Analytic Pseudodifferential Operators for the Heisenberg Group and Local Solvability

Analytic Pseudodifferential Operators for the Heisenberg Group and Local Solvability

by

Daryl Geller

Mathematical Notes 37

PRINCETON UNIVERSITY PRESS

PRINCETON, NEW JERSEY

1990

Princeton University Press books are printed on acid-free paper, and meet the guidelines for permanence and durability of the Committee on Production Guidelines for Book Longevity of the Council on Library Resources

Printed in the United States of America
by Princeton University Press, 41 William Street
Princeton, New Jersey

Library of Congress Cataloging-in-Publication Data

Geller, Daryl, 1950-
Analytic pseudodifferential operators for the Heisenberg group and local solvability / by Daryl Geller.
p. cm.
(Mathematical notes ; 37)
Includes bibliographical references.

1. Pseudodifferential operators. 2. Functions of several complex variables. 3. Solvable groups. I. Title. II. Title: Heisenberg group and local solvability. III. Series: Mathematical notes (Princeton University Press) ; 37.
QA329.7.G45 1990 515'.7242 89-24111
ISBN 0-691-08564-1

CB-stk
SIMON

Dedicated to my mother, Libby, and
to the memory of my father, Samuel.

CONTENTS

Analytic Pseudodifferential Operators
for the Heisenberg Group
and Local Solvability

Introduction

Section One: Main Results

The main purpose of this book is to develop a calculus of pseudodifferential operators for the Heisenberg group $\mathbb{H}^n$, in the (real) analytic setting, and to apply this calculus to the study of certain operators arising in several complex variables. Our main new application is the following theorem (Theorem 10.2 and Corollary 10.3):

1. Suppose M is a smooth, compact CR manifold of dimension $2n + 1$. Suppose $\mathcal{U} \subset M$ is open and is a real analytic strictly pseudoconvex CR manifold. Further suppose:

(i) There is a smooth, bounded pseudoconvex domain $D \subset \mathbb{C}^{n+1}$ with boundary M. (D may be weakly pseudoconvex.)

Let S denote the Szegö projection (onto $\ker \bar{\partial}_b$ in $L^2(M)$). Then:

(a) If $f \in L^2(M)$, $V \subset \mathcal{U}$, V open, and f is analytic on V, then Sf is analytic on V.

(b) Say $f \in L^2(M)$, $p \in \mathcal{U}$. Then there exists $\omega \in (\mathcal{E}')^{0,1}(M)$ with $\bar{\partial}_b^* \omega = f$ near p if and only if Sf is analytic near p.

(b)' Say $f \in L^2(M)$, $p \in \mathcal{U}$. Then there exists $\omega \in (\mathcal{E}')^{0,1}(M)$ with $\bar{\partial}_b^* \omega = f$ near p if and only if Sf can be extended to a holomorphic function on a neighborhood U of p in $\mathbb{C}^{n+1}$.

(c) Let V be an open subset of $\mathcal{U}$. If $f \in L^2(M)$ and $\bar\partial_b f$ is analytic on V, then $(I-S)f$ is analytic on V. ∎

Even if D is strictly pseudoconvex everywhere, the result is new for $n = 1$. (If $n > 1$, D strictly pseudoconvex, it follows from the results in [77], [81].)

The theorem we shall present is actually more general than what we have stated. For example, instead of (i), for (a), (b), (c) one need only assume:

(i)' The range of $\bar\partial_b : C^\infty(M) \to \Lambda^{0,1}(M)$ is closed in the C^∞ topology.

(This is weaker than (i) by the results of Kohn [58].)

Further, (a), (b) and (c) hold for $f \in \mathcal{E}'(\mathcal{U}) + L^2(M)$ (not just $f \in L^2$.) As we shall discuss, in many situations, (a), (b) and (c) hold for $f \in \mathcal{E}'(M)$.

The importance of knowing that S preserves analyticity was first recognized by Greiner-Kohn-Stein [35]. They proved the analogue of our result for $M = \mathbb{H}^n$. In the case of $\mathbb{H}^1$, $\bar\partial_b^*$ is the unsolvable operator of Lewy and (b) gives necessary and sufficient conditions on f for local solvability of the Lewy operator.

Greiner, Kohn and Stein showed that (b) follows readily from (a). In fact, from (i)', one sees that one can always globally solve $\bar\partial_b^*\omega_1 = (I-S)f$; so we need only understand when we can solve

$\bar{\partial}_b^* \omega_2$ = Sf near p. If Sf is analytic near p, this can clearly be done by Cauchy-Kowalewski. If, on the other hand, $\bar{\partial}_b^* \omega_2$ = Sf near p, and $\omega_2 \in \mathcal{E}'(\mathcal{U})$, then Sf = $S(Sf - \bar{\partial}_b^* \omega_2)$ is analytic near p, by (a) for $\mathcal{E}'(\mathcal{U}) + L^2(M)$.

We prove (a) by use of the work of Henkin [43]. On strictly pseudoconvex domains, Kerzman-Stein [52] found a simple relationship between S and the Henkin projector H (which also projects onto ker $\bar{\partial}_b$). S is the product of H and the inverse of a singular integral operator. In order to prove that, in appropriate situations, inverses of singular integral operators also preserve analyticity, we need a calculus of analytic pseudodifferential operators. Boutet de Monvel and Kree [9] developed an analytic calculus which is suitable for dealing with elliptic operators on $\mathbb{R}^n$; we need a calculus which is suitable for dealing with analogous operators on $\mathbb{H}^n$.

The simplest operators that our calculus is intended to deal with — the analogues of the Laplacian on R^n — are the Folland-Stein operators $\mathcal{L}_\alpha$ $(\alpha \in \mathbb{C})$. These second-order differential operators on $\mathbb{H}^n$ are intimately connected to the Kohn Laplacian $\square_b$. Folland and Stein [20] showed that if $\alpha, -\alpha \notin \{n, n+2, \ldots, \}$, then

$$\mathcal{L}_\alpha \phi_\alpha = \delta \tag{0.1}$$

where ϕ_α is homogeneous with respect to the parabolic dilations which are automorphisms of $\mathbb{H}^n$, and is (real) analytic away from

$0 \in \mathbb{H}^n$. They used this fact to study $\Box_b$ on nondegenerate CR manifolds. They speculated that there ought to be a calculus of pseudodifferential operators modelled on the L_α and parabolically homogeneous distributions on $\mathbb{H}^n$, just as the usual calculus is modelled on Δ and homogeneous distributions on $\mathbb{R}^n$. Such a calculus would then be appropriate for the intrinsic (non-isotropic) Sobolev and Lipschitz spaces on a nondegenerate CR manifold.

We present such a calculus here, and we do so in the analytic setting. Our calculus is analogous to the C^∞ calculus of Taylor [79], but our outlook is quite different from his, and our proofs — of necessity — are much more elaborate, since we are working in the analytic setting.

Besides application #1 above (to the Szegö projection) we obtain a number of other new results in this book. Here is a summary of our main results:

2. A very precise form of an analytic parametrix for $\Box_b$ on any nondegenerate analytic CR manifold (Theorem 9.6). From it, one can read off simultaneously the analytic regularity and the (nonisotopic) Sobolev and Lipschitz regularity for $\Box_b$;

3. An analytic calculus on $\mathbb{H}^n$, natural for dealing with $\Box_b$ and operators like it. Simple, explicit formulae for products and adjoints (Theorem 7.11). Simple and natural representation-theoretic conditions, analogous to ellipticity, for determining if operators in the calculus are analytic hypoelliptic, having

parametrices in the calculus (Theorem 8.1). The calculus may be transplanted to provide a very natural calculus on non-degenerate analytic CR manifolds, and more generally, on analytic contact manifolds.

4. Generalization of the theory of operators like the L_α beyond the study of differential operators, analogous to the way in which the usual theory of pseudodifferential operators introduces one to the notion of elliptic operators which are not differential operators. In this way we find a large new, natural class of analytic hypoelliptic operators (Theorem 8.1). Those which are not differential operators were not previously known to be analytic hypoelliptic. Our calculus is the first analytic calculus modelled on parabolic homogeneity, instead of the usual isotropic notion of homogeneity.

5. A characterization of the Fourier transform of the space $\{K \in \mathcal{S}'(\mathbb{R}^n)$: K is homogeneous (with respect to a given dilation structure) and analytic away from $0\}$ (Theorem 1.3). (The dilations are to be of the type $D_r x = (r^{a_1}x_1,\ldots,r^{a_n}x_n)$ for $x \in \mathbb{R}^n$, $r > 0$, where $a_1,\ldots,a_n$ are positive rationals). Of course, $\hat{K}$ is homogeneous. In the isotropic case $(a_1=\ldots=a_n=1)$ it is known that $\hat{K}$ must be analytic away from 0. This does not hold in general for other dilation weights. In the parabolic case, for instance, where $a_1 = 2$, $a_2=\ldots,=a_n = 1$, with

$(t,x) \in \mathbb{R} \times \mathbb{R}^s = \mathbb{R}^n$, and with dual coordinates (λ,ξ), we have the well-known formula

$$[(|x|^2+it)^{-1}]\hat{} = (4\pi\lambda)^{-s/2} e^{-|\xi|^2/4\lambda} H(\lambda) = J_0(\lambda,\xi), \text{ say,} \quad (0.2)$$

where H is the characteristic function of $(0,\infty)$. J_0 is the familiar kernel of the heat operator $\Delta_\xi + \partial/\partial\lambda$, and is not analytic at $\lambda = 0$.

Our characterization of $\{\hat{K} : K \text{ as above}\}$ is most easily understood in the case $a_1 = p$, $a_2 = \ldots = a_n = 1$, $p \in \mathbb{Q}$, $p > 1$ (Theorem 2.11).

In addition, we lay the groundwork for the following further studies:

6. Generalization of our calculus to a wide class of nilpotent Lie groups ("graded homogeneous groups"—see e.g. [12] for the definition). We see no difficulty in doing this, but it would then usually make more sense to work in the C^∞ setting, since analytic hypoelliptic operators on most other groups are rare.

7. Generalization to the study of operators like the L_α, but for $\alpha \in \pm\{n,n+2,\ldots\}$. In this case, one has not (0.1), but

$$L_\alpha \phi_\alpha = \delta - P_\alpha \quad (0.3)$$

with ϕ_α, P_α homogeneous and analytic away from 0, and with $f \to f * P_\alpha$ $(f \in L^2)$ being the projection in $L^2(\mathbb{H}^n)$ onto $[L_\alpha S(\mathbb{H}^n)]^\perp$.

ϕ_α is called a relative fundamental solution for L_α.

When $\alpha = n$, L_α is the same as $\square_b$ on functions on $\mathbb{H}^n$, P_α is the Szegö projection for $\mathbb{H}^n$, and (0.3) is one of the results of Greiner, Kohn and Stein alluded to above. As in (b) of application #1 above, for $f \in \mathcal{E}'$, $L_\alpha g = f$ is locally solvable near $u \in \mathbb{H}^n$ if and only if $f*P_\alpha$ is analytic near u.

Peter Heller and this author have obtained a generalization of (0.3) to left-invariant differential operators on $\mathbb{H}^n$ which are "transversally elliptic." This generalization will be presented in a future paper. Historical references and further discussions are in Section Two of this Introduction.

Local Solvability and Analytic Pseudolocality of the Szegö Projection

We now explain in more detail our results on the Szegö projection. Again suppose that M is a smooth compact CR manifold of dimension 2n + 1; suppose $\mathcal{U} \subset M$ is open, real analytic, and strictly pseudoconvex, and that the range of $\bar{\partial}_b : C^\infty(M) \to \Lambda^{0,1}(M)$ is closed in the C^∞ topology. Let S be the Szegö projection on M. Our hypotheses are too weak to imply that $S : C^\infty(M) \to C^\infty(M)$, so we cannot necessarily extend $S : \mathcal{E}'(M) \to \mathcal{E}'(M)$. What we shall show is that S is analytic pseudolocal on $\mathcal{U}$ when restricted to its "natural domain." That is: first we shall show $S : C^\infty(M) \to C^\infty(\mathcal{U})$ continuously. Fix $\mathcal{V} \subset \mathcal{U}$, $\mathcal{V}$ open, so that $S : C^\infty(M) \to C^\infty(\mathcal{V})$ continuously. We define $\mathcal{D}(S)$, the domain of S, to be $\mathcal{E}'(\mathcal{V}) + L^2(M)$. We shall show that if $f \in \mathcal{D}(S)$, $W \subset \mathcal{U}$, W open, f analytic on W, then Sf is analytic on W. (By the results of Kohn [57], we may take $\mathcal{V} = M$ if $M = \partial D$, D a smooth bounded pseudoconvex domain in $\mathbb{C}^n$, provided D is of finite type if $n = 2$, or D is of finite ideal type if $n > 2$.)

If $n > 1$, and M is in addition globally strictly pseudoconvex, analytic pseudolocality of S follows from the work of Treves [81] and Tartakoff [77]. Indeed, $\Box_b$ on (0,1)-forms has a good "Hodge theory." Let N invert $\Box_b$ on (0,1)-forms on the orthocomplement of its kernel. Kohn's formula

$$I - S = \bar{\partial}_b^* N \bar{\partial}_b \tag{0.4}$$

shows that the analytic pseudolocality of S may be deduced from that of N, which in turn follows from the results of [81], [77].

When $n = 1$, there is no good Hodge theory for $\Box_b$ on 1-forms, so this method cannot be used in this case, even if M is globally strictly pseudoconvex.

Our proof of analytic pseudolocality of S begins with a reduction to a local analogue, and it is in this reduction that we use the "closed range" hypothesis for $\bar{\partial}_b$. The local analogue is this: Say $p \in \mathcal{U}$. We shall show that p has a neighborhood $V \subset\subset \mathcal{U}$, and there exist operators

$$S_1 : \mathcal{E}'(V) \to \mathcal{D}'(V),\ K_1 : (\mathcal{E}')^{0,1}(V) \to \mathcal{D}'(V),\ S_1, K_1 \text{ analytic pseudolocal on } V \tag{0.5}$$

$$K_1\bar{\partial}_b = (I - S_1),\ \bar{\partial}_b S_1 = 0,\ S_1\bar{\partial}_b^* = 0 \text{ on } V \text{ mod analytic regularizing errors.} \tag{0.6}$$

Precisely: if $f \in \mathcal{E}'(V)$, $\omega \in (\mathcal{E}')^{0,1}(V)$, then

$$[K_1\bar{\partial}_b - (I - S_1)]f,\ \bar{\partial}_b S_1 f,\ S_1\bar{\partial}_b^*\omega \text{ are all analytic on } V. \tag{0.7}$$

If V were M and the equations in (0.6) were exact, then S_1 would have to equal S. In general, though, we can show this: say $p \in V_1 \subset\subset V$, V_1 open, $\zeta \in C_c^\infty(V)$, $\zeta = 1$ on V_1. Then we claim:

p has an open neighborhood $V_2 \subset\subset V_1$, so that if $f \in \mathcal{D}(S)$,

$$Sf - S_1(\zeta f) \text{ is analytic on } V_2. \tag{0.8}$$

This would then clearly show analytic pseudolocality of S on $\mathcal{U}$.

To illustrate the method of proof, let us examine the analogue in the C^∞ setting (replace "analytic" by "smooth" in (0.5), (0.7), (0.8); call the new statements (0.5)', (0.7)', (0.8)'). In (0.8)', we may as well assume V_1, supp ζ are as small as we like, by the pseudolocality of S_1. We may also assume $S_1 : \mathcal{E}'(M) \to \mathcal{D}'(V)$, $K_1 : (\mathcal{E}')^{0,1}(M) \to \mathcal{D}'(V)$ and (0.7)' holds for $f \in \mathcal{E}'(M)$, $\omega \in (\mathcal{E}')^{0,1}(M)$. Indeed, to achieve these relations we need only shrink V slightly and multiply the Schwartz kernels of S_1, K_1 by an appropriate smooth bump function ϕ

supported near the diagonal of $M \times M$. If supp ζ, V_1, V_2 are sufficiently small, the value of $S_1(\zeta f)$ on V_2 is unaltered if we change S_1 in this manner, so (0.8)' is unaffected. To establish (0.8)', we just have to look at S_1Sf. On the one hand, by the closed range property of $\overline{\partial}_b$, we have $(I-S)f = \overline{\partial}_b^* u$ for some $u \in \mathcal{E}'(M)$; thus, on V, $S_1Sf = S_1(f-\overline{\partial}_b^* u) = S_1 f + g_1 = S_1(\zeta f) + g_2$ where g_1 is smooth on V, g_2 is smooth on V_1; while $S_1Sf = (I-K_1\overline{\partial}_b)Sf + g_3 = Sf + g_3$, g_3 smooth on V. (0.8)' follows at once. (This part of the argument is similar to certain reasoning in [10].)

In the analytic setting, the bump function ϕ cannot be chosen to be analytic, of course; so we instead carry out this procedure with a sequence of special bump functions, due to Ehrenpreis. These bump functions, and the errors g_1, g_2, g_3 which result in the above argument, satisfy the conditions for analyticity (i.e. conditions like $|\partial^\alpha F| < CR^{\|\alpha\|}\alpha!$) for $\|\alpha\|$ less than or equal to a number N. We may then let $N \to \infty$ to establish (0.8).

To establish (0.6), since $\mathcal{U}$ is analytic and strictly pseudoconvex, we may in fact assume $\mathcal{U} \subset M'$, $M' = \partial D'$, $D' \subset \mathbb{C}^{n+1}$ a smooth, bounded strictly <u>convex</u> domian. If $n > 1$, we are done by (0.4). If $n = 1$, we use the work of Henkin and a method of Kerzman-Stein.

Henkin [43] shows that there are operators R,H on M' so that

$$R\bar{\partial}_b = I-H, \quad \bar{\partial}_b H = 0. \tag{0.9}$$

H is the Henkin projection onto the kernel of $\bar{\partial}_b$ in L^2. These relations apparently give us part of (0.6). But H, the generalization from $\mathbb{C}$ of the Cauchy projection, need not be orthogonal. Thus we need not have $H\bar{\partial}_b^* = 0$ near p modulo an analytic regularizing error.

Kerzman and Stein, however, observed [52] that if S' is the Szegö projection on M', then S'(I+H-H*) = H (since S'H = H, HS' = S' ⇒ S'H* = S'), so that

$$S' = H(I+H-H^*)^{-1} \tag{0.10}$$

(I + H - H* is invertible on L^2 since H - H* is skew-adjoint.) Since (I-S')(I-H) = I - S', we would like to obtain (0.6) from (0.9) by putting $K_1 = (I-S')R$. The problem, then, is to obtain analytic pseudolocality on $\mathcal{U}$ for S' from (0.10); and for this we need to develop an analytic calculus on the Heisenberg group.

Overview

Here now is an overview of our calculus. Let A be a classical pseudodifferential operator of order j on $\mathbb{R}^n$, so that

$$(Af)(u) = (2\pi)^{-n}\iint e^{-i(u-v)\cdot\xi}a(u,\xi)f(v)\,dv\,d\xi. \tag{0.11}$$

Here let us say the symbol $a(u,\xi) \sim \sum a^m(u,\xi)$ as $\xi \to \infty$, where each a^m is smooth in u and homogeneous of degree j - m in ξ.

We may then also formally write

$$(Af)(u) = (K_u * f)(u) \tag{0.12}$$

where $$K_u(w) = (2\pi)^{-n} \int e^{-iw\cdot\xi} a(u,\xi)\,d\xi, \tag{0.13}$$

the inverse Fourier transform of a in the ξ variable. Let us call $K_u(w) = K(u,w)$ the <u>core</u> of the operator A; the <u>kernel</u> of A is then $K(u,u-v)$. We have

$$K_u(w) \sim \sum K_u^m(w) \quad \text{near } w = 0 \tag{0.14}$$

where K_u^m is smooth in u and homogeneous of degree $k + m$ $(k=-n-j)$, at least if $k + m \notin \mathbb{Z}^+ = \{0,1,2,\ldots\}$. (If $k + m \in \mathbb{Z}^+$, $K_u^m(w)$ may in addition contain a "log term" of the form $p_u(w)\log|w|$, where p_u is a homogeneous polynomial in w of degree $k + m$.) A is a differential operator with analytic coefficients if all the K_u^m are supported at 0. A is elliptic if $\hat{K}_u^0(\xi) \neq 0$ for $u \in U$, $\xi \neq 0$; then it has a parametrix of the same type on any relatively compact open subset V of U.

We intend to develop a calculus on $\mathbb{H}^n$ which is analogous to (0.12), (0.14). In (0.12), * will be replaced by group convolution on $\mathbb{H}^n$, and in (0.14), the $K_u^m(w)$ will now be homogeneous in w (or homogeneous plus a "log term") in the <u>parabolic</u> sense. The condition analogous to ellipticity is that $\pi(K_u^0)$ be injective on Schwartz vectors for all non-trivial irreducible unitary representations π of $\mathbb{H}^n$.

The idea of working with operators of type (0.12) in the C^∞ category and in the nilpotent group situation is due to Folland-Stein [20]. Such operators, and generalizations thereof, were also studied by Rothschild-Stein [70] and Nagel-Stein [66]. Taylor [19], in the C^∞ setting on $\mathbb{H}^n$, defines "pseudo-differential operator" by (0.12), and then chooses to pass directly from these operators to define a new kind of symbol. We shall stay instead on the core level, although the Fourier transform will still play an integral role in our work. Also, we will work in the analytic setting. For 3-step nilpotent groups, in the C^∞ setting, a calculus based on cores was also recently constructed by Cummins [14].

Of course, the concept of "core" is far from new. We have introduced the new name since we intend to place absolute emphasis on the core, as opposed to the kernel or the symbol.

It is worth recalling some of the basic reasons why one usually prefers, in the standard situation on $\mathbb{R}^n$, to use definition (0.11) instead of (0.12), when (0.12) has a much simpler appearance. Two elementary reasons are:

(a) The Fourier transform converts convolution to multiplication, which is easier to handle.

(b) The Fourier transform converts the finding of convolution inverses (such as fundamental solutions for constant coefficient differential operators) to division.

Reason (a) loses much of its significance on a nilpotent group, where the Euclidean Fourier transform on the underlying manifold ($\mathbb{R}^n$) converts group convolution to a complicated analogue of multiplication, while the complicated group Fourier transform converts group convolution to multiplication. One could, then, write down formulas analogous to (0.11), but their complexity makes them undesirable to manipulate compared to (0.12). We shall see that, using convolution, we can obtain an analogue of the Kohn-Nirenberg product rule for $\mathbb{H}^n$ (and more generally, for graded homogeneous groups).

As for reason (b), even when one inverts, it is possible in many circumstances, to avoid use of the Fourier transforms. The problem can be reduced to inverting the principal part of the operator (in (0.12), (0.14), the operator with core equal to $K_u^0(w)$ near $w = 0$). The rest can be handled with a Neumann series, after a product rule is established. Thus, the initial problem to be solved is:

> Given K_1 homogeneous and analytic (resp. smooth) away from 0, can we find K_2 homogeneous and analytic (resp. smooth) away from 0, with $K_2 * K_1 = \delta$?

Say we are in the "smooth" setting, and for simplicity, say that K_1 has the same homogeneity as δ. Then, on graded homogeneous groups, a necessary and sufficient condition for the existence of K_2 is that the map $f \to K_1*f$ be left invertible on L^2 ([12]).

The Fourier transform on the group can nevertheless be a valuable tool for proving K_2 exists. Thus, for certain groups, representation-theoretic conditions for the existence of K_2 are known. In [12], it was conjectured that, in the smooth setting, on homogeneous groups, the existence of K_2 is equivalent to this condition:

> $\pi(K_1)$ is injective on Schwartz vectors for all irreducible unitary representations π (except $\pi \equiv 1$).

This is the analogue of the Helffer-Nourrigat theorem [41] for singular integral operators. The conjecture was proved for $\mathbb{H}^n$ in [12], generalized to graded groups of step less than or equal to three, by Moukaddem [65], and has now been proved in full generality by Glowacki [32].

In the analytic setting, one could only expect to find such a K_2 on a small number of groups (the "H-groups" [62], to which the $\mathbb{H}^n$ belong). One of our objectives is to prove that, in the analytic setting on $\mathbb{H}^n$, the representation-theoretic criterion given above implies the existence of K_2 (Theorem 5.1).

Thus the Fourier transform will still play an important role in our work; but the methods we have discussed will enable us to limit its use to the study of homogeneous distributions. In conclusion, we will frequently take the Fourier transforms of homogeneous distributions, but we will never have to deal with complicated analogues of (0.11).

The Core Approach on $\mathbb{R}^n$

Since the approach to pseudodifferential operators using cores instead of symbols is unfamiliar even on $\mathbb{R}^n$, let alone on $\mathbb{H}^n$, at this point we wish to discuss the $\mathbb{R}^n$ situation in some detail—even though that is not part of the actual subject matter of this book. Many of the methods that can be used on $\mathbb{R}^n$ go through with only minor modification on $\mathbb{H}^n$. When it comes to inversion, however, new methods must be developed for $\mathbb{H}^n$, and that is the main subject matter of the book.

Using cores instead of symbols on $\mathbb{R}^n$, there is of course the annoyance of having to deal with convolution instead of multiplication, but there are certain real conceptual advantages. For instance, as Taylor [79] has pointed out, the Kohn-Nirenberg formula for products of pseudodifferential operators can be proved just by extending the proof that convolution is associative. Indeed, let $\mathcal{K}_1, \mathcal{K}_2$ be the operators with cores K_1, K_2; then

$$(\mathcal{K}_2\mathcal{K}_1 f)(u) = \int K_2(u,u-v)\int K_1(v,v-w)f(w)\,dw\,dv. \qquad (0.15)$$

In this expression we formally write

$$K_1(v,v-w) = \sum (v-u)^{\alpha} \partial_u^{\alpha} K_1(u,v-w)/\alpha! \tag{0.16}$$

to see that $K = K_2 K_1$ should have core K, where

$$K_u = \sum_{\alpha} [(-w)^{\alpha} K_{2u}] * [\partial_u^{\alpha} K_{1u}]/\alpha! \tag{0.17}$$

One could then recover the Kohn-Nirenberg formula through (0.12). (A very similar formula holds on $\mathbb{H}^n$—see (0.28) below). This of course, would all be easiest to justify if K_{1u} were analytic in u.

The analytic pseudodifferential operators of Boutet de Monvel and Kree [6] are also most readily motivated if one looks at the cores. (The operators we discuss are slight modifications of those of Boutet de Monvel and Kree.) Roughly speaking, we look at cores K_u as in (0.14) where each $K_u^m(w)$ is analytic in u, analytic in w away from 0, and where K_u^m grows at most on the order of R^m on $|w| = 1$, for some R. Because of the homogeneity of the K_u^m, the series in (0.14) will now converge in a punctured ball about 0, and provide a very natural generalization of a power series.

A little notation will help to make things clearer. If $k \notin \mathbb{Z}^+$, let

$$AK^k = \{K \in \mathcal{D}' : K \text{ is homogeneous of degree } k \text{ and analytic away from } 0\}. \tag{0.18}$$

If $k \in \mathbb{Z}^+$, let

$$\mathcal{AK}^k = \{K \in \mathcal{D}' : K = K' + p(x)\log|x|, \text{ with } K', p \text{ homogeneous of degree } k \text{ and analytic away from } 0, \text{ and } p \text{ a polynomial}\}. \tag{0.19}$$

If $U \subset \mathbb{R}^n$ is relatively compact and open, and $k \in \mathbb{C}$, we let $C^k(U)$ denote the space of all cores of the form

$$K_u(w) = \chi(w)\left[\sum_{m=0}^{\infty} K_u^m(w) + Q(u,w)\right] \tag{0.20}$$

where for some $r, R > 0$,

χ is the characteristic function of the ball $B_r = \{w : |w| < r\}$, (0.21)

$K_u^m \in \mathcal{AK}^{k+m}$ for all $u \in U$, and depends analytically on u; (0.22)

Q is analytic in $U \times B_r$; (0.23)

The series $\sum K^m(v,w)$ converges absolutely for v in a complexified neighborhood of $\overline{U}$ and for $0 < |w| \leqq r$, to a function holomorphic in v and analytic in w. (0.24)

For every compact $E \subset U$ there exists $C_E > 0$ so that

$$|\partial_u^\gamma \partial_w^\rho Q(u,w)| < C_E R^{|\gamma|+|\rho|} \rho!\gamma! \quad \text{for } u \in E,\ w \in B_r; \tag{0.25}$$

$$|\partial_u^\gamma \partial_w^\rho K^m(u,w)| < C_E R^{|\gamma|+|\rho|+m} \rho!\gamma! \text{ for } 1 \leqq |w| \leqq 2, \; u \in E,\ \gamma, \rho \text{ multi-indices.} \tag{0.26}$$

We write $K \sim \sum K_u^m$, and we call r the support radius of K. ($C^k(U)$ can also be defined for U not relatively compact, by modifying the definition slightly.) By (0.26), K_u^m grows on the order of at most R^m on $|w| = 1$, so the convergence of the series legislated by (0.24) is actually a consequence of (0.26)—at least for w sufficiently small ($0<|w|<1/3R$ will do). The series in fact clearly has a holomorphic extension to $\sum K^m(v,\omega)$ where $(v,\omega) \in V \times S$, V being a complexified neighborhood of $\overline{U}$, S being a truncated sector of the form $\{\omega \in \mathbb{C}^n: |\mathrm{Im}\,\omega| < c|\mathrm{Re}\,\omega|, |\omega| < b\}$ for some $c,b > 0$; (0.26) is virtually equivalent to this. Thus $C^k(U)$ is an extremely natural space of cores.

The one arbitrary point in the definition is the use of the characteristic function χ in (0.20) to localize. One might consider using instead a C^∞ cutoff function ϕ, or the characteristic function χ' of an arbitrary neighborhood of 0. Recall, however, that $(Kf)(u) = (K_u * f)(u)$. We shall always "operate locally"— that is, we shall assume supp f has small diameter and we shall only be interested in the behavior of Kf at points in or very near supp f. Operating locally, it cannot matter whether one uses χ, ϕ or χ' to localize; Kf will be the same.

Operating locally, it is easy to believe that operators with cores in $C^k(U)$ are analytic pseudolocal, and that an operator with core in $C^{k_1}(U)$ may be composed with one whose core is in $C^{k_2}(U)$, if their support radii are unequal, to produce one with core in $C^{k_1+k_2+n}(U)$. Operating locally, it is also easy to believe that if $V \subset\subset U$, V open, then an elliptic operator with core in $C^{k-n}(U)$ will have, on V, as a parametrix an operator with core in $C^{-k-n}(V)$. The parametrix inverts the operator up to an analytic regularizing operator $\mathcal{Q}$, with core $Q \sim 0$.

All these statements are true if one operates locally, and the reader can no doubt envision what some of the proofs might look like, using such techniques as (0.15)-(0.17). We still need to clarify several points about the calculus, but let us first pause to see how things look on $\mathbb{H}^n$.

The Core Approach on $\mathbb{H}^n$

$\mathbb{H}^n$ is the Lie group with underlying manifold $\mathbb{R} \times \mathbb{C}^n$ and multiplication $(t,z)\cdot(t',z') = (t+t'+2 \operatorname{Im} z\cdot\bar{z}', z+z')$ where $z\cdot\bar{z}' = \sum z_j\bar{z}_j'$. The dilations $D_r(t,z) = (r^2t, rz)$ are automorphisms. We write $z = x+iy$. The left-invariant vector fields agreeing with $\partial/\partial x_j$, $\partial/\partial y_j$, $\partial/\partial t$ at 0 are $X_j = \partial/\partial x_j + 2y_j\partial/\partial t$, $Y_j = \partial/\partial y_j - 2x_j\partial/\partial t$, $T = \partial/\partial t$. The right-invariant analogues are $X_j^R = \partial/\partial x_j - 2y_j\partial/\partial t$, $Y_j^R = \partial/\partial y_j + 2x_j\partial/\partial t$, and T. Put $Z_j = (X_j - iY_j)/2$, $Z_j^R = (X_j^R - iY_j^R)/2$.

If $u = (t,z) \in \mathbb{H}^n$, we put $|u| = (t^2+|z|^4)^{1/4}$; then $||$ is (parabolically) homogeneous of degree one and is analogous to the norm function on $\mathbb{R}^n$.

With $||$ as above, and using parabolic homogeneity, we define $\mathcal{A}\mathcal{K}^k$ as in (0.18)-(0.19) and $C^k(U)$ as in (0.20)-(0.26) for $U \subset \mathbb{H}^n$ open and relatively compact. The operator with core K is $\mathcal{K}$ where $(\mathcal{K}f)(u) = (K_u * f)(u)$ for $f \in C_c^\infty(U)$, and where we use group convolution.

Operating locally, analytic pseudolocality again holds for these operators. As for composition, we replace (0.16) by

$$K_1(v,vw^{-1}) = \sum_\alpha (vu^{-1})^\alpha \sigma_\alpha(\mathcal{U}) K_1(u,vw^{-1})/\alpha! \tag{0.27}$$

invoking Taylor's theorem for $\mathbb{H}^n$. Here $\mathcal{U} = (T,X_1^R,\ldots,X_n^R,Y_1^R,\ldots,Y_n^R)$ acting on the u variable, and for $\alpha \in (\mathbb{Z}^+)^{2n+1}$, $\sigma_\alpha(\mathcal{U})$ is the <u>symmetrization</u> of $\mathcal{U}^\alpha$, familiar from the Birkhoff-Poincaré-Witt theorem. That is, $\sigma_\alpha(\mathcal{U})$ is the coefficient of τ^α in the multinomial expansion of $(\alpha!/\|\alpha\|!)(\tau\cdot\mathcal{U})^{\|\alpha\|}$. (Here $\tau \in R^{2n+1}$, $\|\alpha\| = \alpha_1+\ldots+\alpha_{2n+1}$.) If the X_j^R, Y_j^R commuted, $\sigma_\alpha(\mathcal{U})$ would be just $\mathcal{U}^\alpha$; as it is, however, the symmetrization of $X_1^R Y_1^R$ (for example) is $(X_1^R Y_1^R + Y_1^R X_1^R)/2$. (0.27) follows easily from right invariance and the observation that the right-invariant operator agreeing with ∂^α at 0 is $\sigma_\alpha(\mathcal{U})$.

Using (0.23), we see that the formal composition rule (0.17) should be replaced by:

$$K_u = \sum_\alpha [(-w)^\alpha K_{2u}] * [\sigma_\alpha(U) K_{1u}]/\alpha! \tag{0.28}$$

In this manner, we develop a natural analytic calculus which contains $\Box_b$ and a parametrix for it, on nondegenerate analytic CR manifolds, in the following sense. A nondegenerate analytic CR manifold M is an example of what is called a contact manifold, and Darboux's theorem implies that M is locally analytically diffeomorphic to $\mathbb{H}^n$ by means of a specific kind of diffeomorphism—a contact transformation. One of our objectives, then, is to prove that after such a contact transformation, $\Box_b$ and a parametrix for it lie in the analogue of our calculus for systems. This will be true under Kohn's hypothesis—if M is k-strongly pseudoconvex, we must be looking at $\Box_b$ on (0,q) forms, where $q \neq k$ or $n - k$.

In the situation of (0.6), K_1 is a (left) relative parametrix for $\bar{\partial}_b$ on functions; it and S are locally in our calculus. We can also obtain a (left or right) analytic parametrix for $\Box_b$ on functions, within the calculus.

The Core Approach on $\mathbb{R}^n$ and the Fourier Transform

Continuing with our description of the core approach on $\mathbb{R}^n$, let us explain in more detail the role played by the Fourier transform.

Given a sequence $K^m \in AK^{k+m}$, satisfying estimates

$$|\partial^\rho K^m(w)| < CR^{m+|\rho|}\rho! \quad \text{for } 1 \leq |w| \leq 2 \tag{0.29}$$

we let $J^m = \hat{K}^m$. Boutet de Monvel and Kree [9] show that there are dual estimates

$$|\partial^\rho J^m(\xi)| \leqq C_o R_o^{m+|\rho|} \rho! m! \quad \text{for} \quad 1 \leqq |\xi| \quad 2. \tag{0.30}$$

We review the proof of this fact, since the method is very important for us. For simplicity we assume Re $k < 0$, in which case (as is known and as we shall verify in much greater generality later) $\hat{}: AK^k \to AK^{-n-k}$; further if $C_1, R_1 > 0$ there are C_2, R_2 depending only on C,R so that if $K \in AK^k$, $J = \hat{K}$ and

$$|\partial^\rho K(w)| < C_1 R_1^{|\rho|} \rho! \quad \text{for} \quad 1 \leqq |w| \leqq 2, \text{ all } \rho, \tag{0.31}$$

then $|\partial^\rho J(\xi)| < C_2 R_2^{|\rho|} \rho! \quad \text{for} \quad 1 \leqq |\xi| \leqq 2, \text{ all } \rho.$ (0.32)

In (0.29), note that if $|\alpha| = m$, we have estimates

$$|\partial^\rho (\partial^\alpha K^m)(w)| < C_3 R_3^{m+|\rho|} \rho! m!$$

and $\partial^\alpha K^m \in AK^k$. By (0.31), (0.32), this leads to

$$|\partial^\rho (\xi^\alpha J^m)(\xi)| < C_4 R_4^{m+|\rho|} \rho! m! \quad \text{if} \quad |\alpha| = m.$$

If the left side were $|(\xi^\alpha \partial^\rho J^m)(\xi)|$ instead, this would give us (0.30). It is, in fact, not hard to reach (0.30) from here.

One can reverse this argument up to a point. Starting with (0.30), if Re $k < 0$, one can deduce (0.29) for $|\rho| \geqq m$. The argument gives no estimate for $\partial^\rho K^m$, $|\rho| < m$. This, in fact, could hardly be expected, since if k is an integer, J^m

is not even, in general, well-determined by its values for $1 \leqq |\xi| \leqq 2$. J^m is very singular near 0 for m large, and we must specify its effect on test functions whose supports include 0. It is well known how one can do this. For instance, if $G \in C^\infty(\mathbb{R}^n - \{0\})$ is homogeneous of degree $-n$, one can define a distribution $J \in (\mathcal{A}\mathcal{K}^0)^\wedge$ agreeing with G away from 0 by

$$J(f) = \int_{|x|<1} G(x)[f(x)-f(0)]dx + \int_{|x|\geqq 1} G(x)f(x)dx$$

for $f \in C_c^\infty$. If instead G is homogeneous of degree $-n-\varkappa$, $\mathrm{Re}\,\varkappa \geqq 0$, one can similarly define a $J \in (\mathcal{A}\mathcal{K}^\varkappa)^\wedge$ agreeing with G away from 0, by use of similar integrals, in one of which one subtracts additional terms of the Taylor series of f at 0, and splits at $|x| = 1$. We can write $J = \Lambda_G$. It is then not hard to show that if the J^m are as in (0.30), $J^m = G^m$ away from 0, and $J^m = \Lambda_{G^m}$, then (0.29) follows even for $|\rho| < m$.

Our second point illustrates the assistance of the Fourier transform. Consider the formal composition law (0.17), as applied to elements of the $\mathcal{C}^k$ spaces. If $K_1 \sim \sum K_1^m$, $K_2 \sim \sum K_2^m$, the support radii are unequal, and $\mathcal{K}$ is the operator obtained by composing $\mathcal{K}_2$ and $\mathcal{K}_1$, then one might imagine $\mathcal{K}$ has core $K \sim \sum K^m$, where

$$K_u^m = \sum_{|\alpha|+a+b=m} [(-w)^\alpha K_{2u}^b] * [\partial_u^\alpha K_{1u}^a]/\alpha! \tag{0.33}$$

This is hardly possible, since if a and b are large, the convolution no longer makes sense, due to the divergence of the integral at ∞. It is necessary, then, to generalize convolution. Change notation, then; say $\mathrm{Re}(k_1+k_2) > -n$, and $K_\nu \in \mathcal{A}K^{k_\nu}$ for $\nu = 1,2$; what replacement is available for K_2*K_1? One could choose K with $\hat{K} = \hat{K}_2\hat{K}_1$ away from 0; but it is also necessary to specify $\hat{K}$ near 0. In fact, if $\hat{K}_\nu = G_\nu$ away from 0, $\nu = 1,2$, we put

$$K_2\underline{*}K_1 = (\Lambda_{G_2G_1}) \tag{0.34}$$

with Λ as before. One can then see that (0.33) should be replaced by

$$K_u^m = \sum_{|\alpha|+a+b=m} [(-w)^\alpha K_{2u}^b]\underline{*}[\partial_u^\alpha K_{1u}^a]/\alpha! \tag{0.35}$$

In fact, it follows easily from our earlier comments about Λ that if K_u^m is as in (0.35) then the K_u^m satisfy estimates as in (0.26).

[It is less clear that $\mathcal{K}$, the composition of $\mathcal{K}_2$ and $\mathcal{K}_1$, has a core K asymptotic to $\sum K^m$ (operating locally). One must see why, if Q_u^m is the difference between the right side of (0.35) and

$$\sum_{|\alpha|+a+b=m} [(-w)^\alpha \chi_2 K_{2u}^b] * [\partial_u^\alpha \chi_1 K_{1u}^a]/\alpha!$$

then Q_u^m is analytic in (u,w) and small enough that $\sum Q_u^m(w)$ converges to an analytic function of (u,w) ($u \in U$, $|w|$ small). (χ_1, χ_2 are characteristic function of balls about 0 of radii

r_1, r_2, with $r_1 \neq r_2$.) changing notation, this comes down to a study of

$$Q = K_2 * K_1 - \chi_2 K_2 * \chi_1 K_1 \tag{0.36}$$

near 0, if $K_\nu \in AK^{k_\nu}$ ($\nu = 1,2$), $\mathrm{Re}(k_1+k_2) > -n$. Say $r_1 > r_2$. If $|\gamma| > M = [\mathrm{Re}(k_1+k_2)] + n$, one can rather easily estimate $\partial^\gamma Q = (1-\chi_2)K_2 * \partial^\gamma K_1$ near 0, and thereby also prove Q is analytic near 0. It is then necessary to estimate the Taylor polynomial of Q about 0 of degree M. Since this is a polynomial, it is enough to estimate this on a small sphere about 0, and this can be done rather easily directly from (0.36).]

If $K_1 \sim \sum K_1^m$, $K_2 \sim \sum K_2^m$ are in the C^k spaces, we define the formal composition of formal series

$$(\sum K_2^m) \# (\sum K_1^m) = \sum K^m \tag{0.37}$$

with K^m as in (0.35).

Our third and last point about the core approach on $\mathbb{R}^n$ concerns the finding of parametrices for elliptic operators in the calculus, and how the Fourier transform is again of assistance. If $K_1 \sim \sum K_1^m$ is elliptic, we begin by composing with a core asymptotic to $(1/\hat{K}_1^0)^\vee$ to reduce (by (0.35)) to the case $K_1^0 \equiv \delta$. One then solves this case by use of a formal Neumann series, in which $\sum_{m>0} K_1^m$ is repeatedly #'d with itself. To estimate the resulting terms, one uses the Fourier transform and the duality between the estimates (0.29) and (0.30). This requires some simple combinatorics. Once the formal inverse is found, the composition theorem is invoked to rigorously produce a parametrix.

<u>The Core Approach on $\mathbb{H}^n$ and the Fourier Transforms</u>

On $\mathbb{H}^n$, (0.35) should be replaced by

$$K^m = \sum_{|\alpha|+a+b=m} [(-w)^\alpha K^b_{2u}] \underline{*} [\sigma_\alpha(u) K^a_{1u}]/\alpha! \qquad (0.38)$$

where $\underline{*}$ is "generalized group convolution." (0.34), of course, cannot be used. One must first adapt Λ to the parabolic setting. Next note that $K_2 * T^N K_1$ makes sense for N large. Using F to denote Euclidean Fourier transforms on $\mathbb{R}^{2n+1} = \mathbb{H}^n$, and with (λ,p,q) dual to (t,x,y), put $G(\lambda,p,q) = F(K_2 * T^N K_1)/(-i\lambda)^N$ for $\lambda \neq 0$; this ought to be $F(K_2 \underline{*} K_1)$ for $\lambda \neq 0$. In fact G has a C^∞ extension for $(\lambda,\xi,\eta) \neq 0$, and one defines $K_2 \underline{*} K_1 = F^{-1}\Lambda_G$.

This description of $K_2 * K_1$ is the simplest, but it is not clear how to generalize this construction to other groups. An alternative procedure, however, will generalize to other groups, at least in the C^∞ setting; and in fact we shall be using this procedure on $\mathbb{H}^n$ (Chapter 7, Section 2). It is a "Poincaré-lemma" type of procedure; let us illustrate it on $\mathbb{R}^n$ in the C^∞ situation. Say $K_\nu \in K^{k_\nu}$ for $\nu = 1,2$. If N is sufficiently large, we can surely define $K_\alpha = K_2 * \partial^\alpha K_1$ for all $\alpha \in (\mathbb{Z}^+)^n$, $\|\alpha\| \geq N$; and surely $\partial^\beta K_\alpha = \partial^\alpha K_\beta$ for all α,β. Then we need only find a K with

$\partial K^{\alpha} = K_{\alpha}$ for all α; such a K, though only well-defined up to a polynomial, could serve as $K_2 * K_1$. In the analytic setting we would uniquely determine K by requiring $\hat{K} = \Lambda_G$ for some G. (For 3-step nilpotent groups, in the C^{∞} setting, a similar procedure was also recently followed by Cummins [14].)

Back on $\mathbb{H}^n$, the arguments needed to establish analytic pseudolocality, and the composition and adjoint theorems are given in Chapter 7. However, the main problem is in finding analytic parametrices under the hypothesis which is analogous to ellipticity--that $\pi(K_u^o)$ is injective on Schwartz vectors for all $u \in U$ and all non-trivial irreducible unitary representations π of $\mathbb{H}^n$. This necessitates that we obtain a complete understanding of $F(\mathcal{A}K^k)$ (Chapters 1 and 2) as well as a partial understanding of the group Fourier transform of $\mathcal{A}K^k$ (Chapters 4 and 5). The two Fourier transforms will be used, as in the Euclidean case, to invert the principal core (Chapter 5) and then to estimate the terms of the Neumann series (Chapter 8).

Using parabolic homogeneity, neither step would be possible if we used standard Euclidean convolution. $\mathbb{H}^n$ convolution must be used. This is an indication of how subtle the problem is.

The Fourier Transform of $\mathcal{A}K^k$

Again we write ^ for Euclidean Fourier transform; for the reasons indicated above, we seek to understand $(\mathcal{A}K^k)\hat{}$.

A characterization of this space can be given for general dilations $D_r x = (r^{a_1}x_1,\ldots,r^{a_n}x_n)$ on $\mathbb{R}^n$, $a_1,\ldots,a_n$ positive rationals (Theorem 1.3). Our characterization is by far clearest, however, if $n = s + 1$, $(a_1,\ldots,a_{s+1}) = (p,1,\ldots,1)$, $p > 1$, and we restrict to this case here. We use coordinates $(t,x) \in \mathbb{R} \times \mathbb{R}^s$ with dual coordinates (λ,ξ). We define AK^k as in (0.18), (0.19), where now $|\;|$ is a fixed homogeneous norm function which is analytic away from 0. (Thus $|\;|$ is homogeneous of degree 1, $|x| \geq 0$ for all x, and $|x| = 0 \Leftrightarrow x = 0$.)

If $K \in AK^k$, then $J = \hat{K}$ is homogeneous and smooth away from 0, so that it is determined away from 0 by two functions,

$$J_+(\xi) = J(1,\xi), \text{ and } J_-(\xi) = J(-1,\xi).$$

Looking at (0.2), one would suspect that J_+ and J_- must be restrictions to $\mathbb{R}^s$ of entire functions on $\mathbb{C}^s$. This turns out to be correct. Let us denote the entire extensions by $J_+(\zeta)$, $J_-(\zeta)$ for $\zeta \in \mathbb{C}^s$. The salient question is: what is the class of entire functions which arises in this way? Again, looking at (0.2), one might guess that if $p = 2$, then $J_+(\zeta)$ and $J_-(\zeta)$ must have exponential order 2. (An entire functions F on $\mathbb{C}^s$ is said to have exponential order m if for some B, $C > 0$, $|F(\zeta)| < Ce^{B|\zeta|^m}$ for all $\zeta \in \mathbb{C}^s$.) This also turns out to be correct. Further, if $p \in \mathbb{Q}$ is general, we shall show that J_+ and J_- must have exponential order q, where $(1/p) + (1/q) = 1$.

The behavior of $J_+(\xi)$, $J_-(\xi)$ as $\xi \to \infty$ ($\xi\epsilon\mathbb{R}^s$) is more subtle. One might guess from (0.2) that these functions must have Gaussian decay when p = 2. This turns out to be not true in general, but it is true of an important subclass. To be precise, let us define

$$Z_q^q = Z_q^q(\mathbb{C}^s) = \{\text{entire functions } f \text{ on } \mathbb{C}^s | \text{ for some } B_1, B_2, C > 0$$
$$\text{we have } |f(\xi)| < Ce^{B_1|\xi|^q} \text{ for } \xi \in \mathbb{C}^s, \text{ while}$$
$$|f(\xi)| < Ce^{-B_2|\xi|^q} \text{ for } \xi \in \mathbb{R}^s\}.$$

We say $f \in Z_q^q(\mathbb{R}^s)$ if f is the restriction to $\mathbb{R}^s$ of a function $f \in Z_q^q(\mathbb{C}^s)$. This space was first investigated by Gelfand-Silov ([21], [22]); we shall say much more about it presently. For now, let us simply state that <u>if</u> J_+, J_- are two functions in $Z_q^q(R^s)$, <u>then</u> there exists $K \in AK^k$ such that $\hat{K}(1,\xi) = J_+(\xi)$, $\hat{K}(-1,\xi) = J_-(\xi)$.

However, the converse is not true; if $K \in AK^k$, $\hat{K}(1,\xi) = J_+(\xi)$ and $\hat{K}(-1,\xi) = J_-(\xi)$ then $J_+(\xi)$ and $J_-(\xi)$ need not decay so rapidly as $\xi \to \infty$. One observation which will be crucial to us in understanding the behavior of J_+ and J_- as $\xi \to \infty$ was made by Taylor [79], who studied homogeneous distributions which are only assumed to be smooth away from 0. The observation is that, since $\hat{K}(\lambda,\xi)$ is homogeneous, its behavior as $\lambda \to 0$ fixed $\xi \neq 0$ is reflected in its behavior as $\xi \to \infty$ for fixed $\xi \neq 0$. Now, for fixed $\lambda \neq 0$, $\hat{K}(\lambda,\xi)$ has a Taylor series expansion in λ

about $\lambda = 0$. This then implies that J_+ and J_- must have asymptotic expansions in ξ at infinity. It is rare, then, that $J_+(\xi)$ and $J_-(\xi)$ will decay rapidly at infinity; this will only happen when the aforementioned Taylor series vanish identically.

The assumption that K is analytic away from 0 implies, as we shall see, that there are in addition certain growth restrictions on the terms of the asymptotic series for J_+ and J_-. Precisely, we can describe a new space in which J_+ and J_- must lie. Let

$Z^q_{q,j}(\mathbb{C}^s) = \{$entire functions $f(\xi)$ on $\mathbb{C}^s \mid$ for some B,C,R,c:

(i) $|f(\zeta)| < Ce^{B|\zeta|^q}$ for all ζ

(ii) In the sector $S = \{\zeta = \xi + i\eta : |\eta| < c|\xi|\}$ there exist holomorphic functions $g_\ell(\zeta)$, homogeneous of degree $j - p\ell$, which satisfy $|g_\ell(\zeta)| < CR^\ell \ell!^{p-1}$ for $|\zeta| = 1, \zeta \varepsilon S$, and such that for all $L > 0$, if $|\zeta| > 1$, $\zeta \varepsilon S$, then

$$|f(\zeta) - \sum_{\ell=0}^{L-1} g_\ell(\zeta)| < CR^L L!^{p-1} |\zeta|^{\mathrm{Re}\, j - pL}.\} \qquad (0.39)$$

In the situation above, we write $f \sim \sum g_\ell$. The relation with Z^q_q is that $f \varepsilon Z^q_q \iff f \varepsilon Z^q_{q,j}$ and $f \sim 0$.

We say $f \varepsilon Z^q_{q,j}(\mathbb{R}^s)$ if f is the restriction to $\mathbb{R}^s$ of a $Z^q_{q,j}$ function on $\mathbb{C}^s$. Also, let $Q = p+s$.

The main result of Chapters 1 and 2 of this book is:

<u>Theorem 2.11</u>. Suppose $J_+, J_- \in C^\infty(\mathbb{R}^s)$ and $j \in \mathbb{C}$. Then there exists $K \in \mathcal{A}\mathcal{K}^{-Q-j}$ such that $\hat{K}(1,\xi) = J_+(\xi)$ and $\hat{K}(-1,\xi) = J_-(\xi)$ for all ξ if and only if:

$$J_+, J_- \in Z^q_{q,j}(\mathbb{R}^s) \text{ and if } J_+ \sim \sum g_\ell \text{ then } J_- \sim \sum (-1)^\ell g_\ell .$$

If this is the case, then

$$g_\ell(\xi) = (\partial_\lambda^\ell J)(0,\xi)/\ell!. \tag{0.40}$$

Let us explain one of the main ideas in the proof of Theorem 2.11. This idea is fully exploited in Chapter one. For $k \notin \mathbb{Z}^+$, let

$$\mathcal{K}^k = \{\text{distributions K which are homogeneous of degree k and smooth away from } 0\}.$$

Say for simplicity that $p = 2$ and $-Q < \mathrm{Re}\, k < 0$. Then, not only does $\hat{}: \mathcal{K}^k \to \mathcal{K}^{-Q-k}$, but it does so continuously, in the following sense. Let $\|\ \|_\infty$ denote sup norm on the "unit sphere" $\{x : |x| = 1\}$, and let $\|\ \|_{C^N}$ denote the C^N norm on the unit sphere. Then for some N, we have

$$\|\hat{K}\|_\infty < C\|K\|_{C^N}$$

for all $K \in \mathcal{K}^k$. Say now $K \in \mathcal{A}\mathcal{K}^k$. If $\alpha \in (\mathbb{Z}^+)^s$, let us write $|\alpha| = \alpha_1 + \ldots + \alpha_s$. Then, if $|\alpha|$ is even, one has that $\partial_t^{|\alpha|/2} x^\alpha K$ and $\partial_x^\alpha t^{|\alpha|/2} K$ are in $\mathcal{A}\mathcal{K}^k$. It is then not hard to see:

$$\|\lambda^{|\alpha|/2}\partial_\xi^\alpha \hat{K}\|_\infty < C\|\partial_t^{|\alpha|/2} x^\alpha K\|_{C^N} < C_1 R_1^{|\alpha|}(|\alpha|/2)! \qquad (0.41)$$

$$\|\xi^\alpha \partial_\lambda^{|\alpha|/2}\hat{K}\|_\infty < C\|\partial_x^\alpha t^{|\alpha|/2} K\|_{C^N} < C_2 R_2^{|\alpha|}|\alpha|! \qquad (0.42)$$

(0.41), and a variant for $|\alpha|$ odd, implies that if $J_+(\xi) = \hat{K}(1,\xi)$, then for ξ in any compact subset of $\mathbb{R}^s$,

$$|\partial_\xi^\alpha J_+(\xi)| < C_3 R_3^{|\alpha|}(|\alpha|!)^{1/2}$$

and similarly for J_-. This then implies easily that J_+ and J_- have extensions to entire functions of exponential order 2, as claimed before.

Further, for $|\xi| = 1$, (0.42) implies that as $\lambda \to 0$

$$|\partial_\lambda^m \hat{K}(\lambda,\xi)| < C_4 R_4^m (2m)!$$

which shows that $\hat{K}$ will not, in general, be analytic as $\lambda \to 0$. Rather, the terms of its Taylor expansion about $\lambda = 0$ in general grow on the order of $R^m m!$, for $|\xi| = 1$. This in turn implies, by homogeneity, that the terms of the asymptotic expansions of J_+ and J_- in general grow on the order of $R^m m!$, in the sense explained by Theorem 2.11 and the definition of $Z^2_{2,j}$.

Analytic Parametrices on $\mathbb{H}^n$

We briefly explain how our understanding of the Fourier transform of AK^k helps us to find analytic parametrices for operators in our calculus. We write F for Euclidean Fourier transform on $\mathbb{H}^n$, thought of as $\mathbb{R}^{2n+1}$. We use coordinates

(t,x,y) on $\mathbb{H}^n$, with dual coordinates (λ,p,q).

The first step is to find a convolution inverse for the principal core, under natural hypotheses. We have:

<u>Theorem</u> (Part of Theorem 5.1). Say $K_1 \in AK^{k-2n-2}$ for some $k \in \mathbb{C}$. Then the following are equivalent:

(a) There exists $K_2 \in AK^{-k-2n-2}$ so that $K_2 * K_1 = \delta$.

(b) The operator $f \to K_1 * f$ $(f \in E')$ is analytic hypoelliptic.

(c) $\pi(K_1)$ is injective on Schwartz vectors for all irreducible unitary representations π of $\mathbb{H}^n$ (except $\pi \equiv 1$).

We explain the meaning of $\pi(K_1)$ precisely in Chapter five. Suffice it to say now, however, that as a consequence of (c), $FK_1(0,p,q) \neq 0$ for $(p,q) \neq 0$. Say $FK_1 = J_1$, $J_{1+}(p,q) = J_1(1,p,q)$, $J_{1+} \sim \sum g_\ell$ in $Z^2_{2,-k}$. By (0.40),

$$g_0(p,q) \neq 0 \text{ for } (p,q) \neq 0 \qquad (0.43)$$

Using this, we construct a "first guess" K_2' so that if

$$K_0 = K_2' * K_1 - \delta \qquad (0.44)$$

then

$$FK_0(1,p,q) \in Z_2^2(\mathbb{R}^{2n}); \qquad (0.45)$$

in other words, at least the asymptotic series for $FK_0(1,p,q)$ in $Z^2_{2,0}$ is zero. Part of what is involved in finding the needed

K_2' is to work on the level of asymptotic series, using (0.43). Next, by (0.44), it suffices for (a) to find $K_2'' \in AK^{-k-2n-2}$ with $K_2''*K_1 = -K_o$. It turns out that (0.45) says that K_o is of such a simple form that the group Fourier transform now becomes a very effective tool. It enables one to rapidly find K_2'', using the rest of the hypothesis (c).

Now say $K_1 \sim \sum K_{1u}^m \in C^{k-2n-2}(U)$ and K_{1u}^o satisfies the condition (c) above for all $u \in U$. Operating locally, to find an analytic parametrix in $C^{-k-2n-2}(V)$ (V a relatively compact open subset of U), we first use (c) $\Longrightarrow$ (a) of the above Theorem, and the rule (0.38) for composition to reduce to the case $K_{1u}^o(w) = \delta(w)$. The parametrix is constructed by use of a Neumann series, as in the Euclidean case; the terms are estimated by use of <u>both</u> the Euclidean and group Fourier transforms.

We have now completed our discussion of the main results and the main methods of this book. We turn next to a discussion of historical perspectives and future directions. After that, we will discuss ancillary results, and give a more detailed exposition of some of the methods which go into the main results.

Prerequisites

This book is long, but the proofs are very detailed, and the prerequisites are few. Thus it should be pleasant reading for a wide audience. We do not assume that the reader necessarily has any prior acquaintance with this area of analysis. In particular, it is certainly not necessary for the reader to have read any of this author's earlier papers on this subject. At a few points, however, we do refer to these and other papers for proofs of very standard or "obvious-looking" facts.

Acknowledgements

I would like to thank E. M. Stein for many helpful discussions. In particular, in response to my questions, he introduced me to the space Z_q^q, and showed me the Fourier series analogues of the implication (A') $\Longrightarrow$ (A)(II) of Chapter one and Theorem 2.3(c) $\Longleftrightarrow$ (e). He also suggested I approach the Szegö projection through the Henkin kernel. These contributions greatly influenced my thinking on the subject.

I would also like to thank Michael Taylor for many helpful discussions and for his encouragement.

I also want to thank Peter Heller for his contributions to the development and his encouragement. He enthusiastically built upon my work in his Princeton thesis [42] after only the first five chapters of this book were complete. Among other results, he proved a weaker form of Theorem 7.1, which opened the door for me to the entire calculus.

Finally, I would like to thank Estella Shivers for typing the manuscript so beautifully.

Research on Chapters 1-9 supported in part by NSF Grant DMS 8503790. Most of the work on Chapter 10 was done in Spring

1988 while I was a member of the Mathematical Sciences Research Institute in Berkeley. I would like to thank M.S.R.I. for its hospitality.

Section Two: Historical Perspectives and Future Directions

Ideally, a pseudodifferential operator calculus on $\mathbb{H}^n$ should have three properties:

A. Analyticity: It should be an analytic calculus. It should contain analytic parametrices for operators K with principal core K_u^o satisfying

$$\pi(K_u^o) \text{ is injective on Schwartz vectors for all non-trivial irreducible representations } \pi \text{ and all } u. \tag{0.46}$$

B. Homogeneity: It should be modelled on homogeneous distributions and group convolution on $\mathbb{H}^n$.

C. Relative Parametrices: Operators like the L_α in (0.3) should have relative parametrices in the calculus. In (0.3), we say ϕ_α is a relative fundamental solution. More generally, if K, with core in $C^k(\mathbb{H}^n)$, has principal core K_u^o satisfying

$$\pi(K_u^o) \neq 0 \text{ for all one-dimensional non-trivial irreducible unitary representations } \pi \text{ and all } u \tag{0.47}$$

we would like K to have a relative parametrix in the calculus, an operator Φ with

$$K\Phi = \delta - P + Q \quad \text{(operating locally)},$$

where P projects in L^2 onto $[KS(\mathbb{H}^n)]^\perp$ and Q has core χQ, with χ, Q as in (0.21), (0.23).

There should also be generalizations to the situation where K operates on functions on a compact analytic contact manifold, and is locally in the calculus after an analytic contact transformation.

The calculus described in this book is the first to have both properties A and B. Further, as we explain below, we have, with Peter Heller, established C if K is a right-invariant differential operator.

Many researchers have proved results on property A and on property B, and there are a few results on property C. We now review some of the history of each, and attempt to place our work in perspective.

A. After Kohn's breakthrough [54], proving the hypoellipticity of $\Box_b$ under his condition Y(q), it was a number of years before analytic hypoellipticity was established if the manifold is in addition real analytic. This was done independently, and virtually simultaneously, by Tartakoff [77], [78] and Treves [81], in remarkable papers. Métivier [63] adapted Treves' methods, and in particular proved analytic hypoellipticity of differential operators satisfying (0.46). An analytic calculus which contains these differential operators was developed by Sjöstrand [72].

Treves, Métivier and Sjöstrand did construct analytic parametrices; we wish to retract, with apologies to them, a statement made near the beginning of our announcement [30]. (The statement was: "We study subelliptic operators, for which one rarely knows how to construct an analytic parametrix.") The parametrices of these authors, however, did not satisfy property B; in particular, Métivier's parametrices were only known to lie in the $OpS_{1/2,1/2}$ spaces. (Operating locally, our calculus is also contained in $OpS_{1/2,1/2}$.) Further, the operators in our calculus which are not differential operators are not covered by the earlier theorems. (Indeed, these operators are not classical analytic pseudodifferential operators. Further, the operators $\pi(K_u^o)$, mentioned in (0.46), for representations π on $L^2(\mathbb{R}^n)$, are not Grušin operators. For these two reasons, the theorems of [81] and [63] do not apply to these operators.)

Using the FBI transform, Treves, Baouendi, Rothschild and others have obtained a series of fine results about extendability of CR functions and analyticity of CR mappings, for weakly pseudoconvex real analytic domains. (A few references are: Baouendi-Chang-Treves [3], Baouendi-Jacobowitz-Treves [4], Baouendi-Bell-Rothschild [6], Baouendi-Rothschild [5].) Our techniques do not deal with the weakly pseudoconvex case. Perhaps some of our methods of dealing with general dilation weights $(a_1,\ldots,a_n)$ may ultimately be of some assistance in understanding the analytic regularity of the Kohn Laplacian on weakly pseudoconvex domains-which is presently not understood except in certain cases ([15]).

B. Homogeneity. After the initial work of Folland and Stein [20] on the L_α, a number of researchers obtained C^∞ regularity results for homogeneous differential operators. The work of Rockland [68], together with that of Miller [64], showed that if R is a right-invariant differential operator on $\mathbb{H}^n$ satisfying (0.46) (with R in place of K_u^o) then R is hypoelliptic. This result was extended to general nilpotent Lie groups with dilations by Helffer and Nourrigat [41]. That such an R will have a fundamental solution which is homogeneous, or homogeneous plus a log term, and C^∞ away from 0, follows from a theorem of Folland and the author [29].

An outline of a pseudodifferential calculus on $\mathbb{H}^n$ modelled on homogeneous distributions was given by Dynin [17]. Taylor [79] constructed such a calculus, as did Beals and Greiner [8]. Taylor shows, in particular, that operators satisfying (0.46) will have parametrices in the C^∞ calculus. Melin [61] also constructed a C^∞ calculus, and Cummins [14] recently constructed a calculus based on cores for 3-step nilpotent groups in the C^∞ setting.

Let us then summarize our new regularity theorems for our pseudodifferential operators. For the differential operators in our calculus satisfying (0.46), we have the new result that they have parametrices in the calculus. From this one can, all at once, read off regularity in the analytic, C^∞ and (non-isotropic) Sobolev and Lipschitz senses. These regularity properties are, however, themselves not new, since they can be obtained by combining the results of the authors listed in the discussions of A and B above.

For the operators in our calculus satisfying (0.46) which are not differential operators, we have the new results that they are analytic hypoelliptic and have parametrices in the calculus. The C^∞, Sobolev and Lipschitz regularity could already have been read off from Taylor's calculus.

C. <u>Relative Parametrices</u>. An operator K in the calculus which satisfies (0.47) is called transversally elliptic. Any right invariant differential operator on $\mathbb{H}^n$ can be expressed as $p(Z^R,\bar{Z}^R)$ for some non-commuting polynomial p. (Here $Z^R = (Z^r_1,\ldots,Z^R_n)$. For instance, $T = \frac{i}{2}[\bar{Z}^R_1, Z^R_1]$.) If $K = p(Z^R,\bar{Z}^R)$ is a right-invariant differential operator on $\mathbb{H}^n$, then in fact K is transversally elliptic if and only if $p(\partial/\partial z,\partial/\partial\bar{z})$ is elliptic on $\mathbb{C}^n$. (Here $\partial/\partial z = (\partial/\partial z_1,\ldots,\partial/\partial z_n)$.) Similarly, if $p(Z,\bar{Z})$ is a left-invariant differential operator, we say it is transversally elliptic if $p(\partial/\partial z,\partial/\partial\bar{z})$ is elliptic on $\mathbb{C}^n$.

It is known that transversally elliptic left invariant differential operators on $\mathbb{H}^n$ are not necessarily locally solvable. The most famous example is $\bar{Z}_1$ on $\mathbb{H}^1$, which is the unsolvable operator of Hans Lewy [59]. However, one does have the following result, which was first observed for the Lewy operator by Greiner, Kohn and Stein in [35].

Theorem 4.1. Suppose L is a left invariant differential operator on $\mathbb{H}^n$ which is homogeneous of degree k and transversally elliptic. Then there exist $K \in \mathcal{AK}^{-2n-2+k}$ and $P \in \mathcal{AK}^{-2n-2}$ such that $LK = \delta - P$, and such that map $f \to f*P$ is the projection in $L^2(\mathbb{H}^n)$ onto $[L\mathcal{S}(\mathbb{H}^n)]^{\perp}$. If $f \in L^2$ or $\mathcal{E}'$ and if $q \in \mathbb{H}^n$, then there exists a distribution u with $Lu = f$ near q if and only if $f*P$ is real analytic near q.

All these assertions, save one, were proved by the author in [26]. What was not shown was that K could be chosen to lie in $\mathcal{AK}^{-2n-2+k}$; there one only knew that K could be found in $\mathcal{K}^{-2n-2+k}$. Rothschild and Tartakoff [71] later showed that one could write $L\mathcal{K} = I - \mathcal{P}$ where $\mathcal{K}$ and $\mathcal{P}$ were operators which preserved local analyticity; however, they did not show that they were given by convolution with kernels in the $\mathcal{AK}$ spaces.

Theorem 4.1 is "best possible" in the sense that our kernels do have the right form. If $f*P$ is analytic near q, and if f has some regularity near q, one can find local solutions u of $Lu = f$ with the appropriate regularity properties. Thus the operators can be studied and understood even if they are not locally solvable. We call K a relative fundamental solution.

Theorem 4.1 has been extended to inhomogeneous L by the work of Peter Heller and this author. These results will appear elsewhere [31], but we state them here, in order to indicate further applications of the techniques of this book.

Suppose L is a transversally elliptic left invariant differential operator of degree k on $\mathbb{H}^n$, not necessarily homogeneous. Then Grigis and Rothschild [36] have derived representation-theoretic criteria for the analytic hypoellipticity of L*. In his Princeton thesis, Heller uses many of the techniques introduced in this book in showing that if L* is analytic hypoelliptic, then L has an analytic parametrix built out of homogeneous distributions. More precisely, if L* is analytic hypoelliptic, Heller shows that one can write $LK = \delta + Q$ where, for some $s \in \mathbb{Z}^+$, $K \in \sum_{j=k-s-2n-2}^{\infty} AK^j$, and where Q is analytic on $\mathbb{H}^n$. We define $\sum_{j=\ell}^{\infty} AK^j = \{$distributions K on $\mathbb{H}^n | K = \sum_{j=\ell}^{\infty} K_j$ for certain $K_j \in AK^j$, with the sum of the analytic extensions of the K_j converging absolutely and uniformly on compact subsets of a complex neighborhood of $\mathbb{H}^n \setminus \{0\}$.) In particular K is analytic away from 0. As an example of Stein [75] shows, one cannot in general take $s = 0$. One can do so if the "top order" part of L^* is hypoelliptic.

The situation in which L^* is not hypoelliptic is dealt with in an upcoming joint paper of Heller and this author. We shall show that one can always write

$$LK = \delta - P + Q$$

where K, P, and Q are certain tempered distributions, with the following properties:

1. Q is analytic on $\mathbb{H}^n$. Further, for certain analytic functions Q_1, Q_2 on $\mathbb{H}^n$, and for some $s \in \mathbb{Z}^+$:

$$K - Q_1 \in \sum_{j=k-s-2n-2}^{\infty} \mathcal{A}K^j, \quad P - Q_2 \in \sum_{j=-2n-2}^{\infty} \mathcal{A}K^j.$$

If the top order part of L^* is hypoelliptic, we may take $s = 0$.

2. The map $f \to f*P$ extends from $\mathcal{S}$ to a bounded operator on L^2. This operator is the projection onto $[L\mathcal{S}(\mathbb{H}^n)]^{\perp}$

3. If $f \in \mathcal{E}'$, then $Lu = f$ is locally solvable near a point $q \in \mathbb{H}^n$ if and only if $f*P$ is real analytic near q.

The case where $P \neq 0$ corresponds exactly to the situation in which L^* is not hypoelliptic, and in this case it was already known (Corwin-Rothschild [13]) that L was then not locally solvable. The preceding theorem gives more precise information.

The preceding theorem, probably, has an analogue for general differential operators in our calculus, on compact real analytic contact manifolds under certain restrictions. Proving this analogue is a fundamental problem which has not yet been solved. We hope that the techniques presented in this book will be of assistance in the solution.

We close this section by noting that our approach is meant to be an affirmation of the importance of cores, not a rejection of the notion of symbol. For the questions that we discuss, there is no need to generalize the notion of "symbol" as Taylor does; but for other questions, this may well be necessary. In particular, note that Treves [81] and Métivier [63] prove microlocal analytic hypoellipticity for their operators. We have not investigated this question for our operators. We conjecture that there is a new notion of analytic wave front set appropriate to our operators which is "conic," where parabolic homogeneity is used to define "cones."

Section Three: Ancillary Results and Further Discussion of Methods

In this section, we explain several additional results and go into more detail about our methods. The topics we shall discuss are:

1. The structure of the spaces Z_q^q and $Z_{q,j}^q$.
2. A general obstruction to analytic hypoellipticity of homogeneous partial differential equations (dilation weights $(p,1,\ldots,1)$.)
3. The method of proof of Theorem 4.1 (on local solvability) and 5.1 (on inversion of singular integral operators).
4. A natural analytic Weyl calculus on $\mathbb{R}^n$ which is modelled on the Hermite operator.

The Space Z_q^q and $Z_{q,j}^q$

The main result of Chapters 1 and 2 is Theorem 2.11, stated above in the Introduction, which relates the Fourier transform of AK^k to the space $Z_{q,j}^q$. Several non-trivial facts must be proved about Z_q^q and $Z_{q,j}^q$ before one is able to establish Theorem 2.11 or apply it. We do this in Chapter 2.

Many of the facts that we shall use about Z_q^q were established by Gelfand-Šilov ([21], [22]). They used these, and other, spaces in studying the Cauchy problem ([23]). (A sample of one of the types of problems they studied is this: For which classes Φ of functions or distributions is the following statement true? If $u(x,t) = u_t(x)$ is a solution of the heat equation such that $u_o = 0$ and $u_t \in \Phi$ for all t, then $u \equiv 0$).

The main properties of Z_q^q that we shall be using are given in the theorem which follows. In (c) we use this notation: Let S = Schwartz space on $\mathbb{R}^s$, $M_\ell, \partial_\ell : S \to S$ by

$$(M_\ell f)(\xi) = \xi_\ell f(\xi), \ \partial_\ell f = \partial f/\partial\xi \ .$$

Also, let

$$U = \{M_\ell | 1 \leq \ell \leq n\}, \ V = \{\partial_\ell | 1 \leq \ell \leq n\}.$$

Theorem 2.3. For a function f on $\mathbb{R}^s$, the following conditions are equivalent:

(a) $f \in Z_q^q$.

(b) f is the restriction to $\mathbb{R}^s$ of an entire function $\mathbb{C}^s$, also denoted f, which satisfies for some B_1, B_2, C, c, $|f(\zeta)| < Ce^{B_1|\zeta|^q}$ for all $\zeta \in \mathbb{C}^s$, while for ζ in the sector $S = \{\zeta = \xi + i\eta \in \mathbb{C}^s : |\eta| < c|\xi|\}$, $|f(\zeta)| < Ce^{-B_2|\zeta|^q}$.

(c) $f \in S$, and there exist C,R as follows. For every set of operators $\{X_1, \ldots, X_{m+n}\} \in U \cup V$ such that $\#\{X_i | X_i \in U\} = m$ and $\#\{X_i | X_i \in V\} = n$, we have $||X_1 \ldots X_{m+n} f||_\infty < CR^{m+n} m!^{1/q} n!^{1/p}$.

(d) $\hat{f} \in Z_p^p$.

Also, if $p = 2$, the statements are equivalent to

(e) $f \in L^2(\mathbb{R}^s)$ and has the Hermite expansion

$\sum_{\alpha \in (\mathbb{Z}^+)^n} c_\alpha H_\alpha(\zeta)$, where for some $C > 0$, $0 < r < 1$,

we have $|c_\alpha| < Cr^{||\alpha||}$.

Much of this was proved by Gelfand-Šilov. However, (a) $\Longrightarrow$ (c) appears to be sharper than in Gelfand-Šilov (who required all the elements of U to precede all the elements of V), and the equivalence with (e) appears to be new. (e) is an analogue of the well-known fact that if $g \in L^2(\mathbb{R}^s)$, then $g \in S$ if and only if its Hermite coefficients decay rapidly.

(a) $\Longrightarrow$ (b) of this theorem states that the decay of f, which is initially assumed only on $\mathbb{R}^s$, must persist into a sector. This is a consequence of Phragmen-Lindelöf. A related fact holds for $Z^q_{q,j}$. Thus, suppose a function f satisfies all the conditions for membership in $Z^q_{q,j}$, except that (0.2) is only known for $\zeta = \xi \in \mathbb{R}^s$, $|\xi| > 1$, rather than for ζ in a complex sector. Then f can be shown to be in $Z^q_{q,j}$. This relatively deep fact is shown in Theorem 2.6.

Whenever one is dealing with a space defined by asymptotics, like $Z^q_{q,j}$, it is common to ask whether, given a formal series, there is necessarily an element of the space which is asymptotic to it. Precisely, let us make the following definition.

Definition. We say that $Z^q_{q,j}$ is ample if the following condition holds. Given any set of functions $\{g_\ell(\zeta)\}$ as in the definition of $Z^q_{q,j}$ (i.e. g_ℓ is holomorphic in a sector $\{\zeta |\ |\eta| < c|\xi|\}$, homogeneous of degree $j - p\ell$ and satisfies $|g_\ell(\zeta)| < CR^\ell \ell!^{p-1}$ for $\zeta \in S$, $|\zeta| = 1$) then there exists $f \in Z^q_{q,j}$ such that $f \sim \sum g_\ell$.

We have proved the following:

Theorem (a) (Theorem 2.12). If q is an even integer and $j \in \mathbb{C}$, then $Z^q_{q,j}$ is ample.

(b) If $q \in \emptyset$, $q > 1$ and $j \in \mathbb{R}$ then $Z^q_{q,j}$ is ample.

(b) will not be proved or used here. (Probably our proof of (b) works equally well if $j \in \mathbb{C}$, but we have not checked the details.)

Homogeneous Partial Differential Equations

Chapter 3 contains a result about a rather general class of partial differential operators. As before, we assume $(a_1,\ldots,a_{s+1}) = (p,1,\ldots,1)$ and we use coordinates $(t,x) \in \mathbb{R} \times \mathbb{R}^s$ with dual coordinates (λ,ξ). We study differential operators L satisfying these conditions:

L is homogeneous, with homogeneous degree k; the degree of L is also k; the coefficients of L are polynomials in the x_ℓ's; and we can write (0.48)

$L = L' + (\partial/\partial t)L''$, where L' is a constant-coefficient differential operator in the x_ℓ's only, and where L' is elliptic.

An example (with $s = 1$, $p = 3/2$, $k = 4$) is $(\partial/\partial x)^4+x^2(\partial/\partial t)^4$. On the other hand, $(\partial/\partial x)^4+x^5(\partial/\partial t)^6$ does not satisfy (0.48); its degree is 6.

Note that L' must be homogeneous with homogeneous degree k, and degree k. Given all the other assumptions, the hypothesis that L' is elliptic is evidently necessary for L to be analytic hypoelliptic, as one sees easily by considering functions of x alone.

This class of operators includes:

(I) constant coefficient homogeneous differential operators in (t,x);

(II) transversally elliptic homogeneous left invariant differential operators on the Heisenberg group $\mathbb{H}^n$;

(III) Grušin operators in (t,x) (for these, L satisfies (0.48) and is also elliptic away from $\{(t,x)|\, x=0\}$.)

If $L = \sum c_{\alpha\beta m}(\partial/\partial x)^\alpha x^\beta(\partial/\partial t)^m$, let $\hat{L} = \sum c_{\alpha\beta m}(-i\lambda)^m(i\xi)^\alpha(i\partial/\partial\xi)^\beta$, the Fourier transform of L. We then have the following result.

<u>Theorem 3.2</u>. Suppose L satisfies (0.48) and $\sigma \in \mathbb{C}$. Suppose

that $Z^q_{q,-(\sigma+k+Q)}$ is ample. Then of the following conditions, (a) implies (b), and (b) is equivalent to (c):

(a) L is analytic hypoelliptic and L and L^t are hypoelliptic.

(b) For any $K_1 \in AK^\sigma$ there exists $K \in AK^{\sigma+k}$ with $LK = K_1$.

(c) For all λ, if $F \in Z^q_q(R^s)$, there exists $G \in Z^q_q$ such that $\hat{L}_\lambda G = F$.

In particular, (c) is necessary for (a) to hold. As we discuss in Chapters 3 and 4, condition (c) is often satisfied for operators of type (II) and (III). As is evident, it can never be satisfied for operators of type (I). We hope that our methods will ultimately be the starting point for the construction of analytic parametrices of the "right type" for operators of type (III).

The most difficult part of the proof, and the only place where the ampleness assumption is used, is (c) $\Rightarrow$ (b). The key fact used is the following:

<u>Theorem 3.5</u>. Suppose $j \in \mathbb{C}$ and that $Z^q_{q,j}$ is ample. Suppose L satisfies (0.48). Say $K_1 \in AK^{-(j+k+Q)}$. Then there exist $K' \in AK^{-(j+Q)}$ and $K_2 \in AK^{-(j+k+Q)}$ such that $LK' = K_1 - K_2$ and such that $J_{2\lambda}(\xi) = \hat{K}_2(\lambda,\xi) \in Z^q_q(\mathbb{R}^s)$ for all $\lambda \neq 0$.

Once this is known, (c) $\Longrightarrow$ (b) follows easily. One has only to construct $K_0 \in AK^{\sigma+k}$ with $LK_0 = K_2$, for then $L(K'+K_0) = K_1$. By Theorem 2.11, it suffices to construct $J_{0,1}, J_{0,-1} \in Z_q^q$ with $\hat{L}_1 J_{0,1} = J_{2,1}$, $\hat{L}_{-1} J_{0,1} = J_{2,-1}$; this can be done by (c). (This argument must be modified slightly if $\sigma \in \mathbb{Z}^+$.)

We illustrate the method of proof of Theorem 3.5 with an example. Say $s = 1$, $p = 3/2$, $L = (\partial/\partial x)^4 + x^2(\partial/\partial t)^4$, $K_1 = \delta$. Thus $\hat{L}_{\pm 1} = \xi^4 - (\partial/\partial\xi)^2$. By Theorem 2.11 it suffices to construct $F \in Z^3_{3,-4}$ such that $[\xi^4-(\partial/\partial\xi)^2]F \sim 1$ in $Z^3_{3,0}$; for then we can choose K with $\hat{K}(\pm 1,\xi) = F(\xi)$. It is easy to see that $[\xi^4-(\partial/\partial\xi)^2]F \sim 1$ has a formal asymptotic solution of the form $\xi^{-4} + a_1\xi^{-10} + a_2\xi^{-16} + \ldots$. Estimating the a_ℓ and using the ampleness condition, one finds that there does exist $F \in Z^3_{3,-4}$ such that $F \sim \xi^{-4} + a_1\xi^{-10} + \ldots$, as desired.

In the above example, L is a Grušin operator, and was proved by Grušin [39] to be analytic hypoelliptic. (However, one still does not know how to construct an analytic parametrix of the "right type" for L.) In particular, Theorem 3.2(c) holds for this L. As we show in Proposition 3.7, this may also be shown simply and directly by use of Theorem 2.3(a) $\Longleftrightarrow$ (c), the "basic estimate" for Grušin operators, and an extension of a method of Métivier from [62].

The Group Fourier Transform on $\mathbb{H}^n$

The main result of Chapter 4 is Theorem 4.1, which we

restate now.

Theorem 4.1. Suppose L is a left invariant differential operator on $\mathbb{H}^n$ which is homogeneous of degree k and transversally elliptic. Then there exist $K \in AK^{-2n-2+k}$ and $P \in AK^{-2n-2}$ such that $LK = \delta - P$, and such that the map $f \to f*P$ is the projection in $L^2(\mathbb{H}^n)$ onto $[LS(\mathbb{H}^n)]^{\perp}$. If $f \in L^2$ or E' and if $q \in \mathbb{H}^n$, then there exists a distribution u with $Lu = f$ near q if and only if $f*P$ is real analytic near q.

To prove Theorem 4.1, we begin by observing that L satisfies the condition (0.48). Thus Theorem 3.5 holds for L; we are interested in the case $K_1 = \delta$ of Theorem 3.5. The question is how far one can go in eliminating K_2. For this, we shall be using the group Fourier transform. It is convenient, then, to change notation; from now on, in this introduction, F means Euclidean Fourier transform while ^ means group Fourier transform.

Let us use coordinates (t,x,y) on $\mathbb{H}^n$, where $z = x+iy$. Then, if $f \in L^1(\mathbb{H}^n)$,

$$(Ff)(\lambda,p,q) = \int_{\mathbb{H}^n} \exp[i(\lambda t+x\cdot p+y\cdot q)]f(t,x,y)\,dt\,dx\,dy \qquad (0.49)$$

where we are using (λ,p,q) as dual coordinates to (t,x,y). The group Fourier transform is given by a rather similar-looking formula. If $f \in L^1(\mathbb{H}^n)$, $\hat{f}$ is a family of bounded operators

$(\hat{f}(\lambda))$; here $\hat{f}(\lambda)$ acts on a separable Hilbert space H_λ, and λ ranges over $\mathbb{R}\setminus\{0\}$. The formula for $\hat{f}(\lambda)$ is

$$\hat{f}(\lambda) = \int_{\mathbb{H}^n} \exp[i(\lambda t + x\cdot P_\lambda + y\cdot Q_\lambda)] f(t,x,y)\,dt\,dx\,dy \qquad (0.50)$$

where $P = (P_{1\lambda},\ldots,P_{n\lambda})$ and $Q = (Q_{1\lambda},\ldots,Q_{n\lambda})$ are certain n-tuples of unbounded operators on H_λ. H_λ is commonly realized as $L^2(\mathbb{R}^n)$. In this realization, the operators $P_{k\lambda}$ and $Q_{k\lambda}$ take this form:

$$(P_{k\lambda}\varphi)(p) = 4\lambda p_k\varphi(p);\ Q_{k\lambda}\varphi = -i\partial\varphi/\partial p_k \text{ for } \varphi \in L^2(\mathbb{R}^n). \qquad (0.51)$$

(We are using coordinates $p = (p_1,\ldots,p_n)$ on $\mathbb{R}^n$.)

The significance of definition (0.50) is that the map

$$(t,x,y) \to U^\lambda_{(t,x,y)} = \exp(i(\lambda t + x\cdot P_\lambda + y\cdot Q_\lambda)]$$

is an irreducible unitary representation of $\mathbb{H}^n$ for $\lambda \in \mathbb{R}\setminus\{0\}$; further all infinite-dimensional irreducible unitary representations arise in this way up to equivalence. (If H_λ is realized as $L^2(\mathbb{R}^n)$, this is called the Schrödinger representation). A consequence is the rule $\widehat{f*g}(\lambda) = \hat{f}(\lambda)\hat{g}(\lambda)$; also, one has Plancherel and inversion formulas for $\hat{}$.

Let us write $(F_\lambda f)(p,q) = (Ff)(\lambda,p,q)$. Of great interest to us is the map W_λ which makes the following diagram commute:

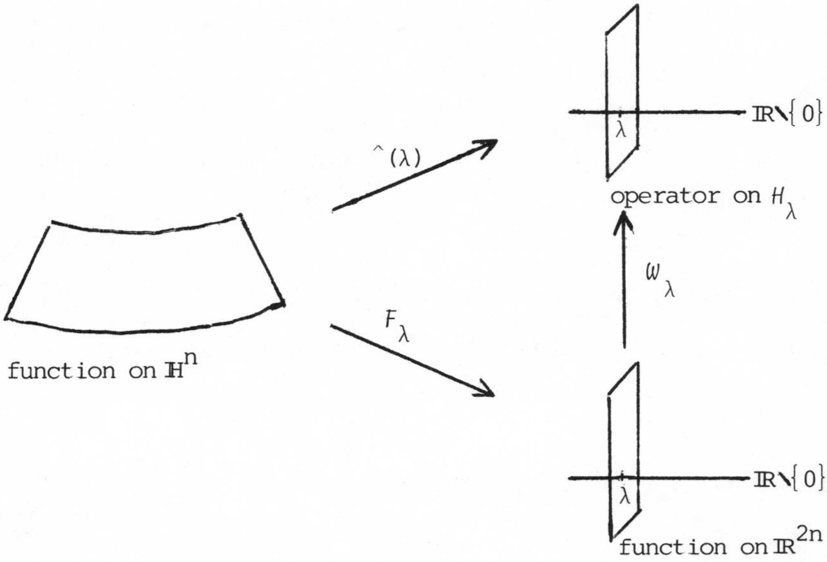

Figure 1

W_λ is called the Weyl correspondence; let us explain some of its properties.

It is often extremely useful to consider the matrix $(R_{\alpha\beta})$ of an operator on H_λ with respect to a certain distinguished orthonormal basis $\{E_{\alpha,\lambda}\}_{\alpha\varepsilon(\mathbb{Z}^+)^n}$. (In the Schrödinger representation, $\{(-i)^{|\alpha|}E_{\alpha,1/4}\}$ are the Hermite functions.) Let us put

$$S(H_\lambda) = \{R\varepsilon B(H_\lambda) \text{ for all N there exists } C_N \text{ with}$$
$$|(RE_{\alpha,\lambda},E_{\beta,\lambda})| < C_N(|\alpha|+1)^{-N}(|\beta|+1)^{-N}\}.$$

$Z_2^2(H_\lambda) = \{S\varepsilon B(H_\lambda)\,|\,\text{for some } C > 0,\ 0 < r < 1, \text{ we have}$

$$|(SE_{\alpha,\lambda}, E_{\beta,\lambda})| < Cr^{|\alpha|+|\beta|}\}.$$

It is well known that $\mathcal{W}_\lambda(S(\mathbb{R}^{2n})) = S(H_\lambda)$. We also have:

<u>Proposition 4.2.</u> $\mathcal{W}_\lambda(Z_2^2(\mathbb{R}^{2n})) = Z_2^2(H_\lambda)$.

(Again, $Z_2^2(\mathbb{R}^{2n})$ is the space of functions which are restrictions to $\mathbb{R}^{2n}$ of $Z_2^2(\mathbb{C}^{2n})$ functions.)

Returning to the situation of Theorem 4.1, we invoke Theorem 3.5 to find K', K_2 as in Theorem 3.5 with $LK' = \delta - K_2$. Let $J_{2\lambda}(p,q) = (FK_2)(\lambda,p,q) \in Z_2^2(\mathbb{R}^{2n})$ for $\lambda \neq 0$. If we attempt to eliminate K_2—that is, if we seek K_0 with $LK_0 = K_2$—we are reduced to producing $J_{0\lambda} \in Z_2^2(\mathbb{R}^{2n})$ with

$$(F_\lambda L)J_{0\lambda} = J_{2\lambda}$$

for $\lambda \neq 0$. ($F_\lambda L$ is what we previously called $\hat{L}_\lambda$.) By homogeneity we need only do this when $\lambda = \pm 1$. We apply $\mathcal{W}_\lambda$ to the equation. Set $S_\lambda = \mathcal{W}_\lambda J_{2\lambda} \in Z_2^2(H_\lambda)$. We seek to solve

$$H_\lambda \pi_\lambda(L) = S_\lambda$$

with $H_\lambda \in Z_2^2(H_\lambda)$, for $\lambda = \pm 1$. Here $\pi_\lambda(L)$ is a certain unbounded operator on H_λ; in fact one could write $\pi_\lambda(L) = (L\delta)\hat{}(\lambda)$, where δ is the Dirac distribution on $\mathbb{H}^n$. This is the same as solving

$$\pi_\lambda(L)^* H_\lambda^* = S_\lambda^*. \tag{0.52}$$

We seek to produce the matrix of H_λ^* one column at a time. Let

$$\mathcal{H}_\lambda^\infty = \{v=\sum v_\alpha E_{\alpha\lambda} \in \mathcal{H}_\lambda \mid \text{for all } N>0 \text{ there exists } C_N > 0 \text{ such that}$$
$$|v_\alpha| < C_N(|\alpha|+1)^{-N}\} \quad \text{and}$$

$$\mathcal{H}_\lambda^\omega = \{v \in \mathcal{H}_\lambda \mid \text{for some } C > 0,\ 0 < r < 1,\ |v_\alpha| < Cr^{|\alpha|}$$
$$\text{for all } \alpha\}.$$

(In the Schrödinger representation, with $\mathcal{H}_\lambda = L^2(\mathbb{R}^n)$, one has $\mathcal{H}_\lambda^\infty = \mathcal{S}(\mathbb{R}^n)$ and $\mathcal{H}_\lambda^\omega = Z_2^2(\mathbb{R}^n)$, by Theorem 2.3 (a) $\Longleftrightarrow$ (c).) The column-by-column solvability of (0.16) is then essentially reduced to two questions. First, does $\pi_\lambda(L)^*$ map $\mathcal{H}_\lambda^\infty$ onto $\mathcal{H}_\lambda^\infty$? Second, if $\pi_\lambda(L)^*v=w$ where $w \in \mathcal{H}_\lambda^\omega$, need $v \in \mathcal{H}_\lambda^\omega$? The answer to the second question is always affirmative; the answer to the first is affirmative only if $\pi_\lambda(L)$ is injective on $\mathcal{H}_\lambda^\infty$. These regularity results for $\pi_\lambda(L)^*$ may be derived as consequences of Theorem 3.5, which already comes very close to solving $LK = \delta$.

If $\pi_\lambda(L)$ is not injective on $\mathcal{H}_\lambda^\infty$, then (0.52) cannot in general be solved, and the projection P of Theorem 4.1 enters as a measure of the solvability of (0.52).

Homogeneous Singular Integral Operators on $\mathbb{H}^n$

We now state more precisely our result on inversion of homogeneous singular integral operators. For $K_1 \in AK^{k-2n-2}$, define $C_{K_1} : \mathcal{E}'(\mathbb{H}^n) \to \mathcal{D}'(\mathbb{H}^n)$ by $C_{K_1}(f) = K_1 * f$; we want to know when C_{K_1} is analytic hypoelliptic. The case where K_1 is supported at 0 has long been understood ([81], [77], [78], [62]). In that case, $K_1 = L\delta$ for some homogeneous left invariant differential operator L, and the question is the same as asking when L* is analytic hypoelliptic.

Before stating our result, let us note that there is always an operator family $(J_1(\lambda))$ which may be thought of as the group Fourier transform of K_1, in a certain sense. (If $k < 2n+2$, one has $\hat{K}_1 = (J_1(\lambda))$ in the sense of tempered distributions in a very natural way. If $k > 2n+2$, $\hat{K}_1 = (J_1(\lambda))$ in a more restricted sense, which we shall call the "$\mathcal{Q}$ sense.") We put $J_{1+} = J_1(1)$ and $J_{1-} = J_1(-1)$. Our result is:

Theorem 5.1. The following are equivalent:

(a) C_{K_1} is hypoelliptic.

(b) C_{K_1} is analytic hypoelliptic.

(c) There exists $K_2 \in \mathcal{AK}^{-k-2n-2}$ so that $K_2 * K_1 = \delta$.

(d) $(\mathcal{F}K_1)(0,p,q) \neq 0$ for $(p,q) \neq 0$; $J_{1+}v = 0$, $v \in \mathcal{H}_1^\infty \Rightarrow v = 0$; and $J_{1-}v = 0$, $v \in \mathcal{H}_{-1}^\infty \Rightarrow v = 0$.

(e) $(\mathcal{F}K_1)$ $(0,p,q) \neq 0$ for $(p,q) \neq 0$; $J_{1+}v = 0$, $v \in \mathcal{H}_1^\omega \Rightarrow v = 0$; and $J_{1-}v = 0$, $v \in \mathcal{H}_{-1}^\omega \Rightarrow v = 0$.

(d) may be stated loosely in an appealing way, which we now describe. If π is any irreducible unitary representation of $\mathbb{H}^n$, one can always define

$$\pi(f) = \int_{\mathbb{H}^n} f(u)\pi(u)du \quad \text{for } f \in L^1.$$

If π is infinite-dimensional, then $\pi(t,x,y) = U^\lambda_{(t,x,y)}$ for some $\lambda \in \mathbb{R}\setminus\{0\}$, and $\pi(f) = \hat{f}(\lambda)$. There are also one-dimensional representations $\pi_{p,q}(t,x,y) = e^{i(x\cdot p+y\cdot q)}$ for $(p,q) \in \mathbb{R}^n \times \mathbb{R}^n$; this leads to $\pi_{p,q}(f) = (\mathcal{F}f)(0,p,q)$. Then (d), loosely stated, is the assertion that $\pi(K_1)$ is injective on Schwartz vectors for all irreducible unitary representations π of $\mathbb{H}^n$, except $\pi \equiv 1$.

The case, in Theorem 5.1, where $K = L\delta$ is supported at 0 was known before ([81], [77], [78], [62], [29]). This case in fact follows at once from Theorem 4.1 and its proof--it is the case $P = 0$ of that theorem.

The proof of Theorem 5.1 follows the pattern of that of Theorem 4.1. The key lemma is a replacement for Theorem 3.5; it is this.

Lemma 5.6. Suppose $k \in \mathbb{C}$, $K_1 \in \mathcal{A}\mathcal{K}^{k-2n-2}(\mathbb{H}^n)$, and $\mathcal{F}K_1(0,p,q) \neq 0$ for $(p,q) \neq 0$. Then there exists $K_3 \in \mathcal{A}\mathcal{K}^{-k-2n-2}$ and $K_4 \in \mathcal{A}\mathcal{K}^{-2n-2}$ so that $K_3 * K_1 = \delta - K_4$ and so that $F_{4\lambda}(p,q) = (\mathcal{F}K_4)(\lambda,p,q) \in Z_2^2(\mathbb{R}^{2n})$ for all $\lambda \neq 0$.

Once this is known, the group Fourier transform can be used just as in Section 4 to complete the proof.

In proving Lemma 5.6, one uses the following lemma. It has appeared in the C^∞ situation in various guises before; it is a form of the product formula for the Weyl calculus [37], [49], [60], [79]). In particular, [79] uses this formula in the C^∞ situation for the same purposes that we use it in the real analytic situation.

We change our notation slightly. $\mathbb{H}^n$ has coordinates $(t,x,y) \in \mathbb{R} \times \mathbb{R}^n \times \mathbb{R}^n$ with dual coordinates $(\lambda,\xi) \in \mathbb{R} \times \mathbb{R}^{2n}$. We use this notation for differential operators on $\mathbb{R}^{2n}_\xi$:

$$\partial = (\partial/\partial\xi_1,\ldots,\partial/\partial\xi_{2n}),$$

$$\tilde{\partial} = (\partial/\partial\xi_{n+1},\ldots,\partial/\partial\xi_{2n},-\partial/\partial\xi_1,\ldots,-\partial/\partial\xi_n).$$

If $\alpha \in (\mathbb{Z}^+)^{2n}$ is a multi-index, ∂^α and $\tilde{\partial}^\alpha$ will have the obvious meanings.

<u>Lemma</u>. Suppose $K_1, K_2 \in AK^{-2n-2}$, $FK_j(1,\zeta) \sim \sum g_{j\ell}(\zeta)$ in $Z^2_{2,0}$ $(j=1,2)$. Let $K = K_2 * K_1$, say $FK(1,\zeta) \sim \sum g_\ell(\zeta)$ in $Z^2_{2,0}$. Then

$$g_\ell = \sum_{a+b+|\alpha|=\ell} [(2i)^{|\alpha|}/\alpha!]\,\partial^\alpha g_{2a}\,\tilde{\partial}^\alpha g_{1b}.$$

A simple proof of this lemma, arguing directly from the group law, is included in the text; it is a special case of Lemma 5.7.

An Analytic Weyl Calculus

In Chapter 6, we discuss a calculus of pseudodifferential operators on $\mathbb{R}^n$ which is intimately related (through representation theory) to the algebra of singular integral operators on $\mathbb{H}^n$. The C^∞ version of the calculus was first investigated by Grossman, Loupias and Stein [37]. Its connection with the matters discussed in Chapters 4 and 5 was noted previously by Howe [49], Melin [60] and Taylor [79]. Further results about this calculus follow as special cases of the results of Beals [7]. We shall introduce a new analytic analogue.

Whereas the standard pseudodifferential calculus is the natural one for dealing with "Laplacian-like" (elliptic) operators, the calculus which we describe in Chapter 6 is the natural one for working with "Hermite-like" operators. A Hermite-like operator R in the C^∞ calculus will have these properties:

$$R: \mathcal{S}'(\mathbb{R}^n) \to \mathcal{S}'(\mathbb{R}^n)$$

$$R: \mathcal{S}(\mathbb{R}^n) \to \mathcal{S}(\mathbb{R}^n)$$

and if $f \in \mathcal{S}'$ and $Rf \in \mathcal{S}$, then $f \in \mathcal{S}$.

If R is a Hermite-like operator in the analytic calculus, it will have the additional properties:

$$R: Z_2^2(\mathbb{R}^n) \to Z_2^2(\mathbb{R}^n) \tag{0.53}$$

and $$\text{if } f \in S' \text{ and } Rf \in Z_2^2, \text{ then } f \in Z_2^2. \tag{0.54}$$

(It was previously known that the Hermite operator itself does have these properties; see [26], [62], [60].)

Our operators are defined through the usual formula for pseudodifferential operators in the Weyl calculus (see [48] for a general study of the Weyl calculus). The symbols here, however, are of a more restricted type than usual, and this is why the very strong properties above can be proved. To be specific, we say that a function r(p,q) is an analytic symbol of order k if it is in $Z_{2,k}^2(\mathbb{R}^{2n}) + Z_{2,k-1}^2(\mathbb{R}^{2n})$. In particular, r must have an asymptotic expansion at infinity. The advantage of adding $Z_{2,k}^2$ to $Z_{2,k-1}^2$ is to permit the homogeneity degrees in the expansion to decrease by one each time, as opposed to the "two each time" forced by $Z_{2,k}^2$ alone. The symbol of the Hermite operator is $|p|^2 + |q|^2$.

The main results in Chapter 6 are these. First, the composition of two analytic operators is again an analytic operator. Second, if an analytic operator R is "Hermite-like" in the sense that its symbol does not vanish for $(p,q) \neq (0,0)$, then there is an analytic operator S such that

$$SR = I-P; \tag{0.55}$$

here P is an error which is "small" in the sense that $P: \mathcal{S}' \to Z_2^2$. P may be taken to be the projection onto the null space of R; this null space is a finite-dimensional space of $Z_2^2(\mathbb{R}^n)$ functions.

To see the connection with analysis on $\mathbb{H}^n$, we first look at (0.49), (0.50) and Figure 1. It would appear that

$$\mathcal{W}_\lambda(p_k) = P_{k\lambda},\ \mathcal{W}_\lambda(q_k) = Q_{k\lambda} \tag{0.56}$$

This seems even clearer when we realize that, heuristically, from (0.49) and (0.50),

$$\mathcal{F}_\lambda(\partial\delta/\partial x_k) = -ip_k,\ (\partial\delta/\partial x_k)\hat{}\,(\lambda) = -iP_{k\lambda}$$

and similarly for y_k; here δ is the Dirac delta distribution.

In the Schrödinger representation with $\lambda = 1/4$, (0.56) becomes

$$\mathcal{W}_{1/4}(p_k) = p_k;\ \mathcal{W}_{1/4}(q_k) = -i\partial/\partial p_k.$$

It would seem from this equation that the map $\mathcal{W}_{1/4}$ must be a very familiar one: namely, the map which assigns to a symbol $r(p,q)$ its associated pseudodifferential operator. As it turns out, this is true if one uses the Weyl calculus instead of the usual calculus. Thus, if, for instance, $r \in \mathcal{S}(\mathbb{R}^{2n})$, one can show that $\mathcal{W}_{1/4}r = R$ where

$$(Rv)(p) = (2\pi)^{-n}\iint r((p+p')/2,q)e^{i(p-p')\cdot q}v(p')dp'dq \quad \text{for } v \in \mathcal{S}(\mathbb{R}^n). \tag{0.57}$$

(This differs from the usual calculus by our having $r((p+p')/2,q)$ instead of $r(p,q)$.)

One may use larger classes of r than just $\mathcal{S}(\mathbb{R}^{2n})$. Thus, if $K \in AK^k$, we show that if $r = \mathcal{F}_{1/4}K$ and $R = \hat{K}(1/4)$ (taken in the $\mathcal{Q}$ sense), then r and R are related by (0.57). (This is Lemma 6.4.)

If r and R are related by (0.54), let us say $R = Op(r)$.

Theorem 5.1 has much to tell us about operators R of the form $Op(r)$ for $r \in Z^2_{2,j}$. With $k = -2n-2-j$, there exists $K_1 \in AK^k$ with $\mathcal{F}_{1/4}K_1 = r$. The first condition of Theorem 5.1 (e)—that $(\mathcal{F}K_1)(0,p,q) \neq 0$ for $(p,q) \neq 0$—has a simple interpretation in terms of the behavior of r itself. By Theorem 2.11 (see the first part of the Introduction), if $r \sim \sum g_\ell$ in $Z^2_{2,j}$ then $g_0(p,q) = (\mathcal{F}K_1)(0,p,q)$. Thus Theorem 5.1 tells us that if $g_0(p,q) \neq 0$ for $(p,q) \neq 0$, and if $R = \hat{K}_1(1/4)$ and $\hat{K}_1(-1/4)$ are injective on $\mathcal{H}^\omega$ vectors, then there exists $s \in Z^2_{2,-j}$ such that if $S = Op(s)$ then $SR = I$. Actually, there is no need to assume $\hat{K}_1(-1/4)$ is injective on $\mathcal{H}^\omega$ vectors; it is enough to assume $g_0(p,q) \neq 0$ for $(p,q) \neq 0$ and that R is injective on $\mathcal{H}^\omega_{1/4} = Z^2_2(\mathbb{R}^n)$.

We take as our class of analytic symbols $AS^j = Z^2_{2,j} + Z^2_{2,j-1}$, as we indicated above. If $r(p,q) = |p|^2 + |q|^2$, note that Op(r) is the Hermite operator $|p|^2 - \Delta$. In this case, $r = F_{1/4}(4L_o\delta)$ and $R = (4L_o\delta)\hat{}(1/4)$, where $L_o = -(1/2)\sum_{m=1}^{n}(Z_m\bar{Z}_m + \bar{Z}_m Z_m)$ is the sublaplacian of Folland-Stein [20].) If $r \in AS^j$, we say Op(r) is <u>Hermite-like</u> if $r \sim \sum_{\ell=0}^{\infty} r_\ell$ where $r_o(p,q) \neq 0$ for $(p,q) \neq 0$.

In Theorem 6.1 and 6.2 we prove the properties of the calculus alluded to above, such as (0.53) for any operator R in the calculus, and (0.54), (0.55) if R is in addition Hermite-like. We also show that if $R_i \in Op(AS^{j_i})$ $(i=1,2)$ then $R_2R_1 \in Op(AS^{j_1+j_2})$. To prove these theorems, we use the results and methods of Chapters 4 and 5.

This bring us to the end of the Introduction and to the main text. A somewhat more detailed précis of the first five chapters is given in [30].

1. Homogeneous Distributions

We assume given an n-tuple of positive rationals $\underset{\sim}{a} = (a_1,\ldots,a_n)$. We put $Q = \sum a_\ell$. For $x \in \mathbb{R}^n$ we put $D_r x = (r^{a_1}x_1,\ldots,r^{a_n}x_n)$. For f a function on $\mathbb{R}^n$, $r > 0$, we define the functions $D^r f, D_r f$ by $D^r f(x) = f(D_r x)$, $D_r f = r^{-Q} D^{1/r} f$. For $F \in S'(\mathbb{R}^n)$ we define $D^r F, D_r F \in S'$ by $(D^r F|g) = (F|D_r g)$, $(D_r F|g) = (F|D^r g)$ for $g \in S$. (Here and elsewhere $(F|g)$ denotes the sequilinear pairing, linear in g). For $k \in \mathbb{C}$, we say that F is homogeneous of degree k if $D^r F = r^k F$ for all $r > 0$. We let $\mathrm{Rhom}_k = \mathrm{Rhom}_k(\underset{\sim}{a}) = \{K \in S' | K$ is homogeneous of degree k and is C^∞ away from $0\}$. Such K are called regular homogeneous distributions. In studying $\mathrm{Rhom}_k(\underset{\sim}{a})$ there is evidently no loss in assuming all a_ℓ are positive integers, for we can always multiply them all by a common demoninator, changing k. For a while we make this assumption. For $x \in \mathbb{R}^n$, we let $|x| = (\sum x_\ell^{2a'_\ell})^{1/2A}$, where $A = \prod_{m=1}^{n} a_m$, $a'_\ell = A/a_\ell$. Then $|\,|$ is homogeneous of degree one, satisfies $|x| \geqq 0$ for all x, $|x| = 0 \iff x = 0$. For these reasons $|\,|$ is called a homogeneous norm function. Note $|\,|$ is also (real) analytic away from 0.

Let G be a homogeneous function of degree j which is locally integrable away from 0; write $G(x) = \Omega(x)|x|^j$ where Ω is homogeneous of degree 0. One then knows [53] that there exists a number $M(\Omega)$ such that for any $0 < A < B$ we have

$$\int_{A<|x|<B} G(x)\,dx = M(\Omega)(Q+j)^{-1}(B^{Q+j}-A^{Q+j}) \qquad \text{if } j \neq -Q$$
$$= M(\Omega)\log(B/A) \qquad \text{if } j = -Q \tag{1.1}$$

We put $M(G) = M(\Omega)$. In particular, $G \in L^1_{loc}$ if $\operatorname{Re} j > -Q$, and G is in L^1 at ∞ for $\operatorname{Re} j < -Q$. It is well-known [53] that if $\operatorname{Re} k > -Q$, $K \in \mathrm{Rhom}_k$ and $K = f$ away from 0, where $f \in C^\infty(\mathbb{R}^n \setminus \{0\})$, then $K = f$ in the S' sense. (Indeed $K - f$ must be supported at 0, and homogeneity considerations show it is zero.) Let $\hat{}$ denote Fourier transform. The following proposition is known [66]; for convenience, we include the proof.

<u>Proposition 1.1</u> (a) $(\mathrm{Rhom}_k)\hat{} = \mathrm{Rhom}_{-Q-k}$

(b) Suppose $\operatorname{Re} k > -Q$. There exists $C_1 > 0$ such that whenever $K \in \mathrm{Rhom}_k$ and $|\xi| = 1$, $|\hat{K}(\xi)| < C_1 \sup|\partial^\gamma K(x)|$ where the sup is taken over $|\gamma| \leqq \operatorname{Re} k + 2Q$, $1 \leqq |x| \leqq 2$. More precisely, given $\varepsilon > 0$ and $1 \leqq \ell \leqq n$, there exists $C_2 > 0$ such that whenever $K \in \mathrm{Rhom}_k$, $|\xi| = 1$ and $|\xi_\ell| > \varepsilon$, we have $|\hat{K}(\xi)| < C_2 \sup|\partial_\ell^r K(x)|$ where the sup is taken over $a_\ell r \leqq \operatorname{Re} k + Q + a_\ell$, $1 \leqq |x| \leqq 2$. (Here $\partial_\ell = \partial/\partial\xi_\ell$.)

<u>Proof</u>. (a) Suppose $K \in \mathrm{Rhom}_k$. Simple considerations show that $\hat{K}$ is homogeneous of degree $-Q - k$. Now select $\varphi \in C_c^\infty$ such that $\varphi = 1$ near 0. Put $K_1 = \varphi K$ and $K_2 = K - K_1$. Since K_1 is a distribution of compact support $\hat{K}_1 \in C^\infty$. (Indeed, let

$g_\xi(x) = e^{ix\cdot\xi}$. Then $\hat{K}_1(\xi) = K_1(g_\xi)$ so that $\hat{K}_1$ is in fact the restriction to $\mathbb{R}^n$ of an entire function of exponential type.) Next, with α an n-tuple, $1 \leqq \ell \leqq n$, put $K_2^{N\alpha\ell}(x) = \partial_\ell^N x^\alpha K_2(x)$ for $Na_\ell > \operatorname{Re} k + |\alpha| + Q$. (Here and elsewhere, $|\alpha| = \underset{\sim}{a}\cdot\alpha$.) Then, for large x, $K_2^{N\alpha\ell}$ is homogeneous of degree m where $\operatorname{Re} m < -Q$, and therefore $K_2^{N\alpha\ell} \in L^1$. Thus $\xi_\ell^N D'\hat{K}_2$, is continuous on $\mathbb{R}^n$ if D' is the differential monomial $D' = (\partial/\partial\xi)^\alpha$ and N is as before. If $\xi \neq 0$, some $\xi_\ell \neq 0$, so $\hat{K}_2$ is smooth away from 0. Thus $\hat{K}$ is as well.

(b) We keep the notation of the proof of (a). Suppose $\varphi = 1$ for $|x| < 1$ and $\varphi = 0$ for $|x| > 2$. If $\operatorname{Re} k > -Q$, $K_1 \in L^1$. Thus $|\hat{K}_1(\xi)| \leqq |K_1|_1 < C(\sup_{|x|=1} |K(x)|) \int_{|x|<2} |x|^{\operatorname{Re} k} dx < C \sup_{|x|=1} |K(x)|$. (Here and elsewhere, C denotes a constant which may be different in different appearances.) Also, if $1 \leqq \ell \leqq n$, pick $N = N_\ell$ with $\operatorname{Re} k + Q < Na_\ell \leqq \operatorname{Re} k + Q + a_\ell$. Then $|\xi_\ell|^N |\hat{K}_2(\xi)| \leqq$
$\|\partial^N K_2\|_1 < C \sup_{1<|x|<2,\, r\leq N} |\partial_\ell^r K(x)| + C \sup_{|x|=1} |\partial^N K(x)| \int_{|x|>2} |x|^{\operatorname{Re} k - Na_\ell} dx <$
$C \sup_{|x|=1,\, r\leq N} |\partial_\ell^r K(x)|$. Adding these estimates for $\hat{K}_1$ and $\hat{K}_2$ we find the "more precise" form of (b); the "less precise" form follows, since for some $\varepsilon > 0$, $|\xi| = 1$ implies $|\xi_\ell| > \varepsilon$ for some ℓ. ∎

If $\operatorname{Re} j \leqq -Q$, we can obtain more insight into Rhom_j by considering the distributions which follow. For G a homogeneous function of degree $j \in \mathbb{C}$ which is smooth away from 0, we define $\Lambda_G \in S'$ by

$$\Lambda_G(\varphi) = \int_{|\xi|\leq 1} [\varphi(\xi) - \sum_{|\alpha|\leq N} \varphi^{(\alpha)}(0)\xi^\alpha/\alpha!]G(\xi)d\xi$$
$$+ \sum_{\substack{|\alpha|\leq N \\ |\alpha|\neq -Q-j}} (Q+|\alpha|+j)^{-1}M(\xi^\alpha G)\varphi^{(\alpha)}(0)/\alpha!$$
$$+ \int_{|\xi|>1} \varphi(\xi)G(\xi)d\xi \tag{1.2}$$

where N is chosen arbitrarily with $N \geq -Q - \text{Re}\, j - 1$. By (1.1), it is easy to see that the choice of N is immaterial. In particular, if $\text{Re}\, j > -Q$, we can choose $N = -1$, thus $\Lambda_G = G$. Suppose instead that $\text{Re}\, j \leq -Q$. Say $-j - Q \notin \mathbb{Z}^+ = \{0,1,2,\ldots\}$. It is then easy to compute, using (1.1), that $\Lambda_G \in \text{Rhom}_j$. Further, if $K \in \text{Rhom}_j$, say $K = G$ away from 0; then $K - \Lambda_G$ is supported at 0 and homogeneous of degree j; it follows easily that $K - \Lambda_G = 0$. Thus, if $-j - Q \notin \mathbb{Z}^+$, it follows that $\text{Rhom}_j = \{\Lambda_G | G \in C^\infty(\mathbb{R}^n \setminus \{0\}),\ G \text{ homogeneous of degree } j\}$. We discuss the case $-j - Q \in \mathbb{Z}^+$ presently.

First, however, we introduce a space of distributions which includes Rhom_k when $k \in \mathbb{Z}^+$. If $k \notin \mathbb{Z}^+$, we let $K^k = \text{Rhom}_k$. If $k \in \mathbb{Z}^+$, we let $K^k = \text{Rhom}_k + \{p(x)\log|x| : p \text{ a homogeneous polynomial of degree } k\}$. We determine $\hat{K}^k$. For this, suppose $j \in \mathbb{C}$. If $-j - Q \notin \mathbb{Z}^+$, we let $J^j = \text{Rhom}_j$. If $-j - Q \in \mathbb{Z}^+$, we let $J^j = \{\Lambda_G | G \in C^\infty(\mathbb{R}^n \setminus \{0\}),\ G \text{ homogeneous of degree } j\} + \{P(\delta)\delta | P$

a homogeneous polynomial of degree $-Q - j\}$. Here, of course, if $P(x) = \sum c_\alpha x^\alpha$, then $P(\partial)\delta = \sum c_\alpha \partial^\alpha \delta$.

Proposition 1.2 (a) $\hat{K}^k = J^j$ for $k = -Q - j$.

(b) Proposition 1.1(b) holds for "K^k" in place of "Rhom_k".

Proof. We may assume $k \in \mathbb{Z}^+$ and set $j = -Q - k$. Let $K \in K^k$, $J \in J^j$, $K = F + p \log|x|$ ($F \in \mathrm{Rhom}_k$), $J = \Lambda_G + P(\partial)\delta$. The same proof as that of Proposition 1.1(a) shows that $\hat{K}$ and $\check{J}$ are each smooth away from 0. Next note that

$$D^r K - r^k K = r^k (\log r) p \tag{1.3}$$

while

$$D_r J - r^k J = r^k (\log r) q(\partial)\delta \tag{1.4}$$

by a computation using (1.1), where $q(\partial) = -\sum_{|\alpha|=k} [\mu(\xi^\alpha G)/\alpha!]\partial^\alpha$.

Put $\hat{K} = J'$; suppose $J' = G'$ away from 0. By (1.3), $D_r J' - r^k J' = r^k (\log r) p(-i\partial)\delta$. Thus if $\varphi \in S$ is supported away from 0, $(D_r J' - r^k J')(\varphi) = 0$. It follows easily that G' is homogeneous of degree j. Put $J'' = J' - \Lambda_{G'}$. By (1.4) for Λ_G in place of J, $D_r J'' - r^k J'' = r^k \log r\, q''(\partial)\delta$ for some homogeneous polynomial q'' of degree k. However, J'' is supported at the origin; say $J'' = \sum c_\alpha \partial^\alpha \delta$. Then $D_r J'' - r^k J'' = \sum c_\alpha (r^{|\alpha|} - r^k)\partial^\alpha \delta$. Putting $r = 2$, we find that $c_\alpha = 0$ unless $|\alpha| = k$. Thus $J' = \Lambda_{G'} + J'' \in J^j$.

Conversely, put $\check{J} = K'$. By (1.4), $D^r K' - r^k K' = r^k(\log r)p'$ for some homogeneous polynomial p' of degree k. Put $K'' = K' - p'(x)\log|x|$. Then $K'' \in \mathrm{Rhom}_k$, so $K' \in \mathcal{K}^k$ as desired. This proves (a).

For (b), let $S = \sup|K(x)|$ for $1 \leq |x| \leq 2$. Choose ϕ as in Proposition 1.1(b), and let $K_1 = \phi K$. Imitating the proof of Proposition 1.1(b), we see that it suffices to show that $\|K_1\|_1 < CS$ for some $C > 0$, depending only on k. For this, write $K = K' + p(x)\log|x|$. If $|x| = 1$, $K(x) = K'(x)$. Thus $|K'(x)| \leq |x|^k S$ for all x. If $|x| = 2$, $p(x) = [K(x) - K'(x)]/\log 2$. So $|p(x)| \leq C_1|x|^k S$ for all x, where C_1 depends only on k. Finally, then,

$$|K(x)| \leq C_2|x|^k(1+\left|\log|x|\right|)S$$

for all x, C_2 depending only on k, and (b) follows.

Suppose now that $j \in \mathbb{C}$, $-j-Q \in \mathbb{Z}^+$. It follows at once from Proposition 1.2 that $\mathrm{Rhom}_j \subset \mathcal{J}^j$. This enables us to clearly understand Rhom_j if $-j-Q \in \mathbb{Z}^+$. Using (1.4) and the explicit formula for q, we see that, in this case, $\mathrm{Rhom}_j = \{\Lambda_G | G \in C^\infty(\mathbb{R}^n\setminus\{0\})$, G homogeneous of degree j, $\mu(\xi^\alpha G) = 0$ whenever $|\alpha| = -j-Q\} + \{P(\partial)\delta | P$ a homogeneous polynomial of degree $-j-Q\}$.

It is often useful, by the way, to know the fact that $\partial_\ell \mathcal{K}^k \subset \mathcal{K}^{k-a_\ell}$ for any k,ℓ. To see this, we may assume $k \in \mathbb{Z}^+$;

say $K \in K^k$. Select $\varphi \in C^\infty(\mathbb{R}^n)$ such that $\varphi(x) = 0$ for $|x| \leq 1$ and $\varphi(x) = 1$ for $|x| \geq 2$. Let $\varphi_\varepsilon = D^{1/\varepsilon}\varphi$. Away from 0, $\partial_\ell K$ equals a smooth function K'; it suffices to show that $\partial_\ell K = K'$ in the S' sense. Simply let $\varepsilon \to 0$ in $(K'|\varphi_\varepsilon g) = -(K|\partial_\ell(\varphi_\varepsilon g))$ and use the fact that $\lim_{\varepsilon\to 0} \varepsilon^{-a_\ell} \int_{\varepsilon<|x|<2\varepsilon} |K(x)|dx = 0$.

We turn now to analyticity. Let $AK^k = \{K \in K^k | K \text{ is real analytic away from } 0\}$, and $AJ^j = (AK^{-Q-j})\hat{}$. We seek to characterize AJ^j. Here is the main idea. For α a multi-index, we write $\|\alpha\| = \alpha_1+\ldots+\alpha_n$. Suppose $K \in AK^k$; we shall examine $K_{\alpha\beta} = \partial^\beta x^\alpha K$ for multi-indices α, β with $|\alpha| = |\beta|$. Note all $K_{\alpha\beta} \in AK^k$. We have that for some $R \geq 0$, $|x^\alpha\partial^\beta K(x)| < CR^{\|\beta\|}\beta!$ whenever $|x| = 1$. It is easy to believe the variant $|K_{\alpha\beta}(x)| < CR^{\|\beta\|}\beta!$ $(|x|=1)$, and that for any fixed N, $\sup|\partial^\gamma K_{\alpha\beta}(x)| < CR^{\|\beta\|}\beta!$ (sup over $|\gamma| \leq N$, $|x| = 1$). By Proposition 1.1(b), then, we have for $|\xi| = 1$ that $|\hat{K}_{\alpha\beta}(\xi)| < CR^{\|\beta\|}\beta!$. That is, if $\hat{K} = J$,

$$|\xi^\beta\partial^\alpha J(\xi)| < CR^{\|\beta\|}\beta! \quad \text{for } |\xi| = 1,\ |\alpha| = |\beta|. \tag{1.5}$$

The analogue of Proposition 1.1(b) for the inverse Fourier transform similarly shows that (1.5) is not only necessary for J to be in AJ^j but is also sufficient, in that one can derive $|K_{\alpha\beta}(x)| < CR^{\|\beta\|}\beta!$ (for $|x| = 1$) from (1.5). We shall prove everything rigorously in a moment.

Actually, we shall be using not (1.5) but rather several equivalent variants of it. These we prove below in Theorem 1.3, and discuss briefly now; particularly we wish to indicate the significance of these conditions.

First, however, here are a few simple inequalities that we shall frequently use. If $|\alpha| = |\beta|$, we have

$$\|\alpha\| \leqq |\alpha| = |\beta| \leqq Q\|\beta\|, \; \|\beta\| \leqq |\beta| = |\alpha| \leqq Q\|\alpha\|. \tag{1.6}$$

Also, by the multinomial theorem, for any multi-index β, we have

$$\beta! \leqq \|\beta\|! \leqq n^{\|\beta\|}\beta! \; . \tag{1.7}$$

Again by the multinomial theorem, if $m,r \in \mathbb{N}$, then

$$(r!)^m \leqq (mr)! \leqq m^{mr}(r!)^m. \tag{1.8}$$

Returning now to (1.5), we examine a special case, when $\xi^\beta = \xi_\ell^r$ for some ℓ, $1 \leqq \ell \leqq n$. In that case $\|\beta\| = r = |\alpha|/a_\ell$. Consequently, using (1.6) and (1.8) we have that for some C,R,

$$|\xi_\ell|^{|\alpha|/a_\ell}|\partial^\alpha J(\xi)| < CR_1^{\|\alpha\|}(|\alpha|!)^{1/a_\ell} \text{ for all } \ell, \; |\xi| = 1. \tag{1.9}$$

This argument works for all α, provided $|\alpha|/a_\ell$ is an integer. Even if it is not, it is not hard to prove (1.9), and we do so below. Further, we shall show that the conditions (1.9) for all α are sufficient to imply $J \in \mathcal{AJ}^j$. Using (1.7), (1.8), we conclude

$$|\xi_\ell|^{|\alpha|/a_\ell}|\partial^\alpha J(\xi)| < CR_2^{\|\alpha\|}(\alpha_1!)^{a_1/a_\ell}\cdots(\alpha_n!)^{a_n/a_\ell} \tag{1.10}$$

for all $\ell,\alpha,|\xi| = 1$.

Assuming $a_1 \geqq a_2 \geqq\ldots\geqq a_n$, $a_k/a_1 \leqq 1$ and it follows readily that J is real analytic away from $\xi_1 = 0$. Even more, suppose $a_1 > a_n$ and for fixed $\xi' = (\xi_1,\ldots,\xi_{n-1})$ with $\xi_1 \neq 0$, set $F(\xi_n) = J(\xi',\xi_n)$. Then F is better than real analytic--it is the restriction to $\mathbb{R}$ on an entire function on $\mathbb{C}$. Along $\{\xi_1 = 0\}$, J may well be worse than real analytic. This agrees with the statements in the introduction about the case $\underset{\sim}{a} = (p,1,\ldots,1)$. We study that case in detail after Theorem 1.3.

With this motivation in mind, we turn to the details. (b) below is a restatement of (1.10), while (c) and (d) are variants.

<u>Theorem 1.3</u>. For $J \in J^j$, the following are equivalient:

(a) $J \in AJ^j$

(b) For some C,R,

$$|\xi_\ell|^{|\alpha|/a_\ell}|\partial^\alpha J(\xi)| < CR^{\|\alpha\|}(\alpha_1!)^{a_1/a_\ell}\cdots(\alpha_n!)^{a_j/a_\ell}$$

for all $\ell,\alpha,|\xi| = 1$.

(c) For some C,R, $|\xi_\ell|^{a_m s/a_\ell}|\partial_m^s J(\xi)| < CR^s s!^{a_m/a_\ell}$

for all $s \in \mathbb{Z}^+$, all ℓ,m $(1\leqq m\leqq n)$, and $|\xi| = 1$.

(d) For some C_0,R_0, $|\xi_1|^{\eta_1 s}\cdots|\xi_n|^{\eta_n s}|\partial_m^s J(\xi)| < C_0 R_0^s s!^{(\eta_1+\ldots+\eta_n)}$

for all m,s, $|\xi| = 1$,

and all n-tuples η of nonnegative real numbers with $\underset{\sim}{a}\cdot\eta = a_m$.

(We use the convention $0^0 = 1$ if some $\xi_\ell = \eta_\ell = 0$ in (d); similarly in (b), (c).) Further, with $k = -Q - j$, we have these uniformities:

(1) Suppose Re $k > -Q$. Then, for any $R_1 > 0$ there exist $C, R, C_0, R_0 > 0$ so that the inequalities of (b), (c) and (d) hold whenever $J = \hat{K}$ for a $K \in AK^k$ which satisfies $\sup_{1\leq|x|\leq 2} |\partial^\gamma K(x)| < R_1^{\|\gamma\|}\gamma!$ for all $\gamma \in (\mathbb{Z}^+)^n$.

(2) Suppose Re $j > -Q$. Then for any $C, R, C_0, R_0 > 0$ there exist $C_1, R_1 > 0$ so that if the inequalities of (b), (c) or (d) hold and if $J = \hat{K}$ then $\sup_{1\leq|x|\leq 2} |\partial^\gamma K(x)| < C_1 R_1^{\|\gamma\|}\gamma!$ for all $\gamma \in (\mathbb{Z}^+)^n$.

<u>Proof</u>. We prove (a) $\Rightarrow$ (b) $\Rightarrow$ (c) $\Rightarrow$ (d) $\Rightarrow$ (a). (b) $\Rightarrow$ (c) is evident, if one takes $\partial^\alpha = \partial_m^s$ in (b). Also (d) $\Rightarrow$ (c) is evident, if one takes $\eta_i = \delta_{i\ell} a_m/a_\ell$ in (d). As we said, we would rather have (c) $\Rightarrow$ (d); this is shown as follows. Assuming (c), write $\eta_\ell = \tau_\ell a_m/a_\ell$; then $\sum \tau_\ell = 1$ and

$$|\xi_1|^{\eta_1 s} \cdots |\xi_n|^{\eta_n s} |\partial_m^s J(\xi)| = \prod_{\ell=1}^{n} (|\xi_\ell|^{a_m s/a_\ell} |\partial_m^s J(\xi)|)^{\tau_\ell}$$
$$< \Pi (CR^s s!^{a_m/a_\ell})^{\tau_\ell} = CR^s s!^{(\eta_1+\cdots+\eta_n)}$$

as desired. Thus the heart of the matter is (a) $\Rightarrow$ (b) and (d) $\Rightarrow$ (a). These we have motivated above; we present the tiresome detailed proof now. The reader may skip this proof without loss of continuity. The uniformity assertion will follow from the proofs of (a) ⇒ (b) and of (b) ⇒ (c) ⇒ (d) ⇒ (a). (These, and future uniformity assertions, will be used in Chapters 7 and 8.)

<u>(a) ⇒ (b)</u>. Suppose $J = \hat{K}$, $K \in AK^k$. Fix ℓ.

As we have motivated, we wish to examine distributions that look like $\partial_\ell^{|\alpha|/a_\ell} x^\alpha K$, and apply Proposition 1.1(b). The evident problems—that $|\alpha|/a_\ell$ might not be an integer, and that we might not have $\operatorname{Re} k > -Q$, are easily dealt with as follows. First, fix $p \in \mathbb{Z}^+$ with $a_\ell p > -(\operatorname{Re} k+Q)$; if $\operatorname{Re} k > -Q$, we could choose $p = 0$. We need prove (b) only for those α with $[|\alpha|/a_\ell] \geqq p$. ([ ] denotes greatest integer function.) If $[|\alpha|/a_\ell] \geqq p$, put $K_\alpha = \partial_\ell^{[|\alpha|/a_\ell]-p} x^\alpha K$. Then $K_\alpha \in K^{k(\alpha)}$ where $k(\alpha) = k + a_\ell p + |\alpha| - a_\ell[|\alpha|/a_\ell]$. Thus $k(\alpha) \in \{k + a_\ell p, k + a_\ell p+1, \ldots, k + a_\ell p + a_\ell - 1\}$ and in particular, $\operatorname{Re} k(\alpha) > -Q$. It suffices then to show that

$$|\partial^\gamma K_\alpha(x)| < CR^{\|\alpha\|}(\alpha_1!)^{a_1/a_\ell}\ldots(\alpha_n!)^{a_n/a_\ell} \tag{1.11}$$

for $1 \leqq |x| \leqq 2$ and

$$|\gamma| \leqq N = \operatorname{Re} k + a_\ell p + a_\ell - 1 + 2Q. \tag{1.12}$$

For by Proposition 1.2(b), for $|\xi| = 1$ we would then have $|\xi_\ell|^{[|\alpha|/a_\ell]-p}|\partial^\alpha J(\xi)|$ is less than or equal to the right side of (1.11). Since $|\xi_\ell|^{|\alpha|/a_\ell}|\partial^\alpha J(\xi)|$ is less than or equal to the former quantity for $|\xi| = 1$, (b) would then follow.

For (1.11), it suffices to show that

$$|\partial^\gamma K_\alpha(x)| < CR^{\|\alpha\|}[|\alpha|/a_\ell]! \tag{1.13}$$

for $1 \leqq |x| \leqq 2$, $|\gamma| \leqq N$. Indeed, $[|\alpha|/a_\ell]!^{a_\ell} \leqq |\alpha|! \leqq n^{|\alpha|}(a_1\alpha_1)!\ldots(a_n\alpha_n)! \leqq n^{|\alpha|}a_1^{a_1\alpha_1}\ldots a_n^{a_n\alpha_n}\alpha_1!^{a_1}\ldots\alpha_n!^{a_n} \leqq R_1^{\|\alpha\|}\alpha_1!^{a_1}\ldots\alpha_n!^{a_n}$ for some $R_1 > 0$, by (1.8), (1.7), (1.8), (1.6). Thus (1.11) is a consequence of (1.13).

(1.13) would surely be true if K_α were replaced by $x^\alpha\partial_\ell^{[|\alpha|/a_\ell]-p}K$. For the required variant, we proceed as follows. Consider $\mathbb{R}^n \subset \mathbb{C}^n$ in the usual fashion. For $r > 0$, $x \in \mathbb{R}^n$, let $\mathcal{B}_r(x) = \{z \in \mathbb{C}^n |$ for all $\ell, |z_\ell - x_\ell| < r\}$. Also let $\mathcal{B}_r = \bigcup_{1\leqq|x|\leqq2} \mathcal{B}_r(x)$. Now, for some $r > 0$, $K|_{\mathcal{B}_r\cap\mathbb{R}^n}$ has an extension to a bounded analytic function on $\mathcal{B}_r$, which we also denote K. For $z \in \mathcal{B}_r$, $|z_\ell| < C_r$ for all ℓ, for some $C_r > 0$. Accordingly we have an estimate

$$|z^\alpha K(z)| < C(C_r)^{\|\alpha\|} \tag{1.14}$$

for $z \in \mathcal{B}_r$. Thus, for all β,

$$|\partial^\beta x^\alpha K(x)| < C(C_r/r)^{\|\alpha\|}\|\beta\|! \quad \text{if } 1 \leqq |x| \leqq 2 \tag{1.15}$$

by (1.6). Thus if $|\gamma| \leqq N$, and $R_2 = C_r/r$, we have

$|\partial^\gamma K_\alpha(x)| < CR_2^{\|\alpha\|}([|\alpha|/a_\ell]-p+N)!$. But

$([|\alpha|/a_\ell]-p+N)! \leqq ([|\alpha|/a_\ell]+N)! \leqq 2^{|\alpha|/a_\ell+N}[|\alpha|/a_\ell]!N!$

$< CR_3^{\|\alpha\|}[|\alpha|/a_\ell]!$ for some R_3, by (1.6). Thus (1.13) is proved, and we are done. The first uniformity assertion, also, follows from the proof just given.

<u>(d) $\Rightarrow$(a)</u> For each m, fix any $p \in \mathbb{Z}^+$ with $a_m p > -(\text{Re } j + Q)$. For $|\beta| \geqq a_m p$, put $J_{m\beta} = \xi^\beta \partial_m^{[|\beta|/a_m]-p} J$. Then $J_{m\beta} \in J^{j(\beta)}$ where $j(\beta) \in \{j + a_m p, \ldots, j+a_m p + a_m - 1\}$. In particular $\text{Re } j(\beta) > -Q$, so that $J^{j(\beta)} = \text{Rhom}_{j(\beta)}$. Our plan will be to show that for some $C,R > 0$,

$$|\partial_m^q J_{m\beta}(\xi)| < CR^{\|\beta\|}\beta! \tag{1.16}$$

for all m, if $|\xi| \leqq 1$ and $a_m q \leqq N = \text{Re } j + a_m p + 2a_m - 1 + Q$. Once this is known, we shall suppose $\hat{K} = J$. By Proposition 1.1(b), we shall have that for any $\varepsilon > 0$ there exists $B > 0$ such that

$$|\partial^\beta x_m^{[|\beta|/a_m]-p} K(x)| < BR^{\|\beta\|}\beta! \tag{1.17}$$

for $|x| = 1$, $|x_m| > \varepsilon$. The real analyticity of K would follow at once if we had these estimates for $x_m^{[|\beta|/a_m]-p}\partial^\beta K$ (for $|x_m| > \varepsilon$, $|x| = 1$) instead of $\partial^\beta x_m^{[|\beta|/a_m]-p} K$; we shall deal with the desired variant below.

We begin by proving (1.16). For ease in notation, suppose $m = 1$. Write $\xi = (\xi_1,\xi')$, $\beta = (\beta_1,\beta')$. Note that

$$\partial_1^q J_{1\beta}(\xi) = \sum_{i=0}^{q} \binom{q}{i}\beta_1\cdots(\beta_1-i+1)\xi_1^{\beta_1-i}(\xi')^{\beta'}\partial_1^{[|\beta|/a_1]-p+q-i} J(\xi).$$

Now $\beta_1\cdots(\beta_1-i+1) \leqq \beta_1^q$, while $\sum_{i=0}^{q}\binom{q}{i} = 2^q$; then it will suffice to show that for all β with $|\beta| \geqq a_1p$, all i,q with $i \leqq q \leqq [N/a_1]$ and $i \leqq \beta_1$, and all ξ with $|\xi| = 1$, we have $|\xi_1^{\beta_1-i}(\xi')^{\beta'}\partial_1^{[|\beta|/a_1]-p+q-i} J(\xi)| < CR^{\|\beta\|}\beta!$. With $\gamma = (\beta_1-i,\beta')$, $q-p = P$, this is the same as $|\xi^\gamma\partial_1^{[|\gamma|/a_1]+P} J(\xi)| < CR^{\|\gamma\|+i}(\gamma_1+i)!(\gamma')!$ since $|\beta|/a_1-i = |\gamma|/a_1$. It suffices to show this for all P with $-p \leqq P \leqq [N/a_1] - p = M$, say, for all $i \leqq N$, and for all γ with $[|\gamma|/a_1] + P \geqq 0$. It will suffice, then to show that there exist $C,R > 0$ such that for all P with $-p \leqq P \leqq M$ and for all γ with $[|\gamma|/a_1] + P \geqq 0$ we have $|\xi^\gamma\partial_1^{[|\gamma|/a_1]+P} J(\xi)| < CR^{\|\gamma\|}\gamma!$. Note that $|\gamma| - a_1 \leqq a_1[|\gamma|/a_1] \leqq |\gamma|$. Thus it suffices to show that for some C,R, we have $|\xi^\gamma\partial_1^s J(\xi)| < CR^{\|\gamma\|}\gamma!$ for $|\xi| = 1$, for all γ and all s satisfying $s \geqq 0$, $|\gamma| - b_1' \leqq a_1s \leqq |\gamma| + b_2'$ where $b_1' = a_1 + a_1p$, $b_2' = a_1M$. The case $s = 0$ is clear since $|\xi| = 1$. Thus, for (1.16), it suffices to show that if $b_1,b_2 \in \mathbb{Z}^+$, then for some C,R,

$$|\xi^\gamma\partial_1^s J(\xi)| < CR^{\|\gamma\|}\gamma! \tag{1.18}$$

for $|\xi| = 1$, for all γ and all s satisfying $s \geqq 1$, $|\gamma| - b_1 \leqq a_1 s \leqq |\gamma| + b_2$.

To prove (1.18) we consider separately the cases $0 \leqq |\gamma| - a_1 s \leqq b_1$ and $0 < a_1 s - |\gamma| \leqq b_2$. In the first case, let $\sigma = \gamma/s$. Then $\underset{\sim}{a}\cdot\sigma = |\gamma|/s = c$, say, and $c \geqq a_1$. Let $\eta = a_1\sigma/c$; then $\underset{\sim}{a}\cdot\eta = a_1$ and all $\eta_\ell \leqq \sigma_\ell$. Thus $|\xi^\gamma \partial_1^s J(\xi)| = |\xi_1|^{\sigma_1 s} \cdots |\xi_n|^{\sigma_n s} |\partial_1^s J(\xi)| \leqq |\xi_1|^{\eta_1 s} \cdots |\xi_n|^{\eta_n s} |\partial_1^s J(\xi)| < C_o R_o^s s!^{(\eta_1 + \cdots + \eta_n)} \leqq C_o R_o^s s!^{(\sigma_1 + \cdots + \sigma_n)} = C_o R_o^s s!^{\|\gamma\|/s} \leqq C_o R_o^s e^s \|\gamma\|! \leqq C_o (R_o e)^{|\gamma|/a_1} \|\gamma\|! \leqq C_o R_1^{\|\gamma\|}\gamma!$ for some $R_1 > 0$. (We used

$$s!^{m/s} \leqq s^m \leqq m!e^s \quad \text{for } m \in \mathbb{Z}^+; \tag{1.19}$$

we assumed, without loss, that $R_o e \geqq 1$; and we used (1.7).) This concludes the first case.

In the second case, with $0 < a_1 s - |\gamma| \leqq b_2$, we begin by choosing $\varepsilon_o > 0$ so that the sets $T_\ell = \{\xi \mid |\xi_\ell| > \varepsilon_o\}$ cover $\{|\xi| = 1\}$. It suffices to prove (1.18) for $\xi \in T_\ell$ for each ℓ separately. We fix ℓ. Again let $\sigma = \gamma/s$. Then $\underset{\sim}{a}\cdot\sigma = |\gamma|/s = c$, say, and $c < a_1$. Define η by $\eta_i = \sigma_i$ if $i \neq \ell$, $\eta_\ell = \sigma_\ell + (a_1 - c)/a_\ell$. Note $a_1 - c \leqq b_2/s$ so $(\eta_\ell - \sigma_\ell)s \leqq b_2/a_\ell$. Consequently, for $\xi \in T_\ell$, $|\xi^\gamma \partial_1^s J(\xi)| = |\xi_1|^{\sigma_1 s} \cdots |\xi_n|^{\sigma_n s} |\partial_1^s J(\xi) \leqq \varepsilon_o^{-b_2/a_\ell} |\xi_1|^{\eta_1 s} \cdots |\xi_n|^{\eta_n s} |\partial_1^s J(\xi)| < C_1 R_o^s s!^{(\eta_1 + \cdots + \eta_n)} \leqq$

$C_1R_o^s s!^{(\sigma_1+\ldots+\sigma_n)} s!^{b_2/a_\ell s} < C_1R_o^s s!^{\|\gamma\|/s} s^{b_2/a_\ell} < C_2R_1^s s!^{\|\gamma\|/s}$

$< C_2(R_1e)^s \|\gamma\|! < C_3R_2^{\|\gamma\|}\gamma!$ for some $C_3, R_2 > 0$, since $s \leqq (|\gamma|+b_2)/a_1$. We used (1.19). This proves (1.18).

Consequently we have (1.16) and hence (1.17). By homogeneity, for any $\varepsilon > 0$ there exists $B > 0$ such that

$$|\partial^\beta x_m^{[|\beta|/a_m]-p} K(x)| < BR^{\|\beta\|}\beta! \tag{1.20}$$

for $1/2 < |x| \leqq 3$, $|x_m| > \varepsilon/2$. We fix $\varepsilon > 0$ so that the sets $S_{m,\varepsilon} = \{x \mid |x_m| > \varepsilon\}$ cover $A = \{1/2 < |x| < 3\}$. It suffices to prove that K is real analytic on each $S_{m,\varepsilon} \cap \{1 \leqq |x| \leqq 2\}$. For ease in notation, suppose $m = 1$. Write $\underset{\sim}{a} = (a_1, a')$, $\beta = (\beta_1, \beta')$, $\partial = (\partial_1, \partial')$, $x = (x_1, x')$. For fixed x', β', write $K_1(x_1) = (\partial')^{\beta'} x_1^{[\underset{\sim}{a}'\cdot\beta'/a_1]-p} K(x_1, x')/(R^{\|\beta'\|}\beta'!)$ for $(x_1, x') \in A \cap S_{1,\varepsilon/2}$. From (1.20) we have, with $N = \beta_1$,

$$|\partial^N x_1^N K_1(x_1)| < BR^N N!\ . \tag{1.21}$$

Now if a function $f(t)$ is defined and smooth on some interval in $\mathbb{R}$, we have the identity

$$(t^N D^N f)/N! = \sum_{m=0}^{N} (-1)^{N-1} \binom{N}{M} D^M t^M f/M! \tag{1.22}$$

where $D = d/dt$. To see (1.22), observe that since the conclusion is local, we may assume $f \in C_c^\infty(\mathbb{R})$. Let $g = \hat{f}$, and observe that by Leibniz's rule $D^N \tau^N g/N! = \sum_{M=0}^{N} \binom{N}{M} \tau^M D^M g/M!$ where now $D = d/d\tau$. (1.22) follows upon taking the inverse Fourier transform. Applying (1.22) to (1.21) we find $|x_1^N \partial_1^N K_1(x_1)|/N! < B \sum_{M=0}^{N} \binom{N}{M} R^M = B(1+R)^N$. Thus, if $(x_1,x') \in S_{1,\varepsilon/2}$,

$$|\partial_1^N K_1(x_1)| < B[2(1+R)/\varepsilon]^N N! = BR_1^N N! \ . \qquad (1.23)$$

Select $\varepsilon_1 > 0$ so that if $1 \leqq |(y_1,y')| \leqq 2$ and if $|y_1 - u| \leqq \varepsilon_1$, then $(u,y') \in A$. Let $r = \min(1/2R_1, \varepsilon_1, \varepsilon/2)$, suppose $1 \leqq |(x_{10},x')| \leqq 2$, $|x_{10}| > \varepsilon$, and let $\mathcal{B} = \{z \in \mathbb{C} \,|\, |x_{10}-z| < r\}$. Then if $x_1 \in \mathcal{B} \cap \mathbb{R}$, $(x_1,x') \in S_{1,\varepsilon/2} \cap A$. Thus by (1.23), $K_{1|\mathcal{B}\cap\mathbb{R}}$ has an extension to a holomorphic function on $\mathcal{B}$ which we denote $K_1(z_1)$; further, in $\mathcal{B}$, $|K_1(z_1)| \leqq B$. Since $r \leqq \varepsilon/2$ and $|x_{10}| \geqq \varepsilon$, $|z_1| \geqq \varepsilon/2$ in $\mathcal{B}$. Accordingly, if $z_1 \in \mathcal{B}$, $|z_1^{p-[\underset{\sim}{a}'\cdot\beta'/a_1]} K_1(z_1)| \leqq B(2/\varepsilon)^{\underset{\sim}{a}'\cdot\beta'/a_1-p} < B_1 R_2^{\|\beta'\|}$ for some $B_1, R_2 > 0$. By the Cauchy estimates, $|\partial_1^{\beta_1} x_1^{p-[\underset{\sim}{a}'\cdot\beta']/a_1]} K_1(x_1)| \leqq B_1 R_2^{|\beta'|} r^{-\beta_1} \beta_1!$ for all $\beta_1 \in \mathbb{Z}^+$, when $x_1 = x_{10}$. But this then says precisely that $|\partial^\beta K(x)| < B_1 (R\,R_2)^{\|\beta'\|} r^{-\beta_1} \beta!$ for any multi-index β and $|x| = 1$; so K is real analytic in $S_{1,\varepsilon} \cap \{1 \leqq |x| \leqq 2\}$, as desired. The second uniformity assertion follows from the proof; this proves Theorem 1.3.

Note, incidentally, that the proof shows that the condition (d) for any <u>fixed</u> m is sufficient to prove that $\check{J}$ is real analytic away from $\{x_m = 0\}$.

The following corollary was known.

Corollary 1.4. If $\underset{\sim}{a} = (1,\ldots,1)$, $AJ^j = \{J \in J^j | J$ is real analytic away from $0\}$.

Proof. This is immediate from Theorem 1.3(b). ∎

Up to now, it has been convenient to require that all $a_\ell \in \mathbb{N}$. We now intend to specialize the a_ℓ, and it will be best if we allow them to be rational. This situation is immediately reducible to the preceding if we multiply all a_ℓ by a common denominator, changing j and k correspondingly. Here are the definitions and results in this situation. Let b be the least positive integer for which all $a_\ell b \in \mathbb{N}$. If $k \notin \mathbb{Z}^+/b$, we put $K^k = \mathrm{Rhom}_k$; if $k \in \mathbb{Z}^+/b$, we put $K^k = \mathrm{Rhom}_k + \{p(x)\log|x| :$ p a homogeneous polynomial of degree $k\}$. Here $|x| = (\sum x_\ell^{2a_\ell'})^{1/2A}$, where $A = \prod_{m=1}^{n} (a_m b)$, and where $a_\ell' = A/a_\ell$. Next, if $j - Q \notin \mathbb{Z}^+/b$, we put $J^j = \mathrm{Rhom}_j$. If $-j - Q \in \mathbb{Z}^+/b$, we put $J^j =$ $\{\Lambda_G | G \in C^\infty(\mathbb{R}^n \setminus \{0\})$, G homogeneous of degree $j\} + \{P(\partial)\delta | P$ a homogeneous polynomial of degree $-Q - j\}$. Here Λ_G is defined as in (1.2). Then Proposition 1.2 still holds. Further, let $AK^k = \{K \in K^k | K$ is real analytic away from $0\}$, and, let $AJ^j =$ $(AK^{-Q-j})^\wedge$; then Theorem 1.3 still holds.

For the rest of the book, unless stated otherwise, we confine ourselves to the case $\underset{\sim}{a} = (a_1,\ldots,a_{r+s})$ where $p = a_1 = \ldots = a_r$, $1 = a_{r+1} = \ldots = a_{r+s}$. Here $p \in \mathbb{Q}$, $p > 1$, and $p = a/b$ in lowest terms. Clearly, this takes in the general case where there are only two different a_ℓ's. Later, we shall even restrict to the case r = 1. We seek to elucidate Theorem 1.3 in this case. What can be said in general is an interesting question. Rather than using the letter n to denote r + s, we free this letter for other uses.

For $\gamma = (\gamma_1, \ldots, \gamma_{r+s}) \in (\mathbb{Z}^+)^{r+s}$, we write $|\gamma| = \underset{\sim}{a} \cdot \gamma$, $\|\gamma\| = \sum \gamma_i$. If $\alpha = (\alpha_1, \ldots, \alpha_r) \in (\mathbb{Z}^+)^r$, we write $|\alpha| = \|\alpha\| = \sum \alpha_i$; similarly for $\beta \in (\mathbb{Z}^+)^s$, we use $|\beta|$ and $\|\beta\|$ interchangeably.

On $\mathbb{R}^{r+s}$ we use coordinates $(t_1, \ldots, t_r, x_1, \ldots, x_s)$ with dual coordinates $(\lambda_1, \ldots, \lambda_r, \xi_1, \ldots, \xi_s)$. We write $|t| = (\sum t_i^2)^{1/2}$, $|x| = (\sum x_i^2)^{1/2}$, $|(t,x)| = (|t|^{2b} + |x|^{2a})^{1/2a}$; the latter is a homogeneous norm function, analytic away from 0. Similarly we have $|\lambda|$, $|\xi|$, $|(\lambda,\xi)|$. We write $\partial_\lambda = (\partial/\partial\lambda_1, \ldots, \partial/\partial\lambda_r)$; similarly ∂_ξ. We now reformulate Theorem 1.3 in the present case, to suit our needs.

<u>Corollary 1.5</u>. Suppose $J \in J^j$. Then $J \in AJ^j$ if and only if there exist C,R so that all of the following conditions hold: for all r-tuples α, and all s-tuples β:

(A) If $|\xi| = 1$, (I) $|\lambda|^{\|\alpha\|} |\partial_\lambda^\alpha J(\lambda,\xi)| < CR^{|\alpha|} \alpha! (1+|\lambda|)^{\mathrm{Re}\, j/p}$

(II) $|\partial_\lambda^\alpha J(\lambda,\xi)| < CR^{|\alpha|} \alpha!^p (1+|\lambda|)^{\mathrm{Re}\, j/p}$

(B) If $|\lambda| = 1$, (I) $|\xi|^{\|\beta\|} |\partial_\xi^\beta J(\lambda,\xi)| < CR^{\|\beta\|} \beta! (1+|\xi|)^{\mathrm{Re}\, j}$

(II) $|\partial_\xi^\beta J(\lambda,\xi)| < CR^{\|\beta\|} \beta!^{1/p} (1+|\xi|)^{\mathrm{Re}\, j}$.

Further, we have these uniformities:

(1) Let $k = -Q-j$, suppose $\mathrm{Re}\, k > -Q$. Then for any $R_1 > 0$ there exist $C, R > 0$ so that (A) and (B) hold whenever $J = \hat{K}$ for a $K \in AK^k$ which satisfies $\sup_{1 \leq |x| \leq 2} |\partial^\gamma K(x)| < R_1^{\|\gamma\|}\gamma!$ for all $\gamma \in (\mathbb{Z}^+)^n$.

(2) Suppose $\mathrm{Re}\, j > -Q$. Then for any $C, R > 0$ there exist $C_1, R_1 > 0$ so that if (A) and (B) hold and if $J = \hat{K}$ then $\sup_{1 \leq |u| \leq 2} |\partial^\alpha K(u)| < C_1 R_1^{\|\gamma\|}\gamma!$ for all $\gamma \in (\mathbb{Z}^+)^n$.

<u>Proof</u>. By homogeneity, the conditions are equivalent to this: there exist C, R so that all of the following conditions hold for all $(\lambda, \xi) \neq 0$, all α and all β:

(C) (I) $|\lambda|^{\|\alpha\|} |\partial_\lambda^\alpha J(\lambda,\xi)| < CR^{\|\alpha\|}\alpha! \, |(\lambda,\xi)|^{\mathrm{Re}\, j}$

(II) $|\xi|^{p\|\alpha\|} |\partial_\lambda^\alpha J(\lambda,\xi)| < CR^{\|\alpha\|}\alpha!^p \, |(\lambda,\xi)|^{\mathrm{Re}\, j}$

(D) (I) $|\xi|^{\|\beta\|} |\partial_\xi^\beta J(\lambda,\xi)| < CR^{\|\beta\|}\beta! \, |(\lambda,\xi)|^{\mathrm{Re}\, j}$

(II) $|\lambda|^{\|\beta\|/p} |\partial_\xi^\beta J(\lambda,\xi)| < CR^{\|\beta\|}\beta!^{1/p} \, |(\lambda,\xi)|^{\mathrm{Re}\, j}$

Indeed, suppose we have (A) and (B). Note that for some $C > 0$, $(1+|\lambda|)^{1/p} < C|(\lambda,\xi)|$ for all $|\xi| = 1$, while $(1+|\xi|) < C|(\lambda,\xi)|$ for all $|\lambda| = 1$. Thus the conditions (C) follow by homogeneity for (λ,ξ) with $\xi \neq 0$, hence by continuity when $\xi = 0$ as well. Similarly (D) follows, and conversely (A) and (B) follow from (C) and (D).

To prove the corollary, suppose first we have (C) and (D). Then the conditions of Theorem 1.3(c) obviously hold, so $J \in AK^j$. Conversely, suppose $J \in AJ^j$. Note this fact: suppose $q \in \mathbb{R}$, $q > 0$. Then there exists C so that for all $y \in \mathbb{R}^n$ and $N \in \mathbb{Z}^+$ we have

$$|y|^{qN} < C^N \sum |y_\ell|^{qN}. \tag{1.24}$$

(Here $|y| = (\sum y_\ell^2)^{1/2}$.) Indeed, we may neglect the finitely many cases $qN \leq 1$ and assume $qN > 1$. For some C_1, $|y| < C_1 \sum |y_\ell| < C_1 n(\sum |y_\ell|^{qN})^{1/qN}$ by Hölder's inequality. (1.24) follows at once. Thus $|\lambda|^{||\alpha||} < C^{||\alpha||} \sum |\lambda_\ell|^{||\alpha||}$, $|\xi|^{p||\alpha||} < C^{||\alpha||} \sum |\xi_\ell|^{p||\alpha||}$ etc., and the condition (C) and (D) follow immediately from Theorem 1.3(b). The uniformity assertions, also, follow from this and from the uniformity assertion of Theorem 1.3. This completes the proof. ∎

Next we consider $\mathbb{R}^r \subset \mathbb{C}^r$, $\mathbb{R}^s \subset \mathbb{C}^s$ in the usual fashion, and write $\chi_\ell = \lambda_\ell + i\tau_\ell$, $\zeta_m = \xi_m + i\eta_m$. Define q by $(1/p) + (1/q) = 1$.

<u>Proposition 1.6</u>. Suppose $J \in J^j$.

(a) (A) is equivalent to the following:

(A') there exist $C_1, c, R_1 > 0$ such that for any ξ with $|\xi| = 1$, J has an extension to a function $J(\chi,\xi)$ analytic in the sector $S_1 = \{\chi : |\tau| < c|\lambda|\}$, satisfying these properties:

(I) $|J(\chi,\xi)| < C_1(1+|\chi|)^{\mathrm{Re}\, j/p}$

(II) There exist functions $F_\alpha(\xi)$ and $E_N(\chi,\xi)$ for $\alpha \in (\mathbb{Z}^+)^r$, $N \in \mathbb{Z}^+$, $|\xi| = 1$, $\chi \in S_1$, such that for any $N \in \mathbb{Z}^+$, $J(\chi,\xi) = \sum_{|\alpha| \leqq N-1} F_\alpha(\xi)\chi^\alpha + E_N(\chi,\xi)$, and such that $|F_\alpha(\xi)| < C_1 R_1^{\|\alpha\|}\alpha\,!^{p-1}$, $|E_N(\chi,\xi)| < C_1 R_1^N N!^{p-1}|\chi|^N$ for $|\chi| \leqq 1$. (If $N = 0$, we define the summation above to be zero.) Further, if (A) holds, we can take $F_\alpha(\xi) = (\partial_\lambda^\alpha J)(0,\xi)/\alpha!$.

(b) (B) is equivalent to the following:

(B') There exist $C_1, c, B > 0$ such that for any λ with $|\lambda| = 1$, $J(\lambda,\xi)$ has an extension to an entire function $J(\lambda,\zeta)$, satisfying:

(I) In the sector $S_2 = \{\zeta : |\eta| < c|\xi|\}$, $|J(\lambda,\zeta)| < C_1(1+|\zeta|)^{\mathrm{Re}\, j}$

(II) $|J(\lambda,\zeta)| < C_1 e^{B|\zeta|^q}$.

Further, we have these uniformities:

(1) For any fixed $j, C, R > 0$ there exist fixed $C_1, c, R_1 > 0$ such that if (A) and (B) hold for a $J \in AJ^j$ then (A') and (B') hold with those C_1, c, R_1.

(2) For any fixed $j, C_1, c, R_1 > 0$ there exist fixed $C, R > 0$ so that if (A') and (B') hold with those C_1, c, R_1 for a $J \in AJ^j$, then (A) and (B) hold.

Proof. We first note (A)(I) $\Longleftrightarrow$ (A')(I), (B)(I) $\Longleftrightarrow$ (B')(I). Indeed, assuming (A)(I), for each $\lambda_0 \in \mathbb{R}\setminus\{0\}$ we can, by use of power series, extend $J(\lambda,\xi)$ analytically to $J(\chi,\xi)$ for $|\chi-\lambda_0| < |\lambda_0|/2R$, and hence to the sector S_1 with $c = 1/2R$. In that sector we have $|J(\chi,\xi)| < C_1'(1+|\mathrm{Re}\ \chi|)^{\mathrm{Re}\ j/p} < C_1(1+|\chi|)^{\mathrm{Re}\ j/p}$. Conversely, assume (A')(I). We may select $c_0 > 0$ so that, for any $\lambda \in \mathbb{R}\setminus\{0\}$, if $\mathcal{D}_{c_0|\lambda|}(\lambda)$ denotes the polydisc with center λ and all polyradii $c_0|\lambda|$, then $\mathcal{D}_{c_0|\lambda|}(\lambda) \subset S_1$. (Indeed, we need only do this for $|\lambda| = 1$, and this is easy.) Applying the Cauchy estimates to $J(\chi,\xi)$ in any such polydisc, we obtain (A)(I) at once. Similarly (B)(I) $\Longleftrightarrow$ (B')(I). So we need only establish the conditions labeled (II) under the appropriate hypotheses.

<u>(A) $\Rightarrow$ (A')(II)</u> We first claim this improvement on (A)(II): for some c', $0 < c' < c$, there exist $C_2, R_2 > 0$ such that for all $\chi \in S_1' = \{\lambda+i\tau \,|\, |\tau|<c'|\lambda|\}$, and for all $|\xi| = 1$ and all α,

$$|\partial_\chi^\alpha J(\chi,\xi)| < C_2R_2^{\|\alpha\|}\alpha!^p(1+|\chi|)^{\mathrm{Re}\ j/p}. \tag{1.25}$$

Indeed, since $\partial_\chi^\alpha J(\chi,\xi) = \sum_\gamma \partial_\lambda^{\alpha+\gamma}J(\lambda,\xi)(\chi-\lambda)^\gamma/\gamma!$ for $\lambda = \mathrm{Re}\ \chi$, it suffices to show that there exist $C_3, R_2 > 0$ such that for all $|\xi| = 1$ and all α,γ,

$$|\lambda|^{\|\gamma\|}|\partial_\lambda^{\alpha+\gamma}J(\lambda,\xi)| < C_3R_2^{\|\alpha\|+\|\gamma\|}\alpha!^p\gamma!(1+|\lambda|)^{\mathrm{Re}\ j/p}. \tag{1.26}$$

For if (1.26) is known, (1.25) follows for $c' = 1/2R_2$, just as in the proof of (A)(I) => (A')(I). Note, by the way, that $\|\alpha\| + \|\gamma\| = \|\alpha+\gamma\|$. Now (1.26) follows form (A) in much the same way as Theorem 1.3(d) followed from Theorem 1.3(c). Indeed, replace α by $\alpha + \gamma$ in (A)(I) and (II); call the resulting inequalities (*) and (**). Raise (*) to the power $\|\gamma\|/(\|\alpha+\gamma\|)$ and raise (**) to the power $\|\alpha\|/(\|\alpha+\gamma\|)$. Multiply the two resulting inequalities together, to find

$$|\lambda|^{\|\gamma\|}|\partial_\lambda^{\alpha+\gamma}J(\lambda,\xi)| < CR^{\|\alpha+\gamma\|}(\alpha+\gamma)!^{(p\|\alpha\|+\|\gamma\|)/\|\alpha+\gamma\|}(1+|\lambda|)^{\mathrm{Re}\ j/p}.$$

But $(\alpha+\gamma)! \leqq \|\alpha+\gamma\|!$, and by (1.19), $\|\alpha+\gamma\|!^{(p\|\alpha\|+\|\gamma\|)/\|\alpha+\gamma\|} =$

$$(\|\alpha+\gamma\|!^{(a\|\alpha\|+b\|\gamma\|)/\|\alpha+\gamma\|})^{1/b} \leqq e^{\|\alpha+\gamma\|/b}(a\|\alpha\|+b\|\gamma\|)!^{1/b}.$$

We thus have a bound of $R_3^{\|\alpha+\gamma\|}(a\|\alpha\|)!^{1/b}(b\|\gamma\|)!^{1/b} \leqq R_4^{\|\alpha+\gamma\|}\alpha!^p\gamma!$ for some $R_3, R_4 > 0$ by (1.7), (1.8), (1.7). This proves (1.26) and hence (1.25).

Next, for $\chi \in S_1'$, define $f(y) = f_\chi(y) = J(y\chi,\xi)$ for $y \in \mathbb{R}$, $y > 0$. By (1.25), $f^{(n)}(y)$ is bounded for each n, for $0 < y < 1$. Thus f has a smooth extension to $\{y \geqq 0\}$, which we also denote f. The Taylor expansion of f about 0, evaluated when y = 1, yields $J(\chi,\xi) = \sum_{M=0}^{N-1} f^{(M)}(0)/M! + f^{(N)}(u)/N!$ for some u between 0 and 1. Let $\partial^\alpha = (\partial/\partial\chi)^\alpha$. For $|\chi| \leqq 1$, we have $|f^{(N)}(u)/N!| =$

$$\Big|\sum_{|\alpha|=N}(\partial^\alpha J)(u\chi,\xi)\chi^\alpha/\alpha!\Big| < C_2R_2^N|\chi|^N\sum_{|\alpha|=N}\alpha!^{p-1} < C_4R_4^N N!^{p-1}|\chi|^N$$

for some $C_4, R_4 > 0$, since all $\alpha! \leqq N!$ and $\#\{\alpha : |\alpha| = N\} \leqq (N+1)^r$.

Also $f^{(M)}(0)/M! = \lim_{\varepsilon\to 0} \sum_{|\alpha|=M} (\partial^\alpha J)(\varepsilon\chi,\xi)\chi^\alpha/\alpha!$, while if ε is so small that $|\varepsilon\chi| < 1$, $|\partial^\alpha J(\varepsilon\chi,\xi)|/\alpha! < C_2 R_2^{\|\alpha\|}\alpha!^{p-1}$. Thus (A')(II) follows for $\chi \in S_1'$, with $E_N(\chi,\xi) = f_\chi^{(N)}(u)/N!$, $F_M(\xi) = \lim_{\varepsilon\to 0^+}(\partial^\alpha J)(\varepsilon\chi,\xi)/\alpha!$, once we verify that the latter limit exists and is independent of χ.

It is easy to see that the limit exists. Indeed, let $g(y) = g_\chi(y) = (\partial^\alpha J)(y\chi,\xi)$. Observe again that $g'(y)$ is bounded for $0 < y < 1$, so the limit does exist. To show it is independent of χ, let us first assume $r > 1$. Then S_1' is connected, since $\mathbb{R}^r\setminus\{0\}$ is, so we need only show the limit is locally constant. For $\chi_0 \in S_1'$, suppose $B_\rho(\chi_0) = \{\chi \mid |\chi-\chi_0| < \rho\} \subset S_1'$; we need only show the limit is independent of $\chi \in B_\rho(\chi_0)$. But if $\chi \in B_\rho(\chi_0)$, $|\partial^\alpha J(\varepsilon\chi,\xi) - \partial^\alpha J(\varepsilon\chi_0,\xi)| < \varepsilon|\chi-\chi_0|\sup|\nabla\partial^\alpha J(\chi_1,\xi)|$ where ∇ is in the λ,τ variables and the sup is taken over all $\chi_1 \in B_{\varepsilon\rho}(\varepsilon\chi_0)$. By (1.26) this sup is bounded as $\varepsilon \to 0^+$, hence $\lim_{\varepsilon\to 0^+}[g_\chi(\varepsilon)-g_{\chi_0}(\varepsilon)] = 0$ as desired. If $r = 1$, S_1' has two components and this argument shows only that the limit is constant for χ in each component. However, we are assuming $J \in J^j$, so $\lim_{\varepsilon\to 0^+}\partial^\alpha J(\varepsilon\lambda,\xi) = \lim_{\varepsilon\to 0^+}\partial^\alpha J(-\varepsilon\lambda,\xi)$, so the limits in each component must be equal, as desired.

<u>(A') = (A)(II)</u>: With $|\alpha| = N$, if $|\lambda| < 1$, we have $\partial_\lambda^\alpha J(\lambda,\xi) = \partial_\lambda^\alpha E_N(\lambda,\xi)$. This we estimate with the Cauchy estimates in the polydisc $\mathcal{D}_{C_0|\lambda|}(\lambda)$ (see the first paragraph of the proof).

We assume $|\lambda| \leqq \varepsilon$, where ε is small enough that $|\lambda| \leqq \varepsilon$ implies that $|\chi| < 1$ everywhere in this polydisc. We find $|\partial_\lambda^\alpha E_N(\lambda,\xi)| < C_1 R_1^N N!^{p-1}\alpha!(|\chi|/C_0|\lambda|)^{N+1}$. Since $|\chi| < C\,|\lambda|$ in the polydisc, and since $\alpha! \leqq N!$, (A)(II) is immediate for $|\lambda| \leqq \varepsilon$. For $|\lambda| \geqq \varepsilon$, (A)(II) is trivally weaker than (A)(I).

<u>(B)(II) $\Rightarrow$ (B')(II):</u> (B)(II) guarantees that $J(\lambda,\xi)$ has an entire extension to $J(\lambda,\zeta)$ if $|\lambda| = 1$. Let $\boldsymbol{|}\zeta\boldsymbol{|} = (|\zeta_1|,\ldots,|\zeta_n|)$. Expanding in power series, about $\zeta = 0$, and using Hölder's inequality, we find that $|J(\lambda,\zeta)| < C\sum R^{\|\alpha\|}\boldsymbol{|}\zeta\boldsymbol{|}^\alpha(\alpha!)^{(1/p)-1} = C\sum(2R)^{\|\alpha\|}\boldsymbol{|}\zeta\boldsymbol{|}^\alpha\alpha!^{-1/q}2^{-\|\alpha\|} \leqq C(\sum(2R)^{q\|\alpha\|}\boldsymbol{|}\zeta\boldsymbol{|}^{q\alpha}\alpha!^{-1})^{1/q}(\sum 2^{-p\|\alpha\|})^{1/p} = C\exp[(2R)^q(|\zeta_1|^q+\ldots+|\zeta_s|^q)/q] \times (\sum_{n=0}^{\infty}(n+1)^s 2^{-pn})^{1/p}$ since $\#\{\alpha:|\alpha|=n\} \leqq (n+1)^s$. (B')(II) is immediate.

<u>(B') $\Rightarrow$ (B)(II):</u> We consider two cases, $|\xi| \geqq \|\beta\|^{1/q}$ and $|\xi| < \|\beta\|^{1/q}$. In the first case, by (B)(I), $|\partial_\xi^\beta J(\lambda,\xi)| < CR^{\|\beta\|}(\beta!/\|\beta\|^{\|\beta\|/q})(1+|\xi|)^{\mathrm{Re}\,j}$. However $\beta! \leqq \|\beta\|! \leqq \|\beta\|^{\|\beta\|}$; so $\beta!/\|\beta\|^{\|\beta\|/q} \leqq \beta!/\beta!^{1/q} = \beta!^{1/p}$ and we have (B)(II) in this case. If, instead, $|\xi| < \|\beta\|^{1/q}$, we apply the Cauchy estimates to J, using a polydisc centered at ξ and with all polyradii $\|\beta\|^{1/q}$. In this polydisc $|\zeta| \leqq (s+1)\|\beta\|^{1/q}$, so $|\zeta|^q \leqq B_1\|\beta\|$ for some B_1. By (B')(II) we find $|\partial_\xi^\beta J(\lambda,\xi)| < C_1 e^{B_1\|\beta\|}\beta!/\|\beta\|^{\|\beta\|/q} < CR_0^{\|\beta\|}\beta!^{1/p}$. If $\mathrm{Re}\,j < 0$, we are done; otherwise, write $R_0^{\|\beta\|} \leqq R_0^{\|\beta\|}(1+\|\beta\|^{1/q})^{-\mathrm{Re}\,j}(1+|\xi|)^{\mathrm{Re}\,j} < R^{\|\beta\|}(1+|\xi|)^{\mathrm{Re}\,j}$ to complete the proof.

The uniformity assertions follow from the proofs. ∎

We now can obtain a clear understanding of $J(\lambda,\xi)$ for fixed $\lambda \neq 0$. In order to incorporate the information in (A'), we make the key observation that, by homogeneity, the behavior of $J(\lambda,\xi)$ for fixed $\xi \neq 0$ as $\lambda \to 0$ is reflected in the behavior of $J(\lambda,\xi)$ for fixed $\lambda \neq 0$ as $\xi \to \infty$. This fact was exploited by Taylor [79], who observed (a) of Theorem 1.7 below, and the partial converse Theorem 1.8(a).

Theorem 1.7(a). Suppose $J \in J^j$. Fix $\lambda \neq 0$. Then there exist smooth functions $\{g_\ell(\xi)\}$ on $\mathbb{R}^s \setminus \{0\}$, homogeneous (in the usual sense) of degree $j - p\ell$, so that for all $L \in \mathbb{Z}^+$, $\beta \in (\mathbb{Z}^+)^s$, there exists $C_{L,\beta}$ so that $\left|\partial_\xi^\beta[J(\lambda,\xi) - \sum_{\ell=0}^{L-1} g_\ell(\xi)]\right| < C_{L,\beta}|\xi|^{\mathrm{Re}\, j-pL-|\beta|}$ for $|\xi| > 1$. In fact, $g_\ell(\xi) = \sum_{|\gamma|=\ell} [(\partial_\lambda^\gamma J)(0,\xi)/\gamma!]\lambda^\gamma$.

(b) Suppose $J \in AJ^j$. Fix $\lambda \neq 0$. Then J has an extension to an entire function $J(\lambda,\zeta)$ which is in the space $RZ^q_{q,j}$, which we define as follows:

Definition. $RZ^q_{q,j} = \{$entire functions $f(\zeta)$ on $\mathbb{C}^s \mid$ for some B,C,R:

(i) $|f(\zeta)| < Ce^{B|\zeta|^q}$ for all ζ.

(ii) There exist real analytic functions $\{g_\ell(\xi)\}$ on $\mathbb{R}^s \setminus \{0\}$, homogeneous (in the usual sense) of degree $j - p\ell$, which satisfy $|\partial_\xi^\alpha g_\ell| < CR^{\|\alpha\|+\ell}\|\alpha\|!\ell!^{p-1}$ for $|\xi| = 1$, such that for all $L \geq 0$, if $|\xi| > 1$, $|f(\xi) - \sum_{\ell=0}^{L-1} g_\ell(\xi)|$ $CR^L L!^{p-1} |\xi|^{\text{Re } j-pL}\}$.

Notation. In the above situation, we often write $f \sim \sum g_\ell$. (It is easy to see, by use of the homogeneity of the g_ℓ, that the g_ℓ are uniquely determined by f.) If we need to make the constants B,C,R explicit, we shall very occasionally write $f \in RZ^q_{q,j}(B,C,R)$.

As in (a), $J(\lambda,\xi) \sim \sum g_\ell$ (in $RZ^q_{q,j}$) where

$$g_\ell(\xi) = \sum_{|\gamma|=\ell} [(\partial_\lambda^\gamma J)(0,\xi)/\gamma!]\lambda^\gamma.$$

Further, we have the following uniformity in (b). Suppose $-Q < \text{Re } j < 0$. For any $C,R > 0$ and any fixed $\lambda \neq 0$ there exist B,C',R' so that $J(\lambda,\zeta) \in RZ^q_{q,j}(B,C',R')$ whenever $J \in AJ^j$ and J satisfies (A) and (B).

Proof. For (a), define, for $\xi \neq 0$ and $\gamma \in (\mathbb{Z}^+)^r$, $F_\gamma(\xi) = (\partial_\lambda^\gamma J)(0,\xi)/\gamma!$. Also, for $\xi \neq 0$ and $N \in \mathbb{Z}^+$, define $E_N(\lambda,\xi)$ by the equation

$$J(\lambda,\xi) = \sum_{|\gamma|\leq N-1} F_\gamma(\xi)\lambda^\gamma + E_N(\lambda,\xi). \tag{1.27}$$

Note that $E_N(\lambda,\xi)$ is homogeneous of degree j in (λ,ξ), since $J(\lambda,\xi)$ and $\sum F_\gamma(\xi)\lambda^\gamma$ are. Also, by Taylor's theorem, we have that for any $M > 0$ there exists $C_N' > 0$ so that if $|\xi| = 1$,

$|\lambda| \leqq M$, then $|E_N(\lambda,\xi)| = \sum_{|\alpha|=N} (\partial^\alpha J)(u\lambda,\xi)\lambda^\alpha/\alpha! < C_N'|\lambda|^N$. Here $0 < u < 1$. Now fix $\lambda \neq 0$ and put, for $\ell \in \mathbb{Z}^+$, $g_\ell(\xi) = \sum_{|\gamma|=\ell} F_\gamma(\xi)\lambda^\ell$. Then g_ℓ is smooth on $\mathbb{R}^s\setminus\{0\}$ and homogeneous (in the usual sense) of degree $j - p\ell$. Further, $|J(\lambda,\xi) - \sum_{\ell=0}^{L-1} g_\ell(\xi)| = |E_L(\lambda,\xi)| = |\xi|^{\mathrm{Re}\, j}\, E_L(\lambda/|\xi|^p,\xi/|\xi|)| <$ $C_L'|\xi|^{\mathrm{Re}\, j}(|\lambda|/|\xi|^p)^L < C_L|\xi|^{\mathrm{Re}\, j-pL}$ for some $C_L > 0$, if $|\xi| \geqq 1$. This proves (a), in case $\beta = 0$. In general, note $(\partial_\xi^\beta J)(\lambda,\xi) = \sum_{|\gamma|\leqq N-1} [\partial_\xi^\beta F_\gamma(\xi)]\lambda^\gamma + (\partial_\xi^\beta E_N)(\lambda,\xi)$, and apply the above argument to $\partial_\xi^\beta J$.

For (b), (i) of the definition of $RZ_{q,j}^q$ follows from (B')(II). For (ii), we observe that the expansion in (A')(II) is the same as that of (a) of this proof when $\chi = \lambda \in \mathbb{R}^n\setminus\{0\}$. Let us show the estimate in (ii) for g_ℓ. Since $\#\{\gamma:|\gamma|=\ell\} \leqq (\ell+1)^r$, it suffices to show that $|\partial_\xi^\alpha F_\gamma(\xi)| < CR^{\|\alpha+\gamma\|}\|\alpha\|!\,\|\gamma\|!^{p-1}$ for each γ, for $|\xi| = 1$. However, $F_\gamma(\xi) = [\partial_\lambda^\gamma J](0,\xi)/\gamma!$. Note that, if $|\xi| = 1$, then $|(0,\xi)| = 1$. Thus, by (1.24) and Theorem 1.3(b), if $|\xi| = 1$, $|\partial_\xi^\alpha F_\gamma(\xi)| = |\xi|^{|\alpha|+p|\gamma|}|\partial_\xi^\alpha F_\gamma(\xi)| \leqq$ $C^{|\alpha|+p|\gamma|}\sum|\xi_\ell|^{|\alpha|+p|\gamma|}|\partial_\xi^\alpha\partial_\lambda^\gamma J(0,\xi)|/\gamma! < CR^{\|\alpha+\gamma\|}\alpha!\gamma!^{p-1}$ as claimed. Further, arguing as in (a), but using the estimate of (A')(II), we find $|J(\lambda,\xi) - \sum_{\ell=0}^{L-1} g_\ell(\xi)| = |\xi|^{\mathrm{Re}\, j}\,|E_L(\lambda/|\xi|^p,\xi/|\xi|)| <$ $|\xi|^{\mathrm{Re}\, j}\, C_1R_1^L L!^{p-1}(|\lambda|/|\xi|^p)^L < CR^L L!^{p-1}|\xi|^{\mathrm{Re}\, j-pL}$ for $|\xi| > |\lambda|^{1/p}$. This establishes (ii) at once if $|\lambda| \leqq 1$.

Otherwise, the fact that $|J(\lambda,\xi) - \sum_{\ell=0}^{L-1} g_\ell(\xi)| < CR^L L!^{p-1}|\xi|^{\mathrm{Re}\, j-pL}$ for $1 < |\xi| \leq |\lambda|^{1/p}$ follows trivially from the triangle inequality applied to the left side of this inequality. The uniformity assertion in (b) follows from the proof and from the uniformity assertions of Theorem 1.3, Corollary 1.5 and Proposition 1.6.

Note that, by homogeneity, the functions g_ℓ above satisfy $|\xi|^{||\alpha||}|\partial_\xi^\alpha g_\ell| < CR^{|\alpha||+\ell}||\alpha||!\ell!^{p-1}|\xi|^{\mathrm{Re}\, j-p\ell}$ for all $\xi \neq 0$. As in the proof of (A)(I)<==>(A')(I), this is the same as saying that the $g_\ell(\xi)$ can be extended to holomorphic functions $g_\ell(\zeta)$ in a sector $S_3 = \{|\eta|<c_3|\xi|\}$ in which $|g_\ell(\zeta)| < CR^\ell \ell!^{p-1}|\zeta|^{\mathrm{Re}\, j-p\ell}$. It is easily seen that $g_\ell(\zeta)$ must also be homogeneous of degree $j - p\ell$.

It is of interest that if $f \in RZ^q_{z,j}$ then the error estimate in (ii) of the definition of $RZ^q_{q,j}$ must necessarily hold in a sector. Precisely, we make the following modified definition.

<u>Definition</u>. $Z^q_{q,j} = \{$entire functions $f(\zeta)$ on $\mathbb{C}^s|$ for some B,C,R,c:

(i) $|f(\zeta)| < Ce^{B|\zeta|^q}$ for all ζ

(ii) In the sector $S = \{\zeta| |\eta|<c|\xi|\}$, there exist holomorphic functions $\{g_\ell(\zeta)\}$, homogeneous of degree $j - p\ell$, which satisfy $|g_\ell(\zeta)| < CR^\ell \ell!^{p-1}$ for $|\zeta| = 1$, $\zeta \in S$, such that for all $L \geq 0$, if $|\zeta| > 1$, $\zeta \in S$, then $|f(\zeta) - \sum_{\ell=0}^{L-1} g_\ell(\zeta)| < CR^L L!^{p-1}|\zeta|^{\mathrm{Re}\, j-pL}\}$

Notation. In the above situation, we often write $f \sim \sum g_\ell$. (It is easy to see that the g_ℓ are uniquely determined by f.) If we need to make the constants B,C,R,c explicit, we shall very occasionally write $f \in Z^q_{q,j}(B,C,R,c)$.

Evidently, $Z^q_{q,j} \subset RZ^q_{q,j}$. We assert, and will prove in the next chapter (Theorem 2.6), that $Z^q_{q,j} = RZ_{q,j}$.

For the rest of this paper, unless stated otherwise, we restrict to the case r = 1, that is, $\underset{\sim}{a} = (p,1,\ldots,1)$. We free the letter r for other uses. We write t for t_1, λ for λ_1. With this restriction, Theorem 1.7 has the following converse.

Theorem 1.8(a). Suppose $J_+, J_- \in C^\infty(\mathbb{R}^s)$ and that $\{g_\ell(\xi)\}$ are smooth functions on $\mathbb{R}^s\setminus\{0\}$, homogeneous of degree $j - p\ell$. For $L \in \mathbb{Z}^+$, let $E_{L+}(\xi) = J_+(\xi) - \sum_{\ell=0}^{L-1} g_\ell(\xi)$, $E_{L-}(\xi) = J_-(\xi) - \sum_{\ell=0}^{L-1} (-1)^\ell g_\ell(\xi)$. Suppose that for all $L \in \mathbb{Z}^+$, $\beta \in (\mathbb{Z}^+)^s$, there exists $C_{L\beta}$ so that $|\partial^\beta_\xi E_{L+}(\xi)| < C_{L\beta}|\xi|^{\mathrm{Re}\, j-pL-\|\beta\|}$ for $|\xi| > 1$ and $|\partial^\beta_\xi E_{L-}(\xi)| < C_{L\beta}|\xi|^{\mathrm{Re}\, j-pL-\|\beta\|}$ for $|\xi| > 1$. Then there exists $J \in J^j$ so that $J(1,\xi) = J_+(\xi)$, $J(-1,\xi) = J_-(\xi)$ for all ξ.

(b) Suppose that $J_+(\xi), J_-(\xi) \in C^\infty(\mathbb{R}^s)$ and that they have extensions to entire functions $J_+(\zeta)$, $J_-(\zeta) \in Z^q_{q,j}$. Say $J_+ \sim \sum g_\ell$ in $Z^q_{q,j}$, and suppose that $J_- \sim \sum (-1)^\ell g_\ell$. Then there exists $J \in AJ^j$ so that $J(1,\xi) = J_+(\xi)$, $J(-1,\xi) = J_-(\xi)$ for all ξ.

Further, we have the following uniformity in (b). For any $B,C',R',c > 0$ there exist $C,R > 0$ such that if

$J_+(\zeta)$, $J_-(\zeta) \in Z^q_{q,j}(B,C',R',c)$, then there exists $J \in AJ^j$ as in (b) which satisfies (A) and (B) with that C and R.

Proof (a). For $\lambda > 0$, put $G(\lambda,\xi) = \lambda^{j/p} J_+(\xi/\lambda^{1/p})$; for $\lambda < 0$, put $G(\lambda,\xi) = |\lambda|^{j/p} J_-(\xi/|\lambda|^{1/p})$; finally, for $\xi \neq 0$, put $G(0,\xi) = g_0(\xi)$. Then G is a homogeneous function of degree j on $\mathbb{R}^s \setminus \{0\}$. It suffices to show that $G \in C^\infty(R^s \setminus \{0\})$; for then we can put $J = \Lambda_G$ as in (1.2). It suffices to prove that G is smooth near any point of the form $(0,\xi_0)$, $\xi_0 \neq 0$.

We deifne $E_L(\lambda,\xi)$ for $L \in \mathbb{Z}^+$, as follows. For $\lambda > 0$, put $E_L(\lambda,\xi) = \lambda^{j/p} E_{L+}(\xi/\lambda^{1/p})$; for $\lambda < 0$, put $E_L(\lambda,\xi) = |\lambda|^{j/p} E_{L-}(\xi/|\lambda|^{1/p})$; for $\xi \neq 0$ put $E_L(0,\xi) = g_0(\xi)\delta_{0L}$. Then E_L is a homogeneous function of degree j on $\mathbb{R}^s \setminus \{0\}$. Further, for any $\ell \in \mathbb{Z}^+$, $G(\lambda,\xi) = \sum_{\ell=0}^{L-1} g_\ell(\xi)\lambda^\ell + E_L(\lambda,\xi)$, so it suffices to prove that E_L is C^{L-1} near $(0,\xi_0)$, for $L \geqq 1$. Fix $L \geqq 1$. It suffices to prove by induction on M that if $0 \leqq M \leqq L - 1$, then E_L is C^M near $(0,\xi_0)$ and that if D is a differential monomial of degree $\leqq M$ then for some $C_0 > 0$, $|DE_L(\lambda,\xi)| \leqq C_0|\lambda|^{L-M}$ for (λ,ξ) near $(0,\xi_0)$. This is evident from the definitions if $M = 0$; assume $M \leqq L - 1$ and it is known for $M - 1$ in place of M. Let D be a differential monomial of degree M. The second part of the induction assumption implies at once that if ξ is near ξ_0 then $(DE_L)(0,\xi)$ exists and equals zero; thus E_L is C^M near $(0,\xi_0)$. Further, if $\lambda > 0$, $DE_L(\lambda,\xi)$ is a linear combination of terms of the form $\lambda^{j/p-A-B/p}\xi^{\alpha}(\partial^{\beta}E_{L+})(\xi/\lambda^{1/p})$, where $\|\beta\| = B$ and $0 \leqq A \leqq M$, as is easily seen by induction on M. If ξ is close to $\xi_0 \neq 0$, and $0 < \lambda \leqq 1$, then for such a term we have a bound of $C\lambda^{\mathrm{Re}\ j/p-A-B/p}(\lambda^{-1/p})^{\mathrm{Re}\ j-pL-\|\beta\|} \leqq C\lambda^{L-M}$. Similarly for $\lambda < 0$. This completes the induction step and the proof of (a).

(b) We first check, using (a), that there exists $J \in J^j$ so that $J(1,\xi) = J_+(\xi)$, $J(-1,\xi) = J_-(\xi)$ for all ξ. Define E_{L+}, E_{L-} as in (a); we must check that the required estimates hold for $\partial^{\beta}E_{L\pm}$. By definition of $Z^q_{q,j}$, there exists a sector $S = \{|\eta|<c|\xi|\}$ so that $E_{L+}(\xi)$ has an extension to a holomorphic function $E_{L+}(\zeta)$ in S. Further, for some $C > 0$, $|E_{L+}(\zeta)| < CR^L L!^{p-1}|\zeta|^{\mathrm{Re}\ j-pL}$ for $|\zeta| > 1$, $\zeta \in S$. As in the proof of (A')(I) $\Rightarrow$ (A)(I), the Cauchy estimates show that $|\partial^{\beta}E_{L+}(\xi)| < C_L|\xi|^{\mathrm{Re}\ j-pL-\|\beta\|}$, as desired; similarly for E_{L-}. Thus $J \in J^j$ exists as desired; we must show $J \in AJ^j$. For this, we check (A') and (B').

(B')(II) is immediate, and (B')(I) follows form the error estimate in (ii) of the definition of $Z^q_{q,j}$, in the case $L = 0$. For (A'), note that, for $\lambda > 0$, the function $J(\lambda,\xi) = \lambda^{j/p}J_+(\xi/\lambda^{1/p})$ has an extension $J(\chi,\xi) = \chi^{j/p}J_+(1,\xi/\chi^{1/p})$ which is holomorphic for $\mathrm{Re}\ \chi > 0$ (we use the principal branch of the power functions). Say that $J_+ \in Z^q_{q,j}(B,C,R,c)$. Let $\Theta = \arctan c$. Note that if $|\arg \zeta_\ell| < \Theta$ for all ℓ, then

$\zeta \in S = \{|\eta|<c|\xi|\}$. Let $S_1 = \{\chi||\arg \chi|< \min(p\Theta,\pi/2)$. For $\chi \in S_1$, $|\xi| = 1$, then, we have $\zeta = \xi/\chi^{1/p} \in S$, so that $|J(\chi,\xi)|<C|\chi|^{\mathrm{Re}\, j/p}(1+|\xi/\chi^{1/p}|)^{\mathrm{Re}\, j}$ by (B')(I). Thus, for $|\xi| = 1$, $\chi \in S_1$, $|J(\chi,\xi)| < C(|\chi|^{1/p}+1)^{\mathrm{Re}\, j}$, giving (A')(I) for Re $\chi > 0$. Similarly, for Re $\chi < 0$ we can consider $J(\chi,\xi) = (-\chi)^{j/p}J(-1,\xi/(-\chi)^{1/p})$ to obtain (A')(I) in full. Note that we therefore also have (A)(I).

Finally we show (A')(II). For $|\xi| = 1$, $\chi \in S_1$, Re $\chi > 0$, $|\xi/\chi^{1/p}| \geqq 1$, note that for any $L \geqq 0$, $|J(1,\xi/\chi^{1/p}) - \sum_{\ell=0}^{L-1} g_\ell(\xi/\chi^{1/p})| < CR^L L!|\xi/\chi^{1/p}|^{\mathrm{Re}\, j-pL}$. Observe however that $u(\chi) = g_\ell(\xi/\chi^{1/p}) - \chi^{-j/p+\ell}g_\ell(\xi)$ is holomorphic in χ $(\chi \in S_1, \mathrm{Re}\,\chi >0)$ and vanishes when χ is real; consequently $u \equiv 0$. Multiplying the inequality by $|\chi^{j/p}|$, and noting that for some $C > 0$, $|z^{j/p}| < C|z|^{\mathrm{Re}\, j/p}$ whenever Re $z > 0$, we then find $|J(\chi,\xi) - \sum_{\ell=0}^{L-1} g_\ell(\xi)\chi^\ell| < CR^L L!^{p-1}|\chi|^L$ for $|\xi/\chi^{1/p}| > 1$, i.e., for $|\chi| < 1$. This gives (A')(II) for Re $\chi > 0$; similarly one finds (A')(II) for Re $\chi < 0$, as desired. The uniformity assertion follows from the proof and from the uniformity assertions of Proposition 1.6.

The next section is devoted to a study of $Z^q_{q,j}$, which should elucidate matters. We close this section with a few remarks that will note be used again.

Remarks. Here, briefly, is another approach to proving the results of this chapter. Suppose, for simplicity, that $\underset{\sim}{a} = (p,1)$ and $K \in AK^k$. Let $U(\lambda,x) = (F_t K)(\lambda,x)$, $V(t,\xi) = (F_x K)(t,\xi)$, $J = \hat{K}$; here F_t and F_x denote the Fourier transforms in the t and x variables respectively. If we extend K holomorphically to a sector in $\mathbb{C}^2$, then for ε sufficiently small the change of contour $U(\lambda,x) = \int e^{i\lambda(t+i\varepsilon|x|^p)} K(t+i\varepsilon|x|^p,x)\,dt$ is easily seen to be justified if Re $k < -p$, $x \neq 0$. This formula shows that $|U(\lambda,x)| < Ce^{-\varepsilon\lambda|x|^p}$ for $\lambda > 0$, $|x| \geqq 1$, and in fact shows that for each fixed $\lambda, U(\lambda,x)$ has an extension to a function $U(\lambda,z)$ holomorphic in some sector $\{z=x+iy \mid |y| < C_1|x|\}$. This condition may then be shown to hold even if Re $k \geqq -p$. Further, if Re $k \geqq -p$, then $U(\lambda,z)$ satisfies $|U(\lambda,z)| < Ce^{-\varepsilon|\lambda||z|^p}$ in the sector. Similarly, for fixed ξ, $V(t,\xi)$ has an extension to a function $V(w,\xi)$ holomorphic in some sector $\{w=t+iu \mid |u| < \varepsilon_2|t|\}$. Further, if Re $k \geqq 1$, there exist ε', $C > 0$ so that $V(w,\xi)$ satisfies $|V(w,\xi)| < Ce^{-\varepsilon'|\xi||w|^{1/p}}$ in the sector. These conditions on U and V are then easily seen to be sufficient to prove that $K \in AK^k$. Further, by means of estimates, they may be shown to be essentially equivalent to the conditions (A) and (B). This method of proof, however, also requires numerous technical arguments.

It should be noted that obtaining the conditions on U above for certain kernels, when $p = 2$, is an essential point of the analytic hypoellipticity studies of Treves [81] and Métivier [62]. As we have said, these conditions are basically equivalent to (B) and quickly give the analyticity of K away from x = 0.

The condition (A), or equivalently (A'), is a Gevrey-type condition; this type of condition has been extensively studied in other contexts, and variants of (A) $\Longleftrightarrow$ (A') have been observed before. It is somewhat curious that our methods of studying real analyticity require one to study Gevrey conditions. As a precedent, however, let us observe that condition (ii) of the definition of $Z^2_{2,j}$ is very similar to the definition of analytic pseudodifferential operators of Boutet de Monvel and Kree [9].

It is natural, then, to ask if one can also characterize the Fourier transform of $\{K \in \mathcal{K}^k | \; K$ satisfies a specific Gevrey condition away from $0\}$. This undoubtedly can be done by using the method of Theorem 1.3, although we have carried it out. (I thank E. M. Stein for this suggestion.)

2. The Space $Z^q_{q,j}$

In this section, unless otherwise stated, q is an arbitrary real number which is greater than 1, and $(1/p) + (1/q) = 1$. Also, unless otherwise stated, j will be an arbitrary complex number.

The subspace $\{f \in RZ^q_{q,j} | f \sim 0\}$ is especially important, and we begin by studying it.

We define $Z^q_q = \{$entire functions f on $\mathbb{C}^s |$ for some $B_1, B_2, C > 0$ we have $|f(\zeta)| < Ce^{B_1|\zeta|^q}$ for $\zeta \in \mathbb{C}^s$, while $|f(\xi)| < Ce^{-B_2|\xi|^q}$ for $\xi \in \mathbb{R}^s\}$. If we need to make the constants B_1, B_2, C explicit, we shall very occasionally write $f \in Z^q_q(B_1, B_2, C)$. This space was first investigated by Gelfand and Šilov ([21], [22]). (In the latter, and later, reference the space is usually denoted $S^{1/q}_{1/p}$. We prefer to return to the earlier notation because S is now customarily used to denote a symbol class.) These spaces have been used in related contexts by several authors; more references follow. First of all, however, we have:

<u>Proposition 2.1</u>. $Z^q_q \subset RZ^q_{q,j}$. More specifically, $f \in Z^q_q \Longleftrightarrow f \in RZ^q_{q,j}$ and $f \sim 0$.

Further, we have these uniformities (with fixed j):

(1) Say $B_1, B_2, C > 0$. Then there exist B, C_1, R so that $f \in RZ^q_{q,j}(B, C_1, R)$ whenever $f \in Z^q_q(B_1, B_2, C)$.

(2) Say $B, C_1, R > 0$. Then there exist $B_1, B_2, C > 0$ so that $f \in Z_q^q(B_1, B_2, C)$ whenever $f \in RZ_{q,j}^q(B, C_1, R)$ and $f \sim 0$.

Proof. This is a consequence of the following lemma, which is used many times in [22] and will also be used repeatedly here.

Lemma 2.2. Suppose F is a function defined on a subset $S_0 \subset \mathbb{R}^s$. Then the following are equivalent:

(a) For some $C_a, R_a > 0$, $|F(\xi)| < C_a e^{-R_a|\xi|^q}$ (for all $\xi \in S_0$).

(b) For some $C_b, R_b > 0$, $|F(\xi)| < C_b R_b^L L!^{p-1} |\xi|^{-pL}$ (for all $\xi \in S_0$, $L \in \mathbb{Z}^+$).

(c) For some $C_c, R_c > 0$, $|F(\xi)| < C_c R_c^L L!^{1/q} |\xi|^{-L}$ (for all $\xi \in S_0$, $L \in \mathbb{Z}^+$).

(d) For some $C_d, R_d > 0$, $|F(\xi)| < C_d R_d^L L! |\xi|^{-qL}$ (for all $\xi \in S_0$, $L \in \mathbb{Z}^+$).

Further, we have this uniformity. Say $x \in \{a,b,c,d\}$, and $C_x, R_x > 0$. Then there exist constants $C_y, R_y > 0$ for $y \in \{a,b,c,d\}$, $y \neq x$, as follows: if F is any function defined on any subset $S_0 \subset \mathbb{R}^s$ which satisfies condition (x), then it also satisfies all the conditions (y) for $y \in \{a,b,c,d\}$, $y \neq x$.

Proof of Lemma 2.2. We start with (a) $\Longleftrightarrow$ (d). Suppose (a) holds. For all $L \in \mathbb{Z}^+$, we have $R_a^L|\xi|^{qL}/L! \leq e^{R_a|\xi|^q}$; consequently (d) holds with $R_d = R_a^{-1}$, $C_d = C_a$. Next suppose (d) holds. For all $L \in \mathbb{Z}^+$ we have $[(2R_d)^{-1}|\xi|^q]^L L!^{-1}|F(\xi)| < C_d 2^{-L}$. Summing over L, we find (a) with $C_a = C_d$, $R_a = (2R_d)^{-1}$.

For (a) $\Longleftrightarrow$ (b), apply (a) $\Longleftrightarrow$ (d) to $|F|^{1/(p-1)}$; for (a) $\Longleftrightarrow$ (c) apply (a) $\Longleftrightarrow$ (d) to $|F|^q$. The uniformity assertion follows from the proof.

<u>Proof of Proposition 2.1</u>. Suppose f is an entire function and for some $C, B_1 > 0$, $|f(\zeta)| < Ce^{B_1|\zeta|^q}$ for all $\zeta \in \mathbb{C}^s$. Then: $f \in Z^q_q \Longleftrightarrow$ for some C_1, B_2, $|f(\xi)| < C_1 e^{-B_2|\xi|^q} \Longleftrightarrow$ for some C_2, B_2', $|\xi|^{-\mathrm{Re}\, j}|f(\xi)| < C_2 e^{-B_2'|\xi|^q}$ for $|\xi| \geq 1 \Longleftrightarrow$ for some $C_3, R > 0$, $|\xi|^{-\mathrm{Re}\, j}|f(\xi)| < C_3 R^L L!^{p-1}|\xi|^{-pL}$ for $|\xi| \geq 1$, $L \in \mathbb{Z}^+ \Longleftrightarrow f \in RZ^q_{q,j}$ and $f \sim 0$. In the third equivalence, we used Lemma 2.2(a) $\Longleftrightarrow$ (b). The uniformity assertion follows from this proof and the uniformity assertion of Lemma 2.2. ∎

Accordingly we begin with a study of Z^q_q. Its basic properties are given in the following theorem. If f is a function on $\mathbb{R}^s$, we say $f \in Z^q_q$ if f is the restriction to $\mathbb{R}^s$ of a Z^q_q function on $\mathbb{C}^s$. We shall always, without further comment, denote the extended function on $\mathbb{C}^s$ also by the same letter as the function (in this case, f). Also, let $\mathcal{S}$ denote Schwartz space on $\mathbb{R}^s$. We define $M_\ell : \mathcal{S} \to \mathcal{S}$ by $(M_\ell f)(\xi) = \xi_\ell f(\xi)$, $\partial_\ell : \mathcal{S} \to \mathcal{S}$ by $\partial_\ell f = \partial f/\partial \xi_\ell$. In the statement of the theorem, we let $U = \{M_\ell | 1 \leq \ell \leq s\}$, $V = \{\partial_\ell | 1 \leq \ell \leq s\}$.

Theorem 2.3. For a function f on $\mathbb{R}^s$, the following conditions are equivalent:

(a) $f \in Z_q^q$

(b) f is the restriction to $\mathbb{R}^s$ of an entire function on $\mathbb{C}^s$, also denoted f, which satisfies: for some B_1, B_2, C, c, $|f(\zeta)| < Ce^{B_1|\zeta|^q}$ for all $\zeta \in \mathbb{C}^s$, while for ζ in the sector $S = \{|\eta| < c|\xi|\}$, $|f(\zeta)| < Ce^{-B_2|\zeta|^q}$.

(c) $f \in S$, and there exist C,R as follows. For every set of operators $\{X_1, \ldots, X_{m+n}\} \subset U \cup V$ such that $\#\{X_i | X_i \in U\} = m$ and $\#\{X_i | X_i \in V\} = n$, we have $\|X_1 \ldots X_{m+n} f\|_\infty < CR^{m+n} m!^{1/q} n!^{1/p}$.

(d) $\hat{f} \in Z_p^p$

In addition, the statements are equivalent to $(c)_1$, which is the statement of (c) with $\|\ \|_\infty$ replaced by $\|\ \|_1$, and to $(c)_2$, which is the statement of (c) with $\|\ \|_\infty$ replaced by $\|\ \|_2$.

Also, if p = 2, the statements are equivalent to

(e) $f \in L^2(\mathbb{R}^s)$ and has the Hermite expansion $\sum_{\alpha \in (\mathbb{Z}^+)^n} c_\alpha H_\alpha(\xi)$, where for some $C > 0$, $0 < r < 1$, we have $|c_\alpha| < Cr^{\|\alpha\|}$.

Further, we have these uniformities. Say $f \in Z_q^q(\tilde{B}_1, \tilde{B}_2, C_0)$, so that $|f(\zeta)| < C_0 e^{\tilde{B}_1|\zeta|^q}$ for $\zeta \in \mathbb{C}^s$ while $|f(\xi)| < C_0 e^{-\tilde{B}_2|\xi|^q}$ for $\xi \in \mathbb{R}^s$. Then in (b), we can choose $B_1 = \tilde{B}_1$, and B_2 arbitrary with $0 < B_2 < \tilde{B}_2$, provided c is sufficiently small. The required c depends only on $\tilde{B}_1, \tilde{B}_2, B_2$ and not on C_o or f. In (c), $(c)_1$ or $(c)_2$, we can choose $C = C'C_o$, where C' and R depend only $\tilde{B}_1$ or $\tilde{B}_2$ and not on C_o or f. In (d) we can choose B_1', B_2', C_1 depending only on $\tilde{B}_1, \tilde{B}_2$ and not on C_o or f so that $\hat{f} \in Z^p_p(B_1', B_2', C_1 C_o)$.

Conversely, say $R > 0$; then there exist $\tilde{B}_1, \tilde{B}_2, C_o$ so that if (c) (or $(c)_1$ or $(c)_2$) is satisfied with that R for some C, then $f \in Z^q_q(\tilde{B}_1, \tilde{B}_2, C_o C)$.

Remarks. Here is the history, as best we can determine it. (a) $\Longleftrightarrow$ (b) was almost completely proved by Gelfand and Šilov in [22]; an additional argument, needed to put (b) in its present form for $s > 1$, is given in the last paragraph of (a) $\Rightarrow$ (b) below. (a) $\Longleftrightarrow$ (d) is proved in [22]. The equivalence with (c) appears to be partially new; Gelfand and Šilov have (c) where all the elements of U are required to be to the left of all the elements of V. The condition (c) as stated appears in a related context in Métivier [62] when $q = 2$; more on this in Chapter 3. The equivalence with (e) appears to be new; however, as Goodman essentially remarks at the end of [34], (e) $\Rightarrow$ (a) can be shown as a consequence of Mehler's formula. The condition (e) appeared in a related context in Geller [26]; more on this in Section 4. Goodman appears to have been the first to use Z^q_q in contexts related to those in this paper, as early as 1970 ([33]). Goodman says that E. M. Stein suggested he do so.

For the reader's convenience, and because we shall need certain modifications later, we give a detailed proof of the theorem, without referring back to [22].

Proof of Theorem 2.3. We show (a) $\Rightarrow$ (b) = (c) $\Rightarrow$ (a); afterwards we show (c) $\Rightarrow$ $(c)_2$ $\Rightarrow$ $(c)_1$ $\Rightarrow$ (d) $\Rightarrow$ (c), and, if $p \Rightarrow 2$, $(c)_2 \Longleftrightarrow$ (e).

(a) $\Rightarrow$ (b). The content is that the decay of f, of exponential order q, must persist in a sector. This is an immediate consequence of the following lemma, which will be used again later.

Lemma 2.4. Suppose $q > 0$. Say $B_1, B_2, C_1 > 0$. Suppose f is holomorphic in the sector $S = \{|\eta| < c|\xi|\} \subset \mathbb{C}^s$, is continuous on the closure of S, and satisfies the estimate $|f(\zeta)| \leq C_1 e^{B_1|\zeta|^q}$ for $\zeta \in S$ and $|f(\xi)| \leq C_1 e^{-B_2|\xi|^q}$ for $\xi \in \mathbb{R}^s$. Then for any B_3 with $0 < B_3 < B_2$, there exists $c_1 > 0$ so that if $\zeta \in S_1 = \{|\eta| < c_1|\xi|\}$, then $|f(\zeta)| \leq C_1 e^{-B_3|\zeta|^q}$. c_1 may be chosen so as to depend only on B_1, B_2, B_3 and c, and not on C_1 or f.

Proof. Essentially following [22], page 241, we begin by showing the apparently weaker fact that the conclusion is true not in a sector S_1 but in a domain of the form $D_{\theta'} = \{\zeta \,|\, |\arg \zeta_\ell| < \theta'$ for all $\ell\}$ for some $\theta' > 0$. Precisely, we show:

(*) Say $\varepsilon > 0$, $B_1, B_2', C_1' > 0$, $0 < \theta < \pi$. Suppose f is analytic on a domain D_θ, is continuous on $\overline{D}_\theta$ and satisfies

$|f(\zeta)| \leqq C'e^{B_1'(|\zeta_1|^q+\ldots+|\zeta_s|^q)}$ for $\zeta \in D_\theta$, while $|f(\xi)| \leqq C_1' e^{-B_2'(|\xi_1|^q+\ldots+|\xi_s|^q)}$ for $\xi \in \mathbb{R}^s \cap D_\theta$ (= the first quadrant). Then for any B_3' with $0 < B_3' < B_2'$, there exists θ' with $0 < \theta' < \theta$, depending only on B_1', B_2', B_3', so that $|f(\zeta)| \leqq C_1' e^{-B_3'(|\zeta_1|^q+\ldots+|\zeta_s|^q)}$ for $\zeta \in D_{\theta'}$.

To see (*), first note that we may assume $\theta < \pi/q$. We may immediately reduce to the case $q = 1$ by replacing f by g where $g(\zeta) = f(\zeta_1^{1/q},\ldots,\zeta_s^{1/q})$ for $\zeta \in D_{q\theta}$ (principal branches of the power functions). Accordingly, we prove (*) when $q = 1$, still using f to denote the function. Given an s-tuple σ, with $\sigma_\ell = \pm 1$ for each ℓ, and $0 < \phi < \pi$, let $D_{\phi,\sigma} = \{\zeta \mid 0 \leqq \sigma_\ell \arg \zeta_\ell < \phi$ for all $\ell\}$. Then $D_\phi = \bigcup_\sigma D_{\phi,\sigma}$, where the union is over all 2^s possibilities for σ. It suffices to show that, for each fixed σ, there exists θ' dependiong only on B_1', B_2', B_3' and θ so that

$|f(\zeta)| \leqq C_1' e^{-B_3'(|\zeta_1|+\ldots+|\zeta_s|)}$ for $\zeta \in D_{\theta',\sigma}$.

Consider now the function F on $\mathbb{C}$ defined by $F(z) = e^{(-B_2'+iA)z}$ where A is to be selected in such a way that $|F(x)| = e^{-B_2'x}$ for $x > 0$, while $|F(re^{i\theta})| = e^{B_1'r}$ for $r > 0$. Clearly the first condition holds regardless of the value of A. The second requires $-[B_2' \cos\theta + A \sin\theta] = B_1'$, so it suffices to choose $A = -[B_1 + B_2' \cos\theta]/\sin\theta$. Fixing, then, A as above, let $F_\ell(z) = e^{(-B_2'+iA)z}$ if $\sigma_\ell = 1$, $F_\ell(z) = e^{(-B_2'-iA)z}$ if $\sigma_\ell = -1$.

Note that $|F_\ell(x)| = e^{-B_2'x}$ if $x > 0$, while $|F_\ell(re^{i\sigma_\ell\theta})| = e^{B_1'r}$ if $r > 0$--the latter since $|e^{(-B_2'-iA)z}| = |e^{(-B_2'+iA)\bar{z}}|$.

Let $G(\zeta) = F_1(\zeta_1)\ldots F_s(\zeta_s)$, and $h(\zeta) = f(\zeta)/G(\zeta)$. By construction, $|h(\zeta)| \leqq C_1'$ when each $\arg \zeta_\ell = 0$ or $\sigma_\ell\theta$, while for some number $A' > 0$, $|h(\zeta)| \leqq C_1'e^{A'|\zeta|}$ for $\zeta \in D_{\theta,\sigma}$. In this set-up, the Phragmen-Lindelöf Theorem yields that $|h(\zeta)| \leqq C_1'$ in $D_{\theta,\sigma}$. Accepting this for the moment, it follows that $|f(\zeta)| \leqq C_1'|G(\zeta)|$ in $D_{\theta,\sigma}$. Fix B_3' with $0 < B_3' < B_2'$. It is clear from the form of G that for some θ', depending only on B_1', B_2', B_3' and θ, we have $|G(\zeta)| < e^{-B_3'(|\zeta_1|+\ldots+|\zeta_s|)}$ for $\zeta \in D_{\theta',\sigma}$. Thus (*) will follow once we see how Phragmen-Lindelöf implies that $|h(\zeta)| \leqq C_1'$ in $D_{\theta,\sigma}$.

To use Phragmen-Lindelöf, note that if $\xi_2,\ldots,\xi_s$ are fixed nonnegative reals, the function $u_1(\zeta_1) = h(\zeta_1,\xi_2,\ldots,\xi_s)$ satisfies $|u_1(\zeta_1)| < C_1'$ when $\arg \zeta_1 = 0$ or $\sigma_1\theta$; further $|u_1(\zeta_1)| < Ce^{A''|\zeta_1|}$ in sector $D_1 = \{0 < \sigma_1 \arg \zeta_1 < \theta\}$, for some C, A''. Consequently, the one-dimensional Phragmen-Lindelöf Theorem for a sector— which follows from the more familiar three-lines theorem by use of the transformation $\zeta_1 = e^{\sigma_1\theta w}$ $(w \in \{w|\ 0 < \operatorname{Im} w < 1\})$— gives that $|u_1(\zeta_1)| \leqq C_1'$ for $\zeta_1 \in D_1$. (The growth restriction in the hypothesis of the three-lines theorem is satisfied since $0 < \theta < \pi$.) Next fix $\zeta_1 \in D_1, \zeta_3,\ldots,\zeta_s$ with $\arg \zeta_\ell = 0$ or $\sigma_\ell\theta$, $3 \leqq \ell \leqq s$; the function $u_2(\zeta_2) = h(\zeta_1,\zeta_2,\zeta_3,\ldots,\zeta_s)$

satisfies, by Phragmen-Lindelöf, $|u_2(\zeta_2)| \leqq C_1'$ in $D_2 = \{0 < \sigma_2 \arg \zeta_2 < \theta\}$. Continuing in this way, we find $|h(\zeta)| \leqq C_1'$ in $D_{\theta,\sigma}$ after s steps. This establishes (*).

Finally, we establish the Lemma. Returning to the notation of the statement of the Lemma, note that (*) applies since we may choose θ so that $D_\theta \subset S$. We conclude that there exists θ' so that if f is any function satisfying the hypothesis of the lemma, then $|f(\zeta)| \leqq C_1 e^{-B_2|\zeta|^q}$ for $\zeta \in D_{\theta'}$. Define $c_1 > 0$ by $c_1 = (\arctan \theta')/s^{1/2}$; it will suffice to show that $|f(\zeta)| \leqq C_1 e^{-B_3|\zeta|^q}$ for $\zeta \in S_1 = \{|\eta| < c_1|\xi|\}$. Indeed, if $\zeta' = \xi' + i\eta' \in S_1$ is fixed, select an orthogonal transformation T on $\mathbb{R}^s$ so that $T^{-1}\xi' = (|\xi'|/s^{1/2})(1,\ldots,1)$. Put $f_T(\zeta) = f(T\xi+iT\eta)$ for $\zeta \in S$. Since T may be extended by using the same matrix, to be a unitary transformation T on $\mathbb{C}^s$, and since $f_T(\zeta) = f(T\zeta)$, f_T is evidently holomorphic on S and satisfies all the hypotheses. Since, however, $|(T^{-1}\eta')_\ell| \leqq |T^{-1}\eta'| = |\eta'| < c_1|\xi'| = c_1 s^{1/2}|(T^{-1}\xi')_\ell|$, we have $T^{-1}\zeta' \in D_{\theta'}$, so that $|f(\zeta')| = |f_T(T^{-1}\zeta')| \leqq C_1 e^{-B_3|T^{-1}\zeta'|^q} = C_1 e^{-B_3|\zeta'|^q}$, as desired.

The lemma gives (a) $\Rightarrow$ (b) at once, as well as the uniformity in the implication.

<u>(b) $\Rightarrow$ (c)</u>. First, we establish the following lemma, also to be used again later.

Lemma 2.5. Suppose $z_o \in \mathbb{C}^s$ and let $\mathcal{D}_\rho$ be the polydisc centered at z_o with polyradii $(\rho,\ldots,\rho)$. Let $N \in \mathbb{N}$. Suppose $f_1,\ldots,f_N$ are holomorphic on $\mathcal{D}_\rho$ and that $D_1,\ldots,D_N \in \{\partial/\partial z_1,\ldots,\partial/\partial z_s\}$. Then $|(D_1f_1D_2f_2\ldots D_Nf_N)(z_o)| \leqq (2\pi eN)^{1/2}(e/\rho)^N N!\|f_1\|_\infty\ldots\|f_N\|_\infty$, where $\|\ \|_\infty = \|\ \|_{\infty,\mathcal{D}_\rho}$ is the sup taken over $\mathcal{D}_\rho$.

Remark. This, of course, generalizes the situation where $f_1,\ldots,f_{N-1} \equiv 1$, in which case the right side can be sharpened to $\rho^{-N}N!\|f_N\|_\infty$.

Proof. For $1 \leqq n \leqq N$, $z \in \mathcal{D}_\rho$, let $g_n(z) = (D_nf_nD_{n+1}\ldots D_Nf_N)(z)$; we are to estimate $g_1(z_o)$. By the Cauchy estimates applied on $\mathcal{D}_{\rho/N}(z_o)$, $|g_1(z_o)| \leqq (N/\rho)\|f_1g_2\|_{\infty,\mathcal{D}_{\rho/N}} \leqq (N/\rho)\|f_1\|_{\infty,\mathcal{D}_\rho}\|g_2\|_{\infty,\mathcal{D}_{\rho/N}}$. At any point $z \in \mathcal{D}_{\rho/N}$ we can apply the Cauchy estimates on $\mathcal{D}_{\rho/N}(z) \subset \mathcal{D}_{2\rho/N}(z_o)$, to find $\|g_2\|_{\infty,\mathcal{D}_{\rho/N}} \leqq (N/\rho)\|f_2g_3\|_{\infty,\mathcal{D}_{2\rho/N}} \leqq (N/\rho)\|f_2\|_{\infty,\mathcal{D}_\rho}\|g_3\|_{\infty,\mathcal{D}_{2\rho/N}}$. Continuing in this way, after N steps we find $|g_1(z_o)| \leqq (N/\rho)^N\|f_1\|_\infty\ldots\|f_N\|_\infty$, which is what we wanted, by Stirling's formula.

Proof of (b) ⇒ (c). We modify the proof of (B') ⇒ (B)(II) from Chapter one. We must show that for some C,R and all $\{X_1,\ldots,X_{m+n}\}$ as described in (c), and all $\xi \in \mathbb{R}^s$ we have $|X_1\ldots X_{m+n}f(\xi)| < CR^{m+n}m!^{1/q}n!^{1/p}$. We consider two cases, $|\xi| \geqq n^{1/q}$ and $|\xi| < n^{1/q}$. For the first case, choose b_o with $0 < b_o < 1/2s$

so that if $\xi \in \mathbb{R}^s \setminus \{0\}, \mathcal{D}_{b_o|\xi|}(\xi) \subset S$. Assuming $|\xi| \geq n^{1/q}$, we use Lemma 2.5 with $\mathcal{D} = \mathcal{D}_\rho(z_o) = \mathcal{D}_{b_o|\xi|}(\xi)$ to find $|X_1 \ldots X_{m+n} f(\xi)| \leq (2\pi en)^{1/2}(e/b_o)^n (n!/|\xi|^n)(\sup_{\zeta\in\mathcal{D}}|\zeta|)^m \times \sup_{\zeta\in\mathcal{D}}(Ce^{-B_2|\zeta|^q})$. However, if $\zeta \in \mathcal{D}$, then $(1-sb_o)|\xi| \leq |\zeta| \leq (1+sb_o)|\xi|$; also $|\xi|^n \geq n^{n/q} > n!^{1/q}$; consequently for some $C, R_1, B_2' > 0$, $|X_1 \ldots X_{m+n} f(\xi)| \leq CR_1^{m+n} n!^{1/p}|\xi|^m e^{-B_2|\xi|^q}$. However, by (a) $\Rightarrow$ (c) of Lemma 2.2, for some $C_2, R_2 > 0$,

$$e^{-B_2'|\xi'|^q} < C_2 R_2^m m!^{1/q}|\xi'|^{-m} \tag{2.1}$$

for all $\xi' \in \mathbb{R}^s$, $m \in \mathbb{Z}^+$. This gives the desired estimate in the first case. In case $|\xi| < n^{1/q}$, we use Lemma 2.5 with $\mathcal{D} = \mathcal{D}_{n^{1/q}}(\xi)$ to find $|X_1 \ldots X_{m+n} f(\xi)| \leq (2\pi en)^{1/2} e^n (n!/n^{n/q}) \times (\sup_{\zeta\in\mathcal{D}}|\zeta|)^m \sup_{\zeta\in\mathcal{D}}(Ce^{B_1|\zeta|^q})$. However $(\sup|\zeta|)^m = \sup|\zeta|^m \leq (\sup|\zeta|^m e^{-|\zeta|^q}) \sup e^{|\zeta|^q} \leq C_2 R_2^m m!^{1/q}(\sup e^{|\zeta|^q})$ for some $C_2, R_2 > 0$ by (2.1) (with $\xi' \in \mathbb{R}^s$ chosen so that $|\xi'| = |\zeta|$). Also, for $\zeta \in \mathcal{D}$, $|\zeta|^q \leq (1+s)^q n$. Combining these estimates, we see that the desired estimate follows in this case as well. Thus (b) $\Rightarrow$ (c) is established. The proof, and the already established uniformity in (a) $\Rightarrow$ (b), give the uniformity in (a) $\Rightarrow$ (c).

<u>(c) $\Rightarrow$ (a)</u>. Using (1.24), choose $R_o > 0$ so that for all $M \in \mathbb{Z}^+$,

$|\xi|^M < R_o^M \sum |\xi_\ell|^M$. Then $|\xi|^m |f(\xi)| < C_1 (RR_o)^m m!^{1/q}$, so that $|f(\xi)| < C_2 e^{-B_2|\xi|^q}$ for some $C_2, B_2 > 0$, by Lemma 2.2(c) $\Rightarrow$ (a).

Secondly, we have $|\partial^\beta f(\xi)| < CR^{\|\beta\|}\|\beta\|!^{1/p}$ for any multi-index β. As in the proof of (B)(II) $\Rightarrow$ (B')(II) in Chapter one, this at once shows that f has an entire extension f to $\mathbb{C}^s$ with $|f(\zeta)| < Ce^{B_1|\zeta|^q}$ for some $C, B_1 > 0$. This proves (a) under an apparently much weaker hypothesis than (c)—namely (c) where the X_i are required to <u>all</u> be in U or <u>all</u> in V. This of course is not new ([22]). The uniformity in (c) $\Rightarrow$ (a) follows from the proof.

<u>(c) $\Rightarrow$ (c)$_2$ $\Rightarrow$ (c)$_1$ $\Rightarrow$ (d) $\Rightarrow$ (c).</u> If (c) holds, to show (c)$_2$ it suffices to bound $\|(1+|\xi_1|^s+\ldots+|\xi_s|^s)X_1\ldots X_{m+n}f\|_\infty$ by the right side of (2.2), with a different C and R. But $\|\xi_\ell^s X_1 \ldots X_{m+n} f\|_\infty < CR^{m+n+s+1}(m+s+1)!^{1/q}n!^{1/p}$ and (c)$_2$ follows. If (c)$_2$ holds, then similarly $\|(1+|\xi_1|^s+\ldots+|\xi_s|^s)X_1\ldots X_{m+n}f\|_2 < C_1R_1^{m+n}m!^{1/q}n!^{1/p}$ for some C_1, R_1, and (c)$_1$ follows from Cauchy-Schwarz. If (c)$_1$ holds, the analogue of (c) holds with $(\hat{f},p,q)$ replacing (f,q,p); thus (d) holds. In particular (c) $\Rightarrow$ (d), so by symmetry, (d) $\Rightarrow$ (c). The uniformity in (a) $\Rightarrow$ (d) follows from this proof and the uniformity in (a) $\Rightarrow$ (c). The uniformities in (c)$_2$ $\Rightarrow$ (a) and (c)$_1$ $\Rightarrow$ (a) also follow from this proof and the uniformity in (c) $\Rightarrow$ (a).

$\underline{(c)_2 \Rightarrow (e).}$ (p=2) Let $D = \sum(M_\ell^2 - \partial_\ell^2)$ be the Hermite operator. As is well known,

$$DH_\alpha = (2\|\alpha\|+1)H_\alpha. \tag{2.2}$$

By $(c)_2$ and the simple inequality $(m!n!)^{1/2} \leqq (m+n)!^{1/2}$, we have that for some $C_2, R_2 > 0$,

$$\|X_1 \ldots X_L f\|_2 < C_2 R_2^L L!^{1/2} \tag{2.3}$$

for any $X_1, \ldots, X_L \in U \cup V$. Thus if $M \in \mathbb{Z}^+$, $\|D^M f\|_2 < C_2(2s)^M R_2^{2M}(2M)!^{1/2} < C_3 R_3^M M!$ for some $C_3, R_3 > 0$, and D^M may be written as a sum of $(2s)^M$ expressions of the form $\pm X_1 \ldots X_{2M}$. Since the H_α are orthonormal, if $\|\alpha\| = N$ we have in particular that $|c_\alpha|(2N+1)^M < C_3 R_3^M M!$. Thus $|c_\alpha| < C_3 R_4^M M!/N^M$ if $\|\alpha\| = N$. For $N \in \mathbb{Z}^+$, let $f(N) = \max_{\|\alpha\|=N} |c_\alpha|$. By Lemma 2.2(c) $\Rightarrow$ (a) with $S_o = \mathbb{Z}^+$, we find $f(N) < Ae^{-BN}$ for some $A, B > 0$, as desired.

$\underline{(e) \Rightarrow (c)_2.}$ (p=2) Let $A_\ell^- = (M_\ell + \partial_\ell)/2^{1/2}$, $A_\ell^+ = (M_\ell - \partial_\ell)/2^{1/2}$ be the annihilation and creation operators. As is well known, if $e_\ell = (0, \ldots, 1, \ldots, 0)$ (1 in the ℓth spot), then $A_\ell^- H_\alpha = \alpha_\ell^{1/2} H_{\alpha - e_\ell}$ (zero if $\alpha_\ell = 0$), while $A_\ell^+ H_\alpha = (\alpha_\ell + 1)^{1/2} H_{\alpha + e_\ell}$. The implication will follow readily from the fact that for some $C_1, R_1 > 0$, if $A_1, \ldots, A_L \in \{A_1^-, \ldots, A_s^-, A_1^+, \ldots, A_s^+\}$, then

$$\|A_1 \ldots A_L f\|_2 < C_1 R_1^L L!^{1/2}. \tag{2.4}$$

Let us first show (2.4). Since, for fixed $A_1,\ldots,A_L$, the functions $\{A_1\ldots A_L H_\alpha\}$ are orthogonal, we have $\|A_1\ldots A_L f\|_2^2 = \sum |c_\alpha|^2 \|A_1\ldots A_L H_\alpha\|_2^2 \leq C^2 \sum r^{2\|\alpha\|}(\|\alpha\|+1)\ldots(\|\alpha\|+L+1) \leq C^2 \sum_N (N+1)^s r^{2N}(N+1)\ldots(N+L)$ since $\#\{\alpha \mid |\alpha| = N\} \leq (N+1)^s$. For some $C' > 0$, $0 < r_1 < 1$ we have $(N+1)^s r^{2N} < C' r_1^{2N}$ for all N; thus for some $C_1 > 0$, $L!^{-1}\|A_1\ldots A_L f\|_2^2 \leq C_1^2 \sum r_1^N (N+L)!/(N!L!) = C_1^2 \sum r_1^N (L+1)\ldots(L+N)/N! = C_1^2 (1-r_1)^{-(L+1)}$, yielding (2.4) at once.

(2.3) quickly follows, for some $C_2, R_2 > 0$. Indeed, each $X_\ell = (A_\ell^- \pm A_\ell^+)/2^{1/2}$, so that $X_1\ldots X_L$ can be written as a sum of 2^L expressions of the form $\pm A_1\ldots A_L/2^{L/2}$. Thus (2.3) follows, with $R_2 = 2^{1/2}R_1$; this gives $(c)_2$, since if $L = M + N$, $L! < 2^{M+N}M!N!$. ∎

<u>Remark</u>. (e) does have a generalization for $p > 2$, but the generalization is not for Z_q^q but for Z_q^p. Details are left to the interested reader. For a related result, see Hille [44].

(e) suggests a rough analogy. If S^1 is the unit circle, $C^\infty(S^1)$ consists of those functions with Fourier expansions whose coefficients decay rapidly. $A(S^1) = \{$real analytic functions on $S^1\}$ consists of those functions with Fourier expansions whose coefficients decay exponentially (by Laurent's theorem). $\mathcal{S}(\mathbb{R}^s)$ consists of those functions with Hermite expansions whose coefficients decay rapidly, while Z_2^2 consists of those functions with Hermite expansions whose coefficients decay exponentially.

Thus we can think of Z^2_2 (or its variant, Z^q_q) as "a natural Schwartz space in the real analytic category."

We now show how Lemma 2.4 can also be utilized to show the fact claimed towards the end of the first section, that $RZ^q_{q,j} = Z^q_{q,j}$.

<u>Theorem 2.6</u>. $RZ^q_{q,j} = Z^q_{q,j}$.

Further, we have this uniformity. Suppose $B,C,R > 0$; then there exist C_1,R_1,c so that whenever $h \in RZ^q_{q,j}(B,C,R)$ then $h \in Z^q_{q,j}(B,C_1,R_1,c)$.

<u>Proof</u>. As we remarked after the definition of $Z^q_{q,j}$, we have $Z^q_{q,j} \subset RZ^q_{q,j}$. For the converse, we shall use the following lemma:

<u>Lemma 2.7</u>. Suppose $c,C,R > 0$. Let $\{g_\ell(\zeta)\}$ be a set of functions such that g_ℓ is holomorphic in $S = \{|\eta| < c|\xi|\}$ and homogeneous of degree $j - p\ell$ there, and such that $|g_\ell(\zeta)| < CR^\ell \ell!^{p-1}$ for $|\zeta| = 1$, $\zeta \in S$. Then there exist $c',C_o,R_o > 0$, and a function f which is defined and continuous on the closure of the sector $S' = \{|\eta| < c'|\xi|\}$, holomorphic in S', and satisfies:

$$\left|f(\zeta) - \sum_{\ell=0}^{L-1} g_\ell(\zeta)\right| < C_o R_o^L L!^{p-1}|\zeta|^{\mathrm{Re}\ j-pL} \qquad \text{for } |\zeta| > 1,\ \zeta \in S' \tag{2.5}$$

Further, we have this uniformity. Suppose $c,C,R > 0$; then there exist $c',C_o,R_o > 0$ so that whenever $\{g_\ell(\zeta)\}$ is a set of functions as described in the hypothesis, there then exists f defined and continuous on the closure of $S' = \{|\eta| < c'|\xi|\}$, holomorphic in S', and which satisfies (2.5).

<u>Proof of Theorem 2.6, assuming Lemma 2.7</u>. Suppose $h \in RZ^q_{q,j}$ and $h \sim \sum g_\ell$. As we remarked before the definition of $Z^q_{q,j}$, the g_ℓ can be extended to functions $g_\ell(\zeta)$ which satisfy the hypothesis of Lemma 2.7 for some c,C,R. Select c', C_0, R_0, f as in the conclusion of Lemma 2.7. Then we surely have $|(h-f)(\zeta)| < C_1 e^{B_1|\zeta|^q}$ in S', for some $C_1, B_1 > 0$. Further, for $\xi \in \mathbb{R}^s$, $|\xi| \geqq 1$, we have $|\xi|^{-\mathrm{Re}\, j}|(h-f)(\xi)| < C'R_1^L L!^{p-1}|\xi|^{-pL}$ for some $C', R_1 > 0$. By Lemma 2.2(a) => (b), then there exist $C_2, B_2 > 0$ so that $|(h-f)(\xi)| < C_2 e^{-B_2|\xi|^q}$ for $\xi \in \mathbb{R}^s$. By Lemma 2.4, there exist $c_2, C_3, B_3 > 0$ so that if $\zeta \in S_2 = \{|\eta| < c_2|\xi|\}$, then $|(h-f)(\zeta)| < C_3 e^{-B_3|\zeta|^q}$. Thus for some $C_4, B_4 > 0$, $|\zeta|^{-\mathrm{Re}\, j}|(h-f)(\zeta)| < C_4 e^{-B_4|\zeta|^q}$ for $\zeta \in S_2$, $|\zeta| \geqq 1$, so that by Lemma 2.2(a)<==>(b), for some $C_5, R_5 > 0$, $|(h-f)(\zeta)| < C_5 R_5^L L!^{p-1}|\zeta|^{\mathrm{Re}\, j-pL}$ for $\zeta \in S_2$, $|\zeta| \geqq 1$. If we combine this with (2.5), we find that for some $C_6, R_6 > 0$, $|h(\zeta) - \sum_{\ell=0}^{L-1} g_\ell(\zeta)| < C_6 R_6^L L!^{p-1}|\zeta|^{\mathrm{Re}\, j-pL}$ for $\zeta \in S_2$, $|\zeta| \geqq 1$, so that $h \in Z^q_{q,j}$ as desired. The uniformity assertion follows from the arguments and the uniformity assertions of Lemmas 2.2, 2.4, and 2.7. ∎

<u>Proof of Lemma 2.7</u>. We claim that we may assume $\mathrm{Re}\, j < 0$. Indeed, for any fixed $N \in \mathbb{N}$, put $G_\ell(\zeta) = g_\ell(\zeta)[\zeta_1^2 + \ldots + \zeta_s^2]^{-N}$.

The G_ℓ are then holomorphic in some sector $S_1 = \{|\eta| < c_1|\xi|\}$, are homogeneous of degree $j - 2N - p\ell$, and for some $C_1 > 0$, they satisfy $|G_\ell(\zeta)| < C_1R^\ell \ell!^{p-1}$ for $|\zeta| = 1$, $\zeta \in S_1$. If the construction we seek is possible for Re $j < 0$, then if N is large enough there exists a function F such that for some $c', C_1, R_1 > 0$, F is holomorphic in $S' = \{|\eta| < c'|\xi|\}$, continuous on the closure, and satisfies $|F(\zeta) - \sum_{\ell=0}^{L-1} G_\ell(\zeta)| < C_1R_1^L L!^{p-1}|\zeta|^{\text{Re } j-2N-pL}$ for $\zeta \in S$, $|\zeta| \geqq 1$. The function $f(\zeta) = [\zeta_1^2+\ldots+\zeta_2^2]^N F(\zeta)$ is then as desired. The uniformity assertion for Re $j \geqq 0$ would also follow from these arguments and the uniformity assertion for Re $j < 0$.

We therefore assume Re $j < 0$ and construct f. We free all the notation introduced in the last paragraph. For simplicity, let us first give the argument when $s = 1$. Write the function $g_\ell(\xi)$ as a sum of an even and an odd function. The even part is just a multiple of $|\xi|^{j-p\ell}$; the odd part is a multiple of $\xi|\xi|^{j-p\ell-1}$. Thus $g_\ell(\xi) = A_{\ell,0}|\xi|^{j-p\ell} + A_{\ell,1}\xi|\xi|^{j-p\ell-1}$, where $A_{\ell,0}, A_{\ell,1} \in \mathbb{C}$. The even part is just $[g_\ell(\xi) + g_\ell(-\xi)]/2$, so that $|A_{\ell,\varkappa}| < CR^\ell \ell!^{p-1}$ ($\varkappa=0,1$). g_ℓ has a holomorphic extension to any sector $\{|\eta| < c_1|\xi|\}$, namely $g_\ell(\zeta) = A_{\ell,0}(\pm\zeta)^{j-p\ell} + A_{\ell,1}\zeta(\pm\zeta)^{j-p\ell-1}$, the $\pm$ sign taken according as Re $\zeta > 0$ or Re $\zeta < 0$ (principal branches of the power functions). Next, for $\varkappa = 1$ or 2, note that we have the simple estimate $|\Gamma(p\ell+\varkappa-j)/q)| > C_1R_1^\ell \ell!^{p-1}$ for some $C_1, R_1 > 0$, since

$p/q = p - 1$. (We shall prove a more precise estimate in detail in Lemma 2.10 below.) We next define functions $G_\varkappa(\tau)$ for $\varkappa = 0,1$, and $|\tau|^{p-1} < R_1/R$, as follows:

$$G_\varkappa(\tau) = \sum_{\ell=0}^{\infty} [A_{\ell\varkappa}/\Gamma(p\ell+\varkappa-j)/q)]\tau^{((p\ell+\varkappa-j)/q)-1}.$$

This series converges absolutely for $|\tau|^{p-1} < R_1/R$, as the estimate for $\Gamma((p\ell+k-j)/q)$ together with the estimates for the $A_{\ell,\varkappa}$, show at once. Let $R_2 = (R_1/3R)^{1/(p-1)}$. We claim that we can simply put

$$f(\zeta) = \int_0^{R_2} \exp(-\tau(\zeta^2)^{q/2})[G_0(\tau) + \zeta G_1(\tau)]d\tau$$

for ζ in any sector $S' = \{|\eta| < c'|\xi|\}$ where $c' < \arctan(\pi/2q)$. (We use the principal branch of the $(q/2)$-power function; note ζ^2 cannot be on the negative real axis.) Indeed, f is clearly holomorphic in S' and continuous in the closure of S'. Fix $\varkappa = 0$ or 1. Let

$$e_L(\zeta) = \left|\int_0^{R_2} \exp(-\tau(\zeta^2)^{q/2})G_\varkappa(\tau)d\tau - \sum_{\ell=0}^{L-1} A_{\ell,\varkappa}(\pm\zeta)^{j-p\ell-\varkappa}\right|$$

where the $\pm$ sign is taken as $\operatorname{Re}\zeta > 0$ or $\operatorname{Re}\zeta < 0$. It suffices to show that for some $C_1, R_0 > 0$, $e_L(\zeta) < C_1R_0^L L!^{p-1}|\zeta|^{\operatorname{Re} j-pL-k}$ for $|\zeta| > 1$, $\zeta \in S'$. Clearly we may assume $\operatorname{Re}\zeta > 0$, so that $(\zeta^2)^{q/2} = \zeta^q$. In the expression for e_L, write $G_\varkappa(\tau) = \sum_{\ell=0}^{L-1} [A_{\ell,\varkappa}/\Gamma(p\ell+\varkappa-j)/q)]\tau^{(p\ell+\varkappa-j)/q-1} + G_{L,\varkappa}(\tau)$, where

$|G_{L,\varkappa}(\tau)| < C_2\tau^{(pL+\varkappa-\text{Re }j)/q-1}$ for some C_2. Since

$$\int_0^{R_2}\exp(-\tau\zeta^q)\tau^{(p\varkappa+\ell-j)/q-1}d\tau=\Gamma((p\ell+\varkappa-j)/q)\zeta^{j-p\ell-\varkappa}$$
$$-\int_{R_2}^{\infty}\exp(-\tau\zeta^q)\tau^{(p\ell+\varkappa-j)/q-1}d\tau$$

for $\zeta\in S$, $\text{Re }\zeta > 0$, we find

$$e_L(\zeta) \leqq C_3\Big[\sum_{\ell=0}^{L-1}[|A_{\ell,\varkappa}|/\Gamma((p\ell+\varkappa-j)/q)|]\times$$
$$\int_{R_2}^{\infty}\exp(-\tau\text{Re }\zeta^q)\tau^{(p\ell+\varkappa-\text{Re }j)/q-1}d\tau$$
$$+\int_0^{R_2}\exp(-\tau\text{ Re }\zeta^q)\tau^{(pL+\varkappa-\text{Re }j)/q-1}d\tau\Big].$$

For ζ in the sector S', there exists a constant $B_1 > 0$ such that $\text{Re }\zeta^q > B_1|\zeta|^q$. Thus we may estimate

$$\int_{R_2}^{\infty}\exp(-\tau\text{ Re }\zeta^q)\tau^{(p\ell+\varkappa-\text{Re }j)/q-1}d\tau$$
$$< R_2^{(\ell-L)p/q}\int_{R_2}^{\infty}\exp(-\tau B_1|\zeta|^q)\tau^{(pL+\varkappa-\text{Re }j)/q-1}d\tau. \tag{2.6}$$

Substituting this in the above, and estimating all the integrals by integrals from 0 to ∞, we find that for some $C_4,B_2,R_3,C_5,R_4 > 0$,

$$e_L(\zeta) \leqq C_4\Big[\sum_{\ell=0}^{L-1}R_3^L\Gamma((pL+\varkappa-\text{Re }j)/q)|\zeta|^{\text{Re }j-pL-\varkappa}$$
$$+R_3^L\Gamma((pL+\varkappa-\text{Re }j)/q)|\zeta|^{\text{Re }j-pL-\varkappa}\Big]$$

$< C_5R_4^LL!^{p-1}|\zeta|^{\text{Re }j-pL-k}$, as desired. The uniformity assertion, when $s = 1$, follows also from the above arguments.

When $s > 1$, we must be more precise in our estimates. Again we free all notaiton introduced since the beginning of the proof. Let $\{P_m\}$ $(m \in \mathbb{Z}^+)$ be a basis for the spherical harmonics which is orthonormal on the unit sphere. We define $\varkappa : \mathbb{Z}^+ \to \mathbb{Z}^+$ by letting $\varkappa(m)$ be the degree of P_m; we may suppose the P_m are so ordered that $\varkappa$ is a nondecreasing function. We adopt the convention that in any expression, equality or inequality in which both m and $\varkappa$ occur, $\varkappa$ means $\varkappa(m)$. We consider the spherical harmonic expansion of $g_\ell(\xi)$, say $g_\ell(\xi) = \sum_m A_{\ell m} P_m(\xi)|\xi|^{j-p\ell-\varkappa}$ (this, as we said, is shorthand for $\sum_m A_{\ell m} P_m(\xi)|\xi|^{j-p\ell-\varkappa(m)}$). We shall need the following lemma. (The proof of the forward direction was shown to us by E. M. Stein):

<u>Lemma 2.8</u>. Suppose $R_1 > 0$; then there exist $C_1 > 0$ and $0 < r_1 < 1$ as follows. Let g be a real analytic function on $\mathbb{R}^s \setminus \{0\}$, homogeneous of degree 0, such that $|\partial_\xi^\alpha g(\xi)| < R_1^{\|\alpha\|}\|\alpha\|!$ for $|\xi| = 1$. Let $g(\xi) = \sum A_m P_m(\xi)$ be the spherical harmonic expansion of g, for $|\xi| = 1$. Then $|A_m| < C_1 r_1^{\varkappa}$.

Conversely, given $C_1 > 0$, $0 < r_1 < 1$, and constants $\{A_m\}$ such that $|A_m| < C_1 r_1^{\varkappa}$, then $g(\xi) = \sum A_m P_m(\xi)$ is real analytic in a neighborhood of $\{|\xi| = 1\}$. ∎

Let us accept this lemma temporarily and proceed. It is an easy consequence of the lemma that for some $C_2, R_2 > 0$,

$0 < r < 1$, we have $|A_{\ell m}| < C_2R_2^{\ell}\ell!^{p-1}r^{\varkappa}$. Indeed, let $h_\ell(\xi) = |\xi|^{p\ell-j}g_\ell(\xi)$ for $\xi \neq 0$; this function is homogeneous of degree 0. There exist $C_3,R_3,c_1 > 0$ such that for each ℓ, h_ℓ has a holomorphic extension to the sector $S_1 = \{|\eta| < c_1|\xi|\}$, to the function $h_\ell(\zeta) = (\zeta_1^2+\ldots+\zeta_s^2)^{(p\ell-j)/2}g_\ell(\zeta)$, and such that, if $\zeta \in S_1$, then $|h_\ell(\zeta)| < C_3R_3^{\ell}\ell!^{p-1}$. Accordingly, for some $C_2,R_2 > 0, |\partial_\xi^\alpha h_\ell(\xi)| < C_2R_2^{\ell+|\alpha|}\ell!^{p-1}\|\alpha\|!$ for $|\xi| = 1$. We apply the forward direction of the lemma to $g(\xi) = h_\ell(\xi)/C_2R_2^{\ell}\ell!^{p-1}$, whose spherical harmonic expansion is $\sum[A_{\ell m}/C_2R_2^{\ell}\ell!^{p-1}]P_m(\xi)$ for $|\xi| = 1$. The asserted estimate,

$$|A_{\ell m}| < C_2R_2^{\ell}\ell!^{p-1}r^{\varkappa} \tag{2.7}$$

(for some $0 < r < 1$) follows at once.

Now, there exists a sector $S_2 = \{|\eta| < c_2|\xi|\}$ such that if $\zeta \in S_2$, $g_\ell(\zeta) = \sum_m A_{\ell m}P_m(\zeta)[\zeta_1^2+\ldots+\zeta_s^2]^{(j-p\ell-\varkappa)/2}$. This is a simple consequence of the following lemma:

<u>Lemma 2.9</u>. There exists $C_4 > 0$, as follows.

(a) For any $0 < r_2 < 1$ there exists $c_3 > 0$ such that for all $\zeta \in S_3 = \{|\eta| < c_3|\xi|\}$ with $|\zeta| = 1$, and all m we have $|P_m(\zeta)| < C_4r_2^{-\varkappa}$.

(b) There exists a number $a > 0$ such that for all $\zeta \in \mathbb{C}^s$ with $|\zeta| = 1$, we have $|P_m(\zeta)| < C_4a^{\varkappa}$ for all m.

Accepting this lemma temporarily also, we verify the claim at once, as follows. Select any r_2 with $r < r_2^2 < 1$. (r as in (2.7)). Use (a) of the lemma to select $c_3 > 0$ such that in $S_3 = \{|\eta| < c_3|\xi|\}$, we have $|P_m(\zeta)| < C_4|\zeta|^{\varkappa} r_2^{-\varkappa}$. Further, select c_2 with $0 < c_2 < c_3$ such that if $\zeta \in S_2 = \{|\eta| < c_2|\xi|\}$, we have $|\zeta_1^2+\ldots+\zeta_s^2| > (r_2|\zeta|)^2$. Then if $\zeta \in S_2$,

$$\sum|A_{\ell m}P_m(\zeta)[\zeta_1^2+\ldots+\zeta_s^2]^{(j-p\ell-\varkappa)/2}| < C(\ell)|\zeta|^{\mathrm{Re}\ j-p\ell}\sum_{\varkappa} r^{\varkappa}(\dim H_{\varkappa})r_2^{-2\varkappa}$$

where $C(\ell)$ is a constant depending only on ℓ, and $H_{\varkappa}$ is the space of all spherical harmonics of degree $\varkappa$. As is known ([76], page 140), there exists $N_1 > 0$ such that $\dim H_{\varkappa} < (\varkappa+1)^{N_1}$; thus the series converges absolutely and uniformly on compact subsets of S_2, and therefore coincides with $g_{\ell}(\zeta)$ there.

We next define functions G_m in analogy with the functions $G_{\varkappa}$ of the case $s = 1$. We need a precise estimate for $\Gamma((p\ell+\varkappa-j)/q)$; this is given in the following lemma.

<u>Lemma 2.10</u>. Suppose $\mathrm{Re}\ j < 0$, $p \geqq 1$ and $Q > 0$.

(a) There exist $c_1, c_2 > 0$ such that for all $\ell \in \mathbb{Z}^+$, $c_1^{\ell}\ell!^{1/Q} \leqq [\ell/Q]! \leqq c_2^{\ell}\ell!^{1/Q}$ ([ ] = greatest integer function).

(b) There exist $C_5, N_2, R_3 > 0$ such that for all $\ell, \varkappa \in \mathbb{Z}^+$, $|\Gamma((p\ell+\varkappa-j)/Q)| > C_5(\varkappa+1)^{-N_2}R_3^{\ell}\ell!^{p/Q}[\varkappa/Q]!$

(c) For any $0 < r_3 < 1$, there exist $C_6, R_4 > 0$ such that for all $\ell, \varkappa \in \mathbb{Z}^+$, $|\Gamma((p\ell+\varkappa-j)/Q)| < C_6 r_3^{-\varkappa}R_4^{\ell}\ell!^{p/Q}[\varkappa/Q]!$

The lemma is quite simple to prove. However, we just accept it also for the time being. We apply Lemma 2.9(b) with $Q = q$, and observe that in (b) we can then write $\ell!^{p/Q} = \ell!^{p-1}$. We can thus form

$$G_m(\tau) = \sum_{\ell=0}^{\infty} [A_{\ell m}/\Gamma((p\ell+\varkappa-j)/q)]\tau^{(p\ell+\varkappa-j)/q-1} \tag{2.8}$$

which, as in the case $s = 1$, converges absolutely for $|\tau|^{p-1} < R_3/R_2$, because of (2.7). We put $R_5 = (R_3/3R_2)^{1/(p-1)}$ and wish to put

$$f(\zeta) = \sum_m [\int_0^{R_5} \exp[-\tau(\zeta_1^2+\ldots+\zeta_s^2)^{q/2}]P_m(\zeta)G_m(\tau)\,d\tau] \tag{2.9}$$

for ζ in a sector $S_4 = \{|\eta| < c_4|\xi|\}$, where c_4 is sufficiently small that $\zeta \in S_4 \Rightarrow |\arg(\zeta_1^2+\ldots+\zeta_s^2)| < \pi/q$. To see that f is holomorphic in S_4, let $w = \tau^{1/q}\zeta$. Choose c_1 as in Lemma 2.10(a), so that $\varkappa!^{1/q} \leq [\varkappa/q]!/c_1^{\varkappa}$. Note that elementary estimate $x^{\varkappa} = \varkappa!^{1/q}(x^{q\varkappa}/\varkappa!)^{1/q} \leq \varkappa!^{1/q}\exp(x^q/q)$ for $x > 0$. Observe that for $(\zeta,\tau) \in S_4 \times [0,R_5]$, we have

$$|\exp(-\tau(\zeta_1^2+\ldots+\zeta_s^2)^{q/2})P_m(\zeta)G_m(\tau)| < C_4 e^{|w|^q}\tau^{-\varkappa/q}|w|^{\varkappa}(a/c_1)^{\varkappa}c_1^{\varkappa}|G_m(\tau)|$$

$$< C_4\exp[(1 + (a/c_1)^q/q)|w|^q]\varkappa!^{1/q}c_1^{\varkappa}(\tau^{-\varkappa/q}|G_m(\tau)|)$$

$$< C_7 e^{B_1|\zeta|^q}[\varkappa/q]!((r')^{\varkappa}[\varkappa/q]!^{-1}\tau^{-\mathrm{Re}\, j/q-1})$$

for some $a,B_1,C_7 > 0$, $0 < r' < 1$, by Lemma 2.9(b), (2.7) and Lemma 2.10(b). By Lemma 2.9(a) and the inequality $\dim H_{\varkappa} < (\varkappa+1)^{N_1}$, then, we have that the series for f converges absolutely and locally uniformly, and that for some $C_8, C_9 > 0$ and $0 < r'' < 1$,

$$|f(\zeta)| < C_8 \sum (r'')^{\varkappa} e^{B_1|\zeta|^q} \int_0^{R_5} \tau^{-\mathrm{Re}\, j/q-1} d\tau = C_9 e^{B_1|\zeta|^q}.$$

Evidently, then, f is holomorphic on S_4 and continuous on its closure. For $\zeta \in S_4$, let

$$\begin{aligned} e_{Lm}(\zeta) = \Big| \int_0^{R_5} & \exp(-\tau(\zeta_1^2+\ldots+\zeta_s^2)^{q/2}) G_m(\tau)\, d\zeta \\ & - \sum_{\ell=0}^{L-1} A_{\ell m}[\zeta_1^2+\ldots+\zeta_s^2]^{(j-p\ell-\varkappa)/2} \Big| . \end{aligned} \tag{2.10}$$

It suffices to show that there exist $c_5, C_{10}, R_6 > 0$ and $0 < r_4 < 1$ such that if $\zeta \in S_5 = \{|\eta| < c_5|\xi|\}$, $|\zeta| \geqq 1$, then for all m,

$$e_{Lm}(\zeta) \leqq C_{10} R_6^L (L!)^{p-1} r_4^{\varkappa} |\zeta|^{\mathrm{Re}\, j-pL-\varkappa}. \tag{2.11}$$

Indeed, (2.5) is an immediate consequence of this and Lemma 2.9(a) (where the "r_2" of that lemma is simply chosen with $r_4 < r_2 < 1$). In the expression for e_{Lm}, write

$$G_m(\tau) = \Big[\sum_{\ell=0}^{L-1} A_{\ell m}/\Gamma((p\ell+\varkappa-j)/q) \Big] \tau^{(p\ell+\varkappa-j)/q-1} + G_{Lm}(\tau). \tag{2.12}$$

Observe that by Lemma 2.10(b) and (2.7), there exist $C_{11}, R_7, N_2 > 0$ such that

$$|G_{Lm}(\tau)| \leqq C_{11}(\varkappa+1)^{N_2} R_7^L r^{\varkappa} [\varkappa/q]!^{-1} \tau^{(pL+\varkappa-\mathrm{Re}\, j)/q-1} \tag{2.13}$$

if $\tau \in (0, R_2)$. Since

$$\int_0^{R_5} \exp[-\tau(\zeta_1^2+\ldots+\zeta_s^2)^{q/2}]\tau^{(p\ell+\varkappa-j)/q-1}d\tau$$

$$= \Gamma((p\ell+\varkappa-j)/q)(\zeta_1^2+\ldots+\zeta_s^2)^{(j-p\ell-\varkappa)/2} \qquad (2.14)$$

$$- \int_{R_5}^{\infty} \exp[-\tau(\zeta_1^2+\ldots+\zeta_s^2)^{q/2}]\tau^{(p\ell+\varkappa-j)/q-1}d\tau$$

for $\zeta \in S_4$, we find that for some $C_{12} > 0$,

$$e_{Lm}(\zeta) \leqq C_{12}[\sum_{\ell=0}^{L-1}[|A_{\ell m}/\Gamma((p\ell+\varkappa-j)/q)|]\Big| \times$$

$$\int_{R_5}^{\infty} \exp[-\tau(\zeta_1^2+\ldots+\zeta_s^2)^{q/2}]\tau^{(p\ell+\varkappa-j)/q-1}d\tau \qquad (2.15)$$

$$+ \left|\int_0^{R_5} \exp[-\tau(\zeta_1^2+\ldots+\zeta_s^2)^{q/2}G_{LM}(\tau)d\tau\right|.$$

Select r_5 with $r < r_5^{1/q} < 1$. Select c_5 with $0 < c_5 < c_3$, such that if $\zeta \in S_5 = \{|\eta| < c_5|\xi|\}$, then $\operatorname{Re}([\zeta_1^2+\ldots\zeta_s^2]^{q/2}) > r_5|\zeta|^q$. Then $\zeta \in S_5$, $\exp(-\tau(\zeta_1^2+\ldots+\zeta_s^2)^{q/2}) \leqq e^{-\tau r_5|\zeta|^q}$. We estimate the absolute value of the integrals in (2.15) by the integrals of the absolute values, use (2.13) and also use the last estimate. We make an observation analogous to (2.6) again, and estimate all resulting integrals by integrals from 0 to ∞. We find that for some $C_{13}, R_8, N_2 > 0$ we have

$$e_{Lm}(\zeta)$$

$$\leqq C_{13}([\sum_{\ell=0}^{L-1}[(\varkappa+1)^{N_2}[\varkappa/q]!^{-1}R_8^{\ell}r^{\varkappa}]\ ((pL+\varkappa-\operatorname{Re} j)/q)R_8^{L-\ell}r_5^{-(pL+\varkappa-\operatorname{Re} j)/q}|\zeta|^{\operatorname{Re} j-pL-\varkappa} \qquad (2.16)$$

$$+ [(\varkappa+1)^{N_2}[\varkappa/q]!^{-1}R_7^{L}r^{\varkappa}]\Gamma((pL+\varkappa-\operatorname{Re} j)/q)r_5^{-(pL+\varkappa-\operatorname{Re} j)/q}|\zeta|^{\operatorname{Re} j-pL-\varkappa}).$$

Pick r_3 with $r/r_5^{1/q} < r_3 < 1$ and use Lemma 2.10(c) to estimate $\Gamma((pL+\varkappa-\operatorname{Re} j)/q) < C_6(1/r_3)^{\varkappa}R_4^L L!^{p-1}[\varkappa/q]!$. There then exist

$C_{10}, R_6 > 0$ and $0 < r_4 < 1$ such that $e_{Lm}(\zeta) \leq$ $C_{10}R_6^L L!^{p-1} r_4^{\varkappa} |\zeta|^{\mathrm{Re}\, j-pL-\varkappa}$ for $\zeta \in S_5$. This is (2.11), precisely what we wanted. This gives Lemma 2.7, since the uniformity assertion also follows from the arguments.

We must still demonstrate the technical Lemmas 2.8, 2.9 and 2.10; we turn to this now.

Proof of Lemma 2.9. Let $\| \ \|_2$ denote L^2 norm on S^{s-1}, the unit sphere in $\mathbb{R}^s$. We begin with the identity that if P is a spherical harmonic of degree $\varkappa$,

$$\sum_{i=1}^{s} \|\partial_i P\|_2^2 = \varkappa(2\varkappa+s-2)\|P\|_2^2 . \tag{2.17}$$

It suffices to show that if Q is a homogeneous polynomial of degree $2\varkappa$, $\int_{S^{s-1}} \Delta Q dS = 2\varkappa(2\varkappa+s-2)\int_{S^{s-1}} Q dS$, since we may then apply this identity to $Q = |P|^2/2$ to obtain (2.17). To see the latter identity, observe that $\int_{|\xi|\leq 1} (\Delta Q)(\xi)d\xi =$ $(\int_{S^{s-1}} \Delta Q dS)\int_0^1 r^{s-1+2\varkappa-2} dr = (2\varkappa+s-2)^{-1}\int_{S^{s-1}} \Delta Q ds$ since ΔQ is homogeneous of degree $2\varkappa - 2$. On the other hand, since $\Delta = \mathrm{div}\,\mathrm{grad}$, by Stokes' theorem, if ν denotes unit normal, $\int_{|\xi|\leq 1} (\Delta Q)(\xi)d\xi =$ $\int_{S^{s-1}} (\mathrm{grad}\, Q)(\xi)\cdot\nu(\xi)dS(\xi) = 2\varkappa\int_{S^{s-1}} Q dS$ since $\nu(\xi)\cdot(\mathrm{grad}\, Q)(\xi) =$ $\xi\cdot(\mathrm{grad}\, Q)(\xi) = 2\varkappa Q(\xi)$ by Euler's identity. The identities follow at once.

From (2.17) we evidently have that if $A = (s-2)/2$, $\|\partial_i P\|_2^2 \leq$ $\varkappa(2\varkappa+s-2)\|P\|_2^2 \leq 2(\varkappa+A)^2 \|P\|_2^2$ for each i. Proceeding inductively, we have that if ∂^α is any monomial with $\|\alpha\| = M$, $\|\partial^\alpha P\|_2 \leq 2^{M/2}(\varkappa+A)^M\|P\|_2$. It is known, however, ([73], appendix C.4) that there exist $C,N > 0$ such that for any spherical harmonic R of degree $\varkappa'$, $\|R\|_\infty < C(\varkappa'+1)^N\|R\|_2$--here $\| \ \|_\infty$ denotes L^∞ norm on S^{s-1}. Accordingly, if $\|\alpha\| = M$, then $\|\partial^\alpha P_m\|_\infty \leq C2^{M/2}(\varkappa+A')^{M+N}$ where $A' = \max(A,1)$. Using this and a power series expansion about a point on S^{s-1}, we find the following. Suppose $\zeta \in \mathbb{C}^s$, $d > 0$, and $\operatorname{dist}(\zeta, S^{s-1}) \leq d$. Then, for some $C_4' > 0$,

$$|P_m(\zeta)| \leq C(\varkappa+A')^N \sum_\alpha d^{\|\alpha\|}[2^{1/2}(\varkappa+A')]^{\|\alpha\|}/\alpha! = C(\varkappa+A')^N e^{d2^{1/2}(\varkappa+A')s} < C_4'(\varkappa+1)^N e^{c\varkappa},$$ where $c = 2^{1/2}sd$.

The lemma is now immediate. Indeed, for (a), given r_2 with $0 < r_2 < 1$, we may select $d > 0$ such that if $c = 2^{1/2}sd$, then $e^c < 1/r_2$. Next we may select $c_3 > 0$ such that if $\zeta \in S_3 = \{|\eta| < c_3|\xi|\}$ and $|\zeta| \geq 1$, then $\operatorname{dist}(\zeta, S^{s-1}) \leq d$. Then for some $C_4 > 0$, if $\zeta \in S_3$, then $|P_m(\zeta)| < C_4 r_2^{-\varkappa}$, as desired.

For (b), observe that there certainly exists $d > 0$ such that for all $\zeta \in \mathbb{C}^s$ with $|\zeta| = 1$, we have $\operatorname{dist}(\zeta, S^{s-1}) \leq d$. (b) follows with $a = e^c$ where $c = 2^{1/2}sd$.

Proof of Lemma 2.8. The converse is immediate from Lemma 2.9. Indeed, with data as in the hypotheses of the converse, select $r_2 > 0$ with $r_1 < r_2 < 1$, and c_3, S_3 as in Lemma 2.9. Then $\sum A_m P_m(\zeta)$ converges to a holomorphic function in $S_3 \cap \{\zeta : |\zeta| < r_2/r_1\}$.

For the forward direction, let Δ_S be the spherical Laplacian. Since Δ_S is a second order differential operator on the sphere, it is easy to believe that if R_1, g are as in the statement, then there exist $C, R_2 > 0$, depending only on R_1, such that for all L,

$$\|\Delta_S^L g\|_\infty \leqq C_2 R_2^L L!^2. \tag{2.18}$$

Let us accept (2.18) for the moment and proceed. As is known ([73], Section 3.14), $\Delta_S P_m = \varkappa(\varkappa+s-2)P_m$; consequently $\Delta_S^L g = \sum[\varkappa(\varkappa+s-2)]^L A_m P_m$. Thus $|\varkappa(\varkappa+s-2)|^L |A_m| \leqq \|\Delta_S^L g\|_2 \leqq C' R_2^L L!^2$ for some $C' > 0$. For $\varkappa' \in \mathbb{Z}^+$, let $A^{\varkappa'} = \max_{\varkappa(m)=\varkappa'} |A_m|$. Then if $\varkappa' \geqq 1$, $|A^{\varkappa'}|^{1/2} \leqq C'' R_3^L L!/(\varkappa')^L$ for some $C'' > 0$, where $R_3 = R_2^{1/2}$. (Of course, we are assuming $s \geqq 2$.) By Lemma 2.2(a) $\Longleftrightarrow$ (d), then with $q = 1$, for some $C_1, B > 0$, $|A^\varkappa|^{1/2} \leqq C_1 e^{-B\varkappa}$. This is what we wanted, since as is immediately checked, C_1 and B can be chosen to depend only on R_1.

(2.18) is not quite obvious, and we give a direct proof. Let $W_{jk} = \xi_j \partial_k - \xi_k \partial_j$ where $\partial_k = \partial/\partial\xi_k$; then $\Delta_S = \sum_{j<k} W_{jk}^2$. (This is shown in [20], Section 1]. For a quick proof, one can easily compute that $(\sum_{j<k} W_{jk}^2)(|\xi|^{2a} P_m) = \varkappa(\varkappa+s-2)|\xi|^{2a} P_m$ for any $a \in \mathbb{Z}^+$. Since every polynomial is a linear combination of polynomials of the form $|\xi|^{2a} P_m$ ([73], Theorem IV.2.1), $\Delta_S P = (\sum W_{jk}^2)P$ for any polynomial P, and in particular for every

polynomial of degree less than or equal to 2. Thus the two differential operators must be equal, since both are second order.) Now Δ_S^L can be written as a sum of $[s(s-1)/2]^L$ terms of the form W, where W is a product of 2L factors, each of the form W_{jk} for some j,k. it suffices, then, to show that for some $C_2, R_4 > 0$, depending only on R_1, we have $\|Wg\|_\infty < C_2 R_4^M M!$ whenever W is a product of M of the W_{jk}'s. (Indeed, we can then use this for M = 2L.) Such a W, however, can be written as a sum of 2^M terms of the form D, where D is a product of M factors, each of the form $x_j \partial_k$ for some j,k. It suffices, then, to show that there exist $C_3, R_5 > 0$, depending only on R_1, such that for every such D, $\|Dg\|_\infty < C_3 R_5^M M!$. This, however, is easy from Lemma 2.5. Indeed, let $\rho = 1/2R_1$, and for any fixed ξ with $|\xi| = 1$, let $\mathcal{D}_\rho = \mathcal{D}_\rho(\xi) \subset \mathbb{C}^S$. There exists $C_4 > 0$ such that g, restricted to $\mathcal{D}_\rho \cap \mathbb{R}^S$, can be extended to a holomorphic function g on $\mathcal{D}_\rho$ such that $|g(\zeta)| < C_4$ for $\zeta \in \mathcal{D}_\rho$. For $\zeta \in \mathcal{D}_\rho$, however, $|\zeta_j| \leqq 1 + \rho$ for all j. lemma 2.5 applies directly to show $\|Dg\|_\infty \leqq C_4 (2\pi e M)^{1/2} (e/\rho)^M (1+\rho)^{M-1} M!$, as desired.

<u>Proof of Lemma 2.10</u>. (a) follows at once from the elementary inequalities $N^N e^{-N} \leqq N! \leqq N^N$, applied to $N = \ell$ and to $N = [\ell/Q]$.

To prove (b) and (c), we observe several facts about the gamma function:

(i) If $x > 0$, $\tau \in \mathbb{R}$, then $|\Gamma(x+i\tau)| \leqq \Gamma(x)$. This follows from

$$|\Gamma(x+i\tau)| \leqq \int_0^\infty |e^{-u}u^{x+i\tau-1}|\,du.$$

(ii) Fix $\tau \in \mathbb{R}$ and $\varepsilon > 0$. Then there exists $C_0 > 0$ so that $\Gamma(x) \leqq C_0|\Gamma(x+i\tau)|$ for all $x \geqq \varepsilon$. To see this, observe that we can obviously select $C_0 > 0$ so that $\Gamma(x) \leqq C_0|\Gamma(x+i\tau)|$ for $\varepsilon \leqq x \leqq 1 + \varepsilon$. Now if $x \geqq \varepsilon$, we may select r,N so that $\varepsilon \leqq r < 1 + \varepsilon$, $N \in \mathbb{Z}^+$ and so that $x = r + N$. Then

$\Gamma(x) = (r+N-1)\ldots r\Gamma(r) \leqq C_0|(r+N-1+i\tau)\ldots(r+i\tau)\Gamma(r+i\tau)| = C_0|\Gamma(x+i\tau)|$ as claimed.

(iii) $\Gamma(x+1) = x\Gamma(x)$ for $x > 0$.

(iv) If $1 \leqq x \leqq y \leqq x + 1$, then $\Gamma(x) \leqq 1 + \Gamma(y) \leqq 2 + \Gamma(x+1)$ as the inequality $\Gamma(x) \leqq 1 + \int_1^\infty u^{x-1}e^{-u}du$ shows.

To prove (b) and (c), we may assume $(\ell, \varkappa) \neq (0,0)$. By (i) and (ii) we see that we may assume j is real. By (iii) and (iv), we may further assume $0 \leqq -j/Q \leqq 1$, and in fact we may assume $j = 0$. (iii) and (iv) then further show that it suffices to prove (b) and (c) with $\Gamma((p\ell+\varkappa-j)/Q)$ replaced by $([p\ell/Q] + [\varkappa/Q])!$. (b) is now immediate from (a), since by (1.7), $([p\ell/Q]+[\varkappa/Q])! \geqq [p\ell/Q]![\varkappa/Q]!$ For (c), note that, because of (a), it suffices to show that if $0 < r_3 < 1$, there exists $R_4 > 0$ such that $([p\ell/Q]+[\varkappa/Q])! \leqq R_4 r_3^{-\varkappa}[p\ell/Q]![\varkappa/Q]!$ It suffices then to show that if $0 < r < 1$, there exists $R > 0$ such that for all $m, n \in \mathbb{Z}^+$, $(m+n)! \leqq R^m r^{-n} m!n!$. (We could then apply this with $r = r_3^Q$, for then $r^{-[\varkappa/Q]} = r_3^{-Q[\varkappa/Q]} \leqq r_3^{-\varkappa}$. We claim that we can choose $R = (1-r)^{-1}$. Indeed, $\binom{m+n}{n} r^n (1-r)^m \leqq 1$ by the binomial theorem. This completes the proof. ■

This completes the proof of Theorem 2.6. As we indicated in Chapter One, Theorem 2.6 implies the following result, which is our main result so far. We use the notation of Chapter One with $\underset{\sim}{a} = (p,1,\dots,1)$. We let $Z^q_{q,j}(\mathbb{R}^s) = \{\text{functions } f \text{ on } \mathbb{R}^s \mid f$ is the restriction to $\mathbb{R}^s$ of a $Z^q_{q,j}$ function on $\mathbb{C}^s\}$. We state only one of the uniformities here.

<u>Theorem 2.11</u>. Suppose $J_+, J_- \in C^\infty(\mathbb{R}^s)$. Then there exists $J \in AJ^j$ so that $J(1,\xi) = J_+(\xi)$ and $J(-1,\xi) = J_-(\xi)$ for all ξ if and only if: $J_+, J_- \in Z^q_{q,j}(\mathbb{R}^s)$, and if $J_+ \sim \sum g_\ell$ then $J_- \sim (-1)^\ell g_\ell$. If this is the case,

$$g_\ell(\xi) = (\partial_\lambda^\ell J)(0,\xi)/\ell!. \tag{2.19}$$

Further, we have this uniformity. Let $k = -Q - j$; suppose $\operatorname{Re} k > -Q$. Then for any $R_1 > 0$ there exist $B,C,R,c > 0$ so that $J_+, J_- \in Z^q_{q,j}(B,C,R,c)$ whenever $J = \hat{K}$ for a $K \in AK^k$ which satisfies $\sup_{1 \leq |u| \leq 2} |\partial^\gamma K(u)| < R_1^{\|\gamma\|}\gamma!$ for all $\gamma \in (\mathbb{Z}^+)^{s+1}$.

<u>Proof</u>. This is an immediate consequence of Theorems 1.7, 2.8, and 2.6. The uniformity follows from the uniformities in Corollary 1.5 and Theorems 1.7 and 2.6, in the case $-Q < \operatorname{Re} k < 0$. If $\operatorname{Re} k \geqq 0$, for the uniformity we argue as follows. Select

$m \in \mathbb{Z}^+$ with $-Q < \mathrm{Re}\, k - pm < 0$ and, with K as above, consider $K' = i^m \partial_t^m K$. For some C_2, R_2 depending only on R_1, we have $\sup_{1 \leq |u| \leq 2} |\partial^\gamma K'(u)| < C_2 R_2^{\|\gamma\|} \gamma!$ for all $\gamma \in (\mathbb{Z}^+)^{s+1}$. Thus if $J' = \hat{K}'$, we have $J'_+, J'_- \in Z^q_{q,j+pm}(B,C,R,c)$ with B,C,R,c depending only on R_1. But $J'_+ = J_+$ and it is clear, either directly from this or through (2.19), that if $J'_+ \sim \sum g'_\ell$ in $Z_{q,j+pm}$, then $g'_\ell = 0$ for $\ell < m$, and $g'_\ell = g_{\ell-m}$ for $\ell \geq m$. From these relations, and from $J'_- = \pm J_-$, the uniformity in the case $\mathrm{Re}\, k \geq 0$ follows easily from that in the case $-Q < \mathrm{Re}\, k < 0$.

As we said after the proof of Theorem 2.3, one can think of Z^q_q as "a natural Schwartz space in the real analytic category." $Z^q_{q,j}$ might then be thought of as "a natural asymptotic symbol class in the real analytic category." Its C^∞ analogue would simply be $S^j_{(q)} = \{f \in C^\infty(\mathbb{R}^s) \mid$ there exist C^∞ functions $\{g_\ell(\xi)\}$ on $\mathbb{R}^s \setminus \{0\}$ homogeneous of degree $j - p\ell$ and constants $C_{L\beta}$ (for $L \in \mathbb{Z}^+$, $\beta \in (\mathbb{Z}^+)^n$) such that for all L, β, if $|\xi| > 1$, $|\partial_\xi^\beta (f(\xi) - \sum_{\ell=0}^{L-1} g_\ell(\xi))| < C_{L\beta} |\xi|^{\mathrm{Re}\, j - pL - \|\beta\|}\}$. Write $f \sim \sum g_\ell$ (in $S^j_{(q)}$). By Theorem 1.7(a) and Theorem 1.8(a), one can characterize J^j in terms of these spaces. It is an important, and easy, result in the theory of pseudodifferential operators that given <u>any</u> collection of C^∞ functions $\{g_\ell(\xi)\}$ on $\mathbb{R}^s \setminus \{0\}$ such that g_ℓ is homogeneous of degree $j - p\ell$, then there exists $f \sim \sum g_\ell$ (in $S^j_{(q)}$). In order to apply our methods to study partial differential equations, it is important to have the analogue of this result for $Z^q_{q,j}$. To be precise, let us make the following definition.

<u>Definition</u>. We say $Z^q_{q,j}$ is <u>ample</u> if the following condition holds: Given any set of functions $\{g_\ell(\zeta)\}$ which are holomorphic in a sector $S = \{|\eta| < c|\xi|\}$, homogeneous of degree $j - p\ell$ and which satisfy $|g_\ell(\zeta)| < CR^\ell \ell!^{p-1}$ for $\zeta \in S$, $|\zeta| = 1$, there exists $f \in Z^q_{q,j}$ such that $f \sim \sum g_\ell$. ∎

Clearly, if it were in general known that $Z^q_{q,j}$ is ample, it would be an improvement on Lemma 2.7. In fact, if q is an even integer, the proof of Lemma 2.7 gives this at once. Indeed, if Re $j < 0$, we can simply choose f as in (2.9). In this case, $\exp[-\tau(\zeta_1^2+\ldots+\zeta_s^2)^{q/2}]$ is entire. Thus the argument after (2.9) which showed that f is holomorphic also shows that f is entire and that there exist $C_9, B_1 > 0$ so that $|f(\zeta)| < C_9 e^{B_1|\zeta|^q}$. Indeed, in that argument, we can replace the sector S_4 by $\mathbb{C}^n$. It is then evident that $f \in Z^q_{q,j}$ and $f \sim \sum g_\ell$. If instead Re $j \geq 0$, we can proceed as in the first paragraph of the proof of Lemma 2.7. This latter argument then shows that if q is such that $Z^q_{q,j}$ is ample for all j with Re $j < 0$, it is then ample for all $j \in \mathbb{C}$.

We state the result as a theorem.

<u>Theorem 2.12</u>. Say q is an even integer. Then $Z^q_{q,j}$ is ample for all $j \in \mathbb{C}$. Further, we have the following uniformity.

Suppose $c, C, R > 0$; then there exist $B, C_1, R_1, c_1 > 0$ as follows. If $\{g_\ell(\zeta)\}$ are holomorphic in $S = \{|\eta| < c|\xi|\}$, are homogeneous of degree $j - p\ell$ and satisfy $|g_\ell(\zeta)| < CR^\ell \ell!^{p-1}$ for $\zeta \in S$, $|\zeta| = 1$, then there exists $f \in Z^q_{q,j}(B, C_1, R_1, c_1)$ such that $f \sim \sum g_\ell$.

The uniformity assertion follows from the proof of Lemma 2.7.

If q is not an even integer, $\exp[-\tau(\zeta_1^2 + \ldots \zeta_s^2)^{q/2}]$ is no longer entire. However, it is possible to replace it by similar suitable functions which are entire—in fact in Z^q_q—and to prove the following fact.

<u>Theorem</u>. If $q > 1$ and q is rational, then $Z^q_{q,j}$ is ample for all $j \in \mathbb{R}$.

This will not be proved or used here, but will be shown in a later paper. The key point is the construction of the replacements of $\exp[-\tau(\zeta_1^2 + \ldots + \zeta_s^2)^{q/2}]$. Our construction is lengthy. To see the difficulty, note that it is not even evident that $Z^q_q \neq \{0\}$. This was, however, proved in [22] in an argument attributed to B. IA. Levin. (Šilov has an earlier, less elementary proof.) We, however, need an element of Z^q_q with very specific properties, and must use a very different construction. For a sketch of our construction, see [30].

Our construction probably works for all $j \in \mathbb{C}$, but we have not checked all the details. Probably the construction could be carried out for q irrational as well. (However, we are only concerned with the case of q rational, as we said in Chapter One.)

From Theorem 2.12, we now find the following fact and uniformity, which is a counterpart to Theorem 2.11.

Corollary 2.13. Say q is an even integer, Re $j > -Q$; let $k = -Q - j$. Say $c, R > 0$; then there exist $C_1, R_1 > 0$ as follows. Let $\{g_\ell(\zeta)\}$ be a set of functions such that g_ℓ is holomorphic in $S = \{|\eta| < c|\xi|\}$ and homogeneous of degree $j - p\ell$, and such that $|g_\ell(\zeta)| < R^\ell \ell!^{p-1}$ for $|\zeta| = 1$, $\zeta \in S$. Then there exists $K \in \mathcal{A}\mathcal{K}^k$ which satisfies $\sup_{1 \leq |u| \leq 2} |\partial^\gamma K(u)| < C_1 R_1^{\|\gamma\|} \gamma!$ for all $\gamma \in (\mathbb{Z}^+)^{s+1}$, and such that $\hat{K}(1,\xi) \sim \sum g_\ell(\xi)$ in $Z^q_{q,j}$.

Proof. This is an immediate consequence of Theorem 2.12, Theorem 1.8(b) and Corollary 1.5.

We close this chapter with some generalizations that will be needed in Chapter 7 and 8. Say $\Omega \subset \mathbb{C}^n$ for some n, Ω open, and suppose $k \in \mathbb{C}$. Suppose that for each $\omega \in \Omega$ we are given $K_\omega \in \mathcal{K}^k$. We say that K_ω depends holomorphically on the parameter ω if the following two conditions hold:

(1) For each $x \neq 0$, $K_\omega(x)$ is a holomorphic function of ω.

(2) For each $f \in \mathcal{S}$, $K_\omega(f)$ is a holomorphic function of ω.

Here $K_\omega(f)$ means the tempered distribution K_ω as applied to f. If $-k-Q \notin \mathbb{Z}^+$, then clearly the second condition follows from the first. If $-k-Q \in \mathbb{Z}^+$, we obtain an equivalent condition by replacing condition (2) by:

(2)' $K_\omega = \Lambda_{G_\omega} + \sum_{|\alpha|=-k-Q} c_\alpha(\omega)\partial^\alpha\delta$ where each $c_\alpha(\omega)$ is holomorphic in $\omega \in \Omega$.

That (1) + (2)' $\Rightarrow$ (1) + (2) is trivial, while (1) + (2) $\Rightarrow$ (1) + (2)' follows form testing on appropriate f.

If instead $K_\omega \in J^k$ for each $\omega \in \Omega$, we still say K_ω depends holomorphically on the parameter ω if (1) and (2) hold. Again, (1) implies (2) if $-k-Q \notin \mathbb{Z}^+$. If $-k-Q \in \mathbb{Z}^+$, (1) + (2) is equivalent to (1) + (2)'.

If $K_\omega \in K^k$ (or J^k, respectively), and depends holomorphically on ω, and if $1 \leqq m \leqq n$, note that there exists $K_\omega^m \in K^k$ (or J^k, respectively) depending holomorphically on ω, with $\partial/\partial\omega_m[K_\omega(f)] = K_\omega^m(f)$. In fact, if $K_\omega = G_\omega$ away from 0 where G_ω is C^∞. then $K_\omega^m = \partial G_\omega/\partial\omega_m$ away from 0. If $K_\omega = \Lambda_{G_\omega}$, then $K_\omega^m = \Lambda_{\partial G_\omega/\partial\omega_m}$, and if K_ω is as in (2)', then $K_\omega^m = \Lambda_{\partial G_\omega/\partial\omega_m} + \sum_{|\alpha|=-k-Q} [\partial c_\alpha(\omega)/\partial\omega_m]\partial^\alpha\delta$.

Say now $W \subset \mathbb{R}^n$ for some n, W open, $k \in \mathbb{C}$, and we are given $K_w \in K^k$ for each $w \in W$. We naturally say K_w depends analytically on the parameter w if there is a complexified neighborhood Ω of W in $\mathbb{C}^n$, and an extension of the family $\{K_w\}$ to $\{K_\omega\}$, where

K_ω depends holomorphically on $\omega \in \Omega$. We have a similar definition of analytic dependence for the J^k spaces.

From (1) and (2), and the proof of Proposition 1.1(a), it is easy to see this:

Say $K_\omega \in K^k$ for each $\omega \in \Omega$. Let $J_\omega = \hat{K}_\omega \in J^{-Q-k}$. Then K_ω depends holomorphically on the parameter ω if and only if J_ω does. (2.20)

Say now $K_\omega \in K^k$ depends holomorphically on the parameter $\omega \in \Omega$, and that for each $\gamma \in (\mathbb{Z}^+)^{s+1}$ there is a constant C_γ so that

$$|\partial^\gamma K_\omega(u)| < C_\gamma \quad \text{for } 1 \leqq |u| \leqq 2,\ \omega \in \Omega.$$

Let $J_\omega = \hat{K}_\omega$. Then the proof of Proposition 1.1(a) shows that for each $\gamma \in (\mathbb{Z}^+)^{s+1}$ there is a C'_γ so that

$$|\partial^\gamma J_\omega(v)| < C'_\gamma \quad \text{for } 1 \leqq |v| \leqq 2,\ \omega \in \Omega.$$

In this case it is easy to see that, for any γ, $K^\gamma_\omega = \partial^\gamma K_\omega \in K^{k-|\gamma|}$ and $J^\gamma_\omega = \partial^\gamma J_\omega \in J^{-k-Q-|\gamma|}$ depend holomorphically on ω. Indeed, we need only check the case where there is a single differentation. Property (2) for K^γ_ω is clear, while property (1) follows from arguments involving difference quotients and uniform convergence on compact subsets of Ω. Similarly for J^γ_ω.

In Chapters 7 and 8, we will frequently examine the following situation:

$K_\omega \in AK^k$ depends holomorphically on $\omega \in \Omega$; and (2.21)

For some $C, R > 0$, $\sup_{1 \leq |u| \leq 2} |\partial^\gamma K_\omega(u)| < CR^{\|\gamma\|}\gamma!$

for all $\omega \in \Omega$, $\gamma \in (\mathbb{Z}^+)^{s+1}$ (2.22)

In that case, we have the following result:

Say $C, R > 0$; then there exist $B, C_1, R_1, c > 0$ as follows. Suppose we have (2.21) and (2.22), and in addition that $\operatorname{Re} k > -Q$. Suppose $\hat{K}_\omega = J_\omega$, $J_{\omega+}(\xi) = J_\omega(1,\xi)$, $J_{\omega-}(\xi) = J_\omega(-1,\xi)$ and $J_{\omega+} \sim \sum g_{\omega\ell}$ in $Z^q_{q,j}(\mathbb{R}^s)$. Then $J_{\omega+}, J_{\omega-} \in Z^q_{q,j}(B, C_1, R_1, c)$. Further, (2.23)

for each $\ell \in \mathbb{Z}^+$ and $\zeta \in S = \{\xi + i\eta : |\eta| < c|\xi|\}$, $g_{\omega\ell}(\zeta)$ is a holomorphic function of $\omega \in \Omega$.

Indeed, since the statement about $J_{\omega\pm}$ follows at once from the uniformity in Theorem 2.11, we have only to check the last statement. By (2.19), (2.20) and the remarks following (2.20), we see that if $\alpha \in (\mathbb{Z}^+)^s$, $\ell \in \mathbb{Z}^+$, $\xi \neq 0$, then $\partial^\alpha_\xi g_{\omega\ell}(\xi)$ is a holomorphic function of ω. Further, we know that for some $C_2, R_2 > 0$,

$$|\partial^\alpha_\xi g_{\omega\ell}(\xi)| < C_2 R_2^{\|\alpha\|+\ell}\|\alpha\|!\,\ell!^{p-1} \quad \text{for } |\xi| = 1,\ \omega \in \Omega.$$

(2.23) now follows from an argument involving power series whose coefficients depend on ω and which converge uniformly for ω in a compact subset of Ω.

We shall also at times need a version of (2.23) which holds when $k = -Q$:

$$\text{(2.23) holds for } k = -Q, \text{ provided that } g_{\omega 0} = 0 \text{ for all } \omega \in \Omega. \tag{2.24}$$

To see this, note that by (2.19), $J_\omega(0,\xi) = 0$ for $\xi \neq 0$. We may therefore define $J'_\omega \in J^{-p}$ as follows:

$$J'_\omega(\lambda,\xi) = J_\omega(\lambda,\xi)/(-i\lambda) \text{ if } \lambda \neq 0; \; J'_\omega(0,\xi) = i(\partial_\lambda J_\omega)(0,\xi).$$

Select $K'_\omega \in K^{-Q+p}$ with $\hat{K}'_\omega = J_\omega$, and observe that $TK'_\omega = K_\omega$. Thus if $(t,x) \neq 0$,

$$K'_\omega(t,x) = \int_{-\infty}^{t} K_\omega(s,x)\,ds \quad \text{or} \quad -\int_{t}^{\infty} K_\omega(s,x)\,ds.$$

(Either integral gives $K'_\omega(t,x)$ if $x \neq 0$. If $x = 0$, the first works for $t < 0$, the second for $t > 0$.) These formulae show clearly that $K'_\omega \in AK^{-Q+p}$, depends holomorphically on ω, and satisfies a condition analogous to (2.22). We may therefore apply (2.23) to K'_ω. We then may use simple relations, comparing $J'_{\omega+}$ and its asymptotics to $J_{\omega+}$ and its asymptotics, to complete the proof of (2.24).

Having listed all these elementary properties of holomorphic dependence on a parameter, we now have a non-trivial assertion to make. It is the generalization of Theorem 2.12 to the case where the functions depend holomorphically on a parameter.

Corollary 2.14. Say q is an even integer. Suppose $c,C,R > 0$; then there exist $B,C_1,R_1,c > 0$ as follows. Say $\Omega \subset \mathbb{C}^n$ for some n, $j \in \mathbb{C}$. For each $\omega \in \Omega$, $\ell \in \mathbb{Z}^+$, suppose $g_{\omega\ell}(\zeta)$ is holomorphic in $S = \{|\eta| < c|\xi|\}$, is homogeneous of degree $j - p\ell$ and satisfies $|g_{\omega\ell}(\zeta)| < CR^{\ell}\ell!^{p-1}$ for $\zeta \in S$, $|\zeta| = 1$. Suppose, further, that for each $\zeta \in S$, $g_{\omega\ell}(\zeta)$ is a holomorphic function of $\omega \in \Omega$. Then there exists $f_\omega \in Z^q_{q,j}(B,C_1,R_1,c_1)$ such that $f_\omega \sim \sum g_{\omega\ell}$, and such that for each $\zeta \in \mathbb{C}^s$, $f_\omega(\zeta)$ is a holomorphic function of $\omega \in \Omega$.

Proof. This is actually a Corollary of the proof of Theorem 2.12. We reduce to the case Re $j < 0$ as in that proof. One has then only to check that if $f_\omega(\zeta)$ is defined by the method evidenced in (2.9), the result will be holomorphic in ω. This follows easily from the estimates proving the convergence of the series in (2.9), and the fact that the parameterized versions of the $A_{\ell m}$ in (2.8) will surely be holomorphic in ω.

From Corollary 2.14 we now extract the parameterized generalization of Corollary 2.13.

Corollary 2.15. Say q is an even integer, Re $j > -Q$: let $k = -Q-j$. Say $c,C,R > 0$; then there exist $C_1,R_1 > 0$ as follows. Say $\Omega \subset \mathbb{C}^n$. For each $\omega \in \Omega$, $\ell \in \mathbb{Z}^+$, suppose $g_{\omega\ell}(\zeta)$ is holomorphic in $S = \{|\eta| < c|\xi|\}$, is homogeneous of degree $j - p\ell$ and satisfies $|g_{\omega\ell}(\zeta)| < CR^{\ell}\ell!^{p-1}$ for $\zeta \in S$, $|\zeta| = 1$. Suppose, further, that for each $\zeta \in S$, $g_{\omega\ell}(\zeta)$ is a holomorphic function of $\omega \in \Omega$. Then there exist $K_\omega \in AK^k$ which satisfy $\sup_{1 \leq |u| \leq 2} |\partial^\gamma K_\omega(u)| < C_1 R_1^{\|\gamma\|}\gamma!$ for all $\gamma \in (\mathbb{Z}^+)^{s+1}$, such that $\hat{K}_\omega(1,\xi) \sim \sum g_\ell(\xi)$ in $Z^q_{q,j}$, and such that K_ω depends holomorphically on ω.

Proof. This is an immediate consequence of Corollary 2.14, Theorem 1.8(b), Corollary 1.5, and (2.20).

3. Homogeneous Partial Differential Equations

We proceed to examine the implications of the theory we have developed to the study of analytic hypoellipticity of homogeneous partial differential operators. As our first (well known) proposition shows, the constant-coefficient case is not the place to begin.

Proposition 3.1. Let $P(\partial)$ be a constant-coefficient differential operator on $\mathbb{R}^n$ which is homogeneous with respect to the dilations $D_r x = (r^{a_1}x_1,\ldots,r^{a_n}x_n)$ $(a_1,\ldots,a_n \in \mathbb{N})$. Then $P(\partial)$ is analytic hypoelliptic if and only if $a_1=\ldots=a_n$ and $P(\partial)$ is elliptic.

Remark. In fact, by [45], Vol. II, Corollary 11.4.13, one can drop the "homogeneous" hypothesis; if $P(\partial)$ has constant coefficients, $P(\partial)$ is analytic hypoelliptic if and only if $P(\partial)$ is elliptic.

Proof. If $P(\xi_0) = 0$ for any $\xi_0 \neq 0$, $P(\partial)$ has non-smooth functions in its kernel. Indeed, $g(x) = \int_0^\infty e^{ix\cdot D_r\xi_0} f(r)dr$ is in the kernel for any $f \in L^1$ which decays sufficiently rapidly at ∞. For appropriate f, g is not smooth in x. In fact, say ξ_{01} (the first component of ξ_0) is not zero; let $q = \xi_{01}$. Then $g(x_1,0) = \hat{F}(qx_1/2\pi)$ where $F(s) = (1/a_1)f(s^{1/a_1})s^{1/a_1-1}$ $(s>0)$. It suffices, then to show that there exists, for each $N > 0$, a function F supported in $[0,\infty)$ with $|F(s)| < C(1+s)^{-N}$ such that $\hat{F}$ is not smooth at 0; for from F, the appropriate f could be found. Now $F_N(s) = (1+s^2)^{-(N+1)}$ $(s>0)$ has the desired property. Were $\hat{F}_N$ smooth at 0, $(1+(2\pi i\partial)^2)^N\hat{F}_N$ would be also, so $\hat{F}_0$ would be also. If $G_0(s) = F_0(-s)$, $\hat{G}_0$ would also be smooth at 0, and $[(1+s^2)^{-1}]^\wedge = (F_0+G_0)^\wedge$ would be also. It is not; it equals $Ce^{-2\pi|x|}$.

If $P(\xi) \neq 0$ for any $\xi \neq 0$, then put $G(\xi) = P(-2\pi i\xi)^{-1}$, $J = \Lambda_G$ as in (1.2). Then $K = \check{J}$ is a fundamental solution of $P(\partial)$, so that $P(\partial)$ is C^∞ hypoelliptic. It suffices to show that K is real analytic away from 0 if and only if $a_1=\ldots=a_n$. If $a_1=\ldots=a_n$, use Corollary 1.4. If not, we show that J cannot satisfy the conclusion of Theorem 1.3. Assume it did. Suppose $a_1\geq\ldots\geq a_n$. Fix $\xi' = (\xi_1,\ldots,\xi_{n-1})$ with $\xi_1 \neq 0$, and let us examine the real analytic function $H(\xi_n) = J(\xi',\xi_n)$. We have an estimate $|\partial^m H(\xi_n)| < CR^m m!^{a_n/a_1}$ at $\xi_n = 0$. Since $a_n < a_1$, H is the restriction to $\mathbb{R}$ of an entire function in ξ_n. But it is the inverse of a polynomial in ξ_n; consequently, it is constant. Write $P(\xi) = \sum P_k(\xi')\xi_n^k$. The polynomials $P_k(\xi')$ vanish if $k > 0$, $\xi_1 \neq 0$; by continuity, they vanish identically. So $P(\xi) = P_0(\xi')$. Thus $P(0,\xi_n) = 0$; this contradicts our hypothesis and completes the proof. ∎

To study analytic hypoellipticity for homogeneous partial differential operators, it is therefore necessary to let the coefficients be polynomials. We restrict attention to the case $\underset{\sim}{a} = (p,1,\dots 1)$, $p \in \mathbb{Q}$, $p > 1$. We use coordinates $(t,x) \in \mathbb{R} \times \mathbb{R}^s$, with dual coordinates (λ,ξ), and we put $Q = p + s$ as in Chapter One. We restrict attention to those partial differential operators L with the following properties:

> L is homogeneous, with homogeneous degree k; the degree of L is also k; the coefficients of L are polynomials in the x_ℓ's; and we can write $L = L' + (\partial/\partial t)L''$, where L' is a constant-coefficient differential operator in the x_ℓ's only, and where L' is elliptic. (3.1)

Note that L' must be homogeneous, with homogeneous degree k, and degree k. Given all the other assumptions, the hypothesis that L' is elliptic is evidently necessary for L to be analytic hypoelliptic, as one sees easily considering functions of x alone.

If $L = \sum c'_{\alpha\beta m}(\partial/\partial x)^\alpha x^\beta (\partial/\partial t)^m$, put $\hat{L}_\lambda = \sum c'_{\alpha\beta m}(-i\xi)^\alpha(-i\partial/\partial\xi)^\beta(-i\lambda)^m$, the "Fourier transform of L". We shall often impose the following additional conditions on L:

For all $\lambda \neq 0$, if $F \in Z^q_q(\mathbb{R}^s)$, there exists $G \in Z^q_q$ such that $\hat{L}_\lambda G = F$. (3.2)

We shall see many examples later of cases in which (3.2) holds. The main result of this section is the following.

Theorem 3.2. Suppose L satisfies (3.1) and $\sigma \in \mathbb{C}$. Suppose that $Z^q_{q,-(\sigma+k+Q)}$ is ample. Then of the following conditions (a) implies (b), and (b) is equivalent to (c):

(a) L is analytic hypoelliptic and L and L^t are hypoelliptic.

(b) For any $K_1 \in AK^{\sigma}$, there exist $K \in AK^{\sigma+k}$ such that $LK = K_1$.

(c) (3.2) holds. ∎

In particular, (3.2) is necessary for (a) to hold. As we said at the end of Chapter 2, $Z^q_{q,j}$ is ample for all $j \in \mathbb{R}$, and we shall demonstrate this in a later paper. For now we just include it as a hypothesis. We will in fact not use this hypothesis in proving (a) ⇒ (b) (for $\sigma = -Q$) ⇒ (c). (Of course, by Theorem 2.12, if q is an even integer then $Z^q_{q,j}$ is ample for all $j \in \mathbb{C}$.)

In (b), the special case where $K_1 = \delta$ and where L is left invariant on the Heisenberg group will be studied in detail in Chapter 4. In this case, (b) implies $LK = \delta$ and L^t is hypoelliptic and analytic hypoelliptic.

For (a) ⇒ (b), we can work in a more general setting.

Lemma 3.3. Suppose D_o is a differential operator with polynomial coefficients which is homogeneous of degree k with respect to the dilations $D_r x = (r^{a_1}x_1,\ldots,r^{a_n}x_n)$. Let $Q = \sum a_i$. Suppose D_o and D_o^t are hypoelliptic. Then for all $K_1 \in \mathcal{K}^{-Q}$ there exists $K_o \in \mathcal{K}^{-Q+k}$ such that $D_o K_o = K_1$. ($\mathcal{K}^{-Q}, \mathcal{K}^{-Q+k}$ are as in Chapter One.) In particular, in Theorem 3.2, (a) $\Rightarrow$ (b) if $\sigma = -Q$.

Proof. By Corollary 1, page 540 of [80], D_o has a fundamental kernel $J(x,y)$ in some neighborhood V of 0; the map $u \to \int J(x,y)u(y)dy$ takes $\mathcal{E}'$ to $\mathcal{D}'$. Considering, then, $\int J(x,y)\varphi(y)K_1(y)dy$ where $\varphi \in C_c^\infty$, $\varphi = 1$ near 0, we see that there exists a neighborhood U of 0 and $K \in \mathcal{D}'(U)$ such that $D_o K = K_1$ in U. It is now necessary to modify K to obtain $K_o \in \mathcal{K}^{-Q+k}$. In case $K_1 = \delta$, it is a theorem of Folland and the author that this can be done ([29], Theorem 3.) In case $K_1 \in \mathcal{K}^{-Q}$ is general, the procedure in [29] can still be followed, almost word for word, to obtain K_o. ∎

In fact, by making minor modifications in the proof in [29], one could obtain, for any ℓ and any $K_1 \in \mathcal{K}^{\sigma}$, a $K_o \in \mathcal{K}^{\sigma+k}$ with $D_o K_o = K_1$. (a) $\Rightarrow$ (b) could then be shown in general from this. However, the general (a) $\Rightarrow$ (b) will come out automatically later, so we do not explain the modifications needed in [29] here.

Instead we turn at once to the heart of the matter, (b) $\Longleftrightarrow$ (c). We shall work almost entirely on the Fourier transform side, so for later notational ease, we write $-(j+k+Q)$ for σ

in (b), where $j \in \mathbb{C}$ is appropriate. From now on, if $\alpha \in (\mathbb{Z}^+)^s$ is a multi-index, we write $|\alpha| = \alpha_1+\ldots+\alpha_s$.

By Theorem 2.11, if $K_1 \in AK^{-(j+k+Q)}$, then solving $LK = K_1$ with $K \in AK^{-(j+Q)}$ is essentially the same as solving $\hat{L}_\lambda J'_\lambda = J_{1\lambda}$ with $J'_\lambda \in Z^q_{q,j}$ for each $\lambda \neq 0$, where $\hat{K}_1(\lambda,\xi) = J_{1\lambda}(\lambda)$ for $\lambda \neq 0$. Let us then begin by remarking the following proposition:

Proposition 3.4. Suppose L satisfies (3.1). Then for all $j \in \mathbb{C}$, $\hat{L}_\lambda : Z^q_{q,j} \to Z^q_{q,j+k}$.

Proof. $\hat{L}_\lambda$ is a sum of terms of the form $c\xi^\alpha(\partial/\partial\xi)^\beta\lambda^m$, where $|\alpha| - |\beta| = k\text{-}pm$. (We reiterate that $|\alpha| = \alpha_1+\ldots+\alpha_s$.) Since $Z^q_{q,j+k-pm} \subset Z^q_{q,j+k}$ for all j, it suffices to show that for all j,ℓ, $\partial/\partial\xi_\ell : Z^q_{q,j} \to Z^q_{q,j-1}$, and (multiplication by ξ_ℓ) : $Z^q_{q,j} \to Z^q_{q,j+1}$. The second fact is trivial; the first follows at once from the Cauchy estimates. ∎

We now prove (c) ⇒ (b) in Theorem 3.2. The key fact which we use is the following.

Theorem 3.5. Suppose $j \in \mathbb{C}$ and that $Z^q_{q,j}$ is ample. Suppose L satisfies (3.1). Say $K_1 \in AK^{-(j+k+Q)}$. Then there exist $K \in AK^{-(j+Q)}$ and $K_2 \in AK^{-(j+k+Q)}$ such that $LK = K_1 - K_2$ and such that $J_{2\lambda}(\xi) = \hat{K}_2(\lambda,\xi) \in Z^q_q$ for all $\lambda \neq 0$.

Remark. Once this is known, (c) ⇒ (b) follows at once. Indeed, it suffices, under the hypothesis of (c), to construct

$K_0 \in AK^{-j-Q}$ with $LK_0 = K_2$; then $L(K+K_0) = K_1$. To do this, observe that by (3.2) there exist $J_{0,1}, J_{0,-1} \in Z_q^q$ with $\hat{L}_1 J_{0,1} = J_{2,1}$, $\hat{L}_{-1} J_{0,-1} = J_{2,-1}$. By Theorem 2.11 we can select $K_0' \in AK^{-(j+Q)}$ such that $\hat{K}_0'(1,\xi) = J_{0,1}(\xi)$, $\hat{K}_0'(-1,\xi) = J_{0,-1}(\xi)$. Thus $(LK_0'-K_2)^\wedge$ is supported at 0, so $LK_0' = K_2 - p_0$ for some polynomial p_0. We may assume $p_0 \neq 0$, otherwise we are done. Then since $p_0 \in AK^\sigma$ where $\sigma = -(j+k+Q)$, we must have $\sigma \in \mathbb{Z}^+$, and that p_0 is homogeneous of degree σ. Using the Cauchy-Kowalewski theorem, we can find f, real analytic in a neighborhood U of 0, so that $Lf = p_0$. If p_1 is the polynomial which equals the sum of the terms in the Taylor expansion of f about 0 which are homogeneous of degree $\sigma + k$, then evidently $Lp_1 = p_0$, so we can put $K_0 = K_0' + p_1$.

<u>Proof of Theorem 3.5</u>. Let $J_{1\lambda}(\xi) = \hat{K}_1(\lambda,\xi) \in Z^q_{q,j+k}$ for $\lambda \neq 0$. Let $J_{1+} = J_{1,1}$, $\hat{L}_+ = \hat{L}_1$. Suppose $J_{1+} \sim \sum h_\ell$. It suffices to construct $J_+ \in Z^q_{q,j}$ so that $\hat{L}_+ J_+ - J_{1+} \in Z^q_q$. For, say this is known. Say $J_+ \sim \sum g_\ell$. Using the ampleness of $Z^q_{q,j}$, select $J_- \in Z^q_{q,j}$ so that $J_- \sim \sum (-1)^\ell g_\ell$. Using Theorem 2.11, select $K \in AK^{-(j+Q)}$ such that $\hat{K}(1,\xi) = J_+(\xi)$, $\hat{K}(-1,\xi) = J_-(\xi)$. Then $LK - K_1 \in AK^{-(j+k+Q)}$; also $(LK-K_1)^\wedge(1,\xi) \in Z^q_q$. So $(LK-K_1)^\wedge(1,\xi) \sim 0$ in $Z^q_{q,j+k}$; by Theorem 2.11, $(LK-K_1)^\wedge(-1,\xi) \sim 0$ also, as desired.

Write $\hat{L}_\lambda = P(\xi) + \sum_{m>0,\alpha,\beta} c_{\alpha\beta m} \lambda^m \xi^\alpha (\partial/\partial\xi)^\beta$ where P is a homogeneous polynomial of degree k such that $P(\xi) = 0$ only when

$\xi = 0$. It suffices to construct an asymptotic series $\sum g_\ell$ such that g_ℓ is holomorphic in a sector $S = \{|\eta| < c|\xi|\}$, homogeneous of degree $j - p\ell$ and satisfying, for some $C,R > 0$, $|g_\ell(\zeta)| < CR^\ell \ell!^{p-1}$ for $|\zeta| = 1$, $\zeta \in S$, with the property that

$$(P(\xi) + \sum c_{\alpha\beta m}\xi^\alpha(\partial/\partial\xi)^\beta)\sum g_\ell = \sum h_\ell \text{ formally.} \tag{3.3}$$

That is, the left side of (3.3) is a formal sum of terms homogeneous of degree $j + k - p\ell$, as the proof of Proposition 3.4 shows. If this sum formally equals $\sum h_\ell$, we can, by the ampleness of $Z^q_{q,j}$, select $J_+ \in Z^q_{q,j}$ with $J_+ \sim \sum g_\ell$, and then the proof of Proposition 3.4 shows that $\hat{L}_+ J_+ \sim \sum h_\ell$, as desired.

Define $D: C^\infty(\mathbb{R}^n \setminus \{0\}) \to C^\infty(\mathbb{R}^n \setminus \{0\})$ by $D = -\sum c_{\alpha\beta m}[\xi^\alpha/P(\xi)](\partial/\partial\xi)^\beta$. Put $H_\ell = h_\ell/P$. We need to construct $\{g_\ell\}$ as before (3.3), with $(1-D)\sum g_\ell = \sum H_\ell$ formally. To do this, we shall make sense of $(1+D+D^2+\ldots)\sum H_\ell$. H_ℓ is holomorphic in a sector $S_1 = \{|\eta| < c_1|\xi|\}$, homogeneous of degree $j - p\ell$, and satisfies $|H_\ell(\zeta)| < C_1 R_1^\ell \ell!^{p-1}$ for $|\zeta| = 1$, $\zeta \in S_1$. We may assume $P(\zeta) \neq 0$ for $\zeta \in S_1$. Note that $c_{\alpha\beta m}[\xi^\alpha/P(\xi)](\partial/\partial\xi)^\beta$ reduces the homogeneity of a function by $pm \geq p$, since $|\alpha| - |\beta| - k = -pm$ as in Proposition 3.4. Accordingly, for any N,L, we can write $\sum_{n=0}^{N} D^n \sum_{\nu=0}^{M} H_\nu = \sum_\ell G_\ell^{NM}$ where G_ℓ^{NM} is homogeneous of degree $j - p\ell$ and the sum is finite. Let $g_\ell = G_\ell^{\ell\ell}$. We check (3.3). Observe that $g_\ell = G_\ell^{NM}$ if $N \geq \ell$ and $M \geq \ell$. Thus, for any I, we can write

$$\sum_{n=0}^{I} D^n \sum_{\nu=0}^{I} H_\nu = \sum_{\ell=0}^{I} g_\ell + \sum_{\ell > I} G_\ell^{II} \tag{3.4}$$

Therefore, $(1-D)\sum_{\ell=0}^{I} g_\ell = \sum_{\nu=0}^{I} H_\nu - D^{I+1}\sum_{\nu=0}^{I} H_\nu - (1-D)\sum_{\ell>I} G_\ell^{II} = \sum_{m=0}^{I} H_m + \sum_{\ell>I} F_\ell^I$ where the last sum is finite and F_ℓ^I is homogeneous of degree $j - p\ell$. Now, if $(1-D)\sum g_\ell$ is formally computed, the term which is homogeneous of degree $j - pI$ is the same as the term which is homogeneous of degree $j - pI$ in the expansion of $(1-D)\sum_{\ell=0}^{I} g_\ell$. This is just H_I, so (3.3) has been established.

Now g_ℓ is holomorphic in $S_1 = \{|\eta| < c_1|\xi|\}$ and homogeneous of degree $j - p\ell$. Let $c = c_1/2$, $S = \{|\eta| < c|\xi|\}$. To finish the proof it suffices to show that for some $C,R > 0$, $|g_\ell(\zeta)| < CR^\ell \ell!^{p-1}$ for $|\zeta| = 1$, $\zeta \in S$. We have written $D = -\sum c_{\alpha\beta m}[\xi^\alpha/P(\xi)](\partial/\partial\xi)^\beta$. Now write i for the tuple (α,β,m), and write $D = -\sum_{i\in\mathcal{I}} c_i D_i$ where $\mathcal{I}$ is an index set. If $I = (i_1,\dots,i_n)$, with $i_1,\dots,i_n \in \mathcal{I}$, write length $(I) = n$, $D_I = D_{i_1}\dots D_{i_n}$, $c_I = (-1)^n c_{i_1}\dots c_{i_n}$. Now we can write $D_I H_\nu = \sum f_\ell^{I\nu}$ where $f_\ell^{I\nu}$ is homogeneous of degree $j - p\ell$ and the sum has only one non-zero term. Further $g_\ell = \sum_{\text{length}(I)\leqq\ell}\ \sum_{\nu=0}^{\ell} c_I f_\ell^{I\nu}$. The number of terms in this sum is no more than $\ell\cdot[\#(\mathcal{I})]^\ell(\ell+1)$. Thus it suffices to show that for some $C_2,R_2 > 0$, $|f_\ell^{I\nu}(\zeta)| < C_2R_2^\ell\ell!^{p-1}$ for $|\zeta| = 1$, $\zeta \in S$, whenever $0 \leqq \nu \leqq \ell$ and length$(I) \leqq \ell$. If $f_\ell^{I\nu} \neq 0$, we can write, with $r = \text{length}(I)$,

$$f_\ell^{I\nu}(\zeta) = [\zeta^{\alpha_1}/P(\zeta)](\partial/\partial\zeta)^{\beta_1}\dots[\zeta^{\alpha_r}/P(\zeta)](\partial/\partial\zeta)^{\beta_r}H_\nu(\zeta). \qquad (3.5)$$

By Lemma 2.5, then, there exist $C_0,R_0 > 0$ such that

$$|f_\ell^{I\nu}(\zeta)| < C_0 R_0^{|\beta_1|+\ldots+|\beta_r|}(|\beta_1|+\ldots+|\beta_r|)!\,\|\zeta^{\alpha_1}/P\|_\infty \cdots \|\zeta^{\alpha_r}/P\|_\infty \|H_\nu\|_\infty .$$

Here $\| \ \|_\infty$ denotes L^∞ norm on $\{\zeta \in S_1 | 1/2 \leq |\zeta| \leq 2\}$. Noting $\nu, r \leq \ell$ and that for some N, we have $|\alpha_1|, \ldots, |\alpha_r|$, $|\beta_1|, \ldots, |\beta_r| < N$, we find that for some $C_3, R_3 > 0$, if $\zeta \in S$, then $|f_\ell^{I\nu}(\zeta)| < C_3 R_3^\ell (|\beta_1|+\ldots+|\beta_r|)!\nu!^{p-1}$. It suffices then to show that

$$|\beta_1|+\ldots+|\beta_r| \leq (p-1)(\ell-\nu) \tag{3.6}$$

for then $(|\beta_1|+\ldots+|\beta_r|)!\nu!^{p-1} \leq [(p-1)(\ell-\nu)]!\nu!^{p-1} \leq R_4^\ell (\ell-\nu)!^{p-1}\nu!^{p-1} \leq R_4^\ell \ell!^{p-1}$. (We have used (1.7), (1.8), and the fact $p - 1 \in \emptyset$, and we have let [ ] denote greatest integer function.)

To demonstrate (3.6), first note that by (3.5) $\sum_{n=1}^{r} (|\alpha_n|-|\beta_n|-k) + j - p\nu = j - p\ell$. Now for $1 \leq n \leq r$ we define m_n by $|\alpha_n| - |\beta_n| = k - pm_n$. Then $c_{\alpha_n \beta_n m_n} \lambda^{m_n} \xi^{\alpha_n} (\partial/\partial\xi)^{\beta_n}$ occurs as a term in $\hat{L}_\lambda$. We have $\sum_{n=1}^{r} m_n = \ell - \nu$. Now, for the first time, we use another of the hypotheses of (3.1), that the degree of L is k. As a consequence, $|\alpha_n| + |m_n| \leq k$, so that $|\beta_n| = pm_n + |\alpha_n| - k \leq (p-1)m_n$. (3.6) follows and the proof is complete. ■

The following lemma will complete the proof of Theorem 3.2.

Lemma 3.6. Suppose L satisfies (3.1). Further suppose that there exists $\sigma \in \mathbb{C}$ as follows: for all $K_1 \in AK^{\sigma}$, there exists $K \in AK^{\sigma+k}$ such that $LK = K_1$. Then (3.2) follows.

Remark. Once this is known, Theorem 3.2 follows at once. Indeed (c) $\Rightarrow$ (b) by Theorem 3.5; (b) $\Rightarrow$ (c) by Lemma 3.6; and (a) $\Rightarrow$ (c) by combining Lemmas 3.3 and 3.6.

Proof. We prove (3.2) for $\lambda = 1$. The case $\lambda = -1$ is similar, and the case of general λ follows from simple scaling considerations. Suppose $F \in Z_q^q$. By Theorem 2.11, we can select $K_1 \in AK^{\sigma}$ such that $\hat{K}_1(1,\xi) = F(\xi)$, $\hat{K}_1(-1,\xi) = 0$. By hypothesis, we can select $K \in AK^{\sigma+k}$ such that $LK = K_1$. Let $G(\xi) = \hat{K}(1,\xi)$; then $G \in Z_{q,j}^q$ for $j = -Q - k - \sigma$, and $\hat{L}_1 G = F$. We have only to show that G must be in Z_q^q. Suppose it is not. Suppose $G \sim \sum_{\ell=0}^{\infty} g_\ell$ in $Z_{q,j}^q$; sleect M such that $g_\ell = 0$ if $\ell < M$, $g_M \neq 0$. Write $\hat{L}_1 = \hat{L}_+ = P(\xi) + \sum_{m>0,\alpha,\beta} c_{\alpha\beta m}\xi^{\alpha}(\partial/\partial\xi)^{\beta}$ as in the proof of Theorem 3.5. In the sum for $m > 0$, $|\alpha| - |\beta| = k - pm < k$. Thus, in $Z_{q,j+k}^q$, $\hat{L}_+ G \sim \sum_{\ell=0}^{\infty} h_\ell$ where $h_\ell = 0$ for $\ell < M$, but $h_M = Pg_M$. In particular, $\hat{L}_+ G \notin Z_q^q$, contradiction. ∎

By the way, the proofs we have given show a C^∞ version of Theorem 3.2. Namely, suppose L satisfies (3.1). Then of the following conditions, (a) implies (b) and (b) is equivalent to (c): (a) L and L^t are hypoelliptic; (b) for any $\sigma \in \mathbb{C}$, if

$K_1 \in K^{\sigma}$, there exists $K \in K^{\sigma+k}$ such that $LK = K_1$; (c) for all λ, if $F \in S(\mathbb{R}^n)$ then there exists $G \in S(\mathbb{R}^n)$ such that $\hat{L}_{\lambda} G = F$. One can even drop the hypothesis in (3.1) that the degree of L is k.

Three groups of interesting operators which satisfy the hypotheses (3.1) are:

(I) constant coefficient homogeneous hypoelliptic operators in (t,x);

(II) transversally elliptic homogeneous left invariant differential operators on the Heisenberg group $\mathbb{H}^n$;

(III) Grušin operators in (t,x).

In case (I), it is evident that (3.2) could never hold. This gives an alternate proof of Proposition 3.1 in the case $\underset{\sim}{a} = (p,1,\ldots,1)$. (The ampleness hypothesis is not used in proving (a) $\Rightarrow$ (b) (for $\sigma = -Q$) $\Rightarrow$ (c).)

Case (II) will be discussed in Chapter Four. We close this section with a discussion of case (III); this will not be used in the rest of the book.

The operators L in class (III) can be described, with minimal hypotheses, as follows:

$$\begin{aligned}&\text{L is homogeneous, with homogeneous degree } k;\\ &\text{the degree of L is less than or equal to } k;\\ &\text{the coefficients of L are polynomials in the } x_\ell\text{'s};\\ &\text{and L is elliptic for } x \neq 0.\end{aligned} \tag{3.7}$$

(3.1) follows from (3.7). Indeed, write $L = L' + (\partial/\partial t)L''$ where L' is a differential operator in the x_ℓ's. Since L' is homogeneous with homogeneous degree k and degree less than or equal to k, it follows at once that L' has constant coefficients and degree k. In particular, L must have degree k. Further, since L is elliptic for $x \neq 0$, L' must be elliptic.

For these operators, Theorem 3.2 takes a very clean form.

Proposition 3.7. Suppose L satisfies (3.7) and $\sigma \in \mathbb{C}$. Suppose that $Z^q_{q,-(\sigma+k+Q)}$ is ample. Then the following are equivalent:

(a) L^t is analytic hypoelliptic;

(b) If $K_1 \in \mathcal{AK}^\sigma$, there exists $K \in \mathcal{AK}^{\sigma+k}$ with $LK = K_1$;

(c) For all $\lambda \neq 0$, $\hat{L}_\lambda : Z^q_q \to Z^q_q$ is surjective;

(d) For all $\lambda \neq 0$, $\hat{L}^*_\lambda : Z^q_q \to Z^q_q$ is injective;

(a)' L^t is hypoelliptic;

(b)' if $K_1 \in \mathcal{K}^\sigma$, there exists $K \in \mathcal{K}^{\sigma+k}$ with $LK = K_1$;

(c)' For all $\lambda \neq 0$, $\hat{L}_\lambda : \mathcal{S} \to \mathcal{S}$ is surjective;

(d)' For all $\lambda \neq 0$, $\hat{L}^*_\lambda : \mathcal{S} \to \mathcal{S}$ is injective.

Proof. We shall prove the implication (d) $\Rightarrow$ (d)' below in Lemma 3.8; we assume it for now. We now show that the remaining implications follow easily from this, Theorem 3.2, and, especially, the work of Grušin. Indeed, Grušin demonstrated (d)' $\Rightarrow$ (a)' ([38]) and (d)' $\Rightarrow$ (a) ([39]).

We shall show (b) $\Rightarrow$ (c) $\Rightarrow$ (d) $\Rightarrow$ (a) $\Rightarrow$ (b); (b)' $\Rightarrow$ (c)' $\Rightarrow$ (d)' $\Rightarrow$ (a)' $\Rightarrow$ (b)'. This will be enough, since (d) $\Rightarrow$ (d)' $\Rightarrow$ (a) as we have said.

(b) $\Rightarrow$ (c) follows from Theorem 3.2 (c) $\Rightarrow$ (d) since, given (c) and $\hat{L}_\lambda^* F = 0$, $F \in Z_q^q$, we may write $F = \hat{L}_\lambda G$, $G \in Z_q^q$. Since $\hat{L}_\lambda^* \hat{L}_\lambda G = 0$, $\hat{L}_\lambda G = 0$, so $F = 0$. (d) $\Rightarrow$ (a) since, as we have said, (d) $\Rightarrow$ (d)' $\Rightarrow$ (a). Thus (b) $\Rightarrow$ (c) $\Rightarrow$ (d) $\Rightarrow$ (a); similarly (b)' $\Rightarrow$ (c)' $\Rightarrow$ (d)' $\Rightarrow$ (a)'. Thus we need only show (a) $\Rightarrow$ (b) and (a)' $\Rightarrow$ (b)'.

First note (a) $\Rightarrow$ (d). In fact, we can prove more generally that if L is <u>any</u> homogeneous analytic hypoelliptic differential operator in (t,x) with coefficients which are polynomials in the x_ℓ's, then $\hat{L}_{\pm 1}$ must be injective on Z_q^q. Indeed, suppose $\hat{L}_1 F = 0$, $F \in Z_q^q(\mathbb{R}^s)$. By Theorem 1.8 (b) (with all $g_\ell = 0$), there exists $K \in \mathcal{AK}^{-Q}$ so that $\hat{K}(1,\xi) = F(\xi)$, $\hat{K}(-1,\xi) = 0$. Clearly $LK = 0$. If L is analytic hypoelliptic, K is analytic at 0, hence $K = 0$, so $F = 0$. Thus (a) = (d); analogously (a)' = (d)'. (In fact the analogous argument, for (a)' = (d)', was observed by Taylor [79].)

Now assume (a). We have (d), which implies that $\hat{L}_\lambda \hat{L}^*_\lambda : Z^q_q \to Z^q_q$ is also injective. Since (d) $\Rightarrow$ (a) and (a)', LL^* is hypoelliptic and analytic hypoelliptic. By Theorem 3.2, if $K_1 \in AK^\sigma$, there exists $K' \in AK^{\sigma+2k}$ with $LL^*K' = K_1$. Thus (a) $\Rightarrow$ (b). Similarly (a)' = (b)'; this completes the proof, except for the demonstration of (d) = (d)', which we turn to now.

Métivier proved (d) $\Rightarrow$ (d)' in [62] in the case $p = 2$; Helffer [40] adapted Métivier's proof to the case where $p \in \mathbb{Z}$. The proof which we present here for general p is an extension of Métivier's method.

For $m \in \mathbb{Z}^+$, define the Hilbert space

$$H^m = \{f \in L^2(\mathbb{R}^s) \mid \|f\|^2_m = \sum_{|\alpha| \le m} \|(1+|x|^2)^{(p-1)(m-|\alpha|)/2}(\partial/\partial x)^\alpha f\|^2_{L^2} < \infty\}$$

Write $\partial_j = \partial/\partial x_j$. Note that if $m > 0$, then ∂_j (where the derivatives are taken in the sense of distributions) and multiplication by $(1+|x|^2)^{(p-1)/2}$ map H^m to H^{m-1}. In fact, $H^m = \{f \in H^{m-1} \mid \partial_j f \in H^{m-1}$ for all j and $(1+|x|^2)^{(p-1)/2} f \in H^{m-1}\}$. Further, for some $c_1, c_2 > 0$,

$$c_1\|f\|_m \le \sum \|\partial_j f\|_{m-1} + \|(1+|x|^2)^{(p-1)/2} f\|_{m-1} \le c_2\|f\|_m \text{ for all } f \in H^m.$$

It is traditional, when dealing with Grušin operators, to study not $\hat{L}_\lambda$ but its inverse Fourier transform in ξ; and we keep with this tradition. Thus, if

$$L = \sum c''_{\alpha\beta m} x^{\beta} (\partial/\partial x)^{\alpha} (\partial/\partial t)^{m}, \tag{3.8}$$

we examine

$$L_{\lambda} = \sum c''_{\alpha\beta m} x^{\beta} (\partial/\partial x)^{\alpha} (-i\lambda)^{m} \tag{3.9}$$

instead of $\hat{L}_{\lambda} = \sum c''_{\alpha\beta m} (-i\partial/\partial\xi)^{\beta} (-i\lambda)^{\alpha} (-i\lambda)^{m}$.

In the expression for L_{λ}, note that for each term, the relations $|\alpha| - |\beta| + pm = k$, $m \leqq k - |\alpha|$ give $(p-1)|\alpha| + |\beta| \leqq (p-1)k$ or $|\beta|/(p-1) + |\alpha| \leqq k$. From this it is easily seen that for any λ, m, we have $L_{\lambda} : H^{m+k} \to H^{m}$. Now, under hypothesis (3.7), Grušin showed that (for some $C' > 0$):

$$\text{For all } u \in H^{k},\ \|u\|_{k} < C'(\|Lu\|_{0} + \|u\|_{0}) \tag{3.10}$$

where $L = L_{\lambda}$. From this we shall prove the following lemma, which clearly (by Theorem 2.3 (a) <==> (d)) implies Proposition 3.7 (d) => (d)'.

Lemma 3.8. Suppose $k \in \mathbb{N}$ and L is a differential operator of degree k on $\mathbb{R}^{s}$ of the form $L = \sum c_{\alpha\beta} x^{\beta} (\partial/\partial x)^{\alpha}$, where in each term $(p-1)|\alpha| + |\beta| \leqq (p-1)k$. Suppose (3.10) holds, and that (3.10) also holds if L is replaced by L^{t}. Then: If $G \in S$ and $LG \in Z^{p}_{p}$, then $G \in Z^{p}_{p}$.

Proof. As in Theorem 2.3(c), define $M_{\ell} : S \to S$ by $(M_{\ell} f)(x) = x_{\ell} f(x)$, $\partial_{\ell} f : S \to S$ by $\partial_{\ell} f = \partial f/\partial x_{\ell}$. Let $U = \{M_{\ell} | 1 \leqq \ell \leqq s\}$, $V = \{\partial_{\ell} | 1 \leqq \ell \leqq s\}$. We use the notation X for a product of the form $X_1 \ldots X_{m+n}$, where $\{X_1, \ldots, X_{m+n}\} \subset U \cup V$. If $\#\{X_i | X_i \in U\} = m$ and $\#\{X_i | X_i \in V\} = n$, we define $w(X)$, the weight of X, to be $m/(p-1) + n$. Recall $p = a/b$, so that $p - 1 = (a-b)/b$. Put $A = a - b$. Define $\mathbb{Z}^+/A = \{n/A | n \in \mathbb{Z}^+\}$. Then for any X, $w(X) \in \mathbb{Z}^+/A$. For $f \in S$, $j \in \mathbb{Z}^+/A$, put

$$[f]_j = \max\{\|(1+|x|^2)^{r(p-1)/2} Xf\|_{L^2} : r \in \mathbb{Z}^+/A,\ 0 \leq r \leq \max(1,q-1),\ r+w(X) \leq j\}.$$

(Note $q - 1 = (p-1)^{-1} \in \mathbb{Z}^+/A$.) We claim: suppose $f \in S$. Then $f \in Z_p^p$ if and only if there exist $C_0, R_0 > 0$ such that for all $j \in \mathbb{Z}^+/A$, $[f]_j < C_0 R_0^j (Aj)!^{1/Aq}$. For, suppose $f \in Z_p^p$. Using Theorem 2.3 (a)<==>(c)$_2$, select $C, R > 0$ so that if $\{X_1, \ldots, X_{m+n}\} \subset U \cup V$, $\#\{X_i | X_i \in U\} = m$ and $\#\{X_i | X_i \in V\} = n$, then $\|X_1 \ldots X_{m+n}\|_{L^2} < CR^{m+n} m!^{1/p} n!^{1/q}$. Say $j \in \mathbb{Z}^+/A$, $r \in \mathbb{Z}^+/A$, $0 \leq r \leq \max(1,q-1)$, $r + w(X) \leq j$. Pick $N \in \mathbb{N}$ with $N \geq (p-1)/2$. Then for some $C_2, C_1, R_1 > 0$, $\|(1+|x|^2)^{r(p-1)/2} Xf\|_{L^2} \leq$
$\|(1+|x|^2)^N Xf\|_{L^2} \leq C_1 R^{m+n} (m+2N)!^{1/p} n!^{1/q} \leq C_2 R_1^{m+n} [m!^{A/(p-1)} n!^A]^{1/Aq} \leq$
$C_2 R_1^{m+n} (Aj)!^{1/Aq}$.

Conversely, suppose that $f \in S$ and for all $j \in \mathbb{Z}^+/A$, $[f]_j < C_0 R_0^j (Aj)!^{1/Aq}$.

Say $\{X_1, \ldots, X_{m+n}\} \subset U \cup V$, $\#\{X_i | X_i \in U\} = m$, $\#\{X_i | X_i \in V\} = n$; let $X = X_1 \ldots X_{m+n}$. Then $\|Xf\|_{L^2} \leq [f]_{n+m/(p-1)} <$
$C_0 R_1^{n+m} (An+Am/(p-1))!^{1/Aq} < C_0 R_2^{m+n} (An)!^{1/Aq} (Am/(p-1))!^{1/Aq} <$
$C_0 R_3^{m+n} m!^{1/p} n!^{1/q}$. Thus Theorem 2.3(c)$_2$ holds, and $f \in Z_p^p$.

With this in mind, we come to the proof. Say $G \in S$. We prove the lemma by estimating $[G]_j$ in terms of $[G]_{j-q}$ and $[LG]_j$. Suppose then that $j \in \mathbb{Z}^+/A$, $j > k + q$. Then, evidently there exists $r \in \mathbb{N}/A$ with $0 \leqq r < \max(1,q-1)$ and X with $j - q < r + w(X) \leqq j$ such that $[G]_j \leqq [G]_{j-q} + \|(1+|x|^2)^{r(p-1)/2}XG\|_{L^2}$. We can write $X = X^{(1)}X^{(2)}$ where $k \leqq r + w(X^{(1)}) < k + \max(1,q-1)$; $X^{(1)}$ is just the shortest initial substring in X with $r + w(X^{(1)}) \geqq k$. Put $r' = r + w(X^{(1)}) - k$, so that $0 \leqq r' < \max(1,q-1)$, and $r' + w(X^{(2)}) \leqq j - k$. Write $(1+|x|^2)^{r(p-1)/2}XG =$ $[(1+|x|^2)^{r(p-1)/2}X^{(1)}(1+|x|^2)^{-r'(p-1)/2}][(1+|x|^2)^{r'(p-1)}X^{(2)}G]$. This procedure can be carried through for any j, and for each j we obtain $(r,X^{(1)},X^{(2)},r') = (r_j,X_j^{(1)},X_j^{(2)},r_j')$. Note that, despite the infinite number of possibilities for j, there are only finitely many possibilities for $(r_j,X_j^{(1)},r_j')$, because of the restrictions on r_j,r_j' and $w(X_j^{(1)})$. Also, there exists C_0 as follows: For any j, and for all $f \in H^k$,

$$|(1+|x|^2)^{r_j(p-1)/2}X_j^{(1)}(1+|x|^2)^{-r_j'(p-1)/2}f| < C_0\|f\|_k. \qquad (3.11)$$

Indeed, it suffices to prove (3.11) for any fixed $(r,X^{(1)},r') = (r_j,X_j^{(1)},r_j')$. $X^{(1)}$ is a linear combination of terms of the form $x^\beta\partial^\alpha$, each of whose weights do not exceed $w(X^{(1)})$, so it suffices to prove (3.11) with $X^{(1)}$ replaced by $X^{(3)} = x^\beta\partial^\alpha$ where $w(X^{(3)}) \leqq w(X^{(1)})$. This is now elementary, since $r_j + w(X^{(3)}) - r_j' \leqq k$. From (3.11) and (3.10), then

$$[G]_j \leqq [G]_{j-q} + C_o C' \| L(1+|x|^2)^{r'(p-1)/2} X^{(2)} G\|_o$$
$$+ C_o C' \| (1+|x|^2)^{r'(p-1)/2} X^{(2)} G\|_o$$

so

$$[G]_j \leqq [G]_{j-q} + C_o C' [LG]_{j-k} + C_o C' [G]_{j-k}$$
$$+ C_o C' \| [L,(1+|x|^2)^{r'(p-1)/2} X^{(2)}] G\|_o \tag{3.12}$$

since $r' + w(X^{(2)}) \leqq j - k$. L is a linear combination of terms of the form $X^{(4)} = x^\beta \partial^\alpha$ where $w(X^{(4)}) \leqq k$. Thus

$$\| [L,(1+|x|^2)^{r'(p-1)/2} X^{(2)}] G\|_o$$
$$\leqq C_1 \max \| [X^{(4)},(1+|x|^2)^{r'(p-1)/2} X^{(2)}] G\|_o \tag{3.13}$$

where C_1 does not depend on j and where the max is taken over the finite set of all $X^{(4)}$ of the form $x^\beta \partial^\alpha$ with $w(X^{(4)}) \leqq k$.

Fix $X^{(4)} = x^\beta \partial^\alpha$ with $w(X^{(4)}) \leqq k$, but allow $(r',X^{(2)}) = (r'_j, X_j^{(2)})$ to vary with j. Then

$$\| [X^{(4)},(1+|x|^2)^{r'(p-1)/2} X^{(2)}] G\|_o \leqq B_1 + B_2 \tag{3.14}$$

where

$$B_1 = \| x^\beta [\partial^\alpha,(1+|x|^2)^{r'(p-1)/2}] X^{(2)} G\|_o$$
$$B_2 = \| (1+|x|^2)^{r'(p-1)/2} [X^{(4)}, X^{(2)}] G\|_o . \tag{3.15}$$

We study B_1, B_2 separately. For B_1, note that for any $\tau \in \mathbb{R}$,

$$[\partial^\alpha,(1+|x|^2)^\tau] = \sum C_{\rho n \gamma} x^\rho (1+|x|^2)^n \partial^\gamma \tag{3.16}$$

where in each term $|\rho| + 2n < 2\tau$ and $|\gamma| \leqq |\alpha|$. There thus exists C_2 independent of j such that $B_1 \leqq C_2 \max|x^{\beta+\rho}(1+|x|^2)^n \partial^\gamma X^{(2)} G|_0$, the max taken over all (ρ,n,γ) with $\rho + 2n < r'(p-1)$ and $|\gamma| < |\alpha|$. Thus $B_1 \leqq C_2 \max \|(1+|x|^2)^{r'(p-1)/2} x^\beta \partial^\gamma X^{(2)} G\|_0$, the max taken over all γ with $|\gamma| < |\alpha|$. But $r' + w(x^\beta \partial^\gamma X^{(2)}) \leqq w(X^{(4)}) - 1 + r' + w(X^{(2)}) \leqq k - 1 + j - k = j - 1$. Thus $B_1 \leqq C_2[G]_{j-1}$.

We come to the main point, B_2. Write $X^{(2)} = X_1 \ldots X_{m+n}$ where $\#\{X_i | X_i \in U\} = n$, $\#\{X_i | X_i \in V\} = n$. Put $N = m + n$. Note $N \leqq \max(1,p-1) w(X^{(2)}) < C_3 j$, where C_3 is independent of j. Now $[X^{(4)}, X^{(2)}] = \sum_{i=1}^{N} X_1 \ldots X_{i-1}[X^{(4)},X_i]X_{i+1} \ldots X_N$. Noting that $[\partial_j, x_k] = \delta_{jk}$, we see that $[X^{(4)},X_i] = 0$ or $c_i x^{\beta_i} \partial^{\alpha_i}$ for some $c_i \neq 0$, where $w(x^{\beta_i}\partial^{\alpha_i}) = w(X^{(4)}) + w(X_i) - 1 - (p-1)^{-1} = w(X^{(4)}) + w(X_i) - q \leqq k - q + w(X_i)$. Regardless of j, there are only finitely many possibilities for c_i; let $C_4 = \max c_i$. Note that $r' + w(X_1 \ldots X_{i-1} x^{\beta_i} \partial^{\alpha_i} X_{i+1} \ldots X_N) \leqq r' + w(X_1 \ldots X_{i-1}) + k - q + w(X_i) + w(X_{i+1} \ldots X_N) = r' + k - q + w(X^{(2)}) \leqq j - q$. Thus, for some C_4 independent of j, $B_2 \leqq NC_4[G]_{j-q} \leqq C_3C_4 j[G]_{j-q}$.

Thus, from (3.14),

$$\|[X^{(4)}, (1+|x|^2)^{r'(p-1)/2} X^{(2)}]G\|_0 \leqq C_2[G]_{j-1} + C_3C_4 j[G]_{j-q}$$

where C_2, C_3, C_4 depend on $X^{(4)}$ but not on j. In (3.13) there are only finitely many possibilities for $X^{(4)}$. Thus, from (3.12), there exists $C_5 > 0$, independent of j, such that

$[G]_j \leqq C_5(j[G]_{j-q} + [LG]_j) + C_5[G]_{j-1}$, provided $j \in \mathbb{Z}^+/A$, $j > k + q$. Select $M \in \mathbb{N}$ with $M > q$. If $j > k + q + M$, this inequality holds with $j - 1, \ldots, j - M$ in place of j. Thus there exists $C_6 > 0$ and independent of j such that $[G]_j \leqq C_6(j[G]_{j-q} + [LG]_j)$ provided $j \in \mathbb{Z}^+/A$, $j > J = k + q + M$. Equivalently, for some $C_7 > 0$, independent of j,

$$[G]_j \leqq C_7((j-q)[G]_{j-q} + [LG]_j) \tag{3.17}$$

if $j \in \mathbb{Z}^+/A$, $j > J$.

Now suppose $G \in S$, $LG \in Z_p^p$. We proceed to apply the relation (3.17) to show $G \in Z_p^p$. There exist $C_8, R > 0$ so that $[LG]_j \leqq C_8 R^j (Aj)!^{1/Aq}$ for all $j \in \mathbb{Z}^+/A$. Put $R_1 = \max(R, (2C_7)^{1/q})$. Since $G \in S$, we can select $C_9 \geqq 2C_8$ such that $[G]_j \leqq C_9 R_1^j (Aj)!^{1/Aq}$ for all $j \in \mathbb{Z}^+/A$ with $j \leqq J$. It suffices to show that $[G]_j \leqq C_9 R_1^j (Aj)!^{1/Aq}$ for all $j \in \mathbb{Z}^+/A$. We prove this by induction on Aj. Say $j \in \mathbb{Z}^+/A$ and the statement is known for all lesser values of j. Then, writing $Aq = B$,

$$\begin{aligned} [G]_j &\leqq C_7 C_9 R_1^{j-q} (Aj-B)(Aj-B)!^{1/B} + C_8 R^j (Aj)!^{1/B} \\ &\leqq C_7 C_9 R_1^{j-q} (Aj)!^{1/B} + (C_9/2) R_1^j (Aj)!^{1/B} \\ &\leqq C_9 R_1^j (Aj)!^{1/B} \text{ as desired.} \end{aligned}$$

Remarks. 1. If $p = 2$, the proof reduces to Métivier's procedure. This case is simpler since in the above, r and r' must always be 0—the restriction $r \in \mathbb{Z}^+$, $r < 1$ forces this.

2. It is not difficult to show, in Proposition 3.7, that if $F \in \mathcal{S}'$ and $\mathcal{L}F \in \mathcal{S}$, then $F \in \mathcal{S}$.

3. Transversally elliptic operators on the Heisenberg group, treated in the next chapter, satisfy (3.1) and the conclusion of Proposition 3.7 (that (a)-(d), (a)'-(d)' are all equivalent); but they do not satisfy (3.7) or (3.10) for any λ.

4. Homogeneous Partial Differential Operators on the Heisenberg Group

The Heisenberg group $\mathbb{H}^n$ is the Lie group with underlying manifold $\mathbb{R} \times \mathbb{C}^n$ and multiplication $(t,z)\cdot(t',z') = (t+t'+2 \operatorname{Im} z\cdot\overline{z}', z+z')$, where $z\cdot\overline{z}' = \sum_{j=1}^{n} z_j\overline{z}_{j'}$. The dilations D_r are given by $D_r(t,z) = (r^2t, rz)$; thus we are in the case $p = 2$ of the previous chapters. Write $z_j = x_j + iy_j$; then $X_j = \partial/\partial x_j + 2y_j\partial/\partial t$, $Y_j = \partial/\partial y_j - 2x_j\partial/\partial t$ and $T = \partial/\partial t$ give the basis of left-invariant vector fields agreeing with $\partial/\partial x_j$, $\partial/\partial y_j$, $\partial/\partial t$ at 0. The right-invariant analogues are $X_j^R = \partial/\partial x_j - 2y_j\partial/\partial t$, $Y_j^R = \partial/\partial y_j + 2x_j\partial/\partial t$, and T. Put $Z_j = (1/2)(X_j - iY_j) = \partial/\partial z_j + i\overline{z}_j\partial/\partial t$, $\overline{Z}_j = \partial/\partial\overline{z}_j - iz_j\partial/\partial t$; similarly one has $Z_j^R = \partial/\partial z_j - i\overline{z}_j\partial/\partial t$, and $\overline{Z}_j^R$. We have commutation relations

$$[Z_j,\overline{Z}_k] = -2i\delta_{jk}T, \quad [Z_j,Z_k] = [Z_j,T] = 0\ . \tag{4.1}$$

Let L be a left-invariant differential operator on $\mathbb{H}^n$; it can be expressed as $L = p(Z,\overline{Z})$ for some non-commuting polynomial p, where $Z = (Z_1,\ldots,Z_n)$. Suppose L is homogeneous with respect to the dilations (i.e. p is homogeneous). Let $I_C L = p(\partial/\partial z, \partial/\partial\overline{z})$ where $\partial/\partial z = (\partial/\partial z_1,\ldots,\partial/\partial z_n)$. Note that $I_C L$ does not depend on the choice of p—to obtain $I_C L$, write out L as a differential operator in the $\partial/\partial t$, $\partial/\partial z_j$, $\partial/\partial\overline{z}_j$ and cancel any term involving $\partial/\partial t$. In this section, we prove the following result.

<u>Theorem 4.1</u>. Suppose L is a left-invariant differential operator on $\mathbb{H}^n$ which is homogeneous of degree k. Assume $I_c L$ is elliptic. Then there exist $K \in AK^{-2n-2+k}$ and $P \in AK^{-2n-2}$ such that $LK = \delta - P$, and such that the map $f \to f*P$ is the projection in $L^2(\mathbb{H}^n)$ onto $[LS(\mathbb{H}^n)]^{\perp}$. If $f \in L^2$ or E' and if $q \in \mathbb{H}^n$, then there exists a distribution u with $Lu = f$ near q if and only if $f*P$ is real analytic near q.

All these assertions, save one, were proved by the author in [26]. What was not shown was that K could be chosen to lie in $AK^{-2n-2+k}$; there, one only knew that K could be found in $K^{-2n-2+k}$. Rothschild and Tartakoff [71] later showed that one could write $LK = I - P$ where K and P were operators which preserved local analyticity; however, they did not show that they were given by convolution with kernels in the AK spaces.

To prove Theorem 4.1, we begin by observing that L satisfies the conditions (3.1). (Here, of course, $s = 2n$, $p = 2$. The letters s,p,a,b and Q are now freed for other uses.) Thus Theorem 3.5 holds for L; we are interested in the case $K_1 = \delta$. We prove Theorem 4.1 by establishing a form of (3.2) for L. (3.10) does not hold for $L = \hat{L}_\lambda$, and we must now use the group Fourier transform instead of the Euclidean Fourier transform. It is convenient, then, to change notation: from now on, F means Euclidean Fourier transform, while ^ means group

Fourier transform.

We give a rapid summary of the definition and basic properties of the group Fourier transform. For complete proofs we refer to the first section of [25]. (One piece of notation used in [25] has been changed: the meanings of W and W^+ (to be defined presently) have been reversed.)

Here are the basic definitions. For $\lambda \in \mathbb{R}^* = \mathbb{R}\setminus\{0\}$, let $\mathcal{H}_\lambda$ be a separable complex Hilbert space with fixed orthonormal basis $\{(E_{\alpha,\lambda}\}_{\alpha\in(\mathbb{Z}^+)^n}$ (here $\mathbb{Z}^+ = \{0,1,2,...\}$). At times we identify all $\mathcal{H}_\lambda$ so that $E_{\alpha,\mu}$ is identified with $E_{\alpha,\lambda}$ for all λ,μ. On the algebraic span of the basis, we define the weighted shift operators $\tilde{W}_{k\lambda}, \tilde{W}^+_{k\lambda}$ for $1 \leq k \leq n$ as follows:

$$\begin{aligned} \tilde{W}_k E_\alpha &= (2\alpha_k|\lambda|)^{1/2} E_{\alpha-e_k}, \text{ zero if } \alpha_k = 0 \\ \tilde{W}^+_K E_\alpha &= [2(\alpha_k+1)|\lambda|]^{1/2} E_{\alpha+e_k} \end{aligned} \tag{4.2}$$

for $\lambda > 0$. The right sides are to be reversed if $\lambda < 0$. Here and elsewhere, we frequently drop the λ subscript. Also, $e_k = (0,...,1,...,0)$ with the 1 in the kth position. $\tilde{W}_k$ and $\tilde{W}^+_k$ are closable; we denote their closures by $W_{k\lambda}, W^+_{k\lambda}$ (domain: $\mathcal{D}(W_k) = \mathcal{D}(W^+_k) = \{v = \sum v_\alpha E_\alpha | \sum|\alpha_k||v_\alpha|^2 < \infty\}$. If $\mathcal{H}$ is identified with $L^2(\mathbb{R}^n)$ and $\{(-i)^{|\alpha|}E_\alpha\}$ with the Hermite functions, W_k and W^+_k are $i/2^{1/2}$ times the annihilation and creation operators $2^{-1/2}(x_k \pm \partial/\partial x_k)$ when $\lambda = 1$; we discuss this further, not now, but below. One can restrict W_k, W_k^+ to a domain A_λ on which $-z\cdot W^+ + \bar{z}\cdot W$ is essentially skew-adjoint for all $z \in \mathbb{C}^n$, so that $V_z^\lambda = \exp(-z\cdot W^+ + \bar{z}\cdot W)$ is unitary on H. The Weyl transform $G_\lambda : L^1(\mathbb{C}^n) \to B(H_\lambda)$ is defined by $G_\lambda F = \int_{\mathbb{C}^n} V_z^\lambda F(z)\,dzd\bar{z}$. This is very analogous to the transform $F'_{\mathbb{C}^n} : L^1(\mathbb{C}^n) \to L^\infty(\mathbb{C}^n)$, defined by $(F'_{\mathbb{C}^n}F)(\zeta) = \int \exp(-z\cdot\bar{\zeta} + \bar{z}\cdot\zeta)F(z)\,dzd\bar{z}$. (If we think of $\zeta = p + iq$ with (p,q) the dual coordinates to (x,y), then $F'_{\mathbb{C}^n}$ may be thought of as a slightly skewed Fourier transform. It may be identified with the usual Fourier transform if we instead think of $\zeta = (-q+ip)/2$ with (p,q) dual to (x,y). Usually, though, we won't give names to $\mathrm{Re}\,\zeta$ or $\mathrm{Im}\,\zeta$, so it doesn't matter.) If $G(\zeta)$ is a suitable function on $\mathbb{C}^n$, say $G \in S$, Weyl associates to G an operator $W_\lambda G$, defined by $G_\lambda (F'_{\mathbb{C}^n})^{-1} G$. W essentially replaces $(\zeta, \bar{\zeta})$ by (W, W^+). We call W the Weyl correspondence.

The Weyl transform arises naturally in the definition of the Fourier transform on $\mathbb{H}^n$. We wish to define the Fourier transform of a function f to be a family of bounded operators $\hat{f} = (\hat{f}(\lambda))$ (for $\lambda \in \mathbb{R}^*$; $\hat{f}(\lambda)$ is defined on H_λ). We want $\hat{}$ to have the following properties: for $f \in S(\mathbb{H}^n)$, $(Tf)^\wedge(\lambda) = -i\lambda\hat{f}(\lambda)$, $(Z_j f)^\wedge(\lambda) = \hat{f}(\lambda)W_j^+$, $(\bar{Z}_j f)^\wedge(\lambda) = -\hat{f}(\lambda)W_j$, at least when both sides are applied to each E_α. To see that this is a conceivable objective, one must check commutation relations: we have (4.1), while $[W_j^+, -W_k] = 2\delta_{jk}\lambda I$ (on the algebraic span of the E_α), $[W_j, W_k] = 0$. The definition of such a $\hat{}$ is thus seen to be conceivable. We should also have that $(g*f)\hat{}(\lambda) = \hat{g}(\lambda)\hat{f}(\lambda)$ for $f, g \in L^1$. But formally,

$$(g*f)(t,z) = ([\int_{\mathbb{H}^n} \exp(-t'T - z'\cdot Z - \bar{z}'\cdot\bar{Z}) f(t',z')dt'dz'd\bar{z}']g)(t,z),$$

so we expect that we must define $\hat{f}(\lambda) = \int_{\mathbb{H}^n} U_u^\lambda f(u)du$, where $U^\lambda_{(t,z)} = \exp(i\lambda t - z\cdot W_\lambda^+ + \bar{z}\cdot W_\lambda) = e^{i\lambda t}V_z$. Accordingly, for $f \in L^1(\mathbb{H}^n)$, one does define $\hat{f}$ to be the family of bounded operators $(\hat{f}(\lambda))$, where $\hat{f}(\lambda) = \int_{\mathbb{H}^n} U_u^\lambda f(u)du$. It is then not difficult to check the properties of $\hat{}$ given above. To see the connection with the Weyl transform, for $\lambda \in \mathbb{R}^*$ define $F_c^\lambda : L^1(\mathbb{H}^n) \to L^1(\mathbb{C}^n)$ by $(F_c^\lambda f)(z) = \int_{-\infty}^{\infty} e^{i\lambda t} f(t,z)dt$. Then $\hat{f}(\lambda) = G_\lambda(F_c^\lambda f)$. This is also analogous to the Euclidean Fourier transform F', where $(F'f)(\lambda,\zeta) = [F'_{\mathbb{C}^n}(F_c^\lambda f)](\zeta)$. If $f \in S(\mathbb{H}^n)$, define $F'_\lambda f \in S(\mathbb{C}^n)$ by $(F'_\lambda f)(\zeta) = (F'f)(\lambda,\zeta)$; then $\hat{f}(\lambda) = W_\lambda(F'_\lambda f)$. We shall be using the notations $F, F_{\mathbb{R}^{2n}}, F', F'_{\mathbb{C}^n}, F_c^\lambda, F'_\lambda$ frequently in the rest of the book.

If $f \in S$, we also have that $\hat{f}(\lambda) : H \to D(W_j) = D(W_j^+)$ and $(Z_j^R f)\hat{}(\lambda) = W_j^+\hat{f}(\lambda)$, $(\bar{Z}_j^R f)\hat{}(\lambda) = -W_j\hat{f}(\lambda)$. Another important fact is that if $L_o = (-1/2)\sum(Z_j\bar{Z}_j + \bar{Z}_j Z_j)$ is the Folland-Stein operator, and $f \in S(\mathbb{H}^n)$, then (at least when both sides are applied to any E_α) $(L_o f)\hat{}(\lambda) = \hat{f}(\lambda)A_\lambda$ where $A_\lambda E_\alpha = (2|\alpha|+n)|\lambda|E_\alpha$; here $|\alpha| = \sum\alpha_j$. This follows at once from the formulas above. Also, if $R_o = (-1/2)\sum(Z_j^R\bar{Z}_j^R+\bar{Z}_j^R Z_j^R)$, then $(R_o f)^\wedge(\lambda) = A_\lambda \hat{f}(\lambda)$.

For $u \in \mathbb{H}^n$, let $\pi_\lambda(u) = U_u^\lambda$; then π_λ is an irreducible unitary representation of $\mathbb{H}^n$. Other irreducible unitary representations are given as follows: for $\xi = (a,b) \in \mathbb{R}^n \times \mathbb{R}^n$, one defines π_ξ acting on a one-dimensional Hilbert space by $\pi_\xi(t,x,y) = e^{i(a\cdot x+b\cdot y)}$. Any irreducible unitary representation is equivalent to some π_ξ or some π_λ. Further, no two of these are equivalent. The result is due to von Neumann and Stone.

We rapidly outline the concrete representation of π_λ on $L^2(\mathbb{R}^n)$ which we alluded to above. Nothing we say, except in Chapter 6 will depend on use of this concrete representation, but on several occasions it will provide valuable interpretations. We start afresh in our notation, but will quickly justify that the notation is compatible with, and gives a concrete realization of, what we have already defined. For $\lambda \in \mathbb{R}^*$ we define $E_{o,\lambda} \in L^2(\mathbb{R}^n)$ by $E_{o,\lambda}(p) = (|\lambda|/\pi)^{n/4}e^{-2|\lambda|p^2}$. We define $W_{k\lambda} = i(2\lambda p_k+(1/2)\partial/\partial p_k)$, $W_{k\lambda}^+ = i(-2\lambda p_k+(1/2)\partial/\partial p_k)$; they operate at first on $\mathcal{S}(\mathbb{R}^n)$ and $[W_{j\lambda}^+,-W_{k\lambda}] = 2\delta_{jk}\lambda I$. When $\lambda = 1/4$ these are $i/2^{1/2}$ times the annihilation and creation operators. Say $\alpha \in (\mathbb{Z}^+)^n$. For $\lambda > 0$, put $E_{\alpha,\lambda} = (2|\lambda|)^{-|\alpha|/2}\alpha!^{-1/2}(W_\lambda^+)^\alpha E_{o,\lambda}$. Put $E_{\alpha,\lambda} = (2|\lambda|)^{-|\alpha|/2}\alpha!^{-1/2}W_\lambda^\alpha E_{o,\lambda}$, for $\lambda < 0$. Here $W^+ = (W_{1\lambda}^+,\ldots,W_{n\lambda}^+)$, similarly W_λ. The

$\{(-i)^{|\alpha|}E_{\alpha,1/4}\}$ are the Hermite functions. One verifies (4.2) at once for $\lambda > 0$ and the reverse for $\lambda < 0$. Using (4.2) and induction on $|\alpha|$ one sees that $(E_{\alpha,\lambda}, E_{\beta,\lambda}) = \delta_{\alpha\beta}$ if $|\beta| \leqq |\alpha|$; thus the $\{E_{\alpha,\lambda}\}$ are orthonormal in $L^2(\mathbb{R}^n)$. A standard argument shows that they are a basis. Thus in the preceding discussion we could identify $\mathcal{H}_\lambda$ with $L^2(\mathbb{R}^n)$ and the $E_{\alpha,\lambda}, W_{k,\lambda}, W^+_{k,\lambda}$ with the above functions and operators. This concrete realization is called the Schrödinger representation. To understand V^λ_z in this representation, note that if we write $Q_{k\lambda} = -i\partial/\partial p_k$ and $P_{k\lambda} = 4\lambda p_k$ then $W_k = (-Q_k+iP_k)/2$ and $W^+_k = (-Q_k-iP_k)/2$. Thus $V_z = \exp(-z\cdot W^+ + \bar{z}\cdot W) = \exp(i(x\cdot P+y\cdot Q))$ if $z = x + iy$. By the Weyl relations, $V^\lambda_z = e^{2i\lambda x\cdot y}e^{ix\cdot P}e^{iy\cdot Q}$ formally. (The Weyl relations are a special case of the Campbell-Hausdorff formula; note $[P_{k\lambda}, Q_{k\lambda}] = 4i\lambda$.) Continuing to compute formally, one expects that if $\varphi \in L^2(\mathbb{R}^n)$,

$$V^\lambda_z(\varphi)(p) = e^{2i\lambda x\cdot(y+2p)}\varphi(y+p). \qquad (4.3)$$

As approach to the Schrödinger representation, then, is to define V^λ_z by (4.3), put $U^\lambda_{(t,z)} = e^{i\lambda t}V^\lambda_z$, show directly that $u \to U^\lambda_u$ is a unitary representation of $\mathbb{H}^n$, define $G_\lambda : L^1(\mathbb{C}^n) \to \mathcal{B}(\mathcal{H})$ by $G_\lambda F = \int V^\lambda_z F(z)dzd\bar{z}$ and, for $f \in L^1(\mathbb{H}^n)$, define $\hat{f}(\lambda) = G_\lambda(F^\lambda_c f)$. One can then verify directly all the previously stated properties about ^ such as the relation $(Z_j f)^\wedge(\lambda) = \hat{f}(\lambda)W^+_{j\lambda}$. This is carried out in [24]. (For completeness, we outline a proof that, indeed, $V_z^\lambda = \exp(-z\cdot W^+ + \bar{z}\cdot W)$; shorter proofs can be given. One can show (see e.g. [24]) that the Schrödinger and Bargmann representations are unitarily equivalent; the V_z^λ of (4.3) is carried over to an operator in the Bargmann representation which is shown in [25] to be equal to $\exp(-z\cdot W^+ + \bar{z}\cdot W)$.) A straightforward computation with (4.3) shows that if $F(x,y) \in S(\mathbb{C}^n)$, $\varphi \in L^2(\mathbb{R}^n)$, then

$$[G_\lambda(F)(\varphi)](p) = \int K(p,p')\varphi(p')dp', \text{ where}$$
$$K(p,p') = (F_1 F)(2\lambda(p'+p), p'-p). \tag{4.4}$$

Here $F_1 F$ denotes the Fourier transform of F in the x variables.

We reiterate that, until Chapter 6, we shall be using the Schrödinger representation only to give interesting interpretations. Our own preference is to work in the abstract setting, placing the emphasis on the $\{E_\alpha\}$ and the W_k. For our proof of Theorem 4.1, this choice of emphasis will be essential. In Chapter 6, we shall interpret our results in the Schrödinger representation in order to obtain an interesting calculus of analytic pseudodifferential operators on $\mathbb{R}^n$.

We return now to the abstract setting and set out a little more notation. We shall be studying families of operators $(R(\lambda))$, where $R(\lambda)$ acts on H_λ. At all times, we implicitly assume that the domain of each $R(\lambda)$ contains all the $E_{\alpha,\lambda}$; further, we require that for all α,β, the map

$\lambda \to (R(\lambda)E_{\alpha,\lambda}, E_{\beta,\lambda})$ is measurable. We frequently denote such a family of operators $(R(\lambda))$ by a single letter $\underset{\sim}{R}$; thus $\underset{\sim}{R} = (R(\lambda))$. When they make sense, $\underset{\sim}{R} + \underset{\sim}{S}$, $\underset{\sim}{RS}$ and $\underset{\sim}{R}^*$ will have the obvious meanings; e.g. $\underset{\sim}{RS} = (R(\lambda)S(\lambda))$.

Certain spaces of operator families are very useful. Thus, let $\mathcal{B} = \{$operator families $\underset{\sim}{R} | R(\lambda)$ is bounded for all λ, and $||\underset{\sim}{R}|| = \sup_{\lambda} ||R(\lambda)|| < \infty\}$. Evidently, $\hat{}: L^1(\mathbb{H}^n) \to \mathcal{B}$, and $||\hat{f}|| \leq ||f||_1$. Let $\mathcal{R}_2 = \{\underset{\sim}{R}|$ for a.e.λ, $||R(\lambda)||_2 < \infty$ and $||\underset{\sim}{R}||_2^2 = \int_{-\infty}^{\infty} ||R(\lambda)||_2^2 (2|\lambda|)^n d\lambda < \infty\}$. Here $||R(\lambda)||_2$ is the Hilbert-Schmidt norm of $R(\lambda)$. One then has the Plancherel theorem: $\hat{}$, restricted to $\mathcal{S}(\mathbb{H}^n)$, can be extended to a map from $L^2(\mathbb{H}^n)$ onto $\mathcal{R}_2$, and if $f \in L^2$, $||f||_2^2 = c_n ||\hat{f}||_2^2$; here $c_n = 1/2\pi^{n+1}$. Next, for a bounded operator S on $\mathcal{H}_\lambda$, let $||S||_1^E = \sum_\alpha ||SE_\alpha||$. It is easy to see that if $||S||_1^E < \infty$, then S is of trace class. Let $\mathcal{R}_1^E = \{\underset{\sim}{R}|$ for a.e.λ, $||R(\lambda)||_1^E < \infty$ and $\int_{-\infty}^{\infty} ||R(\lambda)||_1^E (2|\lambda|)^n d\lambda < \infty\}$. It is then very easy to see that $\check{}: \mathcal{R}_1^E \to C(\mathbb{H}^n)$ by $\check{\underset{\sim}{R}}(u) = \int_{-\infty}^{\infty} \mathrm{tr}(U_{-u}^\lambda R(\lambda))(2|\lambda|)^n d\lambda$ is well-defined. One has the inversion theorem: if $f \in L^1(\mathbb{H}^n)$ and $\underset{\sim}{R} = \hat{f} \in \mathcal{R}_1^E$, then $f(u) = c_n \check{\underset{\sim}{R}}(u)$ for a.e. u. Further, $\hat{\mathcal{S}} \subset \mathcal{R}_1^E$. Next, if S is an operator on $\mathcal{H}_\lambda$ whose domain contains all E_α, let $\{S_{\alpha\beta}\}$ be its matrix with respect to the $\{E_\alpha\}$; thus $SE_\beta = \sum S_{\alpha\beta} E_\alpha$. If $\underset{\sim}{R} = (R(\lambda))$ is a family of operators, we let $(R_{\alpha\beta}(\lambda))$ denote the matrix of $R(\lambda)$. Let $\mathcal{Q} = \{\underset{\sim}{R} | R_{\alpha\beta}(\lambda) \in C_0^\infty(\mathbb{R}^*)$ for all α, β, and for some $N \in \mathbb{N}$, $R_{\alpha\beta}(\lambda) = 0$ if $|\alpha| + |\beta| > N\}$. (Here $|\alpha| = \sum |\alpha_j|$.) Then if

$\underset{\sim}{R} \in \mathcal{Q}$, there exists $f \in S(\mathbb{H}^n)$ such that $\hat{f} = \underset{\sim}{R}$.

In this way, we see that $\mathcal{B}, \mathcal{R}_2, \mathcal{R}_1^E, \mathcal{Q}$ play the roles that L^∞, L^2, L^1 and C_o^∞ respectively play on the Fourier transform side in Euclidean Fourier analysis.

At times we shall need corresponding facts about the Weyl transform and the Weyl correspondence $\mathcal{W}$. Again, see Section One of [25] for the proofs. F_c, the t-Fourier transform on $\mathbb{H}^n$, carries $X_j, Y_j, X_j^R, Y_j^R, Z_j, \bar{Z}_j, Z_j^R, \bar{Z}_j^R, L_o, R_o$ into

$X_j = \partial/\partial x_j - 2i\lambda y_j$, $Y_j = \partial/\partial y_j + 2i\lambda x_j$, $X_j^R = \partial/\partial x_j + 2i\lambda y_j$, $Y_j^R = \partial/\partial y_j - 2i\lambda x_j$, $Z_j = \partial/\partial z_j + \lambda \bar{z}_j$, $\tilde{Z}_j = \partial/\partial \bar{z}_j - \lambda z_j$, $Z_j^R = \partial/\partial z_j - \lambda \bar{z}_j$, $\tilde{Z}_j^R = \partial/\partial \bar{z}_j + \lambda z_j$, $L_o = (1/2)\sum (Z_j \tilde{Z}_j + \tilde{Z}_j Z_j)$, $R_o = (-1/2)\sum (Z_j^R \tilde{Z}_j^R + \tilde{Z}_j^R Z_j^R)$. One has: if $F \in S(\mathbb{C}^n)$, $G(Z_j F) = (GF)W_j^+$ (when tested on the E_α); $G(Z_j^R F) : H \to \mathcal{D}(W_j)$ and equals $W_j^+ GF$; etc. Also $G|_{S(\mathbb{C}^n)}$ extends to a constant multiple of a unitary operator from $L^2(\mathbb{C}^n)$ onto $I_2 = \{$operators R on $H \mid \|R\|_2^2 < \infty\}$, $\|R\|_2$ being the Hilbert-Schmidt norm. In fact, if $F \in L^2(\mathbb{C}^n)$, $\|F\|_2^2 = (2|\lambda|/\pi)^n \|G_\lambda F\|_2^2$.

Let $S(H_\lambda) = \{R \in \mathcal{B}(H_\lambda) \mid$ for all N there exists C_N with $|(RE_\alpha, E_\beta)| < C_N(|\alpha|+1)^{-N}(|\beta|+1)^{-N}\}$. Then $G(S(\mathbb{C}^n)) = S(H)$. We sketch the simple proof of this found in [12], Lemma 7.9. Say first $G \in S(\mathbb{C}^n)$; then $G(L_o^N R_o^N G) = A^N (GG) A^N$ on the E_α; this is a bounded operator. Thus, for some C_N, $|((GG)A^N E_\alpha, A^N E_\beta)| < C_N$ for all α, β. This, however says that $GG \in S(H)$. Suppose, conversely, that $R_1 \in S(H)$. It is easy to see that $S(H) \subset I_2$, so there exists $G_1 \in L^2$ with $G(G_1) = R_1$. Further, it is not difficult to show that $R_1 W_j$, $R_1 W_j^+$, defined initially on the algebraic span of the E_α, extend to $S(H)$ operators; that $R_1 : H \to D(W_j)$, and $W_j R_1$, $W_j^+ R_1 \in S(H)$; and that $G(Z_j G) = R_1 W_j^+$, $G(\tilde{Z}_j G_1) = -R_1 W_j$, $G(\hat{\tilde{Z}}_j^R G_1) = W_j^+ R_1$, $G(Z_j^R G_1) = -W_j R_1$, where the derivatives (Z_j, etc.) are taken in the S' sense. Thus, if $GG \in S(H)$, then $G(Z_j G)$, $G(\tilde{Z}_j G)$, $G(Z_j^R G)$, $G(\hat{\tilde{Z}}_j^R G) \in S(H)$. However, $X_j = (Z_j + \tilde{Z}_j)$, $X_j^R = (Z_j^R + \hat{\tilde{Z}}_j^R)$, $\partial/\partial x_j = (X_j + X_j^R)/2$, $y_j = (X_j^R - X_j)/4i\lambda$. Thus $G(G) \in S(H) \Rightarrow G(\partial G/\partial x_j)$, $G(x_j G)$, $G(\partial G/\partial y_j)$, $G(y_j G) \in S(H)$. Iterating this we see that if $GG \in S(H)$ then $G(DpG) \in S(H)$ for any differential polynomial D and any polynomial p. In particular, $DpG_1 \in L^2$ for all D, p, so that by Sobolev's lemma, $G_1 \in S$, as desired.

An alternate proof that $G(S(\mathbb{C}^n)) = S(H)$ could be given on the basis of (4.4). In the notation of (4.4), it is apparent that $F \in S(\mathbb{C}^n)$ if and only if $K \in S(\mathbb{R}^{2n})$. But $K \in S(\mathbb{R}^{2n})$ if and only if it has a Hermite expansion $K(\xi,\eta) = \sum a_{\alpha\beta} E_\alpha(\xi) E_\beta(\eta)$ with $|a_{\alpha\beta}| < C_N(|\alpha|+1)^{-N}(|\beta|+1)^{-N}$ for all N. (This standard fact is proved in much the same way as in the first proof we gave that $G(S) = S$.) This is the same as saying that the operator with kernel K is in $S(H)$.

The first new result of this section is the analogue of the preceding for Z_2^2. Let $Z_2^2(H_\lambda) = \{S \in B(H_\lambda) \mid$ for some

$C > 0$, $0 < r < 1$, $|(SE_\alpha, E_\beta)| < Cr^{|\alpha|+|\beta|}\}$. We think of $\mathbb{C}^n$ as $\mathbb{R}^{2n}$ and consider $G(Z_2^2(\mathbb{R}^{2n}))$ where $Z_2^2(\mathbb{R}^{2n})$ is the space of functions which are restrictions to $\mathbb{R}^{2n}$ of Z_2^2 functions on $\mathbb{C}^{2n}$.

Proposition 4.2. $G(Z_2^2(\mathbb{R}^{2n})) = Z_2^2(H_\lambda)$.

Proof. Again one could prove this by using the Schrödinger representation and (4.4). Notation as in (4.4), note first that, using Theorem 2.3(c)$_2$, one can easily see that $F \in Z_2^2 \Longleftrightarrow F_2F \in Z_2^2$. Using the definition of Z_2^2 and the parallelogram law, we see that $F_2F \in Z_2^2 \Longleftrightarrow K \in Z_2^2$. But, by Theorem 2.3(e), $K \in Z_2^2$ if and only if it has a Hermite expansion $K(p,p') = \sum a_{\alpha\beta}E_\alpha(p)E_\beta(p')$ with $|a_{\alpha\beta}| < Cr^{|\alpha|+|\beta|}$ for some $C > 0$, $0 < r < 1$. This is the same as saying that the operator with kernel K is in $Z_2^2(H)$.

Alternatively, one can give a proof in the abstract setting by adapting the proof of Theorem 2.3(c)$_2 \Longleftrightarrow$ (e) as follows. Let $U = \{2\lambda x_j, 2\lambda y_j, \partial/\partial x_j, \partial/\partial y_j \mid 1 \leq j \leq n\}$, $V = \{Z_j, \tilde{Z}_j, Z_j^R, \tilde{Z}_j^R \mid 1 \leq j \leq n\}$. Say $G \in S(\mathbb{R}^{2n})$, and $GG = S$. Any element ν of V can be written as $\nu = \sum_{j=1}^{4} \sigma_j \chi_j$ where $\chi_j \in U$ and $\sigma_j \in \mathbb{C}$, $|\sigma_j| = 1/2$; further each element χ of V can be written as $\chi = \sum_{j=1}^{4} \sigma_j' \nu_j$ where $\nu_j \in V$ and $\sigma_j' \in \mathbb{C}$, $|\sigma_j'| = 1/2$. Thus $G \in Z_2^2$ if and only if there exist $C_0, R > 0$ so that if $\chi_1, \dots, \chi_N \in U$ then $\|\chi_1 \cdots \chi_N G\|_2 < C_0 R^N N!^{1/2}$, if and only if there exist C_2, $R_2 > 0$

so that if $\nu_1,\ldots,\nu_n \in V$ then $\|\nu_1\cdots\nu_N G\|_2 < C_2R_2^N N!^{1/2}$. (Further, if C_0,R are known, we can take $C_2 = C_0$, $R_2 = 2R_0$; if C_2,R_2 are known, we can take $C_0 = C_2$, $R_0 = 2R_2$.) By Plancherel, this is in turn equivalent to the statement that if $T = \{W_j,W_j^+ | 1 \leqq j \leqq n\}$ and if $\omega_1,\ldots,\omega_N \in T$ and k is arbitrary with $1 \leqq k \leqq n$, then

$$\|\omega_1\cdots\omega_k S\omega_{k+1}\cdots\omega_N\|_2 < C_2'R_2^N N!^{1/2}. \tag{4.5}$$

Here $\|\ \ \|_2$ is the Hilbert-Schmidt norm and $C_2' = (\pi/2|\lambda|)^{n/2}C_2$. Now if (4.5) holds, then for all $K,L \in \mathbb{Z}^+$, $\|A^KSA^L\|_2 < C_2'R_3^{K+L}K!L!$ where $R_3 = 4nR_2^2$, since $A = (-1/2)\sum(W_kW_k^+ + W_k^+W_k)$. Thus for all α,β,K,L, $[((|\alpha|+n/2)|\lambda|)^K/R_3^KK!][((|\beta|+n/2)|\lambda|)^L/R_3^LL!] \times |(SE_\alpha,E_\beta)| < 2^{-K-L}C_2'$. Summing over K,L we find $|(SE_\alpha,E_\beta)| < C_2'e^{-(|\alpha|+|\beta|)|\lambda|/R_3}$, so $S \in Z_2^2(H)$. Conversely, suppose $S \in Z_2^2(H)$, so that $|(SE_\alpha,E_\beta)| < C_1r^{|\alpha|+|\beta|}$ for some C_1, $0 < r < 1$; we check (4.5). Now if $\omega_1,\ldots,\omega_K$, $\omega_1',\ldots,\omega_L' \in T$,

$$\|\omega_1\cdots\omega_K S\omega_1'\cdots\omega_L'\|_2^2 = \sum|(S\omega_1'\cdots\omega_L'E_\alpha,\omega_K^+\cdots\omega_1^+E_\beta)|^2$$

$$\leqq C_1^2r^{-2(K+L)}(2|\lambda|)^{K+L}\sum_{\alpha,\beta}(|\alpha|+1)\ldots(|\alpha|+L+1)(|\beta|+1)\ldots(|\beta|+K+1)r^{2(|\alpha|+|\beta|)}$$

$$\leqq C_1^2R_4^{K+L}[\sum_{N=0}^{\infty}(N+1)^n(N+1)\ldots(N+L+1)r^{2N}][\sum_{N=0}^{\infty}(N+1)^n(N+1)\ldots(N+K+1)r^{2N}]$$

where $R_4 = 2|\lambda|r^{-2}$, since $\#\{\alpha||\alpha|=N\} \leqq (N+1)^n$. Now if $r^2 < r_1 < 1$, $(N+1)^nr^{2N} \leqq C_3r_1^N$ for some $C_3 > 0$. Accordingly $\|\omega_1\cdots\omega_kS\omega_1'\cdots\omega_L'\|_2^2 \leqq (C_1C_3)^2R_4^{K+L}(1-r_1)^{-(K+L)}K!L!$; this gives (4.5), so $G \in Z_2^2$. ∎

The proof of course also shows that for any $S \in Z_2^2(H)$ there exist $C_2', R_2 > 0$ so that (4.5) holds whenever $\omega_1, \ldots, \omega_N \in T$, $1 \leq k \leq N$. In addition, the proof of the Proposition, together with the uniformities in Theorem 2.3 (a) $\Leftrightarrow$ (c)$_2$, show the following uniformities, for λ fixed.

(1) For any $B_1, B_2 > 0$, there exist $C_3, R_2 > 0$ as follows: if $F \in Z_2^2(\mathbb{R}^{2n})$, $S = G_\lambda F$, and in fact $F \in Z_2^2(B_1, B_2, C_o)$, then (4.5) holds for all $\omega_1, \ldots, \omega_N$, with $C_2' = C_3 C_o$.

(2) Conversely, for any $R_2 > 0$, there exist $C_4, B_1, B_2 > 0$ as follows: if $F \in Z_2^2(\mathbb{R}^{2n})$, $S = G_\lambda F$, and (4.5) holds for all $\omega_1, \ldots, \omega_N$, then $F \in Z_2^2(B_1, B_2, C_4 C_2')$.

(3) For any $B_1, B_2 > 0$, there exist $C, r > 0$ as follows: if $F \in Z_2^2(\mathbb{R}^{2n})$, $S = G_\lambda F$, and in fact $F \in Z_2^2(B_1, B_2, C_o)$, then $|(SE_\alpha, E_\beta)| < CC_o r^{|\alpha|+|\beta|}$ for all α, β.

(4) Conversely, for any $C, r > 0$, there exist $B_1, B_2, C_2 > 0$ as follows: if $F \in Z_2^2(\mathbb{R}^{2n})$, $S = G_\lambda F$, and if $|(SE_\alpha, E_\beta)| < Cr^{|\alpha|+|\beta|}$ for all α, β, then $F \in Z_2^2(B_1, B_2, C_2 C)$.

Of course, then, $W_\lambda(S(\mathbb{R}^{2n})) = S(H)$ and $W_\lambda(Z_2^2(\mathbb{R}^{2n})) = Z_2^2(H)$. It is possible also to characterize $(S(\mathbb{H}^n))^\wedge$ and $(K^k)^\wedge$; these are much more difficult than any of the preceding characterizations, and are carried out in [25] and [26]. Fortunately, to prove Theorem 4.1 we shall not need these results, but only certain basic properties of $(K^k)^\wedge$, given next.

If $F \in \mathcal{S}'(\mathbb{H}^n)$ and $\underset{\sim}{R}$ is a family of bounded operators, with $\|R(\lambda)\|$ uniformly bounded for λ in any compact subset $\mathbb{R}\setminus\{0\}$, we say that $\hat{F} = \underset{\sim}{R}$ in the sense of tempered distributions if for every $f \in \mathcal{S}(\mathbb{H}^n)$, $(f|F) = c_n(\hat{f}|\underset{\sim}{R})$. Here $(f|F) = F(\bar{f})$ and $(\hat{f}|\underset{\sim}{R}) = \int_{-\infty}^{\infty} \sum_{\alpha} (\hat{f}(\lambda)E_\alpha, R(\lambda)E_\alpha)(2|\lambda|)^n d\lambda$. (We require that the right side of this last equation make sense for all $f \in \mathcal{S}(\mathbb{H}^n)$.) Thus, by Plancherel, if $F \in L^2(\mathbb{H}^n)$, then $\hat{F}$ is the same when taken in the L^2 sense as in the sense of tempered distributions. (The boundedness assumptions on $\underset{\sim}{R}$ guarantee that $\underset{\sim}{R}$ is uniquely determined by F, because of the existence of $\mathcal{Q}$.)

Of particular interest is $[PV(\mathbb{H}^n)]^\wedge$, where $PV(\mathbb{H}^n) = K^{-2n-2}$, the space of principal value distributions. The following fact is basic:

Proposition 4.3. Suppose $K \in PV(\mathbb{H}^n)$. Then there exists $\underset{\sim}{J} \in \mathcal{B}$ such that $\hat{K} = \underset{\sim}{J}$. Further, if λ_1 and λ_2 have the same sign, $J(\lambda_1) = J(\lambda_2)$ (identifying $\mathcal{H}_{\lambda_1}$ and $\mathcal{H}_{\lambda_2}$). Thus $\underset{\sim}{J}$ is completely determined by two bounded operators, $J_+ = J(1)$ and $J_- = J(-1)$. If $f \in L^2$, we have $(f*K)^\wedge = \hat{f}\hat{K}$. Further, $\max(\|J_+\|, \|J_-\|) = C$ where C is the norm of the operator $B : L^2 \to L^2$ defined $B(f) = f*K$. ∎

A proof of this is given in [12], Proposition 7.4, by use of the fact that B is bounded on L^2, in much the same way as the Euclidean analogue can be shown. The equality $\max(\|J_+\|, \|J_-\|) = C$ is not claimed in [12], but it follows from the proof given there. The next proposition gives more information about $[PV(\mathbb{H}^n)]^\wedge$; for a complete description of $[PV(\mathbb{H}^n)]^\wedge$, see [26].

We discuss the regularity properties of J_+ and J_-. For $M \in \mathbb{Z}^+$, let $H_\lambda^M = \{v = \sum v_\alpha E_\alpha \in H_\lambda : \sum (|\alpha|+1)^M |v_\alpha|^2 = \|v\|_M^2 < \infty\}$, $H_\lambda^\infty = \bigcap_{M \geq 0} H_\lambda^M$ (topologized as a Frechet space). In the Schrödinger representation, H^∞ is just $S(\mathbb{R}^n)$. We claim:

Proposition 4.4. Suppose $K \in PV(\mathbb{H}^n)$, $\hat{K} = \underset{\sim}{J}$. Then $J_+ : H^{2M} \to H^{2M}$, $J_- : H^{2M} \to H^{2M}$ continuously for $M \in \mathbb{Z}^+$.

Sketch of Proof. A simple proof of this is given in [12], Proposition 7.5. We sketch it here because we shall be needing part of the proof in a moment. Note that $R_0 = L_0 + iTD$ where D is a differential operator which is homogeneous of degree zero. In fact a computation shows that

$$D = -2\sum(z_j \partial/\partial z_j - \bar{z}_j \partial/\partial \bar{z}_j). \tag{4.6}$$

Further, TD commutes with L_0 and R_0, since L_0 and R_0 commute; thus D also commutes with L_0 and R_0. Accordingly, for any $M \in \mathbb{N}$, $R_0^M K = \sum_{m=0} \binom{M}{m} L_0^m (iT)^{M-m}(D^{M-m}K)$, and $D^{M-m}K \in PV(\mathbb{H}^n)$. Put $J_{M-m} = (D^{M-m}K)^\wedge$. Operating formally, we are led to believe that in some sense, $A^M J_+ = \sum_{m=0}^{M} \binom{M}{m} J_{M-m,+} A^m$, where $A = A_1$.

This is true if interpreted as follows:

$$\text{If } v \in H^{2M},\ J_+v = A^{-M} \sum_{m=0}^{M} \binom{M}{m} J_{M-m,+} A^m v \tag{4.7}$$

This is justified in [12]. (4.7) shows $J_+v \in H^{2M}$, and a similar equation for J_- completes the proof. ■

We need a refinement of Proposition 4.4. On H^M, we place the equivalent norm $|v|_M = \max\|\omega_1 \cdots \omega_M v\|$ where the max is taken over all possible choices of $\omega_1, \ldots, \omega_M \in \{W_j, W_j^+ | 1 \leqq j \leqq n\}$. We put $H_\lambda^\omega = \{v \in H^\infty|$ for some $C, R > 0$, $|v|_M < CR^M M!^{1/2}$ for all $M\} = \{v \in H^\infty|$ for some $C_1 > 0$, $0 < r < 1$, $|v_\alpha| < C_1 r^{|\alpha|}$ for all $\alpha\}$. The equality of these spaces is seen at once in the Schrödinger representation, where the spaces are just $Z_2^2(\mathbb{R}^n)$. Alternatively one could work abstractly and copy the proof of Theorem 2.3(c)$_2$<==>(e). Observe also that $H_\lambda^\omega = \{v \in H^\infty|$ for some $C_2, R_2 > 0$, $\|A^M v\| < C_2 R_2 M!$ for all $M\}$. Indeed, it is evident that H^ω contains this space; the equality follows from the proof of Theorem 2.3(c)$_2$ =>(e). Observe that the proof of Theorem 2.3(c)$_2$<==>(e) shows that for all $R_2 > 0$ there exist $C_1 > 0$, $0 < r < 1$ so that for all $v \in H^\omega$ with $\|A^M v\| < R_2^M M!$ (for all M) we have $|v_\alpha| < C_1 r^{|\alpha|}$ for all α; further for all $0 < r < 1$ there exist $C, R_2 > 0$ so that for all $v \in H^\omega$ with $|v_\alpha| < r^{|\alpha|}$ (for all α) we have $\|A^M v\| < CR_2^M M!$ (for all M).

Now let $APV(\mathbb{H}^n) = A\mathcal{K}^{-2n-2}$. We then have the following key lemma.

<u>Lemma 4.5</u>. Suppose $K \in APV(\mathbb{H}^n)$, $\hat{K} = \underset{\sim}{J}$. Then $J_+ : H^\omega \to H^\omega$, $J_- : H^\omega \to H^\omega$. Further, for any $0 < r < 1$, there exist $C > 0$, $0 < r < 1$ as follows: for all $v \in H^\omega$ with $|v_\alpha| < r^{|\alpha|}$ (for all α) we have $|(J_+v)_\alpha| < Cr_1^{|\alpha|}$ (for all α); similarly for J_-.

<u>Remark</u>. A proof of this was sketched in [26] (see page 510, Proposition 5.3(b), and the definitions on pages 502 and 522). We give a simpler direct proof here, which is closely related to the proof in [26].

<u>Proof</u>. This is evident if $K = \delta$, so we can assume $K = P.V.(f)$ where f is a real analytic function on $\mathbb{H}^n\setminus\{0\}$, is homogeneous of degree -2n - 2 and has mean value zero. We maintain all the notation of the proof of Proposition 4.4. We begin by asserting:

$$\text{for all } m \in \mathbb{N},\ D^m P.V.(f) = P.V.(D^m f). \tag{4.8}$$

To show this, we can evidently assume m = 1. Now $D(P.V.f) = P.V.(Df) + c\delta$ for some $c \in \mathbb{C}$; we need to show c = 0. To see this it suffices to show that both sides of (4.8) give the same result when tested on those $\varphi \in C_c^\infty(\mathbb{H}^n)$ with support in the unit ball $\{u | |u| < 1\}$ (here $|(t,z)| = (t^2+|z|^4)^{1/4}$.) Now $[DP.V.(F)](\varphi) = [-P.V.(f)](D\varphi) = \lim_{\varepsilon\to 0} \int_{\varepsilon<|u|<1} (D\varphi)(u)f(u)dV(u)$, while $[P.V(Df)](\varphi) = \lim_{\varepsilon\to 0} \int_{\varepsilon<|u|<1} \varphi(u)(Df)(u)dV(u)$. Thus it suffices to show that if $g \in C^\infty(\mathbb{H}^n\setminus\{0\})$, $\varepsilon > 0$, and $U = \{\varepsilon<|u|<1\}$, then $\int_U (Dg)(u)dV(u) = 0$. But $\int_U Dg = 2i\int_U [(d/d\theta)g(t,e^{i\theta}z)]|_{\theta=0} dtdzd\bar{z} = 2i[(d/d\theta)[\int_U g(t,e^{i\theta}z)dtdzd\bar{z}]|_{\theta=0}] = 0$ as desired, since $\int_U g(t,e^{i\theta}z)dtdzd\bar{z}$ is independent of θ.

Using (4.8) we can estimate $\|J_{m,+}\|$ as an element of $\mathcal{B}(H)$. In fact we claim that for some $C_o, R_o > 0$,

$$\|J_{m,+}\| < C_o R_o^m m! \tag{4.9}$$

Indeed, by Proposition 4.3, $\|J_{m,+}\|$ does not exceed the norm of the operator of convolution with $P.V.(D^m f)$, acting on L^2. By the proof of the L^2 boundedness of this operator (see, e.g. [65]), the norm of this operator does not exceed $C\|D^m f\|_{C^1}$ where the C^1 norm is taken over the unit sphere. (Precisely: if $g \in C^1(H^n\setminus\{0\})$, we put $\|g\|_{C^1} = \sup_{|u|=1,\partial\in S} |(\partial g)(u)|$ where $S = \{id, \partial/\partial t, \partial/\partial z_j, \partial/\partial \bar{z}_j | 1 \leq j \leq n\}$.) Thus, for $\partial \in S$, it suffices to bound $\|\partial D^m f\|$ by $C_1 R_o^m m!$. Here $\|\ \ \|$ is the sup norm on $\{u: |u| = 1\}$. Now $\|\partial D^m f\|$ is no greater than the sum of 2^m expressions of the form $\|\partial\zeta_1 D_1 \cdots \zeta_m D_m f\|$ where each $\zeta_\ell = z_j$ for some j, and each $D_\ell = \partial/\partial z_j$ or $\partial/\partial\bar{z}_j$ for some j. It suffices to bound any such expression by $C_1 R_1^m m!$. That such a bound exists follows at once from Lemma 2.5, as applied to the holomorphic extension of f into an open neighborhood of $\{u \in \mathbb{H}^n \mid |u| = 1\}$ in $\mathbb{C}^{2n+1}$.

Thus (4.9) holds, and the lemma is an immediate consequence. Indeed say $v \in H^\omega$, and $||A^m v|| < C_2 R_2^M M!$ for all M. Then by (4.7) $||A^M J_+ v|| \leq \sum_{m=0}^{M} \binom{M}{m} ||J_{M-m,+} A^m v|| \leq C_0 C_2 \sum_{m=0}^{M} R_0^{M-m} R_2^m (M-m)!m! \leq C_0 C_2 R_3^M (M+1)!$ where $R_3 = \max(R_0, R_2)$. Similarly for J_-. So $J_+, J_- : H^\omega \to H^\omega$. This, combined with the remarks made just before the lemma, complete the proof. ■

As we said, we do not seek to characterize $APV(\mathbb{H}^n)^\wedge$, and indeed, no such characterization is known. However it is virtually evident now that we can easily give an important subspace of $(APV)^\wedge$, as in the next proposition. (Recall the Fourier transforms $F'_{\mathbb{C}^n}, F', F'_\lambda$; if $f \in S(H^n)$, then $(F'f)(\lambda,\zeta) = (F'_\lambda f)(\zeta) = [F'_{\mathbb{C}^n}(F^\lambda_c f)](\zeta)$.)

Proposition 4.6 (a) Given operators $J_+ \in Z_2^2(H_1)$, $J_- \in Z_2^2(H_{-1})$, define $\underset{\sim}{J}$ by $J(\lambda) = J_+$ for $\lambda > 0$, J_- for $\lambda < 0$. Then there exists $K \in APV(\mathbb{H}^n)$ such that $\hat{K} = \underset{\sim}{J}$. Further, $F'_\lambda K \in Z_2^2(\mathbb{R}^{2n})$ for each $\lambda \neq 0$.

(b) Suppose $K \in APV(\mathbb{H}^n)$. Let $F_+ = F'_1 K$, $F_- = F'_{-1} K$. Assume $F_+, F_- \in Z_2^2(\mathbb{R}^{2n})$. Let $\hat{K} = \underset{\sim}{J}$. Then $J_+ \in Z_2^2(H_1)$, $J_- \in Z_2^2(H_{-1})$. In fact $J_+ = W_1 F_+$, $J_- = W_{-1} F_-$.

Proof. The $PV(\mathbb{H}^n)$ analogue of this is given in [12], Proposition 7.8; the proof here is just the same. Let us prove (a). Using Proposition 4.2, select $F_+, F_- \in Z_2^2(\mathbb{R}^{2n})$ such that

$\mathcal{W}_1 F_+ = J_+$, $\mathcal{W}_{-1} F_- = J_-$. For $\lambda > 0$, let $F_\lambda(\zeta) = F_+(\zeta/\lambda^{1/2})$; for $\lambda < 0$, let $F_\lambda(\zeta) = F_-(\zeta/|\lambda|^{1/2})$. $(\zeta \varepsilon \mathbb{C}^n = R^{2n}.)$ It is then elementary that $\mathcal{W}_\lambda F_\lambda = J(\lambda)$ for all $\lambda \neq 0$. Let $F(\lambda,\zeta) = F_\lambda(\zeta)$ for $\lambda \neq 0$. By Theorem 2.11, $F = \mathcal{F}'K$ for some $K \varepsilon$ APV. It suffices to show that $\hat{K} = \underset{\sim}{J}$. Suppose then $f \varepsilon \mathcal{S}(H^n)$, $g_\lambda(\zeta) = \mathcal{F}_\lambda' f$; then $g_\lambda \varepsilon \mathcal{S}(\mathbb{C}^n)$. We have, for appropriate c_n'', $(f|K) = c_n''(\mathcal{F}'f|F) = c_n''\int_{-\infty}^{\infty}(g_\lambda|F_\lambda)d\lambda = c_n\int_{-\infty}^{\infty}\sum((\mathcal{W}_\lambda g_\lambda)E_\alpha, (\mathcal{W}_\lambda F_\lambda)E_\alpha)(2|\lambda|)^n d\lambda = c_n(\hat{f}|\underset{\sim}{J})$. (We used the Plancherel theorem for $\mathcal{W}_\lambda$, and $(|)$ denotes the sesquilinear pairing.) This proves (a); the proof of (b) is essentially the same. ∎

We shall need only a little information about $(\mathcal{K}^k)^\wedge$ for more general k. In this chapter, we only look at cases in which k is an integer. The case of general k will be considered in Proposition 5.5. A complete characteristic of $(\mathcal{K}^k)^\wedge$ may be found in [26].

Let $\underset{\sim}{J}$ be a family of operators, $J_+ = J(1)$, $J_- = J(-1)$. We shall say that $\underset{\sim}{J}$ is homogeneous of degree $k \varepsilon \mathbb{C}$ if for $\lambda > 0$, $J(\lambda) = \lambda^{-k/2}J_+$, and for $\lambda < 0$, $J(\lambda) = |\lambda|^{-k/2}J_-$.

Proposition 4.7. Suppose $K \in \mathcal{K}^{k-2n-2}$ where $k \in 2\mathbb{Z}^+$. Then there exists a unique family of bounded operators $\underset{\sim}{J}$ such that $(f|K) = c_n(\hat{f}|\underset{\sim}{J})$ whenever $\hat{f} \in \mathcal{Q}$. Further, $\underset{\sim}{J}$ is homogeneous of degree $-k$, and $J_+, J_- : \mathcal{H}^{2M} \to \mathcal{H}^{2M+k}$ continuously for all $M \in \mathbb{Z}^+$.

In addition, if $K \in A\mathcal{K}^{k-2n-2}$, then $J_+, J_- : \mathcal{H}^{\omega} \to \mathcal{H}^{\omega}$. Further, for any $0 < r < 1$, there exist $C > 0$, $0 < r_1 < 1$ as follows: for all $v \in \mathcal{H}^{\omega}$ with $|v_\alpha| < r^{|\alpha|}$ for all α we have $|(J_+v)_\alpha| < Cr_1^{|\alpha|}$ (for all α); similarly for J_-.

Proof. Let $K_1 = \mathcal{L}_0^{k/2}K \in PV(\mathbb{H}^n)$, $\underset{\sim}{J}_1 = \hat{K}_1$, $\underset{\sim}{J} = \underset{\sim}{J}_1 A^{-k/2}$. We check that $\underset{\sim}{J}$ has the required properties.

Say $\hat{f} \in \mathcal{Q}$; then there exists $g \in \mathcal{Q}$ with $f = \mathcal{L}_0^{k/2}g$; in fact $\hat{g} = \hat{f}\underset{\sim}{A}^{-k/2}$. Thus $(f|K) = (\mathcal{L}_0^{k/2}g|K) = (g|K_1) = c_n(\hat{g}|\underset{\sim}{J}_1) = c_n(\hat{f}|\underset{\sim}{J})$ as desired, since A is diagonal.

$\underset{\sim}{J}$ is unique since the matrix of each $J(\lambda)$ (w.r.t. the $\{E_\alpha\}$) is uniquely determined by the condition $(f|K) = c_n(\hat{f}|\underset{\sim}{J})$ whenever $\hat{f} \in \mathcal{Q}$.

From $\underset{\sim}{J} = \underset{\sim}{J}_1 A^{-k/2}$ and Proposition 4.4, it is evident that $\underset{\sim}{J}$ is homogeneous of degree $-k$ and that $J_+, J_- : H^{2M} \to H^{2M+k}$ continuously. The remaining statements about the case when $K \in AK^{k-2n-2}$ follow at once from $K_1 \in \mathrm{APV}$, Lemma 4.5 and $\underset{\sim}{J} = \underset{\sim}{J}_1 \underset{\sim}{A}^{-k/2}$. ∎

The proposition has an extension to all $k \in \mathbb{C}$; see Proposition 5.5. For the time being, however, we shall content ourselves with one more simple observation and with a partial extension to the case $k \in \mathbb{Z}^+$. Note first that if $k \in \mathbb{C}$, J is homogeneous of degree k and $J_+, J_- : H^\infty \to H^\infty$, then we can still make sense of $(\underset{\sim}{R}|\underset{\sim}{J}) = \int_{-\infty}^{\infty} \sum_\alpha (R(\lambda)E_\alpha, J(\lambda)E_\alpha)(2|\lambda|)^n d\lambda$ for all $\underset{\sim}{R} \in \mathcal{Q}$. We can also define the families of operators $\underset{\sim}{J}\underset{\sim}{W}_j$, $\underset{\sim}{W}_j\underset{\sim}{J}$; these are homogeneous of degree $k + 1$.

<u>Proposition 4.8</u>. Suppose $K \in K^{k-2n-2}$ where $k \in \mathbb{C}$. (a) Suppose that there exists a family of operators $\underset{\sim}{J}$ which is homogeneous of degree $-k$, for which $J_+, J_- : H^\infty \to H^\infty$, and such that $(f|K) = c_n(\hat{f}|\underset{\sim}{J})$ whenever $\hat{f} \in \mathcal{Q}$. Then, whenever $\hat{f} \in \mathcal{Q}$, $(f|Z_jK) = -c_n(\hat{f}|\underset{\sim}{J}\underset{\sim}{W}_j^+)$, $(f|\overline{Z}_jK) = c_n(\hat{f}|\underset{\sim}{J}\underset{\sim}{W}_j)$, $(f|Z_j^RK) = -c_n(\hat{f}|\underset{\sim}{W}_j^+\underset{\sim}{J})$ and $(f|\overline{Z}_j^RK) = c_n(\hat{f}|\underset{\sim}{W}_j\underset{\sim}{J})$.

(b) Suppose $k \in \mathbb{Z}^+$. Then there does exist $\underset{\sim}{J}$ as in (a). Further, we can choose J_λ to be defined on all of H, be bounded, and satisfy $J_+, J_- : H^{2M} \to H^{2M+k}$ continuously for all $M \in \mathbb{Z}^+$.

<u>Proof</u> (a) We prove only the first and third conclusions; the others are similar. If $\hat{f} \in \mathcal{Q}$, then $(\overline{Z}_jf)^\wedge \in \mathcal{Q}$, so we have

$(f|Z_j K) = -(\bar{Z}_j f|K) = -c_n \int_{-\infty}^{\infty} \sum (\hat{f}(\lambda) W_j E_\alpha, J(\lambda) E_\alpha)(2|\lambda|)^n d\lambda =$

$-c_n \int_{-\infty}^{\infty} \sum (\hat{f}(\lambda) E_{\alpha-(\operatorname{sgn}\lambda)e_j}, J(\lambda) W_j^+ E_{\alpha-(\operatorname{sgn}\lambda)e_j})(2|\lambda|)^n d\lambda = -c_n(\hat{f}|\underset{\sim}{J}\underset{\sim}{W}_j^+)$. We used (4.2); here $E_\beta = 0$ if $\beta \notin (\mathbb{Z}^+)^n$. Also $(f|Z_j^R K) = -(\bar{Z}_j^R f|K) = -c_n \int_{-\infty}^{\infty} \sum (W_j \hat{f}(\lambda) E_\alpha, J(\lambda) E_\alpha)(2|\lambda|)^n d\lambda = -c_n(\hat{f}|\underset{\sim}{W}_j^+ \underset{\sim}{J})$ as desired.

(b) Observe first that if p is a polynomial and $f \in \mathcal{Q}$ then $(f|p) = 0$. Indeed, as we observed in the proof of Proposition 4.7, for any $N \in \mathbb{Z}^+$ there exists g_N with $\hat{g}_N \in \mathcal{Q}$ so that $f = \mathcal{L}_o^N g$. Thus $(f|p) = (g_N|\mathcal{L}_o^N p)$. If N is sufficiently large, $\mathcal{L}_o^N p = 0$, as desired. (Indeed, if p were a monomial, $\mathcal{L}_o^N p$ would have a negative degree of homogeneity, but be a polynomial.)

Secondly, suppose $j \in \mathbb{C}$, $K_o \in \mathcal{K}^j$. We claim: there exist $K_1,\ldots,K_n,\tilde{K}_1,\ldots,\tilde{K}_n \in \mathcal{K}^{j+1}$ and a polynomial p so that

$$K_o = \sum Z_\ell^R K_\ell + \sum \bar{Z}_\ell^R \tilde{K}_\ell + p. \tag{4.10}$$

Indeed, first observe that, by Proposition 1.2, there exist $K_{11},\ldots,K_{n1} \in \mathcal{K}^{j+1}$, $K_{n+1} \in \mathcal{K}^{j+2}$ and a polynomial p so that $K_o = \sum (\partial/\partial z_\ell) K_{\ell 1} + TK_{n+1} + p$. We have only to use the Euclidean Fourier transform $\mathcal{F}$; say that (λ,ζ) are the dual variables to (t,z). We have only to choose $K_{\ell 1}, K_{n+1}$ with $\mathcal{F}K_{\ell 1} = 2i\zeta_\ell|\zeta|^2 \times (|\zeta|^4+\lambda^2)^{-1}\mathcal{F}K_o$, $\mathcal{F}K_{n+1} = i\lambda(|\zeta|^4+\lambda^2)^{-1}\mathcal{F}K_o$ away from 0. Since $\partial/\partial z_\ell = Z_\ell^R + iTz_\ell$, it is apparent that $K_o = \sum Z_\ell^R K_{\ell 1} + TK_{n+2} + p$ for some $K_{n+2} \in \mathcal{K}^{j+2}$. Since $T = (-i/2)[Z_1^R, \bar{Z}_1^R]$, (4.10) is evident.

Because of Proposition 4.7, we need only prove (b) for k odd. This is now immediate from (4.10), the preceding comments about $(f|p)$, and Proposition 4.7. ∎

We need one more simple proposition before we prove Theorem 4.1.

<u>Proposition 4.9</u>. Suppose $R,S \in Z_2^2(H_\lambda)$ for some λ. Then:

(a) $RS \in Z_2^2(H)$.

(b) $R : H \to H^\omega$. In fact, there exist $C_o > 0$, $0 < r_o < 1$ so that whenever $\|v\| \leq 1$, $w = Rv$, we have $|w_\alpha| < C_o r_o^{|\alpha|}$ for all α.

<u>Proof</u>. Select $C_1, C_2 > 0$ and $0 < r_1, r_2 < 1$ so that $|(RE_\alpha, E_\beta)| < C_1 r_1^{|\alpha|+|\beta|}$ and $|(SE_\alpha, E_\beta)| < C_2 r_2^{|\alpha|+|\beta|}$ for all α, β. For (a), note $|(RSE_\alpha, E_\beta)| = |(SE_\alpha, R^* E_\beta)| \leq \sum_\gamma |(SE_\alpha, E_\gamma)||(E_\gamma, R^* E_\beta)| < C_1 C_2 r_1^{|\alpha|} r_2^{|\beta|} \sum_\gamma (r_1 r_2)^{|\gamma|} = C_3 r_1^{|\alpha|} r_2^{|\beta|}$ for some $C_3 > 0$, so $RS \in Z_2^2(H)$. For (b), say $\|v\| < 1$, $w = Rv$. Then $|w_\alpha| \leq \sum_\beta |v_\beta||(E_\alpha, RE_\beta)| \leq \|v\|(\sum_\beta |(E_\alpha, RE_\beta)|^2)^{1/2} < C_1 r_1^{|\alpha|} (\sum r^{2|\beta|})^{1/2} = C_4 r_1^{|\alpha|}$ for some $C_4 > 0$, as desired. ∎

Without further ado, we pass to the proof of Theorem 4.1.

<u>Proof of Theorem 4.1</u>. We write $L = p(Z,\overline{Z})$ where p is a non-commuting polynomial. Also we write $L^t = p^t(Z,\overline{Z})$, $\overline{L} = \overline{p(Z,\overline{Z})} = \overline{p}(Z,\overline{Z})$, $L^* = \overline{L}^t = \overline{p^t(Z,\overline{Z})} = p^*(Z,\overline{Z})$,

$L^R = p^t(-Z^R,-\overline{Z}^R)$, $\overline{(L^R)^t} = \overline{p}(-Z^R,-\overline{Z}^R)$. Let $\underset{\sim}{R} = p^t(\underset{\sim}{W}^+,-\underset{\sim}{W})$, $\underset{\sim}{R}^* = \overline{p}(\underset{\sim}{W}^+,-\underset{\sim}{W})$. (We just take this as the definition of $\underset{\sim}{R}^*$, even though in fact R^*_λ is indeed the adjoint of the unbounded operator R_λ. Note, however, that if $v,w \in H^\infty_\lambda$, then $(R_\lambda v,w) = (v,R^*_\lambda w)$.). Then if $f \in S$, $(Lf)^\wedge = \hat{f}\underset{\sim}{R}$, $(L^*f)^\wedge = \hat{f}\underset{\sim}{R}^*$, $(L^R f)^\wedge = \underset{\sim}{R}\hat{f}$, $(L^{R^*}f)^\wedge = \underset{\sim}{R}^*\hat{f}$.

By Lemma 3.5, there exist $K_1 \in AK^{k-2n-2}$ and $K_2 \in APV$ such that $L^*K_1 = \delta - K_2$ and such that $F_{2\lambda} = F'_\lambda K_2 \in Z_2^2(\mathbb{R}^{2n})$ for all $\lambda \neq 0$. Let $F_{2,+} = F_{2,1}$, $F_{2,-} = F_{2,-1}$. Let $\underset{\sim}{J}_2 = \hat{K}_2$, $J_{2,+} = J_2(1)$, $J_{2,-} = J_2(-1)$. By Proposition 4.6, $J_{2,\pm} = W_{\pm 1}F_{2,\pm} \in Z_2^2(H_{\pm 1})$. Next, using Proposition 4.8, select $\underset{\sim}{J}_1$ homogeneous of degree $-k$, with $J_{1\pm} : H^{2M} \to H^{2M+k}$ for all $M \in \mathbb{Z}^+$, such that $(f|K_1) = c_n(\hat{f}|J_1)$ whenever $\hat{f} \in Q$. Applying Proposition 4.8(a) repeatedly we see that whenever $\hat{f} \in Q$, $c_n(f|I-\underset{\sim}{J}_2) = (f|\delta-K_2) = (f|L^*K_1) = c_n(\hat{f}|\underset{\sim}{J}_1\underset{\sim}{R}^*)$. It follows that for all α,λ, $(I-J_2(\lambda))E_\alpha = J_1(\lambda)R(\lambda)^*E_\alpha$. But $R(\lambda)^* : H^k \to H^0$ boundedly, and $J_1(\lambda) : H^0 \to H^k$ boundedly; hence

$$J_1(\lambda)R(\lambda)^*v = (I-J_2(\lambda))v \quad \text{for all } v \in H^k. \qquad (4.11)$$

Again using Lemma 3.5, we select $K_3 \in AK^{k-2n-2}$ and $K_4 \in APV$ such that $L^{R^*}K_3 = \delta - K_4$, and such that $\hat{K}_4 = \underset{\sim}{J}_4$, $J_{4,\pm} \in Z_2^2(H_{\pm 1})$. We also select $\underset{\sim}{J}_3$ homogeneous of degree $-k$, with $J_{3,\pm} : H^{2M} \to H^{2M+k}$ for all $M \in \mathbb{Z}^+$, such that $(f|K_3) = c_n(\hat{f}|\underset{\sim}{J}_3)$ whenever $\hat{f} \in Q$. Applying Proposition 4.8(a) repeatedly we see that whenever $\hat{f} \in \mathcal{Q}$, $c_n(\hat{f}|\, I-\underset{\sim}{J}_4) = (f|\,\delta-K_4) = (f|\,L^R\, K_3) = c_n(\hat{f}|\,\underset{\sim}{R}^*\underset{\sim}{J}_3)$. Since $\underset{\sim}{R}^*\underset{\sim}{J}_3 \in \mathcal{B}$, the uniqueness assertion of Proposition 4.7 shows $\underset{\sim}{R}^*\underset{\sim}{J}_3 = I - \underset{\sim}{J}_4 = (L^R\, K_3)\hat{}$. Thus, for all λ,

$$R(\lambda)^* J_3(\lambda) = (I-J_4(\lambda))v \quad \text{for all } v \in \mathcal{H}. \qquad (4.12)$$

By Proposition 4.9(b), $J_2(\lambda) : \mathcal{H}^k \to \mathcal{H}^k$ and $J_4(\lambda) : \mathcal{H} \to \mathcal{H}$ are compact. Also, $J_1(\lambda)$, $J_3(\lambda) : \mathcal{H} \to \mathcal{H}^k$ boundedly. By Atkinson's theorem ([16]), then, $R(\lambda)^* : \mathcal{H}^k \to \mathcal{H}$ is Fredholm; that is, it has closed range and finite dimensional kernel and cokernel. Indeed, the kernel and cokernel are contained in $\mathcal{H}^\omega$. For if $u \in \mathcal{H}^k$, $R(\lambda)^* u = 0$, then $[I-J_2(\lambda)]u = J_1(\lambda)R(\lambda)^* u = 0$, so $u = J_2(\lambda)u \in \mathcal{H}^\omega$ by Proposition 4.9(b). Further if $v \in \mathcal{H}$, $v \in [R(\lambda)^*\mathcal{H}^k]^\perp$, then $v \in \mathcal{H}^\omega$. It suffices to show $(I-J_4(\lambda)^*)v = 0$, or that $((I-J_4(\lambda))w,v) = 0$ for all $w \in \mathcal{H}$. But this follows from (4.12). Also note $R(\lambda)v = 0$; indeed, since $v \in \mathcal{H}^\omega$, $(E_\alpha, R(\lambda)v) = (R(\lambda)^* E_\alpha, v) = 0$ for all α. Thus, in fact, $[R(\lambda)^*\mathcal{H}^k]^\perp = \{v \in \mathcal{H}^\omega | R(\lambda)v = 0\}$.

Now let $Q(\lambda)$ denote the projeciton in $\mathcal{H}_\lambda$ onto $[R(\lambda)^*\mathcal{H}^k]^\perp$, the cokernel of $R(\lambda)^*$. Since the cokernel is contained in $\mathcal{H}^\omega$, by examining the matrix of $Q(\lambda)$ we see that $Q(\lambda) \in Z_2^2(\mathcal{H}_\lambda)$. Indeed, if $\{v_j\}$ is an orthonormal basis for $Q(\lambda)\mathcal{H}$, we have an estimate $|(Q(\lambda)E_\alpha, E_\beta)| \leq \sum |(E_\alpha, v_j)||(v_j, E_\beta)| < Cr^{|\alpha|+|\beta|}$ for some $C > 0$, $0 < r < 1$. Note $\underset{\sim}{Q}$ is homogeneous of degree 0, so by Proposition 4.6, there exists $P \in APV$ such that $\hat{P} = \underset{\sim}{Q}$.

Let $Q_+ = Q(1)$, $Q_- = Q(-1)$. Note $R(\pm 1)^* J_{3,\pm} = (I-Q_\pm)R(\pm 1)^* J_{3,\pm}$, so

$$\begin{aligned} (I-Q_\pm)(I-J_{4\pm}) &= I - J_{4\pm}, \text{ and} \\ R(\pm 1)^* J_3(\pm 1) &= (I-Q_\pm) - (I-Q_\pm)J_{4\pm}. \end{aligned} \tag{4.13}$$

To see the significance of all this, let us first verify that if $B : L^2 \to L^2$ is given by $Bf = f*P$, then B is the projection in $L^2(\mathbb{H}^n)$ onto a subspace of $[L\mathcal{S}]^\perp$. B is idempotent, since $[(f*P)*P]^\wedge = (f*P)^\wedge \underset{\sim}{Q} = \hat{f}\underset{\sim}{Q}\underset{\sim}{Q} = \hat{f}\underset{\sim}{Q} = (f*P)^\wedge$ by Proposition 4.3. B is self-adjoint, since if $f,g \in L^2$, then $(f*P|g) = c_n(\hat{f}\underset{\sim}{Q}|\hat{g}) = c_n\int_{-\infty}^{\infty} \mathrm{tr}(Q(\lambda)\hat{f}(\lambda)^* g(\lambda))(2|\lambda|)^n d\lambda = c_n\int_{-\infty}^{\infty} \mathrm{tr}(\hat{f}(\lambda)^* g(\lambda)Q(\lambda))(2|\lambda|)^n d\lambda = (f|g*P)$. Thus B is a projeciton. Further, $L\mathcal{S} \subset (I-B)\mathcal{H}$. This is the same as saying that if $f \in \mathcal{S}$, then $Lf*P = 0$. To see this, observe that $(Lf*P)^\wedge = (Lf)^\wedge \underset{\sim}{Q}$; also $(Lf)^\wedge(\lambda)E_\alpha = \hat{f}(\lambda)R(\lambda)E_\alpha$ for all α. Since $\hat{f}$, $(Lf)^\wedge \in \mathcal{B}$, and since $R(\lambda) : \mathcal{H}^\infty \to \mathcal{H}^\infty$ boundedly, $(Lf)^\wedge(\lambda)v = \hat{f}(\lambda)R(\lambda)v$ for all $v \in \mathcal{H}^\infty$. Thus $(Lf*P)^\wedge = \hat{f}\underset{\sim}{R}\underset{\sim}{Q} = 0$ as claimed. (Note, as a consequence, that $L^R P = 0$, since $Lf*P = f*L^R P$ for $f \in \mathcal{S}$.) We see, thus, that B projects onto a subspace of $[L\mathcal{S}]^\perp$. In fact, B projects precisely onto $[L\mathcal{S}]^\perp$; this is easy to believe, but will be verified only at the end of the proof.

Let us speak heuristically for a moment and motivate the rest of the proof. In order to prove Theorem 4.1 we are seeking K so that $LK = \delta - P$. If $(f|K) = c_n(\hat{f}|\underset{\sim}{J})$ whenever $\hat{f} \in \mathcal{Q}$, we expect that $J(\lambda)R(\lambda) = I - Q(\lambda)$, or that $R(\lambda)^*J(\lambda)^* = I - Q(\lambda)$. We are trying to find $\underset{\sim}{J}$, and in light of (4.13) it seems reasonable that the theorem would follow if we knew the following:

$$\text{There exist } H_{\pm} \in Z_2^2(H_{\pm 1}) \text{ such that } R(\pm 1)^* H_{\pm} = (I-Q_{\pm})J_{4\pm} \qquad (4.14)$$

Returning now to the formal proof, we shall establish (4.14) in case k is even, and show how the theorem follows from it. (4.14) is indeed the main point.

Assume, then, that k is even. We construct only $H = H_+$; the construction of H_- is similar. We construct H by constructing its matrix columns H^α, where $H^\alpha = HE_\alpha$. We abbreviate $S = (I-Q_+)J_{4+}$; denote the columns of S by S^α, so that $S^\alpha = SE_\alpha$. We would like $R(1)^*H^\alpha = S^\alpha$. Since $S^\alpha \in (I-Q_+)H$ and since $R(1)^*$ has closed range, we can and do select $H^\alpha \in H^k$ so that $R(1)^*H^\alpha = S^\alpha$ and so that for some $C_o > 0$, $\|H^\alpha\|_k < C_o|S^\alpha|_o$ for all α. We need only show that there exists $H \in Z_2^2(H_1)$ so that $HE_\alpha = H^\alpha$. Let $H_\beta^\alpha = (E_\beta, H^\alpha)$; it suffices to show that there exist $C_1 > 0$, $0 < r_1 < 1$ so that $|H_\alpha^\alpha| < C_1 r_1^{|\alpha| + |\beta|}$. Now, $S \in Z_2^2(H)$ by Proposition 4.9(a). Let $S_\beta^\alpha = (E_\beta, SE_\alpha)$; there exist $C_2 > 0$, $0 < r_2 < 1$ so that $|S_\beta^\alpha| < C_2 r_2^{|\alpha|+|\beta|}$. As a consequence there exists $C_3 > 0$ so that $\|S^\alpha\|_o < C_3 r_2^{|\alpha|}$. Now, by (4.11),

$$H^\alpha = (I-J_2(1))H^\alpha + J_2(1)H^\alpha = J_1(1)S^\alpha + J_2(1)H^\alpha. \qquad (4.15)$$

Now $K_1 \in AK^{k-2n-2}$, and k is even, so by Proposition 4.7, $J_1(1) : H^\omega \to H^\omega$. Further, there exists $C_4 > 0$, $0 < r_4 < 1$ so that $|(E_\beta, J_1(1)S^\alpha)| < C_2 r_2^{|\alpha|} C_4 r_4^{|\beta|}$. In addition, $\|H^\alpha\|_0$ $\|H^\alpha\|_k < C_0 \|S^\alpha\|_0 < C_0 C_3 r_2^{|\alpha|}$. Also $J_2(1) \in Z_2^2(H)$, so by Proposition 4.9(b) there exist $C_5 > 0$, $0 < r_5 < 1$ so that $|(E_\beta, J_2(1)H^\alpha)| < C_5 r_5^{|\beta|} C_0 C_3 r_2^{|\alpha|}$. Thus, indeed, by (4.15), for some $C_1 > 0$, $0 < r_1 < 1$, we have $|H_\beta^\alpha| < C_1 r_1^{|\alpha|+|\beta|}$ as claimed, and (4.14) follows if k is even.

Still assume k is even, and define $\underset{\sim}{H}$ by $H(\lambda) = \lambda^{-k/2} H_+$ if $\lambda > 0$, $H(\lambda) = |\lambda|^{-k/2} H_-$ if $\lambda < 0$. Then $\underset{\sim}{H}$ is homogeneous of degree -k, and, by (4.14), $\underset{\sim}{R}^* \underset{\sim}{H} = (I-\underset{\sim}{Q})\underset{\sim}{J}_4$. Select $G_{\pm} \in Z_2^2(\mathbb{R}^{2n})$ so that $W_{\pm 1} G_{\pm} = H_{\pm}$. Let $G_\lambda(\zeta) = \lambda^{-k/2} G_+(\zeta/\lambda^{1/2})$ if $\lambda > 0$, $G_\lambda(\zeta) = |\lambda|^{-k/2} G_-(\zeta/|\lambda|^{1/2})$ if $\lambda < 0$; then it is elementary that $W_\lambda G_\lambda = H(\lambda)$ for all λ. Let $G(\lambda,\zeta) = G_\lambda(\zeta)$. By Theorem 2.10 there exists $K' \in AK^{k-2n-2}$ so that $F'K' = G$. Then $(f|K') = c_n(\hat{f}|\underset{\sim}{H})$ whenever $\hat{f} \in Q$, since for appropriate c_n'', $(f|K') = c_n''(F'f|G) = c_n'' \int_{-\infty}^{\infty} (F_\lambda' f | G_\lambda) d\lambda = c_n(\hat{f}|\underset{\sim}{H})$ just as in the proof of Proposition 4.6(a). Applying Proposition 4.8(a), we see that whenever $\hat{f} \in Q$, $(f|L^{R*}K') = c_n(\hat{f}|\underset{\sim}{R}^*\underset{\sim}{H}) = c_n(\hat{f}|(I-\underset{\sim}{Q})\underset{\sim}{J}_4)$. Now $L^{R*}K' \in$ APV, so by the uniqueness assertion of Proposition 4.7, $(L^{R*}K')^\wedge = (I-\underset{\sim}{Q})\underset{\sim}{J}_4$. As we argued before (4.12), however, $(L^{R*}K_3)^\wedge = \underset{\sim}{R}^*\underset{\sim}{J}_3 = I - \underset{\sim}{J}_4$; by (4.13), $I - \underset{\sim}{J}_4 = (I-\underset{\sim}{Q})(I-\underset{\sim}{J}_4)$; so $[L^{R*}(K'+K_3)]^\wedge = I - \underset{\sim}{Q}$. If $K'' = K' + K_3$, then $K'' \in AK^{k-2n-2}$ and $L^{R*}K'' = \delta - P$. Now consider the map $\sim: S \to S$ given by

$\tilde{f}(u) = \overline{f(u^{-1})}$; extend $\sim$ to S'. Then if $g \in S'$, $(L^{R*}g)^{\sim} = L\tilde{g}$. Further, if $K_o \in PV$, the adjoint of the operator on L^2 given by $f \to f*K_o$ is the operator given by $f \to f*\tilde{K}_o$. Since B is self-adjoint (recall $Bf = f*P$ for $f \in L^2$), $\tilde{P} = P$. Let $K = \tilde{K}'' \in AK^{k-2n-2}$. Then $LK = \delta - P$. Thus, if k is even, there exists $K \in AK^{k-2n-2}$ with $LK = \delta - P$.

If $n = 1$, k may be odd. In this case, however, we can still find $K'' \in AK^{2k-2n-2}$ with $LL^*K'' = \delta - P'$, where if $\hat{P}' = \underset{\sim}{Q}'$, then $Q'(\lambda)$ is the projection in H_λ onto $\{v \in H^{\omega} | R(\lambda)^*R(\lambda)v = 0\}$. This latter space equals $\{v \in H^{\omega} | R(\lambda)v = 0\}$, so $Q'(\lambda) = Q(\lambda)$ and $P' = P$. If we set $K = L^*K''$ then again $K \in AK^{k-2n-2}$ and $LK = \delta - P$.

Using the facts that $LK = \delta - P$ and $L^R P = 0$, one can now verify the local solvability assertion of Theorem 4.1. See [26], page 549, for the argument, which is the same as that which Greiner-Kohn-Stein [35] gave for the particular case $L = L_n$. (On page 549 of [26], our P is called P_1. Lemma 4.6(c) of [26] is used to show that if $f \in L^2 + E'$ and if $f = 0$ on an open set $U \subset \mathbb{H}^n$, then $f*P$ is real analytic on U. This, however, is easily seen without Lemma 4.6(c). On pages 549-551, it is also shown that, if $P \neq 0$, there exists $f \in C^\infty \cap L^2(\mathbb{H}^n)$ such that f is not real analytic near 0, but $f = f*P$; thus, if $P \neq 0$, $Lu = f$ is not solvable near 0.)

To complete the proof, then, it suffices to show that $I - B$ projects $L^2(\mathbb{H}^n)$ onto $\overline{LS}$. (Recall $Bf = f*P$ for $f \in L^2$, and we have already shown $LS \subset (I-B)L^2(\mathbb{H}^n)$.) We must show LS is dense in $(I-B)L^2(\mathbb{H}^n)$. Let $\check{\mathcal{Q}} = \{f | \hat{f} \in \mathcal{Q}\}$. It suffices to show:

(I) $(I-B)\check{\mathcal{Q}}$ is dense in $(I-B)L^2$

(II) $(I-B)\check{\mathcal{Q}} \subset LS$.

(I) is evident, since $\check{\mathcal{Q}}$ is dense in L^2 and $(I-B) : L^2 \to L^2$ continuously. For (II), say $h \in (I-B)\check{\mathcal{Q}}$ so that $h = f*(\delta-P)$ for some $f \in \check{\mathcal{Q}}$. It suffices to show that $f*K \in S$, for then $h = f*(\delta-P) = L(f*K) \in LS$. To show $f*K \in S$, we proceed as follows. As we observed in the proof of Proposition 4.7, for each $N \in \mathbb{Z}^+$, $f = L_o^N f_N$ for some f_N with $\hat{f}_N \in \mathcal{Q}$. Thus $f*K = f_N * K_N$, where $K_N = R_o^N K \in K^{k-2N-2n-2}$. Now $\partial/\partial x_j = (X_j + X_j^R)/2$, $\partial/\partial y_j = (Y_j + Y_j^R)/2$; we thus see that if D is any differential monomial in the $\partial/\partial x_j$, $\partial/\partial y_j$ and T, and if $k - 2N - 2n - 2 = -M < 0$, then $Dg(u) = 0(1+|u|)^{-M}$. Since M is arbitrary, $Dg \in S$; this completes the proof. ∎

The following corollary was known (see [81], [77], [78], [62] for the equivalence of (a), (b), (d) and (e), and [29], Theorem 3 for the equivalence with (c).)

<u>Corollary 4.10</u>. Suppose L is a left-invariant homogeneous differential operator on $\mathbb{H}^n$ which is homogeneous of degree k.

Define $\underset{\sim}{R}$ by $(Lf)^{\wedge} = \hat{f}\underset{\sim}{R}$, for $f \in S$. Then the following are equivalent:

(a) L^* is hypoelliptic

(b) L^* is analytic hypoelliptic

(c) There exists $K \in AK^{k-2n-2}$ so that $LK = \delta$

(d) $I_C L$ is elliptic; $R(1)v = 0$, $v \in H^\infty \Rightarrow v = 0$; $R(-1)v = 0$, $v \in H^\infty \Rightarrow v = 0$.

(e) $I_C L$ is elliptic; $R(1)v = 0$, $v \in H^\omega \Rightarrow v = 0$; $R(-1)v = 0$, $v \in H^\omega \Rightarrow v = 0$.

Proof. It is evident that (c) $\Rightarrow$ (a) and (b). Next, if (a) or (b) holds, observe that $I_C L$ is elliptic (examine functions of z alone.) Thus we can select K,P as in Theorem 4.1. Since $f \to f*P$ is the projection in L^2 onto $(LS)^\perp$, $L^*P = 0$, so P is smooth, so P = 0. Thus (a) $\Rightarrow$ (c) and (b) $\Rightarrow$ (c), so (a) $\Longleftrightarrow$ (b) $\Longleftrightarrow$ (c). If any of these hold, then $I_C L$ is elliptic and P = 0, so (d) and (e) follow. Finally, if we have (d) or (e), we can again choose K,P as in Theorem 4.1, and by hypothesis P = 0, so (c) follows.

Remarks. 1. The main point of the proof of Theorem 4.1, besides Lemma 3.5, was (4.14). The proof of (4.14) shows that if $v \in (I-Q_{\pm})H \cap H^\omega$, then there exists $w \in H^\omega$ so that $R(\pm 1)^* w = v$. This was shown in [26], and also in [62]; in the latter, however, it was assumed $v = 0$. (By the way, the equivalence of the definitions of H^ω was not realized until the present book. [26] and [62] used different definitions.) The method in [62] is that of Lemma 3.8. Indeed, in the Schrödinger representation, $R(\pm 1)^*$ is a differential operator which satisfies the hypotheses of Proposition 3.7 with $p = 2$.

Note also that the estimate (3.10) for $L = R(\pm 1)^*$ is an immediate consequence of our methods, specifically (4.11).

2. If L satisfies the hypotheses of Theorem 4.1, then it also satisfies the conclusion of Proposition 3.7. Indeed (a) $\Rightarrow$ (b), since if (a) holds, LL^* is hypoelliptic and analytic hypoelliptic by Corollary 4.10, so we may use the argument previously used for Proposition 3.7(a) $\Rightarrow$ (b). (b) $\Rightarrow$ (c) $\Rightarrow$ (d) exactly as in the proof of Proposition 3.7. Next, for (d) $\Rightarrow$ (a), assume (a) fails. Then by Corollary 4.10, if K,P are as in Theorem 4.1, we have $P \neq 0$. Since $\hat{P}(\lambda) \in Z_2^2(H_\lambda)$ for all λ, $F_\lambda P \in Z_2^2(\mathbb{R}^{2n})$ for $\lambda \neq 0$ by Proposition 4.6. But $L^*P = 0$, so $\hat{L}_\lambda^* F_\lambda P = 0$. Since $F_\lambda P \neq 0$ for $\lambda = 1$ or -1, (d) fails. Thus (a) $\Rightarrow$ (b) $\Rightarrow$ (c) $\Rightarrow$ (d) $\Rightarrow$ (a) ; similarly (a)' $\Rightarrow$ (b)' $\Rightarrow$ (c)' $\Rightarrow$ (d)' $\Rightarrow$ (a)'; finally (a) $\Longleftrightarrow$ (a)' by Corollary 4.10.

5. Homogeneous Singular Integral Operators on the Heisenberg Group

Corollary 4.10 gives a necessary and sufficient condition for a homogeneous left-invariant differential operator L* on $\mathbb{H}^n$ to be analytic hypoelliptic. What is the same thing, if L has degree k, it gives a necessary and sufficient condition for there to exist $K_2 \in AK^{k-2n-2}$ such that $LK_2 = \delta$. Let $K_1 = L\delta$; then $K_1 \in AK^{-k-2n-2}$, and we have the precise conditions under which there exists $K_2 \in AK^{k-2n-2}$ with $K_2 * K_1 = \delta$ (since $LK_2 = L(K_2 * \delta) = K_2 * L\delta$). In this chapter we answer the more general question: if $k \in \mathbb{C}$, $K_1 \in AK^{-k-2n-2}$, when does there exist $K_2 \in AK^{k-2n-2}$ with $K_2 * K_1 = \delta$? (The meaning of this convolution will be discussed presently.) We shall also show that the existence of such a K_2 is the same as the analytic hypoellipticity of the operator $f \to f * \tilde{K}_1$ (recall $\tilde{}: S \to S$ by $\tilde{f}(u) = \overline{f(u^{-1})}$ and is extended to a map from S' to S'.) Corollary 4.10 answers this question precisely when K_1 is supported at 0, since it is easy to see that any homogeneous distribution supported at 0 has the form $L\delta$ for some homogeneous left-invariant L. (The analogous question on $\mathbb{R}^n$ is easily answered; K_2 exists if and only if $\hat{K}_1$ does not vanish away from 0, i.e. if and only if convolution with K_1 is an elliptic pseudo-differential operator. In fact, by Corollary 1.4, it is necessary and sufficient that $\hat{K}_2(\xi) = \hat{K}_1(\xi)^{-1}$. All of this follows from techniques we shall see soon.)

The theorem on inverting singular integral operators, to be proved in this section, will be a crucial ingredient in the theory of analytic pseudodifferential operators presented in Chapters 7 and 8.

We state the main theorem now. Not all the terms have been defined precisely yet, but we shall do this immediately after stating the theorem.

Theorem 5.1. Suppose $K_1 \in AK^{k-2n-2}$ for some $k \in \mathbb{C}$. Define $C_k : E'(\mathbb{H}^n) \to D'(\mathbb{H}^n)$ by $C_{K_1}(f) = K_1 * f$. Let $\hat{K}_1 = \underset{\sim}{J}_1$ in the Q sense. Then the following are equivalent:

(a) C_{K_1} is hypoelliptic.

(b) C_{K_1} is analytic hypoelliptic.

(c) There exists $K_2 \in AK^{-k-2n-2}$ so that $K_2 * K_1 = \delta$.

(d) $(FK_1)(0,\xi) \neq 0$ for $\xi \neq 0$; $J_{1+}v = 0$, $v \in H^\infty \Rightarrow v = 0$; and $J_{1-}v = 0$, $v \in H^\infty \Rightarrow v = 0$.

(e) $(FK_1)(0,\xi) \neq 0$ for $\xi \neq 0$; $J_{1+}v = 0$, $v \in H^\omega \Rightarrow v = 0$; and $J_{1-}v = 0$, $v \in H^\omega \Rightarrow v = 0$. ■

The equivalence of (a), (c) and (d) in the C^∞ situation will also follow from our methods; see also [12] for this.

We proceed to explain our terms in detail. First we explain what is meant by a convolution $K_2 * K_1$.

Proposition 5.2. Suppose $K_1 \in K^j$, $K_2 \in K^k$, $\mathrm{Re}(j+k) < -(2n+2)$.

(a) Write $K_1 = K_1' + K_1''$, where $K_1' \in E'$, the distributions of compact support, and $K_1'' \in C^\infty$. Then for $f \in S(\mathbb{H}^n)$, $Qf = (f*K_2)*K_1' + (f*K_2)*K_1''$ exists and is independent of the choice of K_1', K_1''.

(b) There exists $K \in K^{j+k+2n+2}$ (independent of f) with $Qf = f*K$. We write $K = K_2*K_1$.

(c) Similarly, we can form $Q^R f = K_2*(K_1*f)$; this then equals $K*f$.

(d) If R is right invariant, $R(K_2*K_1) = RK_2*K_1$. If L is left invariant, then $L(K_2*K_1) = K_2*LK_1$.

(e) Suppose $j = k = -(2n+2)$, so that $K_1, K_2 \in PV(\mathbb{H}^n)$. For $f \in L^2$, let $K_1 f = f*K_1 \in L^2$, $K_2 f = f*K_2 \in L^2$. Then if $f \in L^2$, $f*K = K_1 K_2 f$.

(f) If $K_1 \in AK^j$ and $K_2 \in AK^k$ then $K_2*K_1 \in AK^{j+k+2n+2}$.

(g) Suppose also $K_3 \in K^\ell$, $\mathrm{Re}(k+\ell) < -(2n+2)$, $\mathrm{Re}(j+k+\ell) < -2(2n+2)$. Then $(K_3*K_2)*K_1 = K_3*(K_2*K_1)$.

(h) Select $\phi \in C_c^\infty(\mathbb{H}^n)$ with $\phi(t,z) = 1$ for $|(t,z)| < 1$. Define ϕ_ε by $\phi_\varepsilon(u) = \phi(D_\varepsilon u)$. Still suppose $K_1 \in K^j$, $K_2 \in K^k$, but do not assume $\mathrm{Re}(j+k) < -2n - 2$ necessarily. Let $K_{1\varepsilon} = \phi_\varepsilon K_1 \in E'$, $K_{2\varepsilon} = \phi_\varepsilon K_2 \in E'$. Let $\mathcal{D}_1, \mathcal{D}_2$ be homogeneous differential operators with polynomial coefficients, of homogeneity degrees d_1, d_2 respectively, and suppose $\mathrm{Re}(j+k) - d_1 - d_2 < -2n - 2$. Then, in the sense of distributions,

$$\lim_{\varepsilon \to 0} \mathcal{D}_2 K_{2\varepsilon} * \mathcal{D}_1 K_{1\varepsilon} = \mathcal{D}_2 K_2 * \mathcal{D}_1 K_1. \quad ■$$

Most of these facts were proved in [26], Theorem 4.4, or in [12], Lemma 9.5; we take the latter as our reference since it is easier to read. The basic idea is to also split $K_2 = K_2' + K_2''$ where $K_2' \in \mathcal{E}'$, $K_2'' \in C^\infty$, and to put

$$K = K_2'*K_1' + K_2'*K_1'' + K_2''*K_1' + K_2''*K_1'' \tag{5.1}$$

This is easily seen to exist and satisfy $Qf = f*K$. The arguments for (a)-(e) in [12] were carried out only for $K_1 \in \mathrm{Rhom}_j$, $K_2 \in \mathrm{Rhom}_k$, j,k real, in which case it is shown that $K \in \mathrm{Rhom}_{j+k+2n+2}$. The proofs in the case $j,k \in \mathbb{C}$ are the same. In the more general case $K_1 \in K^j$, $K_2 \in K^k$, the proofs of (a)-(e) go through as before, but they only show $K = K_2*K_1$ is C^∞ away from 0. However, it is very easy to see from (5.1) that if $K_3 \in K^j$, $K_4 \in K^k$, $\mathrm{Re}(j+k) - 1 < -2n-2$, then $Z_j K_4 * K_3 = K_4 * Z_j^R K_3$. Now suppose $K_2 \in K^k$, $\mathrm{Re}\, k > 0$, $K_1 \in K^j$, $\mathrm{Re}(j+k) < 2n - 2$. Using (4.10), write $K = \sum Z_\ell^R K_{5,\ell} + \sum \overline{Z}_\ell^R K_{6,\ell}$ where $K_{5,\ell}, K_{6,\ell} \in K^{j+1}$. Then $K_2*K_1 = \sum Z_\ell K_2 * K_{5,\ell} + \sum \overline{Z}_\ell K_1 * K_{6,\ell}$ and each $Z_\ell K_2$, $\overline{Z}_\ell K_2 \in K^{k-1}$. Continuing in this way we see that we can write K_2*K_1 as a sum of terms of the form K_7*K_8 where $K_7 \in K^a$, $K_8 \in K^b$, $\mathrm{Re}\, a$, $\mathrm{Re}\, b < 0$, $a + b = j + k$. Thus

$K_2 * K_1 \in K^{j+k+2n+2}$ as claimed. (a)-(e) also hold for $\mathbb{C}^n$ in place of $\mathbb{H}^n$ and 2n in place of 2n + 2; the proofs are the same or easier. (g) and (h) were not proved in [12]; we include the proofs in an appendix to this chapter (in part A.1). We also prove there that if $K_1 \in K^j(\mathbb{R}^{2n})$, $K_2 \in K^k(\mathbb{R}^{2n})$ and Re j, Re k, Re(j+k) + 2n < 0, then $F_{\mathbb{R}^{2n}}(K_1 * K_2) = (F_{\mathbb{R}^{2n}} K_1)(F_{\mathbb{R}^{2n}} K_2)$.

Proposition 5.2(f) is a simple consequence of the following lemma which will also be used later.

Lemma 5.3. Suppose $K \in AK^j$ for some $j \in \mathbb{C}$, and $U \subset \mathbb{H}^n$ is open. Suppose that one of (a), (b) or (c) holds:

(a) f is a distribution of compact support which vanishes on U;

(b) $f \in L^1_{loc}(\mathbb{H}^n)$, f vanishes on U, and for some $\varepsilon > 0$, $f(u) = O(1+|u|)^{-2n-2-\text{Re } j-\varepsilon}$ for $u \in \mathbb{H}^n$;

(c) $f \in C_c^\infty$ and f is real analytic on U.

Then $f*K$ and $K*f$ are real analytic on U. (In case (b), we define for $u \in U$, $f*K(u) = \int f(uv^{-1})K(v)\,dv$; this clearly exists for $u \in U$ under the hypotheses on f.)

Remark. Proposition 5.2(f) follows at once from Lemma 5.3, since if U is open and $\overline{U}$ does not contain 0, then we can write K_2 as a sum of 3 distributions f each of type (a), (b) or (c).

Proof of Lemma 5.3. Say $p \in U$; we show $f*K$ is analytic near p; by replacing f by a translate we can and do assume $p = 0$ and U is a neighborhood of 0. Define $J, \tau_v : C_c^\infty \to C_c^\infty$ by $(JF)(u) = F(u^{-1})$, $(\tau_v F)(u) = F(uv)$; extend $J, \tau_v : \mathcal{D}' \to \mathcal{D}'$; put $g = Jf$. Pick U_1 open with $0 \in U_1$, $\bar{U}_1 \subset U$. In (a) or (b), we can and do regard f as an element of $\mathcal{D}'(U_1^c)$. Note $(f*K)(u) = g(\tau_u K)$ for $u \in \mathbb{H}^n$ close to 0. It is then easy to use this formula to extend $f*K(u)$ to a holomorphic function of $u \in \mathbb{C}^{2n+1}$, u close to 0; this proves the theorem for $f*K$, in cases (a) and (b).

(c) is more subtle. For $f*K$, the case $-2n - 2 < \text{Re } j < -n - 1$ was proved in [26], Theorem 4.6(a). Since the proof of Lemma 4.5 and Theorem 4.6(a) in [26] together give a self-contained proof of this case, we do not reprove it here. (Note, however, a difference of notation in [26]; what we refer to as AK^k is there called ${}^{\omega}K^{-k/2}$.) In particular the theorem holds for $f*K$ in case (c), if $-2n - 2 < \text{Re } j < -2n$. To do the same for general j, observe first that L_o is analytic hypoelliptic by Theorem 4.1 (or the Folland-Stein construction of the fundamental solution of L_o). It suffices, then, to prove $f*L_o^r K$ is analytic on U for some fixed $r \in \mathbb{Z}^+$. We can, then, assume $\text{Re } j < -2n$. Further, by Theorem 3.2 and the analytic hypoellipticity of L_o, we can, for any $s \in \mathbb{Z}^+$, write $K = L_o^s K_s$ for some $K_s \in AK^{k+2s}$. Replacing K by K_s for some s, if necessary, we can assume $-2n - 2 \leq j < -2n$, and, by what we know already, that $j = -2n - 2$. Write $K = L_o K_1$, for some $K_1 \in AK^{-2n}$. Observe that $f*K = -(1/2)\sum[Z_j(f*\overline{Z}_j K_1) + \overline{Z}_j(f*Z_j K_1)]$; this is real analytic near 0 by a known case, since $Z_j K_1, \overline{Z}_j K_1 \in AK^{-2n-1}$. This proves that $f*K$ is analytic on U in cases (a), (b) and (c). For $K*f$, just note $K*f = (\tilde{f}*\tilde{K})^{\sim}$, where $\sim$ is as in the first paragraph of this chapter. ■

The reader who does not wish to refer to [26] for the above proof may wait until Chapter 7. In Lemma 7.6 and Proposition 7.7 we will give a self-contained proof of Lemma 5.3, together with uniformities. Uniformities for Proposition 5.2(f) will then be discussed in Lemma 7.8.

Next we explain the analogue of the condition, in Corollary 4.10, that $I_c L$ be elliptic. Let $F, F_{\mathbb{R}^{2n}}$ denote the Euclidean Fourier transforms on $\mathbb{H}^n$, $\mathbb{R}^{2n}$ respectively; then if L is left invariant, we have the relation $F(L\delta)(0,\xi) = F_{\mathbb{R}^{2n}}((I_c L)\delta)(\xi)$. This follows from the representation $L = I_c L + TL'$ for a differential operator L' which satisfies $F(TL'\delta)(0,\xi) = 0$. In analogy, if $K \in K^k(\mathbb{H}^n)$, $\text{Re } k < -2$, we define $I_c K \in K^{k+2}(\mathbb{R}^{2n})$ as follows. Note $F(\xi) = (FK)(0,\xi) \in K^{-2n-2-k}(\mathbb{R}^{2n})$, so we can define $I_c K$ by $F_{\mathbb{R}^{2n}}(I_c K)(\xi) = (FK)(0,\xi)$. In Theorem 5.1, one would then like to require that $F_{\mathbb{R}^{2n}}(I_c K_1)(\xi) \neq 0$ away from 0, or equivalently, that convolution with $I_c K_1$ be an elliptic pseudodifferential operator on $\mathbb{R}^{2n}$. This does not make sense if Re $j \geqq -2$ (here $K_1 \in K^j$). In this case, though, we can still require that $(FK_1)(0,\xi) \neq 0$ for $\xi \neq 0$. This explains the condition in Theorem 5.1(d) and (e).

We can place the map I_c in a somewhat more formal setting, as follows. Suppose $f \in S'(\mathbb{H}^n)$ satisfies:

(I) $Ff \in L^1_{loc}(\mathbb{R}^{2n+1})$

(II) $g(\xi) = Ff(0,\xi) \in L^1_{loc}(\mathbb{R}^{2n})$

(III) Ff has at worst polynomial growth at infinity.

Then we can define $I_c f \in S'(\mathbb{R}^{2n})$ by the rule $F_{\mathbb{R}^{2n}}(I_c f)(\xi) = (Ff)(0,\xi)$. Important special cases when this is possible are (identifying $\mathbb{C}^n = \mathbb{R}^{2n}$):

1. $f \in L^1(\mathbb{H}^n)$; then $I_c f \in L^1(\mathbb{C}^n)$. In fact $(I_c f)(z) = \int_{-\infty}^{\infty} f(t,z)\,dt$.

2. $f \in K^k(\mathbb{H}^n)$, Re $k < -2$. This case was discussed above. If also Re $k > -2n-2$, then indeed $(I_c K)(z) = \int_{-\infty}^{\infty} K(t,z)\,dt\,(z\neq 0)$; see the remark after Proposition 5.4 below.

3. $f \in E'(\mathbb{H}^n)$. Then $I_c f \in E'(\mathbb{C}^n)$, by the Paley-Wiener theorem.

Let us discuss $I_c f$ in more detail in the case $f \in E'$. First note that if $h \in L^1(\mathbb{H}^n)$, $g \in S(\mathbb{C}^n)$ and $G(t,z) = g(z)$, then $\int_{\mathbb{H}^n} hG\,dV = \int_{\mathbb{C}^n} (I_c h) g\,dV$. Thus it is natural to think that if $f \in E'(\mathbb{H}^n)$, $g \in S(\mathbb{C}^n)$, then $(I_c f)(g) = f(G)$ where

$G(t,z) = g(z)$. To see this, simply choose $\varphi \in C_c^\infty(\mathbb{H}^n)$, $\int\varphi = 1$, and let $\varphi_\varepsilon(u) = \varepsilon^{-(2n+2)}\varphi(D_{1/\varepsilon}u)$. Let $F(\zeta) = (\mathcal{F}f)(0,\zeta)$, $H_\varepsilon(\zeta) = (\mathcal{F}\varphi_\varepsilon)(0,\zeta)$. Then $f(G) = \lim_{\varepsilon\to 0} f*\varphi_\varepsilon(G) = \lim_{\varepsilon\to 0} I_c(f*\varphi_\varepsilon)(g) = \lim_{\varepsilon\to 0} \mathcal{F}^{-1}_{\mathbb{C}^n}(FH_\varepsilon)(g) = \mathcal{F}^{-1}_{\mathbb{C}^n}F(g) = I_c f(g)$ as claimed. From the formula, it is evident that if $f \in \mathcal{E}'$, supp $f \subset \{(t,z) \mid |z| < R\}$, then also supp $I_c f \subset \{z \mid |z| < R\}$. From this it follows easily that if $f \in \mathcal{E}'$ then $(I_c f)(g) = f(G)$ where $G(t,z) = g(z)$ whenever $g \in C^\infty$ (not just $\mathcal{S}$). Indeed, we can write such a $g = g_1 + g_2$ where $g_1 \in C_c^\infty$ and supp $g_2 \subset \{z \mid |z| > R\}$, R as before, and from this the desired statement is evident. It follows at once that if $f \in \mathcal{E}' + L^1$ and g is a bounded C^∞ function on $\mathbb{C}^n$ then for all t,z,

$$(I_c f*g)(z) = (f*G)(t,z), \text{ where } G(t,z) = g(z) \text{ for all } t,z. \tag{5.2}$$

In addition, let us observe the fact that if $f,g \in L^1$, then $I_c(f*g) = I_c f*I_c g$. This is computed at once. The following proposition, then, is entirely natural.

Proposition 5.4. (a) If $f \in \mathcal{S}(\mathbb{H}^n)$, $K \in \mathcal{K}^k$ (Re $k<-2$), then $\int_{-\infty}^{\infty} (f*K)(t,z)\,dt = (I_c f*I_c K)(z)$ for all $z \in \mathbb{C}^n$.

(b) If $K_1 \in \mathcal{K}^j$, $K_2 \in \mathcal{K}^k$ and Re j, Re k, Re$(j+k) + 2n + 2 < -2$, then $I_c(K_2*K_1) = I_c K_2*I_c K_1$.

This was proved in Proposition 7.2 of [12] in case $j,k \in \mathbb{R}$. The proof of the case $j,k \in \mathbb{C}$ is just the same. We do not repeat the proof here. Fubini shows easily that, if $-2n - 2 <$ $\mathrm{Re}\, k < -2$, then $\int_{-\infty}^{\infty} (f*K)(t,z)dt = (I_c f*J)(z)$ where $J(z) = \int_{-\infty}^{\infty} K(t,z)dt$ $(z\neq 0)$. Since every element of $S(\mathbb{C}^n)$ can be written as $I_c f$ for some $f \in S(\mathbb{H}^n)$ (e.g. $f(t,z) = e^{-\pi t^2} g(z)$), we must have $I_c K = J$, as asserted in point II above.

We must still discuss $\hat{K}$ for $K \in K^k$. If $\mathrm{Re}\, k \geq 0$ then $\hat{K}$ is not actually a family of operators, just as FK is not a function. However, we can still form $\hat{K}$ "in the Q sense" much as in Proposition 4.7 or 4.8. The proposition which follows makes this precise. First, for $k \in \mathbb{R}$, we let H_λ^k denote the Hilbert space of all sequences $v = (v_\alpha)_{\alpha\in(\mathbb{Z}^+)^n}$ so that $\sum(|\alpha|+1)^k |v_\alpha|^2 = \|v\|_k^2 < \infty$. We write E_α for the sequence w with $w_\beta = \delta_{\alpha\beta}$; then any $v \in H^k$ can be written as $v = \sum v_\alpha E_\alpha$ in H^k. If $k \in \mathbb{C}$, we let $H^k = H^{\mathrm{Re}\, k}$.

<u>Proposition 5.5</u>. (a) Suppose $K \in K^{k-2n-2}$ wehre $k \in \mathbb{C}$. Then there exists a unique family of operators $\underset{\sim}{J}$ which is homogeneous of degree $-k$, for which J_+, J_- have domain H^∞, map H^∞ to H^∞ boundedly (recall H^∞ is a Frechet space) and such that $(f|K) = c_n(\hat{f}|\underset{\sim}{J})$ whenever $\hat{f} \in Q$. J_+ and J_- have extensions to bounded operators from H^{2M} to H^{2m+k} for any fixed $M \in \mathbb{Z}^+$, or from H^{-k} to H^0. We say $\hat{K} = \underset{\sim}{J}$ in the Q sense. Further, if $K \in AK^{k-2n-2}$, then $J_+, J_- : H^\omega \to H^\omega$. Further, in this case, for any $0 < r < 1$, there exist $C > 0$, $0 < r_1 < 1$ as follows: for all $v \in H^\omega$ with

$|v_\alpha| < r^{|\alpha|}$ (for all α) we have $|(J_+v)_\alpha| < Cr_1^{|\alpha|}$ (for all α); similarly for J_-.

(b) Suppose $K_1 \in K^{j-2n-2}$, $K_2 \in K^{k-2n-2}$ and $\mathrm{Re}(j+k) < 2n + 2$. Put $K = K_1 * K_2$. Suppose $\hat{K}_1 = \underset{\sim}{J}_1$, $\hat{K}_2 = \underset{\sim}{J}_2$, $\hat{K} = \underset{\sim}{J}$, all in the $\mathcal{Q}$ sense. Then if $v \in H^\infty$, $J_+v = J_{2+}J_{1+}v$, $J_-v = J_{2-}J_{1-}v$.

(c) Suppose $R \in Z_2^2(H)$ and $k \in \mathbb{R}$. Then R has an extension to a bounded operator from H^k to H^k; further $R : H^k \to H^\omega$. In fact, there exist $C_o > 0$, $0 < r_o < 1$ so that whenever $\|v\|_k \leq 1$, $w = Rv$, we have $|w_\alpha| < C_o r_o^{|\alpha|}$ for all α.

(d) Suppose $J_+ \in Z_2^2(H_1)$ and $J_- \in Z_2^2(H_-)$, and $k \in \mathbb{C}$. Then there exists $K \in AK^{k-2n-2}$ so that if $\hat{K} = \underset{\sim}{J}$ in the $\mathcal{Q}$ sense, then $J_+ = J(1)$, $J_- = J(-1)$.

(e) Suppose $k \in \mathbb{C}$. Then there exist C,N as follows: if $K \in K^{k-2n-2}$, $K = F \in C^\infty(\mathbb{H}^n \setminus \{0\})$ away from 0, $K = \Lambda_F$ as in (1.2) and $\hat{K} = \underset{\sim}{J}$ in the $\mathcal{Q}$ sense then $\|J_\pm\|_{-k,0} < C\|K\|_{C^N}$. Here $\|\ \|_{-k,0}$ denotes norm as an operator from H^{-k} to H^0, and $\|\ \|_{C^N}$ denotes C^N norm over $\{1 \leq |u| \leq 2\}$. (Of course, if $-k \notin \mathbb{Z}^+$, we may omit the hyopthesis $K = \Lambda_F$.)

Except for the last statement, (a) was proved for the case $k \in \mathbb{Z}^+$ in Proposition 4.8(b). (b) was shown in [26], in greater generality, as was most of (a). We include simpler proofs of (a), (b) and (e) in the appendix to this chapter (in part A.2). For now we content ourselves with proving (c) and (d).

For (c), say $|(E_\alpha, RE_\beta)| < C_1 r_1^{|\alpha|+|\beta|}$. If $v \in H^k$, $v = \sum v_\alpha E_\alpha$, we define $Rv = w = \sum w_\alpha E_\alpha$ by $w_\alpha = \sum_\beta v_\beta (E_\alpha, RE_\beta)$. If $\|v\|_k \leqq 1$, we have $|w_\alpha| \leq \sum |v_\beta||(E_\alpha, RE_\beta)| \leqq C_1 r_1^{|\alpha|} (\sum (|\beta|+1)^k |v_\beta|^2)^{1/2} (\sum (|\beta|+1)^{-k} r_1^{2|\beta|})^{1/2} < C_2 r_1^{|\alpha|}$, as desired.

For (d), say $J_+ = W_1 F_+$, $J_- = W_{-1} F_-$, $F_\lambda(\zeta) = \lambda^{k/2} F_+(\zeta/\lambda^{1/2})$ if $\lambda > 0$, $F_\lambda = |\lambda|^{k/2} F_-(\zeta/|\lambda|^{1/2})$ if $\lambda < 0$. Then $W_\lambda F_\lambda = J_\lambda$ for all λ. If $F(\lambda,\zeta) = F_\lambda(\zeta)$, there exists (by Theorem 2.11) $K \in AK^{k-2n-2}$ so that, away from $0, F'K = F$. If $\hat{f} \in \mathcal{Q}$, then $F'f$ vanishes near 0, since $F_\lambda' f = W_\lambda^{-1} \hat{f}(\lambda)$. Thus $(f|K) = c_n''(F'f|F) = c_n'' \int_{-\infty}^{\infty} (F'f|F_\lambda) d\lambda = c_n(\hat{f}|J)$, as in the proof of Proposition 4.6(a). By the uniqueness assertion of Proposition 5.5(a), $\hat{K} = \underset{\sim}{J}$ in the $\mathcal{Q}$ sense.

Theorem 5.1(d) has the following heuristic interpretation. If π is an irreducible unitary representation of $\mathbb{H}^n$ on Hilbert space H and $f \in L^1$, we can define $\pi(f) \in B(H)$ by $\pi(f) = \int_{\mathbb{H}^n} f(u)\pi(u)du$. If $\pi = \pi_\lambda$, we have $\pi(f) = \hat{f}(\lambda)$. Recall, however, the representation $\pi_{(a,b)}$ $((a,b) \in \mathbb{R}^n \times \mathbb{R}^n)$ which acts on one-dimensional Hilbert space by $\pi_{(a,b)}(t,x,y) = e^{i(a\cdot x + b\cdot y)}$. Then $\pi_{(a,b)}(f) = F_{\mathbb{R}^{2n}}(I_c f)(a,b) = Ff(0,a,b)$. If K_1 is as in Theorem 5.1, we could define $\pi_\lambda(K_1) = J_1(\lambda)$, $\pi_{(a,b)}(K_1) = FK_1(0,a,b)$. (d) then says essentially that $\pi(K_1)$ is to be injective for all irreducible unitary representations π, except $\pi = \pi_{(0,0)}$. ($\pi_{(0,0)}K$ is not defined.

Having explained the terms in Theorem 5.1 and motivated the proof, we give the structure of the proof. We shall show (c) => (a) and (b); (a) => (e) and (b) => (e); (e) => (d); and (d) => (c), the most difficult part. By the way, the arguments, or simpler versions of them, could be used to prove an analogous theorem on $\mathbb{R}^n$.

Proof of Theorem 5.1. (c) => (b) By "C_{K_1} is analytic hypoelliptic" we mean: "Suppose g is a distribution of compact support and $U \subset \mathbb{H}^n$ is open. Suppose $K_1 * g$ is real analytic on U. Then g is real analytic on U."

Suppose then that g is as above. Note that $K_1 * g$ is smooth outside a compact set, and $|(K_1 * g)(u)| = O(1+|u|)^{\text{Re } k-2n-2}$ as $u \to \infty$. (Indeed, K_1 and all its derivatives are $O(1+|u|)^{\text{Re } k-2n-2}$ at ∞.) Thus we can form $K_2 * (K_1 * g)$; by (5.1), it equals $(K_2 * K_1) * g = g$. (Indeed, in the notation of (5.1), we need only show $K_2'' * (K_1'' * g) = (K_2'' * K_1'') * g$; this can be done by approximating K_2'' by appropriate C_c^∞ functions.) Further, with $j = -k - 2n - 2$, we can evidently write $K_1 * g$ as a sum of three distributions f of the form given in Lemma 5.3(a), (b) and (c). Thus $g = K_2 * (K_1 * g)$ is analytic on U, as desired.

(c) ⇒ (a) Replace "analytic" by "smooth" everywhere in the proof of (c) => (b).

(a) ⇒ (e) and (b) ⇒ (e) Assume (e) fails. We show:

(*) There exists $f \in \mathcal{E}'(\mathbb{H}^n)$ which is not smooth at 0 but is such that $K_1 * f$ is real analytic in a neighborhood of 0.

Clearly (*) negates both (a) and (b).

First suppose $\mathcal{F}K_1(0,\xi_o) = 0$ for some $\xi_o \neq 0$; we produce f as above. We may assume $|\xi_o| = 1$. It is convenient to change our notation a bit; we use coordinates (t,x) on $\mathbb{H}^n$ where $x = (x_1,\ldots,x_{2n})$ and where what we used to call y_j is now x_{n+j}. Dual coordinates will still be (λ,ξ) with $\xi \in \mathbb{R}^{2n}$.

We need only consider the case Re $k < 0$. For if this case were known, then for the general case we could select $r \in \mathbb{N}$ so that Re $k - 2r < 0$, and apply the known result to $\mathcal{R}_o^r K_1$ in place of K_1. Indeed, $\mathcal{R}_o = -\Delta + TD$ where Δ is the Laplacian in the x_ℓ's and the coefficients of D are polynomials in the x_ℓ's. Thus $(\mathcal{F}\mathcal{R}_o^r K_1)(0,\xi) = |\xi|^{2r}(\mathcal{F}K_1)(0,\xi)$ for $\xi \neq 0$, and in particular, $(\mathcal{F}\mathcal{R}_o^r K_1)(0,\xi_o) = 0$. We could then select f as in (*) with $\mathcal{R}_o^r J_1$ in place of K_1, and observe that $\mathcal{R}_o^r(K_1 * f) = \mathcal{R}_o^r K_1 * f$ is real analytic near 0. By the analytic hypoelliptici-city of $\mathcal{R}_o$, $K_1 * f$ is real analytic in a neighborhood of 0 also, so that the example of f negates (a) and (b) for K_1 as well.

Thus we assume Re $k < 0$; as a consequence, $K_1 \in \mathcal{E}' + L^1$. Our first objective is to show this:

(**) Suppose $h \in C_c^\infty(\mathbb{R}^+ \setminus \{0\})$. Let $G(t,x) = g(x) = \int_0^\infty e^{-i\tau x\cdot\xi_o} h(\tau)\,d\tau$. Then $K_1 * G = 0$.

To see this, observe first that g is a bounded C^∞ function on $\mathbb{R}^{2n}$; by (5.2) it suffices then to show $K_3 * g = 0$ where $K_3 = I_c K_1$. Thus for (**) we need only show:

(***) Suppose $h \in C_c^\infty(\mathbb{R}^+ \setminus \{0\})$, $\xi' \in \mathbb{R}^{2n}$, $|\xi'| = 1$. Let $g(x) = \int_0^\infty e^{-i\tau x\cdot\xi'} h(\tau)\,d\tau$. Suppose $\text{Re } k < 0$, $K_4 \in \mathcal{K}^{k-2n}(R^{2n})$ and $(\mathcal{F}K_4)(\xi') = 0$. Then $K_4 * g = 0$.

For (***), let $O(\mathbb{R}^{2n})$ denote the orthogonal group on $\mathbb{R}^{2n}$. If $S \in O(\mathbb{R}^{2n})$ and f is a function on R^{2n} define $\rho_S f$ by $\rho_S f(x) = f(Sx)$; extend $\rho_S : C_c^\infty \to C_c^\infty$ to $\rho_S : \mathcal{D}' \to \mathcal{D}'$. Select S so that $S(1,0,\ldots,0) = \xi'$. We need only show that $\rho_S(K_4 * g) = \rho_S K_4 * \rho_S g = 0$. Now $\rho_S g(x) = \int_0^\infty e^{i\tau x_1} h(\tau)\,d\tau$ and $\mathcal{F}(\rho_S K_4)(1,0,\ldots,0) = 0$. In short, to prove (***), we can and do assume $\xi' = (1,0,\ldots,0)$.

In complete analogy to I_c, we define a map I_1 as follows. Suppose $f \in \mathcal{S}'(\mathbb{R}^{2n})$ satisfies:

(I) $\mathcal{F}f \in L^1_{loc}(\mathbb{R}^{2n})$

(II) $f_1(\tau) = (\mathcal{F}f)(\tau,0,\ldots,0) \in L^1_{loc}(\mathbb{R})$

(III) $\mathcal{F}f$ has at worst polynomial growth at infinity.

Then we can define $I_1 f \in \mathcal{S}'(\mathbb{R})$ by the rule $\mathcal{F}_{\mathbb{R}}(I_1 f)(\tau) = (\mathcal{F}f)(\tau,0,\ldots,0)$. The argument leading to (5.2) shows that if $f \in \mathcal{E}' + L^1(\mathbb{R}^{2n})$ and u is a bounded C^∞ function on $\mathbb{R}$ then $I_1 f * u = f * U$ where $U(x_1,\ldots,x_{2n}) = u(x_1)$. Applying this to (***) we find, since $g(x) = \int_0^\infty e^{-i\tau x_1} h(\tau)\,d\tau = v(x_1)$, say, that

$K_4 * g = I_1 K_4 * v = 0$, since $v \in S$, $(Fv)(\tau) = 0$ for $\tau \leq 0$ and $(FI_1K_4)(\tau) = 0$ for $\tau > 0$. This establishes (***) and hence (**).

From (**), it is easy to prove (*) as follows. First write $K_1 = K_1' + K_1''$ where $K_1' \in E_1'$, $K_1'' \in L^1$. Select $M \in \mathbb{Z}^+$ greater than the order of the distribution K_1'. Note that if F is a bounded C^M function on $\mathbb{H}^n$ then $K_1 * F$ is automatically a function. As was shown in the proof of Proposition 3.1, if $h_o(\tau) = (1+\tau^2)^{-(M+1)}$ for $\tau > 0$, $h_o(\tau) = 0$ for $\tau < 0$, then the inverse Fourier transform of h_o is C^M but is not C^∞ near 0. If $G_o(t,x) = g_o(x) = \int_0^\infty e^{-i\tau x\cdot\xi_o} h_o(\tau)\,d\tau$, then G_o is bounded and is in C^M but is not C^∞ near 0. We claim $K_1 * G_o = 0$. Indeed, select a sequence of functions $\{h_m\} \subset C_c^\infty(\mathbb{R}^+ - \{0\})$ so that $0 \leq h_m \leq h_o$ for all m and $h_m(\tau) = h_o(\tau)$ for $1/m \leq \tau \leq m$. Put $G_m(t,x) = \int_0^\infty e^{-i\tau x\cdot\xi_o} h_m(\tau)\,d\tau$. Then $G_m \to G_o$ in sup norm on $\mathbb{H}^n$ and C^M norm on compact subsets of $\mathbb{H}^n$. Hence, for all $u \in \mathbb{H}^n$, $K * G_o(u) = \lim K * G_m(u) = 0$ by (**). Finally, for (*), select $\varphi \in C_c^\infty(\mathbb{H}^n)$ with $\varphi = 1$ near 0. Put $f = \varphi G_o$. Then $K_1 * f = -K_1 * (1-\varphi) G_o$, and this is real analytic near 0 by case (b) of Lemma 5.3. This proves (*) under the assumption $FK(0,\xi_o) \neq 0$. It suffices then to show that (*) holds if there exists $0 \neq v \in H^\omega$ so that $J_{1+}v = 0$ or $J_{1-}v = 0$.

Say $J_{1+}v = 0$; the other case is similar. Let J_+ be the projection onto $\langle v \rangle$ in H_1; then $J_+ \in Z_2^2(H_1)$. For $\lambda > 0$, define

$J_\lambda = \lambda^{k/2}J_+$; for $\lambda < 0$, put $J_\lambda = 0$. By Proposition 5.5(d), there exists $K \in \mathcal{A}K^{-k-2n-2}$ so that $\hat{K} = \underset{\sim}{J}$ in the $\mathcal{Q}$ sense. For all $\lambda \neq 0$ and all $w \in \mathcal{H}_\lambda^\infty$, $J_{1\lambda}J_\lambda v = 0$. By Proposition 5.5(b), then, $(K_1{*}K)\hat{} = 0$ in the $\mathcal{Q}$ sense. But $K_1{*}K \in PV(\mathbb{H}^n)$, so that by the uniqueness assertion of Proposition 5.5(a), $(K_1{*}K)\hat{} = 0$ in the sense of Proposition 4.3. Thus $K_1{*}K = 0$. Now K is homogeneous, but $\hat{K}$, taken in the $\mathcal{Q}$ sense, is not zero; thus K is not a polynomial. (We verified in the proof of Proposition 4.8(b) that if p is a polynomial, $\hat{f} \in \mathcal{Q}$, then $(f|p) = 0$; thus if p is a homogeneous polynomial, $\hat{p} = 0$ in the $\mathcal{Q}$ sense.) Accordingly K is not smooth at 0. Further, $K_1{*}K = 0$. Select $\varphi \in C_c^\infty(\mathbb{H}^n)$ with $\varphi = 1$ near 0; put $f = \varphi K$. Then $K_1{*}f = -K_1{*}(1-\varphi)K$, and this is real analytic near 0 by case (b) of Lemma 5.3. This proves (*) and completes the proof of (a) => (e) and (b) => (e).

We must still show (e) => (d) => (c). We begin by studying carefully those $K_1 \in \mathcal{A}K^{k-2n-2}$ with the property that $\mathcal{F}K_1(0,\xi) \neq 0$ for $\xi \neq 0$. For these, we develop an analogue of Theorem 3.5, as follows.

<u>Lemma 5.6</u>. Suppose $k \in \mathbb{C}$, $K_1 \in \mathcal{A}K^{k-2n-2}(\mathbb{H}^n)$, and $\mathcal{F}K_1(0,\xi) \neq 0$ for $\xi \neq 0$. Then there exists $K_3 \in \mathcal{A}K^{-k-2n-2}$ and $K_4 \in APV$ so that $K_3{*}K_1 = \delta - K_4$ and so that $F_{4\lambda}(\xi) = (\mathcal{F}K_4)(\lambda,\xi) \in Z_2^2(\mathbb{R}^{2n})$ for $\lambda \neq 0$.

Reductions. We make several easy reductions. First, we can and do assume Re k < 2. For if this case were solved, then in general we could find $r \in \mathbb{N}$ with Re k - 2r < 2, find $K_3' \in AK^{-k+2r-2n-2}$ and K_4 as in the statement of the lemma so that $K_3' * R_o^r K_1 = \delta - K_4$; then also $L_o^r K_3' * K_1 = \delta - K_4$ as desired. Next, we can and do assume $0 \leqq$ Re k < 2. For if this case were solved, then in general we could find $r \in \mathbb{Z}^+$ with $0 \leqq$ Re k + 2r < 2, find $K_1' \in AK^{k+2r-2n-2}$ so that $R_o^r K_1' =$ K_1 (by Theorem 3.2) and find $K_3' \in AK^{-k+2r-2n-2}$ and K_4 as in the statement so that $K_3' * K_1' = \delta - K_4$. Finally select $K_3 \in AK^{-k-2n-2}$ so that $L_o^r K_3 = K_3'$; then $K_3 * K_1 = \delta - K_4$ as desired. Now, we can further reduce to the case where

(*) $K_1 \in$ APV and $FK_1(0,\xi) = 1$ for all $\xi \neq 0$ (so that $I_c K_1 = \delta$).

Indeed, suppose this case were solved, and we also knew the following:

(**) There exists $K_5 \in AK^{-k-2n-2}$ so that $FK_5(0,\xi) =$ $[FK_1(0,\xi)]^{-1}$ for all $\xi \neq 0$. (We are in the case $K_1 \in AK^{k-2n-2}$, $0 \leqq$ Re k < 2, and $FK_1(0,\xi) \neq 0$ for $\xi \neq 0$.)

Then $K_5 * K_1 \in$ APV. Also, by Proposition 5.4(b), $F(K_3 * K_1)(0,\xi) =$ $F_{\mathbb{R}^{2n}}(I_c(K_5 * K_1))(\xi) = F_{\mathbb{R}^{2n}}(I_c K_5 * I_c K_1)(\xi) =$ $F_{\mathbb{R}^{2n}}(I_c K_5)(\xi) F_{\mathbb{R}^{2n}}(I_c K_1)(\xi) = 1$ for $\xi \neq 0$. (See the last paragraph of Appendix A.1 for the third equality.)
By the assumed solution in case (*), we can find $K_3' \in$ APV and K_4 as in the statement so that

$K_3'*(K_5*K_1) = \delta - K_4$. Putting $K_3 = K_3'*K_5$ would then solve the general case, by Proposition 5.2(g). Thus, we can restrict to case (*) if we know (**). To prove (**), let us recall this correspondence, which follows at once from Theorem 2.11 and (2.19):

Suppose $K \in AK^{-j-2n-2}$, $FK = F$, $F_+(\xi) = F(1,\xi)$ and $F_+ \sim \sum g_\ell$ in $Z^2_{2,j}$. Then (5.3)

$$g_o(\xi) = F(0,\xi) \text{ for } \xi \neq 0.$$

Accordingly, suppose $FK_1 = F_1$, $F_{1+}(\xi) = F_1(1,\xi)$, $F_{1-}(\xi) = F_1(-1,\xi)$, $F_{1+} \sim \sum g_{1\ell}$, $F_{1-} \sim \sum(-1)^\ell g_{1\ell}$ in $Z^2_{2,-k}$. Then $g_{10}(\xi) = F_1(0,\xi)$ and g_{10} is real analytic away from 0. Using Theorem 2.12, select $F_{5+}, F_{5-} \in Z^2_{2,k}$ so that $F_{5+} \sim \sum g_{5\ell}, F_{5-} \sim \sum(-1)^\ell g_{5\ell}$ where $g_{50}(\xi) = 1/g_{10}(\xi)$ for $\xi \neq 0$ and where $g_{5\ell} = 0$ for $\ell \neq 0$. By Theorem 2.11, there exists $K_5 \in AK^{-k-2n-2}$ so that $F_{5+}(\xi) = FK_5(1,\xi), F_{5-}(\xi) = FK_5(-1,\xi)$. Further $(FK_5)(0,\xi) = 1/g_{10}(\xi)$ for $\xi \neq 0$. This verifies (**). Thus we can indeed work in case (*).

To prove the lemma in case (*), we digress slightly. We need to relate $F(K_2*K_1)$ to FK_1 and FK_2 for $K_2, K_1 \in$ APV; for this we need an additional lemma, Lemma 5.7 below. For the proofs of Lemmas 5.7 and 5.6 only we use the following notation. $\mathbb{H}^n$ has coordinates $(t,x,y) \in \mathbb{R} \times \mathbb{R}^n \times \mathbb{R}^n$ with dual coordinates $(\lambda,\xi) \in \mathbb{R} \times \mathbb{R}^{2n}$. We use this notation for differential operators on $\mathbb{R}^{2n}_\xi$:

$$\partial = (\partial/\partial\xi_1,\ldots,\partial/\partial\xi_{2n}),\ \tilde{\partial} = (\partial/\partial\xi_{n+1},\ldots,\partial/\partial\xi_{2n},-\partial/\partial\xi_1,\ldots,-\partial/\partial\xi_n).$$

If $\alpha \in (\mathbb{Z}^+)^{2n}$ is a multi-index, ∂^α and $\tilde{\partial}^\alpha$ will have the obvious meanings. The following lemma has appeared in the C^∞ situation in various guises before; it is a form of the product formula for the Weyl calculus ([37], [49], [60], [79]). In particular, [79] uses this formula in the C^∞ situation for the same purposes that we use it in the real analytic situation.

<u>Lemma 5.7</u>. Suppose $\mathrm{Re}(k_1+k_2) < -2n-2$. For $\nu = 1,2$, let $j_\nu = -2n-2-k_\nu$, and suppose $K_\nu \in K^{k_\nu}$. Let $FK_\nu = F_\nu$, $F_{\nu+} \sim \sum g_{\nu\ell}$ in Z^2_{2,j_ν} $(\nu=1,2)$. Let $K = K_2 * K_1$, $j = j_1 + j_2$, $FK = F, F_+(\xi) = F(1,\xi), F_+ \sim \sum g_\ell$ in $Z^2_{2,j}$. Then

$$g_\ell = \sum_{a+b+|\alpha|=\ell} [(2i)^{|\alpha|}/\alpha!]\partial^\alpha g_{2a}\tilde{\partial}^\alpha g_{1b} \qquad (5.4)$$

<u>Proof</u>. We define $M_0, M_1,\ldots,M_n, M_1',\ldots,M_n' : S(H^n) \to S(H^n)$ by $(M_0 f)(t,x,y) = tf(t,x,y)$, $(M_j f)(t,x,y) = x_j f(t,x,y)$, $(M_j' f)(t,x,y) = y_j f(t,x,y)$ $(1 \le j \le n)$. A simple computation using the fact that

$$(t,x,y)\cdot(t',-x',-y') = (t-t'-2[(y-y')\cdot x'-(x-x')\cdot y'],x-x',y-y')$$

shows that for all $f,g \in E'$,

$$M_0(f*g) = M_0 f*g + f*M_0 g + 2\sum_{k=1}^{n} (M_k' f*M_k g - M_k f*M_k' g).$$

From this, one sees readily that for all $f,g \in E'$, $\ell \in \mathbb{N}$,

$$M_0^\ell(f*g) = \sum_{a+b+|\alpha|+|\beta|=\ell} 2^{|\alpha+\beta|}(-1)^{|\alpha|}[\ell!/a!b!\alpha!\beta!]M_0^a M^\alpha(M')^\beta f*M_0^b(M')^\alpha M^\beta g. \qquad (5.5)$$

(In the sum, $\alpha,\beta \in (\mathbb{Z}^+)^n$ and $M^\alpha = M_1^{\alpha_1}\ldots M_n^{\alpha_n}$, etc. A good, if pedantic, way to see (5.5) is as follows. On the algebraic tensor product $E'(\mathbb{H}^n) \otimes E'(\mathbb{H}^n)$ we can by universality, define the following linear maps: $P : E'(\mathbb{H}^n) \otimes E'(\mathbb{H}^n) \to E'(\mathbb{H}^n)$ by $P(f\otimes g) = f*g$; and $S_1,\ldots,S_{2n+2}: E'(\mathbb{H}^n) \otimes E'(\mathbb{H}^n) \to E'(\mathbb{H}^n) \otimes E'(\mathbb{H}^n)$ by $S_k(f\otimes g) = -2M_k f \otimes M_k' g$ $(1\le k\le n)$; $S_{k+n}(f\otimes g) = 2M_k' f \otimes M_k(g)$ $(1\le k\le n)$; $S_{2n+1}(f\otimes g) = M_o f \otimes g$; $S_{2n+2}(f\otimes g) = f \otimes M_o g$. $M_o P = P(\sum_{k=1}^{2n+2} S_k)$, so $M_o^\ell P = P(\sum S_k)^\ell = P\sum[\ell!/a!b!\alpha!\beta!]S_1^{\alpha_1}\ldots S_n^{\alpha_n}S_{n+1}^{\beta_1}\ldots S_{2n}^{\beta_n}S_{2n+1}^a S_{2n+2}^b$; this gives (5.5) at once.) From (5.5) and the fact that, if $f,g \in S(\mathbb{H}^n)$, then $F(f*g)(0,\xi) = F_{\mathbb{C}^n}(I_c f * I_c g)(\xi) = Ff(0,\xi)Fg(0,\xi)$, we find that, if $f,g \in S$,

$$(-i\partial/\partial\lambda)^\ell F(f*g)(0,\xi) = \sum_{a+b+|\alpha|=\ell} 2^{|\alpha|}[\ell!/a!b!\alpha!][(-i\partial/\partial\lambda)^a\partial^\alpha Ff(0,\xi)] \times [(-i\partial/\partial\lambda)^b\bar{\partial}^\alpha Fg(0,\xi)] \tag{5.6}$$

In the sum, α is now a multi-index in $(\mathbb{Z}^+)^{2n}$. In light of (5.3), in order to prove (5.4) we need only check that in (5.6) we can replace f,g by K_2,K_1.

To show this, we use the notation of Proposition 5.2(h). Write (5.5) with $K_{2\varepsilon}$, $K_{1\varepsilon}$ in place of f,g, apply $R_o^\ell L_o^\ell$ to both sides and use Proposition 5.2(h) to take $\lim_{\varepsilon\to 0, N\to\infty}$. The result is

$$R_o^\ell L_o^\ell M_o^\ell(K_2*K_1) = \sum_{a+b+|\alpha|+|\beta|=\ell} 2^{|\alpha+\beta|}(-1)^{|\alpha|}[\ell!/a!b!\alpha!\beta!]R_o^\ell M_o^a M^\alpha(M')^\beta K_2 * L_o^\ell M_o^b(M')^\alpha M^\beta K_1.$$

We apply F to both sides and evaluate at $(0,\xi)$ $(\xi\neq 0)$, observing that if $K_1' \in K^{j'}$, $K_2' \in K^{k'}$ and $j', k' \le -2n-2$, then

$F(K_2'*K_1')(0,\xi) = F_{\mathbb{R}^{2n}}(I_cK_2'*I_cK_1')(\xi) = FK_2'(0,\xi)FK_1'(0,\xi)$ by Proposition 5.4(b). The result is (5.6) with K_2, K_1 in place of f,g (actually, this equation multiplied through by $|\xi|^{4\ell}$). This completes the proof. ∎

Before giving the proof of Lemma 5.6, we introduce a little more notation. We use the notation $\underset{\sim}{g} = (g_\ell)$ for a sequence of functions such that g_ℓ is homogeneous of degree -2ℓ and smooth away from 0. If $\underset{\sim}{g}_1 = (g_{1\ell})$ and $\underset{\sim}{g}_2 = (g_{2\ell})$ are two such sequences, we define a new sequence $\underset{\sim}{g} = \underset{\sim}{g}_2 \circ \underset{\sim}{g}_1$ by $g_\ell = \sum_{a+b+|\alpha|=\ell} [(2i)^{|\alpha|}/\alpha!]\partial^\alpha g_{2a}\tilde{\partial}^\alpha g_{1b}$. Thus, in the situation of Lemma 5.7, $\underset{\sim}{g} = \underset{\sim}{g}_2 \circ \underset{\sim}{g}_1$. Also, define the sequence $\underset{\sim}{1}$ by $1_\ell = \delta_{0\ell}$. Then for all $\underset{\sim}{g}$, $\underset{\sim}{1}\circ\underset{\sim}{g} = \underset{\sim}{g}\circ\underset{\sim}{1} = \underset{\sim}{g}$.

With this preparation, we come to the proof of Lemma 5.6, which is the key step in the proof of Theorem 5.1.

<u>Proof of Lemma 5.6</u>. As we already remarked, we can work in the case (*). Write $K_1 = \delta - K$, $FK = F$, $F_+(\xi) = F(1,\xi) = F_-(\xi) = F(-1,\xi)$ and $F_+ \sim \sum g_\ell$ in $Z^2_{2,0}$. Then $g_o = 0$. We attempt to make sense of the Fourier transform of the formal series $K_3 \sim \delta + K + K*K + K*K*K + \ldots$, which should be the solution to our problem. Define sequences $\underset{\sim}{g}^{(m)}$ for $m \in \mathbb{Z}^+$ by $\underset{\sim}{g}^{(o)} = \underset{\sim}{1}$, $\underset{\sim}{g}^{(1)} = \underset{\sim}{g}$, $\underset{\sim}{g}^{(m)} = \underset{\sim}{g}^{(m-1)}\circ\underset{\sim}{g}$ for $m \geq 2$. It is easy to see by induction that $g_\ell^{(m)} = 0$ if $\ell < m$. Thus we can define a sequence $\underset{\sim}{g}_3$ by $g_{3\ell} = \sum_{m=0}^{\infty} g_\ell^{(m)} = \sum_{m=0}^{\ell} g_\ell^{(m)}$. Formally, $\underset{\sim}{g}_3 = \sum_0^\infty \underset{\sim}{g}^{(m)}$, so we

expect $\underset{\sim}{g}_3 \circ (\underset{\sim}{1}-\underset{\sim}{g}) = \underset{\sim}{1}$. This is in fact the case. Indeed, for any ℓ, $(\underset{\sim}{g}_3 \circ \underset{\sim}{g})_\ell = (\sum_{m=0}^{\ell} \underset{\sim}{g}^{(m)} \circ \underset{\sim}{g})_\ell$ since in the expression for $(\underset{\sim}{g}_3 \circ \underset{\sim}{g})_\ell$, g_{3i} appears only for $i \leq \ell$, so that $g_{3i} = \sum_{m=0}^{\ell} g^{(m)}$. Thus $(\underset{\sim}{g}_3 \circ \underset{\sim}{g}) = \sum_{m=1}^{\ell+1} g_\ell^{(m)} = g_{3\ell} - 1_\ell$, so $\underset{\sim}{g}_3 \circ \underset{\sim}{g} = \underset{\sim}{g}_3 - \underset{\sim}{1}$ and $\underset{\sim}{g}_3 \circ (\underset{\sim}{1}-\underset{\sim}{g}) = \underset{\sim}{1}$ as claimed.

We think of $\mathbb{R}^{2n}$ (with coordinates ξ) as contained in $\mathbb{C}^{2n}$ (with coordinates $\zeta = \xi + i\eta$) in the obvious manner. There exists a sector $S = \{|\eta| < c|\xi|\}$ so that all g_ℓ have holomorphic extensions to S, and so that, if g also denotes the holomorphic extension:

$$\text{There exist } C,R > 0 \text{ so that for all } \zeta \in S \text{ with } |\zeta| = 1, \text{ we have } |g_\ell(\zeta)| < CR^\ell \ell!. \tag{5.7}$$

It follows that all $g_\ell^{(m)}$ and all $g_{3\ell}$ have holomorphic homogeneous extensions to S. We still use the notation $g_\ell^{(m)}, g_{3\ell}$ for the holomorphic extensions. Select and fix c' with $0 < c' < c$, and let $S' = \{|\eta| < c'|\xi|\}$. Then the main point is to show the following:

(***) There exist $C_1, R_1 > 0$ so that for all ζ with $\zeta \in S'$, $|\zeta| = 1$, we have $|g_{3\ell}(\zeta)| < C_1 R_1^\ell \ell!$.

If this were known, Lemma 5.6 would follow at once. Indeed, by Theorem 2.12, we could then select $F_{3+}, F_{3-} \in Z^2_{2,0}$ so that $F_{3+} \sim \sum g_{3\ell}$, $F_{3-} \sim \sum (-1)^\ell g_{3\ell}$. By Theorem 2.11 we could select

$K_3 \in$ APV so that $F_{3+}(\xi) = (FK_3)(1,\xi)$, $F_{3-}(\xi) = (FK_3)(-1,\xi)$. Let $\delta - K_4 = K_3 * K_1 = K_3 * (\delta - K)$. By Lemma 5.7 and the fact that $\underset{\sim}{g}_3 \circ (\underset{\sim}{1} - \underset{\sim}{g}) = \underset{\sim}{1}$, it follows at once that $(FK_4)(1,\xi) \sim 0$ in $Z^2_{2,0}$. But $K_4 \in$ APV, so by Theorem 2.11, we have also $(FK_4)(-1,\xi) \sim 0$ in $Z^2_{2,0}$. Thus $FK_4(\pm 1,\xi) \in Z^2_2$, as desired.

Thus we need only show (***). For this, we shall examine an explicit formula for $g_{3\ell}(\zeta)$. To derive this, first put $D = (\partial/\partial\zeta_1,\ldots,\partial/\partial\zeta_{2n})$, $\tilde{D} = (\partial/\partial\zeta_{n+1},\ldots,\partial/\partial\zeta_{2n},$ $-\partial/\partial\zeta_1,\ldots,-\partial/\partial\zeta_n)$ on $\mathbb{C}^{2n}$. Observe that in S we have

$$g_\ell^{(m)} = \sum_{a+b+|\alpha|=\ell} [(2i)^{|\alpha|}/\alpha!]\tilde{D}^\alpha g_a D^\alpha g_b^{(m-1)} = \sum_{a+b+|\alpha|=\ell} g_{a;\alpha} D^\alpha g_b^{(m-1)}$$

where we have set $g_{a;\alpha}(\zeta) = (2i)^{|\alpha|}\tilde{D}^\alpha g_a(\zeta)/\alpha!$ for $\zeta \in S$. This gives at once, by induction, the formula

$$g_\ell^{(m)} = \sum g_{a_{m-1};\alpha^{m-1}} D^{\alpha^{m-1}} g_{a_{m-2};\alpha^{m-2}} D^{\alpha^{m-2}} \cdots g_{a_1;\alpha^1} D^{\alpha^1} g_{a_0} \tag{5.8}$$

in S, where the sum is taken over all $(a_0,\ldots,a_{m-1}) \in \mathbb{N}$ and all 2n-tuples $\alpha^1,\cdots,\alpha^{m-1}$ of nonnegative integers such that $a_0+\ldots+a_{m-1} + |\alpha^1|+\ldots+|\alpha^{m-1}| = \ell$. We estimate $g_\ell^{(m)}(\zeta)$ for $\zeta \in S'$. The number of terms in the summation in (5.8) is no more than the number of ways of writing ℓ as a sum of $m + (m-1)(2n)$ nonnegative integers, which is $\binom{\ell+m+(m-1)2n-1}{\ell} \leq 2^{\ell+(m-1)(2n+1)} < 2^{(2n+2)\ell}$ if $m \leq \ell$. (Recall that if $m > \ell$, then $g_\ell^{(m)} = 0$.) Let us then estimate the size of a typical term in the summation. Select c" with c' < c" < c. By the

Cauchy estimates, (5.7) and the definition of the $g_{a;\alpha}$ we see that there exist $C_2, R_2 > 0$ so that for all a, α and all $\zeta \in S''$ with $1/2 \leq |\zeta| \leq 2$, we have $|g_{a;\alpha}(\zeta)| < C_2 R_2^{a+|\alpha|} a!$. Select $\rho > 0$ so that if $\zeta_0 \in S'$, $|\zeta_0| = 1$, then the polydisc centered at ζ_0 with polyradii $(\rho, \ldots, \rho)$ is contained in $\{\zeta \in S'' \mid 1/2 < |\zeta| < 2\}$. By Lemma 2.5, then if $N = \sum |\alpha^j|$, and $\zeta \in S'$, $|\zeta| = 1$, we have

$$|g_{a_{m-1};\alpha^{m-1}} D^{\alpha^{m-1}} \cdots g_{a_1;\alpha^1} D^{\alpha^1} g_{a_0}(\zeta)| \leq (2\pi e N)^{1/2} (e/\rho)^N N! C_2^m R_2^{\ell} a_0! \ldots a_{m-1}!$$

$< C_3 R_3^{\ell} \ell!$ for some $C_3, R_3 > 0$, if $m \leq \ell$. (Indeed, since $\ell = a_0 + \ldots + a_{m-1} + N$, $a_0! \ldots a_{m-1}! N! \leq \ell!$.) Thus, if $\zeta \in S'$, $|g_{\ell}^{(m)}(\zeta)| < C_3 (2^{(2n+2)} R_3)^{\ell} \ell!$. Since $g_{3\ell} = \sum_{m=0}^{\ell} g_{\ell}^{(m)}$, (***) follows at once. This completes the proof of Lemma 5.6.

<u>Conclusion of the Proof of Theorem 5.1</u>

<u>(e) => (d)</u>. Say K_1 is as in Theorem 5.1(d). Select K_3, K_4 as in Lemma 5.6. Say $\hat{K}_i = \underset{\sim}{J}_i$ in the $\mathcal{Q}$ sense $(i=1,3,4)$. Define $J_{i+} = J_i(1)$ $(i=1,3,4)$; then $J_{4+} \in Z_2^2(H)$ by Proposition 4.6(b). By Proposition 5.5(b), for all $v \in H^{\infty}$, $J_{3+} J_{1+} v = (I - J_{4+})v$. If $v \in H^{\infty}$, $J_{1+} v = 0$, then $v = J_{4+} v \in H^{\omega}$ by Proposition 4.9(b). By (e), $v = 0$. Similarly if $v \in H^{\infty}$, $J_{1-} v = 0$, then $v = 0$. ∎

<u>(d) => (c)</u>. This is the only remaining implication, and is the main point. For this we shall adapt the proof of Theorem 4.1.

Because of Lemma 5.6, only minor changes are needed. For the time being, we use only the hypotheses that $K_1 \in AK^{k-2n-2}$, $(FK_1)(0,\xi) \neq 0$ for $\xi \neq 0$; the hypotheses that J_{1+},J_{1-} have zero kernel in H^∞ will be used only at the end of the proof.

Suppose $\hat{K}_1 = \underset{\sim}{J}_1$ in the Q sense. Define $\tilde{}: S \to S$ by $\tilde{f}(u) = \overline{f(u^{-1})}$ and extend $\tilde{}$ to a map from S' to S'. Say $\hat{\tilde{K}}_1 = \underset{\sim}{\tilde{J}}_1$ in the Q sense. We claim that for all $\lambda \neq 0$, if $v,w \in H_\lambda^\infty$, then $(v,\tilde{J}_1(\lambda)w) = (J_1(\lambda)v,w)$. To see this, suppose $\varphi \in C_c^\infty(\mathbb{R}^*)$, $\alpha,\beta \in (\mathbb{Z}^+)^n$; select f with $\hat{f} \in Q$, $\hat{f}(\lambda)E_\gamma = \delta_{\alpha\gamma}\varphi(\lambda)E_\beta$. Now $\overline{(f|K_1)} = (\tilde{f}|\tilde{K}_1)$; also, $(\tilde{f})^\wedge = (\hat{f})^*$, so that $(\tilde{f})^\wedge(\lambda)E_\gamma = \delta_{\beta\gamma}\overline{\varphi(\lambda)}E_\alpha$. Thus $\int_{-\infty}^\infty \varphi(\lambda)\overline{(E_\beta,J_1(\lambda)E_\alpha)}|\lambda|^n d\lambda =$ $\int_{-\infty}^\infty \varphi(\lambda)(E_\alpha,\tilde{J}_1(\lambda)E_\beta)|\lambda|^n d\lambda$. This is true for all φ, so $(J_1(\lambda)E_\alpha,E_\beta) = (E_\alpha,\tilde{J}_1(\lambda)E_\beta)$ for all α,β. Since $J_1,\tilde{J}_1$ have extensions to bounded operators from H to H^k and H^{-k} to H by Proposition 5.5(a), $(v,\tilde{J}_1(\lambda)w) = (J_1(\lambda)v,w)$ for all $v,w \in H^\infty$, as claimed.

Now, using Lemma 5.6, let us select $\tilde{K}_3,K_5 \in AK^{-k-2n-2}$ and $K_4,K_6 \in$ APV so that $\tilde{K}_3 * K_1 = \delta - \tilde{K}_4$, $K_5 * \tilde{K}_1 = \delta - K_6$, and $(F\tilde{K}_4)(\pm 1,\xi),(FK_6)(\pm 1,\xi) \in Z_2^2(\mathbb{R}^{2n})$. Using (5.1) and the widely-applicable rule $(f*g)^\sim = \tilde{g}*\tilde{f}$, we see that $\tilde{K}_1 * K_3 = \delta - K_4$. Also $(FK_4)(\pm 1,\xi) \in Z_2^2(R^{2n})$. Put $\hat{K}_i = \underset{\sim}{J}_i$ in the Q sense (i=3,4,5,6); then $J_4(\lambda),J_6(\lambda) \in Z_2^2(H)$. By Proposition 5.5(b), if $\lambda \neq 0$, $v \in H^\infty$, then

$$J_5(\lambda)\tilde{J}_1(\lambda)v = (I-J_6(\lambda))v \tag{5.9}$$

$$\tilde{J}_1(\lambda)J_3(\lambda)v = (I-J_4(\lambda))v. \qquad (5.10)$$

By Proposition 5.5(a) and (c), $\tilde{J}_1(\lambda)$ can be extended to a bounded operator from H^{-k} to H^o; $J_3(\lambda)$ and $J_5(\lambda)$ can be extended to bounded operators from H^o to H^{-k}; $J_4(\lambda)$ can be extended to a bounded operator from H^o to H^o, and $J_6(\lambda)$ can be extended to a bounded oeprator from H^{-k} to H^{-k}. From now on, when we write $J_i(\lambda)$ (i=1,3,4,5 or 6) we mean the aforementioned extensions and not the operators with domain H. Then (5.9) holds for all $v \in H^{-k}$ and (5.10) holds for all $v \in H^o$.

By Proposition 5.5(c), $J_4(\lambda)$ and $J_6(\lambda)$ are compact. Thus, by the analogue of the simple argument given after equation (4.12), $\tilde{J}_1(\lambda)$ has closed range, and it also has finite-dimensional kernel and cokernel which are contained in H^ω. In fact, $[\tilde{J}_1(\lambda)H^{-k}]^\perp = \{v \in H^\omega | J_1(\lambda)v = 0\}$; let $Q(\lambda)$ denote the projection in H_λ onto this space. As in the proof of Theorem 4.1, $Q(\lambda) \in Z_2^2(H_\lambda)$, $\underset{\sim}{Q}$ is homogeneous of degree 0, and there exists $P \in$ APV with $\hat{P} = \underset{\sim}{Q}$. Of course, if we have the full hypotheses of (d) (that J_{1+}, J_{1-} have no kernel in H^∞) then $P = 0$. In general, without these hypotheses, we shall show that there exists $K_2 \in AK^{-k-2n-2}$ so that $K_2 * K_1 = \delta - P$. Clearly this will establish (d) => (c). Note that, by the argument following (4.13), $f \to f*P$ is a projection in $L^2(\mathbb{H}^n)$; we postpone discussing what its range is.

We let $J_{i\pm} = J_i(\pm 1)$ $(i=1,3,4,5,6)$, $\tilde{J}_{1\pm} = \tilde{J}_1(\pm 1)$. As in (4.13), we have

$$(I-Q_+)(I-J_{4\pm}) = I - J_{4\pm} \text{ and } \tilde{J}_{1\pm}J_{3\pm} = (I-Q_\pm) - (I-Q_\pm)J_{4\pm} \qquad (5.11)$$

The theorem will follow at once if we can show

$$\text{There exist } H_\pm \in Z_2^2(H_{\pm 1}) \text{ such that } \tilde{J}_{1\pm}H_\pm = (I-Q_\pm)J_{4\pm}. \qquad (5.12)$$

Indeed, say (5.12) were known. By Proposition 5.5(d), select $K_7 \in AK^{-k-2n-2}$ so that if $\hat{K}_7 = \underset{\sim}{J}_7$ in the $\mathcal{Q}$ sense then $H_\pm = J_7(\pm 1)$. Put $\tilde{K}_2 = K_3 + K_7$; by (5.11), (5.12) and homogeneity, $(\tilde{K}_1*\tilde{K}_2)^\wedge$, taken in the $\mathcal{Q}$ sense, equals $I - \underset{\sim}{Q}$. Since $\tilde{K}_1*\tilde{K}_2 \in PV(\mathbb{H}^n)$, $(\tilde{K}_1*\tilde{K}_2)^\wedge = I - \underset{\approx}{Q}$ in the sense of Proposition 4.3. Thus $\tilde{K}_1*\tilde{K}_2 = I - P$. Now $\tilde{P} = P$ since the operator of convolution with P is self-adjoint on L^2; thus $K_2*K_1 = I - P$ as desired. Thus we have only to show (5.12).

To prove (5.12), we construct only $H = H_+$, since H_- is similar. This is done exactly as in the proof of Theorem 4.1, so we will be brief. Let $S = (I-Q_+)J_{4+}$, $S^\alpha = SE_\alpha$; we can select $H^\alpha \in H^{-k}$ so that $\tilde{J}_{1+}H^\alpha = S^\alpha$ and so that for some $C_0 > 0$, $\|H^\alpha\|_{-k} < C_0\|S^\alpha\|_0$ for all α. Let $H_\beta^\alpha = (E_\beta, H^\alpha)$; it suffices to show $|H_\beta^\alpha| < C_1 r_1^{|\alpha|+|\beta|}$ for some $C_1 > 0$, $0 < r_1 < 1$, for then there exists $H \in Z_2^2$ so that $HE_\alpha = H^\alpha$, and H will be as desired. There exist $C_2, C_3 > 0$, $0 < r_2 < 1$ so that $|S_\beta^\alpha| < C_2 r_2^{|\alpha|+|\beta|}$,

$\|S^\alpha\|_0 < C_3 r_2^{|\alpha|}$ (Here $S^\alpha_\beta = (E_\beta, S^\alpha)$.) Now $H^\alpha = J_{5+}S^\alpha + J_{6+}H^\alpha$ by (5.9). By Proposition 5.5(a), there exist $C_4 > 0$, $0 < r_4 < 1$ so that $|(E_\beta, J_{5+}S^\alpha)| < C_2 r_2^{|\alpha|} C_4 r_4^{|\beta|}$. In addition $\|H^\alpha\|_{-k} < C_0\|S^\alpha\|_0 < C_0 C_3 r_2^{|\alpha|}$, so by Proposition 5.5(c) there exist $C_5 > 0$, $0 < r_5 < 1$ so that $|(E_\beta, J_{6+}H^\alpha)| < C_5 r_5^{|\beta|} C_0 C_3 r_2^{|\alpha|}$. Thus, indeed, for some $C_1 > 0$, $0 < r_1 < 1$, we have $|H^\alpha_\beta| < C_1 r_1^{|\alpha|+|\beta|}$ as claimed. This proves (5.12) and completes the proof of Theorem 5.1. ∎

The case $P \neq 0$ in the preceding proof of (d) => (c) is also of some interest. We consider only the case $k = 0$, so that $K_1, K_2 \in APV$.

<u>Corollary 5.8</u>. Suppose $K_1 \in APV$, and $FK_1(0,\xi) \neq 0$ for $\xi \neq 0$.
(a) There exist K_2, $P \in APV$ so that $K_2 * K_1 = \delta - P$ and so that the map $S : L^2 \to L^2$ given by $Sf = f*(\delta - P)$ is the projection onto $L^2 * K_1 = \{f*K_1 | f \in L^2\}$.

(b) Suppose $g \in L^2$, $g*P = 0$, and g is real analytic on an open set $U \subset \mathbb{H}^n$. Then there exists $f \in L^2$ which is real analytic on U, so that $f*K_1 = g$.

<u>Proof</u>. This is a corollary of the proof of (d) =>(c), and we retain all the notation of that proof. For (a), note

$[K_1*(\delta-P)]^\wedge = \underset{\sim}{J}_1(I-Q) = \underset{\sim}{J}_1 = \hat{K}_1$ so $K_1*(\delta-P) = K_1$. Thus, if $f \in L^2*K_1$, then $f*(\delta-P) = f$. On the other hand, if $f \in L^2$ and $f = f*(\delta-P)$, then $f = (f*K_2)*K_1 \in L^2*K_1$. This proves (a). For (b), we have only to set $f = g*K_2$. This is real analytic on U by Lemma 5.3(c) and the proof of Lemma 5.3(b). (That proof shows that if $h \in L^2$ and $h = 0$ on U then $h*K_2$ is real analytic on U.) This completes the proof.

It is rather curious that the projection onto L^2*K_1 is given by convolution with an element of $PV(\mathbb{H}^n)$. The analogous situation on $\mathbb{R}^n$ could happen only if $K_1 = 0$, so that $L^2*K_1 = 0$, or if FK_1 was nonvanishing away from 0, so that $L^2*K_1 = L^2$.

At this point, in preparation for Chapter 8, we would like to discuss the generalization of Theorem 5.1 to the case where K_1 depends analytically on a parameter. (See the discussion at the end of Chapter 2 for the definition.) We shall need to use a lemma from Chapter 7, namely a version of Proposition 5.5(f) with uniformities. (It is alright to use a lemma from Chapter 7, since the theorem we are going to prove will not be used until Chapter 8. Since the essential ideas of the proof of our theorem are those we have just dealt with, we feel it is pedagogically best to present the proof of our theorem now. We continue to use the letter ω for the parameter; this should not be confused with the "ω" appearing in the symbol H^{ω}.

Theorem 5.9. Say $k \in \mathbb{C}$, $\Omega \subset \mathbb{C}^s$ for some s, Ω open. Suppose $K_{1\omega} \in AK^{k-2n-2}$ depends holomorphically on $\omega \in \Omega$. Let $\hat{K}_{1\omega} = \underset{\sim}{J}_{1\omega}$ in the $\mathcal{Q}$ sense. Identify $\mathbb{R}^s = \{\omega \in \mathbb{C}^s : \operatorname{Im} \omega = 0\}$. Suppose $\Omega \cap \mathbb{R}^s \neq \phi$, and for $\omega \in \Omega \cap \mathbb{R}^s$:

$$(FK_{1\omega})(0,\xi) \neq 0 \text{ for } \xi \neq 0; \; J_{1\omega+}v = 0,\; v \in H^{\omega} \Rightarrow v = 0;$$

and $J_{1\omega-}v = 0$, $v \in H^{\omega} \Rightarrow v = 0$.

Suppose also that for some $C_1, R_1 > 0$ we have

$$\begin{aligned} &|\partial^{\gamma} K_{1\omega}(u)| < C_1 R_1^{\|\gamma\|} \gamma! \\ &\text{for all } 1 \leqq |u| \leqq 2, \gamma \in (\mathbb{Z}^+)^{2n+1},\; \omega \in \Omega. \end{aligned} \tag{5.13}$$

Suppose $E \subset \mathbb{R}^s$ is compact, and $E \subset \Omega$. Then there exists an open set $\Omega' \subset \Omega$ with $E \subset \Omega'$, as follows: For $\omega \in \Omega'$, there exist $K_{2\omega} \in AK^{-k-2n-2}$, depending holomorphically on $\omega \in \Omega'$, so that

$$K_{2\omega} * K_{1\omega} = \delta \text{ for all } \omega \in \Omega';$$

and for some $C_2, R_2 > 0$,

$$\begin{aligned} &|\partial^{\gamma} K_{2\omega}(u)| < C_2 R_2^{\|\gamma\|} \gamma! \\ &\text{for } 1 \leqq |u| \leqq 2, \gamma \in (\mathbb{Z}^+)^{2n+1},\; \omega \in \Omega'. \end{aligned}$$

The proof will merely require us to go over the proof of Theorem 5.1(e) => (a) carefully. However, as we said, we do need a lemma from Chapter 7, which is the following:

Fact 5.10. (A consequence of Lemma 7.8(b). A self-contained proof is given in Lemma 7.6, Proposition 7.7 and Lemma 7.8 taken together.) Suppose $\Omega \subset \mathbb{C}^s$ is open, $k_1, k_2 \in \mathbb{C}$, $\mathrm{Re}(k_1+k_2) < -2n - 2$. For $\nu = 1,2$, suppose $K_{\nu\omega} \in AK^{k_\nu}$ depends holomorphically on $\omega \in \Omega$. Further, suppose that there exist $C_1, C_2, R_1, R_2 > 0$ so that

$$|\partial^\gamma K_{\nu\omega}(u)| < C_\nu R_\nu^{\|\gamma\|}\gamma! \quad \text{for } 1 \leq |u| \leq 2, \gamma \in (\mathbb{Z}^+)^{2n+1}.$$

Then $K_\omega = K_{2\omega} * K_{1\omega}$ depends holomorphically on $\omega \in \Omega$. Further, if $\overline{\Omega}' \subset \Omega$, Ω' open, $\overline{\Omega}'$ compact, there exist $C, R > 0$ so that for all $\omega \in \Omega'$,

$$|\partial^\gamma K_\omega(u)| < CR^{\|\gamma\|}\gamma! \quad \text{for } 1 \leq |u| \leq 2, \gamma \in (\mathbb{Z}^+)^{2n+1}. \tag{5.14}$$

That K_ω depends holomorphically on $\omega \in \Omega$ follows easily from the method of (5.1) and from the use of the map Q of Proposition 5.2(a). The real content is the uniform estimate (5.14).

Using this fact for Theorem 5.9, we proceed to retrace through the steps of the proof of Theorem 5.1(e) => (a), seeking holomorphic dependence and uniform estimates at every turn. We begin by generalizing Lemma 5.6. We show:

(***) Suppose $k \in \mathbb{C}$, $\Omega \subset \mathbb{C}^s$ open. Suppose $K_{1\omega} \in AK^{k-2n-2}$ depends holomorphically on $\omega \in \Omega$, satisfies (5.13), and that $FK_{1\omega}(0,\xi) \neq 0$ for $\xi \neq 0$, $\omega \in \Omega \cap \mathbb{R}^s$. Suppose $E \subset \mathbb{R}^s$ is compact, and $E \subset \Omega$. Then there exists an open set $\Omega'' \subset \Omega$ with $E \subset \Omega''$ as follows: For $\omega \in \Omega''$, there exist $K_{3\omega} \in AK^{-k-2n-2}$ and $K_{4\omega} \in APV$ depending holomorphically on ω so that $K_{3\omega} * K_{1\omega} = \delta - K_{4\omega}$ and so that $(FK_{4\omega})(\underline{+}1,\xi) \in Z_2^2(\mathbb{R}^{2n})$. Further, for some $C_3, R_3 > 0$,

$$|\partial^\gamma K_{\nu\omega}(u)| < C_3 R_3^{||\gamma||}\gamma! \quad \text{for } 1 \leqq |u| \leqq 2, \gamma \in (\mathbb{Z}^+)^{2n+1}, \quad \nu = 3 \text{ or } 4, \omega \in \Omega''. \tag{5.15}$$

To prove (***), we imitate the procedure of the Reductions with a few changes. It will be better for us to first reduce to the case $0 < \text{Re } k \leqq 2$ (instead of $0 \leqq \text{Re } k < 2$ as in the Reductions). The reduction to $\text{Re } k \leqq 2$ proceeds just as in the argument in the Reductions which allowed one to assume $\text{Re } k < 2$. The reduction to $0 < \text{Re } k \leqq 2$ proceeds more or less like the argument in the Reductions which allowed one to assume $0 \leqq \text{Re } k < 2$. However, here we do not quote Theorem 3.2 to produce $K'_{1\omega} \in AK^{k+2r-2n-2}$ with $R_o^r K'_{1\omega} = K_{1\omega}$ (r suitable). For in addition, we need $K'_{1\omega}$ to depend holomorphically on ω and to satisfy uniform estimates of the type (5.13). To find $K'_{1\omega}$, what we do is to convolve $K_{1\omega}$ on the left, repeatedly, with the fundamental solution of R_o. This gives what we want, by Fact 5.10, if we replace Ω by a smaller relatively compact open subset which contains E. We next check the analogue of (**) of the Reductions, where now $0 < \mathrm{Re}\, k \leqq 2$:

(**) There exists an open set $\Omega'' \subset \Omega$ with $E \subset \Omega''$ as follows: For $\omega \in \Omega''$, there exists $K_{5\omega} \in AK^{-k-2n-2}$ so that $FK_{5\omega}(0,\xi) = [FK_{1\omega}(0,\xi)]^{-1}$ for all $\xi \neq 0$, so that $K_{5\omega}$ depends holomorphically on $\omega \in \Omega''$, and satisfies estimates analogous to (5.13) uniformly in $\omega \in \Omega''$.

This follows directly from the construction which provides (**) of the Reductions, together with (2.23) and Corollary 2.15. (If $FK_{1\omega} = F_{1\omega}, F_{1\omega+} \sim \sum g_{1\omega\ell}$ in $Z^2_{2,-k}$, note that we cannot necessarily invert $g_{1\omega 0}$ for all $\omega \in \Omega$. But we can do so for ω in some Ω''.) Now we again have $F(K_{5\omega} * K_{1\omega})(0,\xi) = 1$ for $\xi \in \mathbb{R}^{2n}$, $\xi \neq 0$. (We cannot use the map I_c to prove this as we did in the Reductions, for now $\mathrm{Re}\, k - 2n - 2$ just might be -2. It follows, in any event, from Lemma 5.7 in the case $\ell = 0$ together with (5.3). The method of the Reductions, together with Fact 5.10, now allows us to work in case (*):

(*) $K_{1\omega} \in APV$ for all $\omega \in \Omega$ and $FK_{1\omega}(0,\xi) = 1$ for all $\xi \neq 0$.

In this case we may in (***), in fact take Ω'' to be any relatively compact open subset of Ω with $E \subset \Omega''$, $\overline{\Omega''} \subset \Omega$. (**) in case (*) follows easily from the proof of Lemma 5.6, together with (2.24) as applied to $K_{1\omega} - \delta$, and Corollary 2.15. The estimate (5.15) for $\nu = 4$ follows at once from the relation $K_{3\omega} * K_{1\omega} = \delta - K_{4\omega}$ and Fact 5.10.

With (***) established, we are now almost prepared to imitate the proof of Theorem 5.1(d) => (c), to complete the proof of Theorem 5.9. First, however, we need to state a few simple general facts about holomorphic families of vectors and operators.

Say $\Omega \subset \mathbb{C}^s$ for some s. Fix $\lambda \neq 0$. Say $k \in \mathbb{C}$. We shall say that a family of vectors $v_\omega \in H^k_\lambda$ is <u>weakly holomorphic</u> in $\omega \in \Omega$ if (u, v_ω) is a holomorphic function of $\omega \in \Omega$ for every fixed $u \in H^{-k}$. We say v_ω is weakly anti-holomorphic in ω if $v_{\bar{\omega}}$ is weakly holomorphic in ω. The uniform boundedness principle shows at once that if $v_\omega \in H^k$ is weakly holomorphic or anti-holomorphic in $\omega \in \Omega$, and if $\omega_o \in \Omega$, then

$$\begin{array}{l} \|v_\omega\|_k \text{ and } \|(v_\omega - v_{\omega_o})/(\omega-\omega_o)\|_k \text{ are uniformly} \\ \text{bounded for } \omega \text{ in compact subsets of } \Omega. \end{array} \tag{5.16}$$

Next, if $R_\omega : H^j \to H^k$ is a family of operators for $\omega \in \Omega$, we say R_ω is weakly holomorphic in ω if, for each $v \in H^j$, $R_\omega v \in H^k$ is weakly holomorphic in $\omega \in \Omega$. It follows easily from use of adjoints, difference quotients and (5.16) that if $R_\omega : H^j \to H^k$ and $v_\omega \in H^j$ are weakly homomorphic in ω, then

$R_\omega v_\omega \in H^k$ is weakly holomorphic in ω. Further, it follows easily from the usual method of differentiating inverses (5.16), and a use of adjoints, that if $R_\omega : H^j \to H^k$ is weakly holomorphic in $\omega \in \Omega$, and if $R_\omega^{-1} : H^k \to H^j$ exists for all $\omega \in \Omega$, then R_ω^{-1} is weakly holomorphic in $\omega \in \Omega$.

Another general comment. Suppose $k \in \mathbb{C}$, $\Omega \subset \mathbb{C}^s$, $K_\omega \in K^{k-2n-2}$ for all $\omega \in \Omega$, and K_ω depends holomorphically on ω. Suppose also that for all $\gamma \in (\mathbb{Z}^+)^{2n+1}$, and for every compact set $E \subset \Omega$, there exists $C_{E\gamma}$ so that

$$|\partial^\gamma K_\omega(u)| < C_{E\gamma} \text{ for all } \omega \in E,\ 1 \leq |u| \leq 2.$$

Let $\underset{\sim}{J}_\omega = \hat{K}_\omega$. Then we claim that for each λ, $J_\omega(\lambda) : H_\lambda^{-k} \to H_\lambda^0$ and $J_\omega(\lambda) : H_\lambda^0 \to H_\lambda^k$ are weakly holomorphic in ω. Indeed, by use of $\sim$ and adjoints, we see that we need only show $J_\omega(\lambda) : H^{-k} \to H^0$ is weakly holomorphic in ω. Now for f,g with $\hat{f},\hat{g} \in \mathcal{Q}$, we have $(g|f*K_\omega) = c_n(\hat{g}|\hat{f}\underset{\sim}{J}_\omega)$ as the identity $(\hat{f}*g)^\wedge = \hat{f}*\hat{g}$ shows. Now for f with $\hat{f} \in \mathcal{Q}$, we have $(f|K_\omega) = c_n(\hat{f}|J_\omega)$. If we select f as in the second paragraph of the proof of Theorem 5.1(d) => (c), we see that $(E_\beta, J_\omega(\lambda)E_\alpha)$ is holomorphic in ω for all α,β. Use of Proposition 5.5(e) and an argument involving uniform convergence of holomorphic functions on compact subsets of Ω now shows $J_\omega(\lambda) : H^{-k} \to H^0$ is weakly holomorphic in ω.

With these preliminaries complete, we now turn to the proof of Theorem 5.9.

<u>Proof of Theorem 5.9</u>. First note that we may assume Re $k > 0$. Otherwise, select $r \in \mathbb{Z}^+$ with $0 < \text{Re } k + 2r \leqq 2$, and convolve $K_{1\omega}$ on the left r times with the fundamental solution of $\mathcal{R}_0$ to obtain $K'_{1\omega} \in \mathcal{AK}^{k+2r-2n-2}$, depending holomorphically on ω, so that $\mathcal{R}_0^r K'_{1\omega} = K_{1\omega}$. Because of Fact 5.10, if the theorem were known in case Re $k > 0$, then we could find $\Omega', K'_{2\omega} \in \mathcal{AK}^{-k-2r-2n-2}$ with $K'_{2\omega}$ holomorphic in ω and with $K'_{2\omega} * K'_{1\omega} = \delta$ for $\omega \in \Omega'$ and satisfying estimates as in the theorem. We would then obtain $K_{2\omega}$ by convolving $K'_{2\omega}$ on the right r times with the fundamental solution of $\mathcal{R}_0$, using Fact 5.10 (shrinking Ω' a bit) and Proposition 5.2(g). (We have used the relations $F(\mathcal{R}_0 F)(0,\xi) = (|\xi|^2/4)(FK)(0,\xi)$, $(\mathcal{R}_0 K)\hat{} = \underset{\sim}{A}\hat{K}$ in the $\mathcal{Q}$ sense, valid for $K \in \mathcal{AK}^{k'}$ for any k'.)

With $k, \Omega, K_{1\omega}, E$ as in the statement of the theorem, Re $k > 0$, now construct $\Omega'', K_{3\omega}, K_{4\omega}$ as in (***). Clearly we need now only produce Ω' open, $E \subset \Omega' \subset \Omega$, and $K_{7\omega} \in \mathcal{AK}^{-k-2n-2}$ depending holomorphically on $\omega \in \Omega'$, so that

$$K_{7\omega} * K_{1\omega} = K_{4\omega} \quad \text{for all } \omega \in \Omega',$$

so that for some $C_7, R_7 > 0$

$$|\partial^\gamma K_{7\omega}(u)| < C_7 R_7^{\|\gamma\|} \gamma! \qquad (5.17)$$
$$\text{for } 1 \leqq |u| \leqq 2, \gamma \in (\mathbb{Z}^+)^{2n+1}, \ \omega \in \Omega'$$

and so that $F_{7\omega\lambda}(\zeta) = (F'K_{7\omega})(\lambda,\zeta) \in Z_2^2(\mathbb{R}^{2n})$ for $\lambda \neq 0$. (Here $\zeta \in \mathbb{C}^n$, identified with $\mathbb{R}^{2n}$.)

We construct $K_{7\omega}$ by using the group Fourier transform. Let $\hat{K}_{1\omega} = J_{1\omega}$, $(\tilde{K}_{1\omega})^\wedge = \underset{\sim}{J}_{1\omega}$, $\hat{K}_{4\omega} = \underset{\sim}{J}_{4\omega}$, $(\hat{K}_{4\omega})^\wedge = \underset{\sim}{J}_{4\omega}$. We make a slight change from the proof of Theorem 5.1(d) => (c); we wish to extend $\tilde{J}_{1\omega+} : H^0 \to H^k$ as a bounded operator (not from H^{-k} to H^0). $\tilde{J}_{1\omega+}$ is then the adjoint of the corresponding extension $J_{1\omega+} : H^{-k} \to H^0$ as the arguments of the second paragraph of Theorem 5.1(d) => (c) show. Let $S_\omega = \tilde{J}_{4\omega+}$, $S_\omega^\alpha = S_\omega E_\alpha \in H^\omega$; we construct $\hat{\tilde{K}}_{7\omega}(1)$ by solving the equation

$$\tilde{J}_{1\omega+} H_\omega^\alpha = S_\omega^\alpha \qquad (5.18)$$

in an appropriate manner.

Because of (***), we may assume Ω'' is sufficiently small that there exist $K_{5\omega} \in AK^{-k-2n-2}$, $K_{6\omega} \in$ APV with $K_{5\bar{\omega}}$, $K_{6\bar{\omega}}$ depending holomorphically on ω for $\bar{\omega} \in \Omega''$, so that $K_{5\omega} * \tilde{K}_{1\omega} = \delta - K_{6\omega}$, $(FK_{6\omega})(\pm 1,\xi) \in Z_2^2(\mathbb{R}^{2n})$ for $\lambda \neq 0$, and for some $C_4, R_4 > 0$,

$$|\partial^\gamma K_{\nu\omega}(u)| < C_4 R_4^{\|\gamma\|} \gamma! \quad \text{for } 1 \leqq |u| \leqq 2, \gamma \in (\mathbb{Z}^+)^{2n+1}, \qquad (5.19)$$
$$\nu = 5 \text{ or } 6, \ \omega \in \Omega''.$$

Now put $\hat{K}_{\nu\omega} = \underset{\sim}{J}_{\nu\omega}$ for $\nu = 5$ or 6, so that

$$J_{5\omega}(\lambda)\tilde{J}_{1\omega}(\lambda)v = (I-J_{6\omega}(\lambda))v \tag{5.20}$$

for $v \in H^o$. (We are thinking of $J_{5\omega}(\lambda)$ as having been extended to be a bounded operator from H^k to H^o.)

From (5.20), Proposition 5.5(a) and (c), any solution H_ω^α of (5.18) must be in H^ω. We need specific solutions which vary anti-holomorphically in ω and satisfy good estimates. To this end, we shall show:

<u>Claim</u>. There exists Ω' open, with $E \subset \Omega' \subset \Omega''$ and $\mathbb{C}_{\overline{\Omega}'} = \Omega'$, and $C > 0$, so that $\tilde{J}_{1\overline{\omega}+} J_{1\omega+} : H^{-k} \to H^k$ is invertible, and $||(\tilde{J}_{1\overline{\omega}+} J_{1\omega+})^{-1}||_{k,-k} < C$ for all $\omega \in \Omega'$. (Here $\mathbb{C}_{\overline{\Omega}'}$ denotes the complex conjugate of the set Ω', $\{\omega : \overline{\omega} \in \Omega'\}$.)

To see this, for any $r,s \in \mathbb{C}$, let $B(H^r,H^s)$ denote the Banach space of bounded operators from H^r to H^s, with the norm topology. It is easy to see from Proposition 5.5(e), the mean value theorem and the Cauchy estimate for a derivative that $\omega \to J_{1\omega+}$ is continuous from Ω to $B(H^{-k},H^o)$. Thus $\omega \to \tilde{J}_{1\overline{\omega}+}$ is continuous from $\mathbb{C}_{\overline{\Omega}}$ to $B(H^o,H^k)$, and $\omega \to \tilde{J}_{1\overline{\omega}+} J_{1\omega+}$ is continuous from $\Omega \cap \mathbb{C}_{\overline{\Omega}}$ to $B(H^{-k},H^k)$. Since the invertible operators form an open subset of $B(H^{-k},H^k)$, it now clearly suffices to show that $\tilde{J}_{1w+}J_{1w+} : H^{-k} \to H^k$ is invertible for each $w \in \Omega \cap \mathbb{R}^s$. Indeed, by Atkinson's Theorem ([16]) and the analogues of (5.9), (5.10) for J_{1w+} in place of $\tilde{J}_1(\lambda)$ it is clear that $J_{1w+} : H^{-k} \to H^o$ has closed range and kernel contained in H^{ω}. Hence it is injective and bounded below. Since $\tilde{J}_{1w+} : H^0 \to H^k$ is the adjoint of this operator, $\tilde{J}_{1w+}J_{1w+} : H^{-k} \to H^k$ is self-adjoint and injective. Further, it is bounded below, since for some $C > 0$, if $v \in H^{-k}$, $\|v\|_{-k}^2 \leq C\|J_{1w+}v\|_0^2 \leq C\|\tilde{J}_{1w+}J_{1w+}v\|_k\|v\|_{-k}$. Thus $\tilde{J}_{1w+}J_{1w+}$: $H^{-k} \to H^k$ is self-adjoint, injective, and has closed range; hence it is invertible. This establishes the "Claim". As the proof shows, we may assume Ω' is sufficiently small that for some $C' > 0$,

$$\|J_{1\omega+}\|_{-k,0} = \|J_{1\omega+}\|_{0,k} < C' \quad \text{for all } \omega \in \Omega'. \tag{5.21}$$

For (5.18), for $\omega \in \Omega'$, we now put

$$H_{\omega}^{\alpha} = J_{1\bar{\omega}+}(J_{1\omega+}J_{1\bar{\omega}+})^{-1}S_{\omega}^{\alpha} \tag{5.22}$$

(5.18) is then clearly satisfied. Also, $S_{\omega}^{\alpha} = \tilde{J}_{4\omega+}E_{\alpha}$ is weakly anti-holomorphic in $\omega \in \Omega''$, since $J_{4\omega+} : H^0 \to H^0$ is weakly holomorphic in $\omega \in \Omega''$. Thus $H_{\omega}^{\alpha} \in H^0$ is weakly anti-holomorphic in $\omega \in \Omega'$. To complete the proof, we shall have to obtain uniform estimates on H_{ω}^{α} for $\omega \in \Omega'$.

Note that for some $C_5 > 0$, $0 < r_2 < 1$, we have

$$|S_{\omega\beta}^{\alpha}| < C_5 r_2^{|\alpha|+|\beta|} \quad \text{for all } \alpha,\beta \in (\mathbb{Z}^+)^n,\ \omega \in \Omega' \tag{5.23}$$

if $S_{\omega\beta}^{\alpha} = (E_{\beta},S_{\omega}^{\alpha}) = (E_{\beta},\tilde{J}_{4\omega+}E_{\alpha})$. This follows easily from (5.15) with $\nu = 4$, together with (2.24) and the second uniformity of Proposition 2.1, together with the third uniformity in the Remark after Proposition 4.2 and the uniformity in Theorem 2.3(a) =>(d), used in conjunction with $J_{4\omega}(\lambda) = \mathcal{W}_\lambda(F'_\lambda K_{4\omega})$. We need to obtain an analogue of $H^\alpha_{\omega\beta}$ in place of $S^\alpha_{\omega\beta}$. To this end, we write

$$H^\alpha_\omega = J_{6\omega+}H^\alpha_\omega + J_{5\omega+}S^\alpha_\omega \tag{5.24}$$

by (5.20). By (5.23), the "Claim" and (5.21), there exists $C_6 > 0$ so that

$$\|H^\alpha_\omega\|_0 < C_6 r_2^{|\alpha|} \quad \text{for all } \omega \in \Omega'. \tag{5.25}$$

Just as in the argument which verified (5.23), which examined the matrix entries of $J_{4\omega+}$, we see from (5.19) with $\nu = 6$ that for some $C_6 > 0$, $0 < r_3 < 1$,

$$|(E_\beta, J_{6\omega+}E_\alpha)| < C_6 r_3^{|\alpha|+|\beta|} \quad \text{for all } \alpha,\beta \in (\mathbb{Z}^+)^n,\ \omega \in \Omega'. \tag{5.26}$$

From (5.25), (5.26) and the proof of Proposition 5.5(c) we see that for some $C_8 > 0$, $0 < r_4 < 1$,

$$|(E_\beta, J_{6\omega+}H^\alpha_\omega)| < C_8 r_4^{|\alpha|+|\beta|} \quad \text{for all } \alpha,\beta \in (\mathbb{Z}^+)^n,\ \omega \in \Omega'. \tag{5.27}$$

We claim that there also exist $C_9 > 0$, $0 < r_5 < 1$ so that if Ω' is slightly shrunk,

$$|(E_\beta, J_{5\omega+}S^\alpha_\omega)| < C_9 r_2^{|\alpha|} r_5^{|\beta|} \quad \text{for all } \alpha,\beta \in (\mathbb{Z}^+)^n,\ \omega \in \Omega'. \tag{5.28}$$

(5.28) follows from (5.23), (5.19) for $\nu = 5$, and the proof of Proposition 5.5(a). (For the latter, see the Appendix A.2 to this chapter. Note that we need to use a uniform version of the case Re $k < 0$ of that proof. This case in turn follows from a uniform version of the case $0 < \text{Re } k < 2n + 2$, repeated convolution with the fundamental solution of $\mathcal{R}_o$, and from Fact 5.10.) From (5.26), (5.27), (5.28), we see that there exist $C_{10} > 0$, $0 < r_1 < 1$ with

$$|H^{\alpha}_{\omega\beta}| < C_{10} r_1^{|\alpha|+|\beta|} \text{ for all } \alpha,\beta \in (\mathbb{Z}^+)^n,\ \omega \in \Omega'. \tag{5.29}$$

Now define $H_{\omega+} \in Z_2^2(H_1)$ with $(E_\alpha, H_{\omega+}E_\beta) = H^{\alpha}_{\omega\beta}$.

Then, by (5.18), $\tilde{J}_{1\omega+}H_{\omega+} = \tilde{J}_{4\omega+}$. We construct $H_{\omega-} \in Z_2^2(H_{-1})$ in a like manner with $\tilde{J}_{1\omega-}H_{\omega-} = \tilde{J}_{4\omega-}$. If $K_{7\omega}$ in $AK^{-k-2n-2}$ satisfies $\hat{K}_{7\omega}(1) = H^*_{\omega+}, \hat{K}_{7\omega}(-1) = H^*_{\omega-}$, then surely $K_{7\omega} * K_{1\omega} = K_{4\omega}$. To complete the proof we must be certain that $K_{7\omega}$ depends holomorphically on ω and satisfies the estimate (5.17).

For this, put $F_{7\omega+} = W_1^{-1}H^*_{\omega+}, F_{7\omega-} = W_{-1}^{-1}H^*_{\omega-}$; both are in $Z_2^2(\mathbb{R}^{2n})$. It is easy to see from the inversion formula for G_λ and the usual formula for the Fourier transform that $F_{7\omega+}(\zeta)$, $F_{7\omega-}(\zeta)$ are holomorphic in $\omega \in \Omega'$ for each fixed $\zeta \in \mathbb{C}^n$ (identified with $\mathbb{R}^{2n}$). Further, from the fourth uniformity assertion of Proposition 4.2, and (5.29), we see that for some $B_1, B_2, C_{11} > 0$, $F_{7\omega\pm} \in Z_2^2(B_1, B_2, C_{11})$ for all $\omega \in \Omega$. If we now, for the first time, use the fact that we are working in the case Re $k > 0$ (see the beginning of the proof), we see that there is a unique $K_{7\omega} \in AK^{-k-2n-2}$ with $F'K_{7\omega}(\underline{+}1,\zeta) = F_{7\omega\underline{+}}(\zeta)$. Further, by the uniformity assertion of Theorem 1.8(b) and the second uniformity assertion of Corollary 1.5, $K_{7\omega}$ satisfies an estimate of the type (5.17). It also depends holomorphically on ω by (2.20). This completes the proof of Theorem 5.9.

Appendix to Chapter 5

We prove a few facts whose proofs were omitted so as not to interrupt the flow of the presentation.

A.1. We prove Proposition 5.2(g) and (h). For (g), write $K_i = K_i' + K_i''(i=1,2,3)$ where $K_i' \in \mathcal{E}'$, $K_i'' \in C^\infty$. The basic fact we use is (5.1). We start by observing that if $F \in \mathcal{E}'$ then $(F*K_2'')*K_1'' = F*(K_2''*K_1'')$. Indeed, observe that K_2'' and all its derivatives are $O(1+|u|)^{\mathrm{Re}\,k+\varepsilon}$ for any $\varepsilon > 0$, so that $|(F*K_2'')(u)| = O(1+|u|)^{\mathrm{Re}\,k+\varepsilon}$ for any $\varepsilon > 0$. (ε is included just to handle the case where log terms enter.) If χ_N is the characteristic function of $\{|u| < N\}$, then $(F*K_2'')*\chi_N K_1'' = F*(K_2''*\chi_N K_1'')$. As $N \to \infty$, the left side approaches $(F*K_2'')*K_1''$ uniformly on compact sets; for the right side, note $K_2''*\chi_N K_1 \to K_2''*K_1''$ uniformly on compact sets, hence in the distribution sense, so the right side approaches $F*(K_2''*K_1'')$ in the distribution sense. This establishes the claim. Similarly $(K_1''*K_2'')*F = K_1''*(K_2''*F)$.

Using (5.1), these observations, and the associative law if at least two distributions have compact support, we see $K_3*(K_2*K_1) = K_3'*(K_2'*K_1') + K_3'*(K_2'*K_1''+K_2''*K_1'+K_2''*K_1'') + K_3''*(K_2'*K_2') + K_3''*(K_2'*K_1''+K_2''*K_1'+K_2''*K_1'') = (K_3'*K_2')*K_1' + (K_3'*K_2')*K_1'' + (K_3'*K_2'')*K_1' + (K_3'*K_2'')*K_1'' + (K_3''*K_2')*K_1' + (K_3''*K_2'')*K_1' + K_3''*(K_2*K_1'')$, where $K_2*K_1'' = K_2'*K_1''+K_2''*K_1''$ by definition. By (5.1),

$K_2*K_1'' = K-K_2'*K_1'-K_2''*K_1' = 0(1+|u|)^{\mathrm{Re}(j+k)+2n+2+\varepsilon} + 0(1+|u|)^{\mathrm{Re}\,k+\varepsilon}$ for any $\varepsilon > 0$, so that $K_3''*(K_2*K_1'')$ does exist. To complete the proof we must show it equals $(K_3''*K_2)*K_1''$.

To this end we introduce a partition of unity used in the proof of L^2 boundedness of convolution with elements of $PV(\mathbb{H}^n)$ ([65]). We fix $\varphi_0 \in C^\infty[0,\infty)$ such that $\varphi_0(t) = 1$ for $0 \leqq t \leqq 1$, $\varphi_0(t) = 0$ for $t \geqq 2$. Put $\varphi(u) = \varphi_0(|u|)$ and $K_{1M}(u) = [\varphi(D_{2^{-M}}u) - \varphi(D_{2^{-M+1}}u)]K_M(u)$. Then $K_1(u) = \sum_{M=-\infty}^{\infty} K_{1M}(u)$ for $u \neq 0$. Without loss we may assume $K_1'' = \sum_{M=0}^{\infty} K_{1M}(u)$, since in general the funciton on the left differs from K_1'' by a C_c^∞ function. Since $K_3''*K_2$ is $0(1+|u|)^{\mathrm{Re}(k+\ell)+2n+2+\varepsilon} + 0(1+|u|)^{\mathrm{Re}\,k+\varepsilon}$ for any $\varepsilon > 0$, $(K_3''*K_2)*K_1'' = \sum_{M=0}^{\infty}(K_3''*K_2)*K_{1M} = \sum K_3''*(K_2*K_{1M})$, with uniform convergence on compact sets. To complete the proof it suffices to show that for any $\varepsilon > 0$, there exists $C > 0$ so that for all N and all u, $\sum_{M=0}^{N}|(K_2*K_{1M})(u)| < C(1+|u|)^{A+\varepsilon}$ where $A = \max(\mathrm{Re}(j+k)+2n+2, \mathrm{Re}\,k)$.

Now $K_{1M}(u) = 2^{Mj}K_{10}(D_{2^{-M}}u)$. A simple computation shows, then that if $K_2 \in \mathrm{Rhom}_k$, then $(K_2*K_{1M})(u) = 2^{M(j+k+2n+2)}(K_2*K_{10})(D_{2^{-M}}u)$, while if $k \in \mathbb{Z}^+$ and $K_2 = K_{2,hom} + p(x)\log|x|$ ($K_{2,hom} \in \mathrm{Rhom}_k$, p a polynomial) then $(K_2*K_{1M})(u) = 2^{M(j+k+2n+2)}(K_{2,hom}*K_{10})(D_{2^{-M}}u) + 2^{M(j+k+2n+2)}\log(2^M)(p*K_{10})(D_{2^{-M}}u)$. Put $a = \mathrm{Re}\,k$, $b = -[\mathrm{Re}(j+k)+2n+2]$. It follows at once that for any $\varepsilon > 0$ there exists $C > 0$ such that $|(K_2 * K_{1M})(u)| < C(M+1)2^{-Mb}(1+|u|/2^M)^a$. It suffices to show, then, that for any $\varepsilon > 0$ there exists $C > 0$ so that for all N and all $x \geq 0$ one has $\sum(M+1)2^{-Mb}(1+x/2^M)^a < C(1+x)^{A+\varepsilon}$; recall $A = \max(a,-b)$ and $b > 0$. If $a \geq 0$, the relation follows from $1 + x/2^M \leq 1 + x$. Say then $0 > a = -c$. If $-c > -b$, so $b > c$, the relation follows from $\sum(M+1)/[2^{Mb}(1+x/2^M)^a] = \sum(M+1)/[2^{M(b-c)}(2^M+x)^c] < C(1+x)^{-c}$, since $1 + x \leq 2^M + x$. If instead $-c < -b$, so $b < c$, we note that if $0 < \varepsilon < b$, $\sum(M+1)/[2^{M(b-c)}(2^M+x)^c] < [1/(1+x)^{b-\varepsilon}] \times \sum(M+1)2^{M(c-b)}/[2^M+x]^{c-b+\varepsilon} \leq (1+x)^{-b+\varepsilon}\sum(M+1)2^{-M\varepsilon}$ as desired. This proves Proposition 5.2(g).

For Proposition 5.2(h), write

$$\mathcal{D}_1 K_1 = \mathcal{D}_1 K_{1\varepsilon} + \mathcal{D}_1(1-\phi_\varepsilon)K_1,\ \mathcal{D}_2 K_2 = \mathcal{D}_2 K_{2\varepsilon} + \mathcal{D}_2(I-\phi_\varepsilon)K_2.$$

Use the distributive law to write $\mathcal{D}_2 K_2 * \mathcal{D}_1 K_1$ as a sum of four terms. Thereby, write $\mathcal{D}_2 K_2 * \mathcal{D}_1 K_1 - \mathcal{D}_2 K_{2\varepsilon} * \mathcal{D}_1 K_{1\varepsilon}$ as a sum of three terms; we prove all three approach zero in the sense of distributions as $\varepsilon \to 0$.

The terms are

(i) $\mathcal{D}_2 K_{2\varepsilon} * \mathcal{D}_1(1-\phi_\varepsilon)K_1$

(ii) $\mathcal{D}_2(1-\phi_\varepsilon)K_2 * \mathcal{D}_1(1-\phi_\varepsilon)K_1$

(iii) $\mathcal{D}_2(1-\phi_\varepsilon)K_2*\mathcal{D}_1K_{1\varepsilon}$.

For (i), observe that for any compact set C there exist $\varepsilon_o, N_o > 0$ so that, on C, if $\varepsilon \leqq \varepsilon_o$, then $\mathcal{D}_2K_{2\varepsilon}*\mathcal{D}_1(1-\phi_\varepsilon)K_1 = \mathcal{D}_2(\phi_\varepsilon-\phi_{N_o})K_2*\mathcal{D}_1(1-\phi_\varepsilon)K_1$. From this it is evident that (i) approaches 0 uniformly on compact sets as $\varepsilon \to 0$.

(ii) does likewise for obvious reasons, and (iii) is handled in a similar fashion to (i). This proves Proposition 5.2(h).

If $K_1 \in K^j(\mathbb{R}^{2n})$, $K_2 \in K^k(\mathbb{R}^{2n})$ and Re j, Re k, Re(j+k) + 2n < 0, and if $K = K_2*K_1$ is as in (5.1), we show $F_{\mathbb{R}^{2n}}(K_2*K_1) = (F_{\mathbb{R}^{2n}}K_2)(F_{\mathbb{R}^{2n}}K_1)$. Since both sides are homogeneous of degree $-4n - j - k > -2n$, we need only show they agree away from 0. We need only show $F(K_2''*K_1'') = (FK_2'')(FK_1'')$ away from 0, since $K_1', K_2' \in E'$. But this is immediate, since if N is large, $\Delta^{2N}(K_2''*K_1'') = \Delta^N K_2''*\Delta^N K_1''$, and $\Delta^N K_1'', \Delta^N K_2'' \in L^1$. ∎

A.2. We prove Proposition 5.5(a), (b), and (e). First we need some preliminaries. We begin by recalling that there exists, for each $\tau > 0$, $h_\tau \in S(\mathbb{H}^n)$ so that $\hat{h}_\tau(\lambda)E_\alpha = e^{-(2|\alpha|+n)|\lambda|\tau}E_\alpha$. ($h(\tau,u) = h_\tau(u)$ is the heat kernel for L_o; it satisfies $(\partial/\partial\tau+L_o)h = 0$. h can be explicitly computed, and one can thus directly show that it is in S; see, for example, the discussion in the second paragraph preceding Theorem 7.4 of [28].) In fact, simple relations for the

Fourier transform show that $h_\tau(u) = \tau^{-n-1}h_1(D_{\tau^{-1/2}}u)$. From this, it is easy to see that we can, for $0 < \text{Re } k < 2n+2$, define $\Phi_k \in K^{k-2n-2}$ by $\Phi_k(u) = \Gamma(k/2)^{-1}\int_0^\infty h_\tau(u)\tau^{k/2-1}d\tau$.* As $\varepsilon \to 0$, $N \to \infty$, we have $\Gamma(k/2)^{-1}\int_\varepsilon^N h_\tau(u)\tau^{k/2-1}d\tau \to \Phi_k$ in the S' sense, by dominated convergence. It follows immediately that $\hat{\Phi}_k = \underset{\sim}{R}_k$ in the $\mathcal{Q}$ sense, where $R_k(\lambda)E_\alpha = [(2|\alpha|+n)|\lambda|]^{-k/2}E_\alpha$. Next, if $\text{Re } k \leq 0$, define $K_k \in K^{k-2n-2}$ as follows: select $m \in \mathbb{N}$ so that $0 < \text{Re } k + 2m \leq 2$, and put $\Phi_k = L_0^m\Phi_{k+2m}$. Then, by Proposition 4.8(a), $\hat{\Phi}_k = \underset{\sim}{R}_k$ in the $\mathcal{Q}$ sense, where $R_k(\lambda)E_\alpha = [(2|\alpha|+n)|\lambda|]^{-k/2}E_\alpha$.

Suppose $\hat{f} \in \mathcal{Q}$. The argument in the last paragraph of the proof of Theorem 4.1 shows that if $K \in K^k$ for some k then $f*K \in S$. We claim $(f*\Phi_k)^\wedge = \hat{f}\underset{\sim}{R}_k$ (if $\text{Re } k < 2n+2$) so that $(f*\Phi_k)^\wedge \in \mathcal{Q}$ also. To see this, note that whenever $g \in S$ we have $(g|f*\Phi_k) = (\tilde{f}*g|\Phi_k)$. Since $(\tilde{f})^\wedge = (\hat{f})^*$, $(\tilde{f}*g)^\wedge = \hat{f}^*\hat{g}$, so this equals $(\hat{f}^*\hat{g}|\underset{\sim}{R}_k) = c_n(\hat{g}|\hat{f}\underset{\sim}{R}_k) = (g|F)$ where $F \in S$, $\hat{F} = \hat{f}\underset{\sim}{J}_k \in \mathcal{Q}$. Thus, $f*\Phi_k = F$, as claimed.

From this it follows at once that if $-2n-2 < \text{Re } k < 2n+2$, we have $\Phi_{-k}*\Phi_k = \delta$. Indeed, since the left side is in $PV(\mathbb{H}^n)$, and since $\{f|\hat{f} \in \mathcal{Q}\}$ is dense in L^2, it suffices to show that whenever $\hat{f} \in \mathcal{Q}$, $f*\Phi_{-k}*\Phi_k = f$. Since, however, $(f*\Phi_{-k})^\wedge \in \mathcal{Q}$, we have $(f*\Phi_{-k}*\Phi_k)^\wedge = (f*\Phi_{-k})^\wedge\underset{\sim}{R}_k = \hat{f}\underset{\sim}{R}_{-k}\underset{\sim}{R}_k = \hat{f}$, as claimed. Observe that if $\text{Re } k < 0$ and $\hat{f} \in \mathcal{Q}$ then

*A similar construction was given by Folland [18].

$f*\tilde{\Phi}_k = f*\Phi_{\bar{k}}$. Indeed, if $\hat{g} \in \mathcal{Q}$, $(f*\tilde{\Phi}_k|g) = (f|g*\Phi_k) =$ $c_n(\hat{f}|\hat{g}\underset{\sim}{R}_k) = c_n(\hat{f}\underset{\sim}{R}_{\bar{k}}|\hat{g}) = (f*\Phi_{\bar{k}}|g)$, since $R_k(\lambda)$ is diagonal. The claim follows from the density of $\{g|\hat{g} \in \mathcal{Q}\}$ in L^2.

We can now show Proposition 5.5(a) and (e). If $\underset{\sim}{J}$ exists as in Proposition 5.5(a), it must be unique since for any α,β and for any $\varphi \in C_c^\infty(R\setminus\{0\})$ there exists $\underset{\sim}{Q}' \in \mathcal{Q}$ with $Q'(\lambda)E_\gamma = \delta_{\gamma\beta}\varphi(\lambda)E_\alpha$ and since finite linear combinations of the E_α are dense in H^∞. Suppose $0 < \text{Re } k < 2n + 2$. Then if $\hat{f} \in \mathcal{Q}$, $(f|K) = (f|\Phi_k*(\Phi_{-k}*K)) = (\tilde{\Phi}_k*f|\Phi_{-k}*K) = ((\tilde{f}*\Phi_k)^{\sim}|\Phi_{-k}*K)$. Since $\Phi_{-k}*K \in PV(H^n)$, we may form $(\Phi_{-k}*K)^{\wedge}$ in the sense of Proposition 4.3. Let $Q(\lambda)$ denote the restriction of $(\Phi_{-k}*K)^{\wedge}(\lambda)$ to H_λ^∞. Then $\underset{\sim}{J} = \underset{\sim}{R}_k\underset{\sim}{Q}$ makes sense, where the product $R_k(\lambda)Q(\lambda)$ is taken in the sense of bounded operators on H^∞. From the above computation, $(f|K) = c_n(\underset{\sim}{R}_{\bar{k}}\hat{f}|(\Phi_k*K)^{\wedge}) = c_n(\hat{f}|\underset{\sim}{J})$, so $\hat{K} = \underset{\sim}{J}$ in the $\mathcal{Q}$ sense. At the same time, if $S(\lambda)$ denotes the restriction of $(K*\Phi_{-k})^{\wedge}$ to H_λ^∞, we have $(f|K) = (f|(K*\Phi_{-k})*\Phi_k) =$ $(f*\tilde{\Phi}_k|K*\Phi_{-k}) = (f*\Phi_{\bar{k}}|K*\Phi_{-k}) = c_n(\hat{f}R_{\bar{k}}|(K*\Phi_{-k})^{\wedge}) = c_n(\hat{f}|\underset{\sim}{S}\underset{\sim}{R}_k)$ since all $R_{\bar{k}}$ are diagonal. Again the product of $S(\lambda)$ and $R_k(\lambda)$ is taken in the sense of bounded operators on H^∞, and we conclude $\hat{K} = \underset{\sim}{S}\underset{\sim}{R}_k$ in the $\mathcal{Q}$ sense. By uniqueness, we see that $\underset{\sim}{J} = \underset{\sim}{R}_k\underset{\sim}{Q} = \underset{\sim}{S}\underset{\sim}{R}_k$. The first equality and Proposition 4.4 show that J_+, J_- have extensions to bounded operators from H^{2M} to H^{2M+k} for any fixed $M \in \mathbb{Z}^+$; the second equality shows they have extensions to bounded operators from H^{-k} to H^0.

Still assuming $0 < \text{Re } k < 2n + 2$, we prove Proposition 5.5(e). If $G \in C^\infty(\mathbb{H}^n \setminus \{0\})$, we let $\|G\|_{C^N}$ denote the C^N norm of G over $\{1 \leq |u| \leq 2\}$. For Proposition 5.5(e), write $\underset{\sim}{J} = \underset{\sim}{S}\underset{\sim}{J}_k$ as above, and observe that we need only show there exist C,N so that $\|S\| < C\|K\|_{C^N}$. Write $K*\Phi_k = \text{P.V.}F + c\delta$; it suffices to show, by Proposition 4.3, that for some C,N, $\|F\|_{C^1} + |c| < C\|K\|_{C^N}$. (We have used the fact that the norm, on L^2, of the operator of convolution with $K*\Phi_{-k}$ does not exceed a constant times $\|F\|_{C^1} + |c|$.) To see this, write, as in (5.1), $K*\Phi_{-k} = K'*\Phi'_{-k} + K'*\Phi''_{-k} + K''*\Phi'_{-k} + K''*\Phi''_{-k}$. If the supports of K' and Φ'_{-k} are sufficiently small, we have $\|F\|_{C^1} \leq \|K'*\Phi''_{-k}\|_{C^1} + \|K''*\Phi'_{-k}\|_{C^1} + \|K''*\Phi''_{-k}\|_{C^1} \leq C\|K\|_{C^N}$ for some C,N, as one sees by evaluating C^M norms by placing left invariant derivatives on Φ'_{-k}, Φ''_{-k}. Next, select $\varphi \in C_c^\infty(\mathbb{H}^n)$ with $\varphi(0) = 1$, and observe $|c| = |(\varphi|K*\Phi_{-k}-\text{P.V.}F)| \leq$ $|(\varphi|K'*\Phi'_{-k})| + |(\varphi|K*\Phi_{-k}-K'*\Phi'_{-k})| + |(\varphi|\text{P.V.}F)|$. It is easy to estimate that for some C,u,N',M, $|(K*\Phi_{-k}-K'*\Phi'_{-k})(u)| < C\|K\|_{C^{N'}}(1+|u|)^M$, so $|(\varphi|K*\Phi_{-k}-K'*\Phi'_{-k})| < C\|K\|_{C^{N'}}$. Further $|(\varphi|\text{P.V.}F)| < C\|K\|_{C^N}$ by the preceding estimate. Finally, for some N'', $|(\varphi|K'*\Phi'_{-k})| = |(\tilde{K}'*\varphi|\Phi'_{-k})| < C\|\tilde{K}'*\varphi\|_{C^{N''}(\mathbb{H}^n)} <$ $C\|K\|_{C^0}$. This proves (e) in case $0 < \text{Re } k < 2n + 2$.

Suppose now $K \in \mathcal{A}K^{k-2n-2}$, $0 < \text{Re } k < 2n + 2$; we wish to prove the last two sentences of Proposition 5.5(a). It would be simplest to reduce to the case $K \in \text{APV}$ as in the preceding paragraph; however, we do not know that $\Phi_{-k} \in AK^{-k-2n-2}$, so we cannot do this directly. Instead we imitate the proofs of Proposition 4.4 and 4.5. As in Proposition 4.4 we write $R_o^M K = \sum_{m=0}^{M} \binom{M}{m} L_o^m (iT)^{M-m}(D^{M-m}K)$ where D is as in (4.6). Put $J_{\sim M-m} = (D^{M-m}K)^{\wedge}$, taken in the $\mathcal{Q}$ sense. We can take the Fourier transform of both sides in the $\mathcal{Q}$ sense; by uniqueness and Proposition 4.8(a) (recall $T = (i/2)[Z_1,\bar{Z}_1]$) we have that for all $v \in H^{\infty}$, $A^M J_+ v = \sum_{m=0}^{M} \binom{M}{m} J_{M-m,+} A^m v$. Using Proposition 5.5(e), we conclude as in (4.9) that there exist $C_o, R_o > 0$ so that $||J_{m+}||_{-k,0} < C_o R_o^m m!$. Suppose now $v \in H^{\omega}$ and $||A^m v|| < C_2 R_o^m m!$ for all m. Then $||A^m v||_{-k} < C_3 ||A^{m-\text{Re } k/2} v|| < C_3 ||A^{m-[\text{Re } k/2]+1} v|| < C_4 R_4^m m!$ for some C_3, C_4, $R_4 > 0$. ([ ] = greatest integer function.) We find $||A^M J_+ v|| \leqq \sum_{m=0}^{M} \binom{M}{m} ||J_{M-m,+}||_{-k,0} ||A^m v||_{-k} \leqq C_o C_4 \sum \binom{M}{m} R_o^{M-m} R_4^m (M-m)!m! \leqq C_o C_4 R_5^M (M+1)!$ where $R_5 = \max(R_o, R_4)$ as desired. This proves Proposition 5.5(a) in case $0 < \text{Re } k < 2n + 2$.

We next consider Proposition 5.5(a) and (e) in the case $\text{Re } k \leqq 0$. Select $r \in \mathbb{N}$ so that $0 < \text{Re } k + 2r < 2n + 2$. We can then write $K = L_o^r K_1 = R_o^r K_2$ where $K_1, K_2 \in AK^{k+2r-2n-2}$; to obtain K_1, K_2, convolve K with the fundamental solution of L_o r times. If $K \in AK^{k-2n-2}$, then $K_1, K_2 \in AK^{k+2r-2n-2}$ by Proposition 5.2(f). Thus Proposition 5.5(a) follows in this case at once from Proposition 4.8(a) and the previously proved case. For (e), we assume K as in (e), and it suffices to show that for all N, there exist C,N' so that $\|K_1\|_{C^N} < C\|K\|_{C^{N'}}$. Arguing inductively, we see that it suffices to show that for all N, there exist C,N' so that $\|K*\Phi\|_{C^N} < C\|K\|_{C^{N'}}$; here Φ is the fundamental solution of L_0. We can estimate the C^N norm by placing left-invariant derivatives onto Φ and estimating the sup norm of the results. It suffices, then, to show that if $K_3 \in K^j$ for some $j \leq -2$ and if $\mathrm{Re}\, k \leq 0$ then there exist $C,N > 0$ so that for all $K \in K^{k-2n-2}$ as in (e) we have $\|K*K_3\|_{C^0} < C\|K\|_{C^N}$. This follows readily from (5.1). By making the supports of K', K_3' sufficiently small, we see that we need only check that $\|K'*K_3''\|_{C^0} < C\|K\|_{C^M}$ for some C,M where $\|\ \ \|_{C^0}$ denotes sup norm over $\{1 \leq |u| \leq 2\}$. This follows from the fact that if $K = F$ away from 0 then $K = \Lambda_F$ as in (1.2). In fact, by (1.2), we can take $M = 0$. This proves (e) in case $\mathrm{Re}\, k \leq 0$.

We conclude the proof of (a) and (e) by considering the case $\mathrm{Re}\, k \geq 2n + 2$. Select $r \in \mathbb{N}$ so that $0 < \mathrm{Re}\, k - 2r < 2n + 2$. Let $K_1 = L_0^r K$, and let $\underset{\sim}{J}_1 = \hat{K}_1$ taken in the Q sense. If $\hat{f} \in Q$, then $f = L_0^r g$ for some g with $\hat{g} \in Q$. Thus $(f|K) = (g|K_1) = c_n(\hat{g}|\underset{\sim}{J}_1) = c_n(\hat{f}\underset{\sim}{A}^{-r}|\underset{\sim}{J}_1) = c_n(\hat{f}|\underset{\sim}{J}_1\underset{\sim}{A}^{-r})$ since each $A(\lambda)$ is diagonal. Thus $\hat{K} = \underset{\sim}{J}_1\underset{\sim}{A}^{-r}$ in the Q sense. Similarly, let $K_2 = R_0^r K$, $\underset{\sim}{J}_2 = \hat{K}_2$ taken in the Q sense; then $\hat{K} = \underset{\sim}{A}^{-r}\underset{\sim}{J}_2$ in the Q sense. From these facts, (a) and (e) can be immediately read off from a previously proved case.

Next we prove Proposition 5.5(b), by considering various cases.

Case 1: $j=k=0$. We prove the stronger statement that $(K_2*K_1)^\wedge = \hat{K}_2\hat{K}_1$. Indeed, if $f \in L^2$, we have by Proposition 4.3 and Proposition 5.2(e) that $[f*(K_2*K_1)]^\wedge = [(f*K_2)*K_1]^\wedge = (f*K_2)^\wedge\hat{K}_1 = \hat{f}\hat{K}_2\hat{K}_1$. Since $\hat{f}$ could be arbitrary in $\mathcal{R}_2$, the desired relation follows.

Case 2: $0 < \operatorname{Re} j, \operatorname{Re} k < 2n+2$. As in the first part of the proof we can write $K_2 = \Phi_k*K_4$, $K_1 = K_3*\Phi_j$ for some $K_3, K_4 \in PV(\mathbb{H}^n)$; further, if $\underset{\sim}{J}_i = \hat{K}_i$ taken in the $\mathcal{Q}$ sense $(1 \leq i \leq 4)$, we have $\underset{\sim}{J}_2 = \underset{\sim}{R}_k\underset{\sim}{J}_4$, $\underset{\sim}{J}_1 = \underset{\sim}{J}_3\underset{\sim}{R}_j$ (all equalities in the sense of families of operators on $\mathcal{H}^\infty$). Let $K = K_4*K_2$, $\underset{\sim}{J} = \hat{K}$ taken in the $\mathcal{Q}$ sense; then $\underset{\sim}{J} = \underset{\sim}{J}_4\underset{\sim}{J}_2$. Then for all f with $\hat{f} \in \mathcal{Q}$, $(f|K_2*K_1) = (f|\Phi_k*(K*\Phi_j)) = (\tilde{\Phi}_k*f|K*\Phi_j) = ((\tilde{f}*\Phi_k)^\sim*\Phi_j|K) = c_n(\underset{\sim}{R}_{\bar{k}}\hat{f}\underset{\sim}{R}_{\bar{j}}|K) = c_n(\hat{f}|\underset{\sim}{R}_k\underset{\sim}{J}\underset{\sim}{R}_j) = c_n(\hat{f}|\underset{\sim}{J}_2\underset{\sim}{J}_1)$; thus $(K_2*K_1)^\wedge = \underset{\sim}{J}_2\underset{\sim}{J}_1$ as claimed.

In all remaining cases we have $\operatorname{Re} j \leq 0$ or $\operatorname{Re} k \leq 0$. We consider only the case $\operatorname{Re} k \leq 0$, since the case $\operatorname{Re} j \leq 0$ is exactly similar.

Case 3: $\operatorname{Re} k = 0$, $0 < \operatorname{Re} j \leq 2$. Write $K_2 = \mathcal{R}_o K_4$, $K_4 \in \mathcal{K}^{-2n}$; then $K_2 = \sum(Z_\ell^R K_{5\ell} + \bar{Z}_\ell^R K_{6\ell})$ for certain $K_{5\ell}, K_{6\ell} \in \mathcal{K}^{-2n-1}$. Then, in the $\mathcal{Q}$ sense, $(K_2*K_1)^\wedge = [\sum Z_\ell^R(K_{5\ell}*K_1) + \sum \bar{Z}_\ell^R(K_{6\ell}*K_1)]^\wedge = \hat{K}_2\hat{K}_1$ by Case 2 and Proposition 4.8(a).

Case 4: Re k ≦ 0, Re j ≦ 2. Select $r,s \in \mathbb{Z}^+$ with $0 \leqq \text{Re } k + 2r < 2$, $0 < \text{Re } j + 2s \leqq 2$. Write $K_2 = \mathcal{L}_0^r K_4$, $K_1 = \mathcal{L}_0^s K_3$ for some $K_4 \in \mathcal{K}^{k+2r}$, $K_3 \in \mathcal{K}^{j+2s}$. Then, the $\mathcal{Q}$ sense, $(K_2 * K_1)^\wedge = [\mathcal{R}_0^r \mathcal{L}_0^s (K_4 * K_3)]^\wedge = \hat{K}_2 \hat{K}_1$ by Case 2,3 and Proposition 4.8(a).

Case 5: Re k ≦ 0, Re j > 2. Select s with $0 < \text{Re } j - 2s \leqq 2$. Write $K_3 = \mathcal{L}_0^s K_1$. By Proposition 4.8(a) and Case 4, in the $\mathcal{Q}$ sense, $[K_2 * K_1]^\wedge \underset{\sim}{A}^s \underset{\sim}{A}^{-s} = [\mathcal{L}_0^s (K_2 * K_1)]^\wedge \underset{\sim}{A}^{-s} = [K_2 * K_3]^\wedge \underset{\sim}{A}^{-s} = \hat{K}_2 \hat{K}_3 \underset{\sim}{A}^{-s} = \hat{K}_2 \hat{K}_1$. This completes the proof. ∎

6. An Analytic Weyl Calculus

In this chapter, we shall interpret the results of Chapters 4 and 5 in the Schrödinger representation, in order to discuss an interesting calculus of pseudodifferential operators on $\mathbb{R}^n$. The C^∞ version of this calculus was first investigated by Grossman, Loupias and Stein [37]. Its connection with the matters discussed in Chapters 4 and 5 was noted previously by Howe [49], Melin [60] and Taylor [79]. Further results about this calculus follow as special cases of the results of Beals [7]. We shall introduce a new analytic analogue. The results of this chapter will not be used later on.

Whereas the standard pseudodifferential operator calculus is the natural one for dealing with "Laplacian-like" (elliptic) operators, the calculus which we shall describe is the natural one for working with "Hermite-like" operators. Thus, the operators R which this calculus is primarily designed to deal with will, like the Hermite operator, have the following properties:

$$R: \mathcal{S}'(\mathbb{R}^n) \to \mathcal{S}'(\mathbb{R}^n);$$

$$R: \mathcal{S}(\mathbb{R}^n) \to \mathcal{S}(\mathbb{R}^n);$$

and $$\text{if } f \in \mathcal{S}' \text{ and } Rf \in \mathcal{S}, \text{ then } f \in \mathcal{S}. \tag{6.1}$$

We shall also produce an analytic version of this calculus. If R is an operator in the analytic calculus, it will have the additional properties:

$$R\colon Z_2^2(\mathbb{R}^n) \to Z_2^2(\mathbb{R}^n);$$

$$\text{if } f \in \mathcal{S}' \text{ and } Rf \in Z_2^2, \text{ then } f \in Z_2^2. \tag{6.2}$$

(That the Hermite operator has these properties follows from Theorem 2.3, criterion (e).) Part of the point of the calculus is not just to prove that the operators have these properties, but also to exhibit parametrices which are in the calculus. The existence of such parametrices will make properties (6.1) and (6.2) manifest.

Our operators will be defined through the usual formula for pseudodifferential operators in the Weyl calculus:

$$(Rv)(p) = \iint r((p+p')/2,q)\,e^{i(p-p')\cdot q}v(p')\,dp'\,đq \quad \text{for } v \in \mathcal{S}(\mathbb{R}^n). \tag{6.3}$$

Here $đq = dq/(2\pi)^n$.

For the reader unfamiliar with the Weyl calculus, let us observe that if $r((p+p')/2,q)$ is replaced by $r(p,q)$ in the above formula, one obtains the usual definition of pseudodifferential operators with symbol r. As emphasized by Hörmander [48], who has studied the Weyl calculus in great detail, Weyl's definition has several advantages. One major advantage is that in the Weyl calculus, the symbol of R* is simply $\bar{r}$. A good way of thinking of the Weyl calculus is this: if one wants a calculus modelled on differential operators written as $\sum a_\alpha(p)D^\alpha$ (with multiplications to the left of differentiations) then one uses the usual calculus. If one wanted a calculus modelled on operators $\sum D^\alpha a_\alpha(p)$ (differentiations to the left of multiplications) one would use (6.3) with $r((p+p')/2,q)$ replaced by $r(p',q)$. The Weyl calculus is a symmetric compromise between these two.

Since we are seeking a calculus modelled on the Hermite operator, however, our symbols will need to have more restricted properties than are assumed on traditional symbols. The Hermite operator, divided by 4, is

$$A = (|p|^2-\Delta)/4$$

and its symbol is easily seen to be

$$a(p,q) = (|p|^2+|q|^2)/4.$$

This motivates the following definition:

Definition. For $j \in \mathbb{C}$, S^j denotes the set of C^∞ functions $r(p,q)$ on $\mathbb{R}^n$ with the following properties: there exist smooth functions $\{r_\ell\}_{\ell\in\mathbb{Z}^+}$ on $\mathbb{R}^{2n}\setminus\{0\}$, homogeneous of degree $j-\ell$ in (p,q), such that for all $L \in \mathbb{Z}^+, \alpha,\beta \in (\mathbb{Z}^+)^n$ there exist constants $C_{L\alpha\beta}$ such that, if $|(p,q)| = (|p|^2+|q|^2)^{1/2} > 1$, then

$$|\partial_p^\alpha \partial_q^\beta (r - \sum_{\ell=0}^{L-1} r_\ell)(p,q)| < C_{L\alpha\beta}|(p,q)|^{\mathrm{Re}\, j-|\alpha|-|\beta|-L}.$$

Thus this calculus puts the norm $|(p,q)|$ in the central role, and we lose one power every time we differentiate with respect to <u>either</u> p or q. Further, we are restricting to "classical" symbols, i.e. those that have actual asymptotic expansions at infinity. Doubtless, one could obtain corresponding results for "non-classical" symbols (and some such results are obtained by Grossman, Loupias and Stein in [37]). However, these more general symbols do not fall under the scope of the present discussion.

For the analytic analogue of S^j, we prescribe precise growth restrictions on the r_j and also require that r have an entire extension with precise growth restrictions. Precisely, we put

$$AS^j = Z^2_{2,j} + Z^2_{2,j-1}.$$

Thus, for both S^j and AS^j, we are assuming that our symbols have asymptotic expansions at infinity where the homogeneity degrees in these expansions decrease by <u>one</u> each time. One can give a corresponding theory in which the homogeneity degrees decrease by two each time, in which case the symbols would be precisely the type of functions discussed in Theorem

1.8, and in which one would simply define $AS^j = Z^2_{2,j}$. Allowing the degrees to drop by one at a time gives a more general theory of which (as will be easy to see) the "two at a time" theory is a special case.

If $r \in S^j$, we define $R = Op(r)$ by (6.3). The integral is easily seen to converge by a standard method; we need only show that for each p,

$$\int r((p+p')/2,q)e^{i(p-p')\cdot q}v(p')dp'$$

decays rapidly in q. To see this, simply note that for any N we can write

$$e^{ip'\cdot q} = (1+|q|^2)^{-N}(1-\Delta_{p'})^N e^{-ip'\cdot q};$$

we integrate by parts repeatedly to throw the $(1-\Delta_{p'})^N$ onto $r((p+p')/2,q)v(p')$.

Before stating our results about these operators, let us establish some notation (which agrees with the notation used in the paragraph before (4.3), in the case $\lambda = 1/4$). With coordinates $p \in \mathbb{R}^n$, we define the operators

$$W_k = i(\partial/\partial p_k+p_k)/2, \quad W_k^+ = i(\partial/\partial p_k-p_k)/2;$$

these are $i/2^{1/2}$ times the annihilation and creation operators. We define the functions $\{E_\alpha\}_{\alpha\in(\mathbb{Z}^+)^n}$ as follows:

$$E_o(p) = \pi^{-n/4} e^{-|p|^2/2}$$

$$E_\alpha = 2^{|\alpha|/2} \alpha!^{-1/2} (W^+)^\alpha E_o.$$

The $\{(-i)^{|\alpha|} E_\alpha\}$ are the Hermite functions. From the definition we have

$$\begin{aligned} W_k E_\alpha &= (\alpha_k/2)^{1/2} E_{\alpha - e_k}, \text{ zero if } \alpha_k = 0 \\ W_k^+ E_\alpha &= [(\alpha_k+1)/2]^{1/2} E_{\alpha+e_k} \end{aligned} \tag{6.4}$$

where $e_k = (0,\dots,1,\dots,0)$ with the 1 in the kth position.

It follows from (6.4) and a standard argument that the $\{E_\alpha\}$ are an orthonormal basis of $\mathbb{R}^n$. The Hermite operator, divided by 4, is

$$A = (1/2) \sum (W_k^+ W_k + W_k^+ W_k) = (|p|^2 - \Delta)/4$$

and satisfies

$$AE_\alpha = [(2|\alpha|+n)/4] E_\alpha.$$

Using W_k, W_k^+ and A, one easily sees that if $f \in L^2(\mathbb{R}^n)$ then $f \in \mathcal{S}(\mathbb{R}^n)$ if and only if its Hermite coefficients decay rapidly. Now if $g \in \mathcal{S}'(\mathbb{R}^n)$, we can define a sequence $v_\alpha = g(E_\alpha)$. It is easy to see that for some $k \in \mathbb{R}$,

$$\sum (|\alpha|+1)^k |v_\alpha|^2 < \infty; \tag{6.5}$$

this follows from (6.4) and the fact that for some C,N,

$$|g(f)| < C \max_{|\beta|+|\gamma| \leq N} \| W^\beta (W^+)^\gamma f \|_{L^2} .$$

Conversely, we define H^k as the Hilbert space of all sequences $v = (v_\alpha)$ satisfying (6.5) (here $\|v\|_k^2$ is the left side of (6.5).) Then, clearly, any element v of H^k defines an element of S' by the rule $v(\sum w_\alpha E_\alpha) = \sum v_\alpha w_\alpha$. For obvious reasons, then, in H^k we write E_α for the sequence w with $w_\beta = \delta_{\alpha\beta}$; then any $v \in H^k$ can be written as $v = \sum v_\alpha E_\alpha$ in H^k. In this way we obtain

$$\cap H^k = S(\mathbb{R}^n); \ \cup H^k = S'(\mathbb{R}^n).$$

It follows easily from (6.4) that if $k \in \mathbb{Z}^+$ then

$$H^k = \{f \in L^2 | p^\alpha \partial^\beta f \in L^2 \text{ for } |\alpha| + |\beta| \leq k\}.$$

For $k \in \mathbb{C}$, we put $H^k = H^{\text{Re } k}$.

Next, some observations about our symbol classes. Observe that

$$\cap S^j = S(\mathbb{R}^{2n}), \ \cap AS^j = Z_2^2(\mathbb{R}^{2n}).$$

If $R \in \text{Op}(r)$ where $r \in S$ (resp. Z_2^2), we naturally say $R \in \text{Op}(S)$ (resp. $R \in \text{Op}(Z_2^2)$).

If $R = \text{Op}(r) \in \text{Op}(S^j)$, we say that R is <u>Hermite-like</u> if, in the asymptotic expansion $r \sim \sum_{\ell=0}^{\infty} r_\ell$, we have

$$r_0(p,q) \neq 0 \text{ if } (p,q) \neq (0,0).$$

At times, as in Theorem 6.2 below, we write $\xi = (p,q)$ for our coordinates in $\mathbb{R}^{2n}$. We also write $\tilde{\xi} = (q,-p)$. In the statement of Theorem 6.2(b), we write $\partial^{\alpha} = \partial_{\xi}^{\alpha}$ and $\tilde{\partial}^{\alpha} = (\partial_{\tilde{\xi}})^{\alpha}$. For other proofs of the theorems below in the C^{∞} setting, see [37] and [7].

Theorem 6.1(a) (Boundedness). Suppose $R \in Op(S^j)$. Then $R : \mathcal{S}(\mathbb{R}^n) \to \mathcal{S}(\mathbb{R}^n)$ continuously, and extends to an operator $R : \mathcal{S}'(\mathbb{R}^n) \to \mathcal{S}'(\mathbb{R}^n)$. This operator has the property that for all m, $R : H^m \to H^{m-j}$ boundedly. If $R \in Op(AS^j)$, then $R : Z_2^2 \to Z_2^2$, with the uniformity: for any $0 < r < 1$ there exist C and $0 < s < 1$ such that if $v = \sum v_{\alpha}E_{\alpha} \in Z_2^2$, $|v_{\alpha}| < r^{|\alpha|}$, then $Rv = \sum w_{\alpha}E_{\alpha}$ with $|w_{\alpha}| < Cs^{|\alpha|}$.

(b) (Pseudolocality). If $R \in Op(S^j)$ and $v \in \mathcal{S}'$ is smooth on an open set U, then Rv is smooth on U. If $R \in Op(AS^j)$ and $v \in \mathcal{S}'$ is analytic on an open set U, then Rv is analytic on U.

(c) (Smoothing operators). If $R \in Op(\mathcal{S})$, then $R : \mathcal{S}'(\mathbb{R}^n) \to \mathcal{S}(\mathbb{R}^n)$. If $R \in Op(Z_2^2)$, then $R : \mathcal{S}'(\mathbb{R}^n) \to Z_2^2(\mathbb{R}^n)$, and we have this uniformity: say $k \in \mathbb{C}$. Then there exist $C_o > 0$, $0 < r_o < 1$ so that whenever $\|v\|_k \leq 1$, $w = Rv$, we have $|w_{\alpha}| < C_o r_o^{|\alpha|}$ for all α.

Theorem 6.2(a) (Composition). If $R \in Op(S^{j_1})$ and $S \in Op(S^{j_2})$ then $RS \in Op(S^{j_1+j_2})$.

If $R \in Op(AS^{j_1})$ and $S \in Op(AS^{j_2})$ then $RS \in Op(AS^{j_1+j_2})$.

If $R = \mathrm{Op}(r)$ and $S = \mathrm{Op}(s)$ then the symbol of RS is $r \circ s$, the twisted product of r and s, where

$$r \circ s(\xi) = \int_{\mathbb{R}^{2n}} e^{i\xi\cdot\eta} r(\xi - \tilde{\eta}/2)(F^{-1}s)(\eta)\,d\eta \qquad (6.6)$$

$$= \int_{\mathbb{R}^{2n}} e^{i\xi\cdot\eta} (F^{-1}r)(\eta)\, s(\xi + \tilde{\eta}/2)\,d\eta \qquad (6.6')$$

(b) (Asymptotic expansion of a product). The asymptotic expansion of $r \circ s$ at infinity is $r \circ s = \sum (r \circ s)_\ell$, where

$$(r \circ s)_\ell(\xi) = \sum_{a+b+2|\alpha|=\ell} [(i/2)^{|\alpha|}/\alpha!]\,\partial^\alpha r_a \tilde{\partial}^\alpha s_b. \qquad (6.7)$$

(c) (Adjoints). If $R \in \mathrm{Op}(S^j)$ then $R^* \in \mathrm{Op}(S^j)$. (Here R^* is the formal adjoint of R, acting on S.) If $R \in \mathrm{Op}(AS^j)$, then $R^* \in \mathrm{Op}(AS^j)$.

If $R = \mathrm{Op}(r)$ then $R^* = \mathrm{Op}(\bar{r})$ where $\bar{r}(p,q) = \overline{r(p,q)}$.

Theorem 6.3. Suppose $R \in \mathrm{Op}(S^j)$ (resp. $\mathrm{Op}(AS^j)$) and R is Hermite-like.

(a) (Null space). The null space of $R : S' \to S'$ is a finite-dimensional space of Schwartz (resp. Z_2^2) vectors.

(b) (Parametrix). There exists $S \in \mathrm{Op}(S^{-j})$ (resp. $\mathrm{Op}(AS^{-j})$) such that

$$SR = I - P$$

where $P \in \mathrm{Op}(S)$ (resp. $\mathrm{Op}(Z_2^2)$) is the projection (in H^0) onto the null space of R.

(c) (Regularity). Say $f \in S'$. If U is an open set and Rf is smooth (resp. analytic) on U, then f is smooth (resp. analytic) on U.

If $Rf \in S$ (resp. Z_2^2) then $f \in S$ (resp. Z_2^2).

Undoubtedly, these theorems could all be proved directly, and indeed, in the smooth case, direct proofs are given in [37] and [7]. This author himself would in many ways prefer direct proofs to the arguments we are about to give. However, for ease and brevity, we are simply going to explain how these results follow simply by interpreting the results of Chapters 4 and 5 in the Schrödinger representation.

To understand the connection with the Schrödinger representation, let us return to the notation in the paragraph before and after (4.3). Let us also recall that, for $F \in L^1(\mathbb{C}^n)$,

$$(\mathcal{F}'_{\mathbb{C}^n} F)(\zeta) = \int_{\mathbb{C}^n} \exp(-z\cdot\overline{\zeta}+\overline{z}\cdot\zeta)F(z)\,dz d\overline{z}.$$

It will be convenient in this chapter to identify $\mathcal{F}'_{\mathbb{C}^n}$ with the usual Fourier transform. As we said before, this we may do if we think of $\zeta = (-q+ip)/2$ with (p,q) dual to (x,y). Thus, in this chapter, we may as well leave off the primes, as well as the "$\mathbb{C}^n$", and write

$$(\mathcal{F}F)(p,q) = \int_{\mathbb{R}^{2n}} \exp(i(x\cdot p+y\cdot q))F(x,y)\,dxdy\,.$$

Next, in the Schrödinger representation, acting on $L^2(\mathbb{R}^n)$, let us also use coordinates p on $\mathbb{R}^n$. Thus

$$P_{k\lambda} = 4\lambda p_k \text{ and } Q_{k\lambda} = -i\partial/\partial p_k,$$

and we have, for $f \in L^1(\mathbb{H}^n)$,

$$\hat{f}(\lambda) = \int \exp(i(\lambda t + x\cdot P + y\cdot Q))f(t,x,y)\,dtdxdy \qquad (6.8)$$

in analogy to

$$(F_\lambda f)(p,q) = (Ff)(\lambda,p,q) = \int \exp(i(\lambda t+x\cdot p+y\cdot q))f(t,x,y)\,dtdxdy.$$

We can understand the Schrödinger representation better if we look once again at what happens to iX_j and iY_j under $\hat{}$ and F. More precisely, let us examine $(iX_j\delta)^\wedge, (iY_j\delta)^\wedge$, $F_\lambda(iX_j\delta), F_\lambda(iY_j\delta)$. Since $X_j\delta = (\partial/\partial x_j)\delta$ and $Y_j\delta = (\partial/\partial y_j)\delta$, we have this picture:

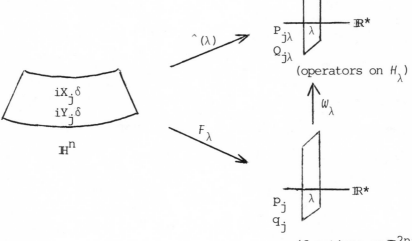

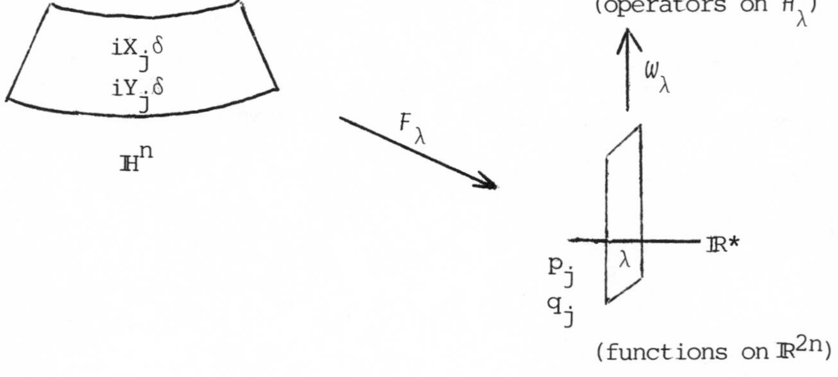

Let us speak heuristically now and focus on the case $\lambda = 1/4$. In that case, we have

$$\mathcal{W}(p_j) = p_j;\ \mathcal{W}(q_j) = -i\partial/\partial p_j.$$

It would seem from these equations that the map $\mathcal{W}$ must be a very familiar one: namely, the map which assigns to a symbol $r(p,q)$ its usual associated pseudodifferential operator. That would, in fact, be the case if we had instead used the definition

$$\hat{f}(\lambda) = \int e^{i\lambda t} e^{ix\cdot P} e^{iy\cdot Q} f(t,x,y)\,dtdxdy$$

for we would then be putting the multiplications P to the left of the differentiations Q instead of adopting the symmetric compromise (6.8). Not surprisingly, then, we assert that when $\lambda = 1/4$, $\mathcal{W}$ is the map which assigns to a symbol $r(p,q)$ its associated pseudodifferential operator in the Weyl calculus.

To check this rigorously, at least when $r \in S$, let us return to (4.4), which states that if $F \in S(\mathbb{R}^{2n})$ and $\lambda = 1/4$, then

$$\begin{aligned}[(GF)(\varphi)](p) &= \int (\mathcal{F}_1 F)((p+p')/2, p'-p)\varphi(p')\,dp' \\ &= \iint (\mathcal{F}F)((p+p')/2, q)\, e^{i(p-p')\cdot q}\varphi(p')\,\bar{d}q\,dp'.\end{aligned}$$

From this it is apparent that if $r \in S(\mathbb{R}^{2n})$ then $R = \mathcal{W}_{1/4} r$

is indeed given by (6.3).

In order to make use of the results of Chapters 4 and 5, we must extend this result to the case where $r \in F_{1/4}(K^k)$ for some k. More precisely, we need to prove the following key lemma.

Lemma 6.4. Suppose $k \in \mathbb{C}$ and $K \in K^k$. Let $r = F_{1/4}K$ and $R = \hat{K}(1/4)$, where $\hat{K}$ is taken in the Q sense. Then, in the Schrödinger representation, r and R are related by (6.3).

Proof. First we shall show that if $F \in S(\mathbb{R}^{2n})$ and $W_{1/4}F = S$ then

$$(R|S) = (2\pi)^{-n}(r|F). \tag{6.9}$$

Letting $(S_{\alpha\beta})$ be the matrix of S with respect to the (E_α), we shall first show (6.9) in the case when $S \in C$, where

$$C = \{S \in S(H) \mid \text{for some } N, S_{\alpha\beta} = 0 \text{ if } |\alpha| + |\beta| > N\}.$$

Select $\varphi \in C_c^\infty(\mathbb{R})$ with $\varphi \geqq 0$, $\int\varphi > 0$, $\text{supp}\,\varphi \subset (0,\infty)$. Then we may define $\underset{\sim}{Q} \in Q$ with $Q(\lambda) = \varphi(\lambda)S$ for all λ. Put $j = -2n - 2 - k$. If $\hat{G} = Q$, then

$$(K|G) = (2\pi^{n+1})^{-1}\int_0^\infty \varphi(\lambda)(4\lambda)^{j/2}(R|S)(2\lambda)^n d\lambda$$

$$= (2\pi^{n+1})^{-1}(R|S)\int_0^\infty \varphi(\lambda)(4\lambda)^{j/2}(2\lambda)^n d\lambda.$$

Further, it is easy to see that, for $\lambda > 0$, $(FK)(\lambda,p,q) = (4\lambda)^{j/2} r(p/2\lambda^{1/2}, q/2\lambda^{1/2})$ and $(FG)(\lambda,p,q) = \varphi(\lambda)F(p/2\lambda^{1/2}, q/2\lambda^{1/2})$;

so $(K|G) = (2\pi)^{-2n-1}\int_0^\infty \varphi(\lambda)(4\lambda)^{j/2}(r|F)(2\lambda^{1/2})^{2n}d\lambda =$

$2^n(2\pi)^{-2n-1}(r|F)\int_0^\infty \varphi(\lambda)(4\lambda)^{j/2}(2\lambda)^n d\lambda.$

This proves (6.9) in case $S \in \mathcal{C}$.

Consider now the general case, where S is not necessarily in $\mathcal{C}$. We may select a sequence $\{S_N\} \subset \mathcal{C}$ with these properties:

(i) for all α,β, $(S_N)_{\alpha\beta} = S_{\alpha\beta}$ or zero;

(ii) for all (α,β), there exists $M > 0$ such that if $N > M$ then $(S_N)_{\alpha\beta} = S_{\alpha\beta}$.

Define F_N by $\mathcal{W}_{1/4}F_N = S_N$; then $(R|S_N) = (2\pi)^{-n}(r|F_N)$. We claim that we may take $\lim_{N\to\infty}$ to obtain (6.9). Indeed, define, H_N, H by $\mathcal{G}_{1/4}H_N = S_N$, $\mathcal{G}_{1/4}H = S$. As is easily seen by Plancherel for $\mathcal{G}$,

$$Z^\alpha \tilde{Z}^\beta (Z^R)^\gamma (\tilde{Z}^R)^\delta H_N \to Z^\alpha \tilde{Z}^\beta (Z^R)^\gamma (\tilde{Z}^R)^\delta H$$

in L^2 for any tuples $\alpha,\beta,\gamma,\delta$. This easily implies that $H_N \to H$ in $\mathcal{S}(\mathbb{R}^{2n})$, whence $\mathcal{F}H_N \to \mathcal{F}H$ in $\mathcal{S}$, i.e. $F_N \to F$ in $\mathcal{S}$. Accordingly $(r|F_N) \to (r|F)$. On the other hand, to show that $\sum(RE_\alpha, S_N E_\alpha) \to \sum(RE_\alpha, SE_\alpha)$, we need only show (by dominated convergence) that for some C, $|(RE_\alpha, S_N E_\alpha)| < C(|\alpha|+1)^{-n-1}$. For this, it suffices to show that for some C', M, $\|RE_\alpha\| < C'(|\alpha|+1)^M$. This follows from Proposition 5.5(a), which states that R can be extended to a bounded operator from $\mathcal{H}^j$ to $\mathcal{H}$, and from the estimate $\|E_\alpha\|_j < C'(|\alpha|+1)^{\mathrm{Re}\, j/2}$. This proves (6.9).

Now for v,w fixed unit vectors in $H^{\infty}_{1/4}$, define $S \in S(H)$ by $Sv' = (v|v')w$ for any $v' \in H$. Also, with $\tilde{}: S \to S$ by $\tilde{f}(u) = \overline{f(u^{-1})}$ and extended to S', say $\hat{\tilde{K}}(1/4) = \tilde{R}$. Then, as in the second paragraph of the proof of Theorem 5.1(d) => (c), we have $(Rv'|w') = (v'|\tilde{R}w')$ for all $v',w' \in H^{\infty}$. Thus

$$\begin{aligned}(R|S) &= \sum(RE_{\alpha},SE_{\alpha}) = \sum(RE_{\alpha},w)(v,E_{\alpha}) = \sum(E_{\alpha},\tilde{R}w)(v,E_{\alpha}) \\ &= (v,\tilde{R}w) = (Rv,w).\end{aligned} \tag{6.10}$$

On the other hand, define F by $W_{1/4}F = S$. Select $\psi \in C_c^{\infty}(\mathbb{R}^{2n})$ with $0 \leq \psi \leq 1$ and $\psi(p,q) = 1$ if $|(p,q)| \leq 1$. For $\varepsilon > 0$, put $\psi_{\varepsilon}(p,q) = \psi(\varepsilon p,\varepsilon q)$. Then

$$\begin{aligned}(2\pi)^{-n}(r|F) &= \lim_{\varepsilon\to 0}(2\pi)^{-n}(\psi_{\varepsilon}r|F) = \lim_{\varepsilon\to 0}(W(\psi_{\varepsilon}r)|S) \\ &= \lim_{\varepsilon\to 0}\operatorname{tr}([W(\psi_{\varepsilon}r)]^{*}S) = \lim_{\varepsilon\to 0}(W(\psi_{\varepsilon}r)v|w).\end{aligned} \tag{6.11}$$

We have noted $\psi_{\varepsilon}r \in S(\mathbb{R}^{2n})$, used Plancherel for $W = W_{1/4}$ and evaluated the trace by using an orthonormal basis including v.

Now, in the Schrödinger representation, we may think of $v = v(p)$, $w = w(p) \in S(\mathbb{R}^{n})$; and, as we remarked just before this proof, $W(\psi_{\varepsilon}r)(v) = g_{\varepsilon}$ where

$$g_{\varepsilon}(p) = \iint(\psi_{\varepsilon}r)((p+p')/2,q)e^{i(p-p')\cdot q}v(p')\,dp'\,\bar{d}q. \tag{6.12}$$

Define also $\psi_{0} \equiv 1$ and g_{0} by (6.12) when $\varepsilon = 0$; we are trying to show that $Rv = g_{0}$. We claim $g_{\varepsilon} \to g_{0}$ uniformly on compact subsets of $\mathbb{R}^{n}$. This is easily seen, by the same method (mentioned earlier) by which one shows that the integral for g_0 converges. Namely, for any M we can write $e^{-ip'\cdot q} = (1+|q|^2)^{-M}(1-\Delta_{p'})^M e^{-ip'\cdot q}$ and integrate by parts repeatedly in the inner integral to throw the $(1-\Delta_{p'})^M$ onto $(\psi_\varepsilon r)((p+p')/2,q)v(p')$. Invoking this technique, we see easily that $g_\varepsilon \to g_0$ uniformly on compact sets.

Finally, then, by (6.9), (6.10) and (6.11) we see that, if $w \in C_c^\infty(\mathbb{R}^n)$, then

$$(Rv|w) = \lim_{\varepsilon\to 0}(\mathcal{W}(\psi_\varepsilon r)v|w) = \lim_{\varepsilon\to 0}(g_\varepsilon|w) = (g_0|w).$$

Thus $Rv = g_0$. Since $v \in \mathcal{S}(\mathbb{R}^n)$ is arbitrary, this completes the proof.

<u>Remark</u>. In the rest of this chapter, if $K \in \mathcal{K}^k$ for some k, then $\tilde{K}$ will also be as in the proof of Lemma 6.4. That is, $\tilde{K}$ is defined through this procedure: define $\tilde{\ }: \mathcal{S}(\mathbb{H}^n) \to \mathcal{S}(\mathbb{H}^n)$ by $\tilde{f}(u) = f(u^{-1})$, and extend to $\mathcal{S}'$. As in the proof of Theorem 5.1(d) => (c), if $K \in \mathcal{K}^k$, $(\hat{K}(\lambda)v|w) = (v|\hat{\tilde{K}}(\lambda)w)$ for all $v,w \in \mathcal{H}_\lambda^\infty$.

<u>Corollary 6.5</u>(a). Suppose $j \in \mathbb{C}$, $r \in S^j$ (resp. AS^j) and $R = Op(r)$. Let $k = -2n - 2 - j$, and let $M \in \mathbb{Z}^+$ be arbitrary. Then there exists $K \in \mathcal{K}^{k-M} + \mathcal{K}^{k-M+1}$ (resp. $A\mathcal{K}^{k-M} + A\mathcal{K}^{k-M+1}$) such that $F_{1/4}K = r$, $\hat{K}(1/4) = R$. ($\hat{K}$ is taken in the $\mathcal{Q}$ sense.)

(b) Say $R \in Op(S)$ (resp. $Op(Z_2^2)$); let $(R_{\alpha\beta})$ be the matrix of R with respect to the $\{E_\alpha\}$. Then for $N > 0$ there exists

$C_N > 0$ (resp. there exist $C > 0$, $0 < r < 1$) such that

$$|R_{\alpha\beta}| < C_N(1+|\alpha|)^{-N}(1+|\beta|)^{-N}$$

$$\text{(resp.} \quad |R_{\alpha\beta}| < Cr^{|\alpha|+|\beta|}. \qquad (**))$$

Conversely, if R is any operator on H whose matrix $(R_{\alpha\beta})$ with respect to the E_α satisfies (*) (resp. (**)), then $R \in Op(S)$ (resp. $Op(Z_2^2)$).

Proof. (a) Since $S^j \subset S^{j+M}$ and $AS^j \subset AS^{j+M}$ for any M, we may assume M = 0. In that case, the Corollary follows at once from Theorem 1.8(a) and Theorem 2.11. (It is necessary to construct an appropriate "J_-" when applying either of these theorems. This is easily done, using Theorem 2.12 or its much simpler, well-known C^∞ analogue.)

(b) If $r \in S$, then $Op(r) = W_{1/4}r$, so this is a consequence of Proposition 4.2 and of $G(S(\mathbb{R}^{2n})) = S(H)$.

Using these facts, we can readily prove Theorems 6.1, 6.2 and 6.3 as follows.

Proof of Theorem 6.1(a) and (c). Most of Theorem 6.1(a) follows from Corollary 6.5, together with Proposition 5.5(a). To obtain the full statement of Theorem 6.1(a), we must extend slightly the result of Proposition 5.5(a). In the notation of Proposition 5.5(a), we only showed there that

$J_+ : H^{2M} \to H^{2M+k}$ boundedly when $M \in \mathbb{Z}^+$. To obtain the full statement of Theorem 6.1(a) it is necessary to show that $J_+ : H^m \to H^{m+k}$ boundedly for all $m \in \mathbb{R}$. We need only show this in the case $k = 0$, i.e. the case $K \in PV(\mathbb{H}^n)$. Indeed, one can then derive the result for general k from this, just as in the proof of Proposition 5.5(a) (see part A.2 of the appendix to Chapter 5). Further, we need only consider the case $m > 0$, for use of $\sim$ and a simple duality argument enables one then to extend the result to $m < 0$.

To complete the proof of Theorem 6.1(a), then, it suffices to prove the following. Suppose $S : H^\infty \to H^\infty$ is linear; that $0 \leq p < m < q$; and that $\|Sv\|_p^2 \leq C_p\|v\|_p^2$, $\|Sv\|_q^2 \leq C_q\|v\|_q^2$ for all $v \in H^\infty$. Then there exists C_m so that $\|Sv\|_m^2 \leq C_m\|v\|_m^2$ for all $v \in H^\infty$.

This follows from a Marcinkiewicz-type interpolation argument. Indeed, for any $u \in H^\infty$, $N \in \mathbb{Z}^+$, write $P_N u = \sum_{|\alpha| \geq N} u_\alpha E_\alpha$; note that for any $s \geq 0$, $\|P_N u\|^2 \leq (N+1)^{-s}\|u\|_s^2$. Now any $v \in H^\infty$, $Sv = w$. For $N \in \mathbb{Z}^+$, put $V_N^2 = \sum_{|\alpha|=N} |v_\alpha|^2$, $W_N^2 = \sum_{|\alpha|=N} |w_\alpha|^2$. Also put $a = \max(2,m)$. Summing by parts, we find:

$$\|Sv\|_m^2 = \sum(|\alpha|+1)^m|w_\alpha|^2 = \sum_{N=0}^{\infty}(N+1)^m W_N^2 = \sum_{N=0}^{\infty}[(N+1)^m - N^m]\,\|P_N w\|^2$$

$$\leq a\sum_{N=0}^{\infty}(N+1)^{m-1}\|P_N Sv\|^2$$

$$\leq 2a\sum_{N=0}^{\infty}(N+1)^{m-1}\|P_N\, SP_N v\|^2 + 2a\sum_{N=0}(N+1)^{m-1}\|P_N S(v-P_N v)\|^2$$

$$\leq 2a\sum_{N=0}^{\infty}(N+1)^{m-p-1}\|SP_N v\|_p^2 + 2a\sum_{N=0}^{\infty}(N+1)^{m-q-1}\|S(v-P_N v)\|_q^2$$

$$\leq 2a\, C_p\sum_{N=0}^{\infty}(N+1)^{m-p-1}\|P_N v\|_p^2 + 2a\, C_q\sum_{N=0}^{\infty}(N+1)^{m-q-1}\|v-P_N v\|_q^2$$

$$= 2a\, C_p\sum_{N=0}^{\infty}(N+1)^{m-p-1}\sum_{M=N}^{\infty}(M+1)^p V_M^2 + 2a\, C_q\sum_{N=0}^{\infty}(N+1)^{m-q-1}\sum_{M=0}^{N-1}(M+1)^q V_M^2$$

$$\leq C_m\sum_{M=0}^{\infty}(M+1)^m V_M^2 = C_m\|v\|_m^2 \quad \text{for some } C_M > 0,$$

as one sees at once by interchanging the $\sum_N$ and $\sum_M$ summations. This completes the proof of Theorem 6.1(a).

Theorem 6.1(c) follows at once from (a) in the case $r \in \mathcal{S}(\mathbb{R}^{2n})$. In the case $r \in Z_2^2(\mathbb{R}^{2n})$, it follows at once from Proposition 5.5(c), Proposition 4.2, and the fact that $r \in \mathcal{S} \Rightarrow \mathrm{Op}(r) = \mathcal{W}_{1/4}r$.

Proof of Theorem 6.2. Theorem 6.2(c) is easy, either from the definition (6.3), or from Corollary 6.5 and use of $\tilde{K}$.

As for Theorem 6.2(a), let us first understand how the integrals in (6.6) and (6.6') are to be interpreted and why they are equal. The interpretation is evident once we recognize that $\mathcal{F}^{-1}r, \mathcal{F}^{-1}s \in \mathcal{E}' + \mathcal{S}$. Indeed, $\xi^\beta\partial_\xi^\alpha r$ and $\partial_\xi^\alpha\xi^\alpha r$ are in L^1 if $|\alpha|$ is sufficiently larger than $|\beta|$. Taking $\mathcal{F}^{-1}$, we see from these facts that $F^{-1}r$ is smooth away from 0 and, outside any given ball centered at 0, agrees with an S function.

If we are even more careful about interpreting the integrals, we can reach an easy understanding of why they are equal. For any symbols r,s, define more generally

$$r\circ_\lambda s(\xi) = \int e^{i\xi\cdot\eta} r(\xi-2\lambda\tilde{\eta})(F^{-1}s)(\eta)d\eta. \qquad (6.13)$$

We claim that $r\circ_\lambda s(\xi)$ equals

$$\int e^{i\xi\cdot\eta}(F^{-1}r)(\eta)s(\xi+2\lambda\tilde{\eta})d\eta.$$

Indeed, for $\xi \in \mathbf{R}^n$, define $d_{2\lambda}, \rho, \rho^t, \tau_\xi, m_\xi : S(\mathbb{R}^n) \to S(\mathbb{R}^n)$ by $(d_{2\lambda}f)(\eta) = f(2\lambda\eta)$; $(\rho f)(\eta) = f(-\tilde{\eta})$; $(\rho^t f)(\eta) = f(\tilde{\eta})$; $(\tau_\xi f)(\eta) = f(\eta+\xi)$; $(m_\xi f)(\eta) = e^{i\xi\cdot\eta}f(\eta)$. Extend these maps to S'. The integral in (6.13) is then $(m_\xi F^{-1}s, d_{2\lambda}\rho\tau_\xi r)$. Since $(F^{-1}F,G) = (F,F^{-1}G)$ for $F,G \in E' + S$ (real inner product), this then equals $(\rho^t d_{2\lambda}\tau_\xi r, m_\xi F^{-1}s)$, as claimed.

Rather than prove (a) directly, we opt to argue via Corollary 6.5. Thus, by Proposition 5.5(b), it is enough to show:

If $K_1 \in K^k$, $K_2 \in K^\ell$, $\mathrm{Re}(k+\ell) < -(2n+2)$ and $\lambda \neq 0$, then

$$F_\lambda(K_1 * K_2) = F_\lambda K_1 \circ_\lambda F_\lambda K_2. \qquad (6.14)$$

Let us first show that if $f,g \in S(\mathbb{H}^n)$, then

$$F_\lambda(f*g) = F_\lambda f \circ_\lambda F_\lambda g. \tag{6.15}$$

Write (t,w) for the coordinates on $\mathbb{H}^n$, $w = (x,y)$; also write $\tilde{w} = (y,-x)$. Then $(t,w)(t',w')^{-1} = (t-t'+2w\cdot\tilde{w}', w-w')$. From this it follows easily that

$$F_c^\lambda(f*g)(w) = F_c^\lambda f \times_\lambda F_c^\lambda g$$

where

$$F \times_\lambda G(w) = \int e^{-2i\lambda w\cdot\tilde{w}'} F(w-w')G(w')\,dw' \tag{6.16}$$

for $F,G \in S(\mathbb{R}^{2n})$. (Recall $(F_c^\lambda h)(w) = \int_{-\infty}^{\infty} e^{i\lambda t} h(t,w)\,dt$ for $h \in S(\mathbb{H}^n)$.) $\times_\lambda$ is called twisted convolution. Thus we need to verify that $F(F\times_\lambda G) = FF\circ_\lambda FG$. We use (6.16) to take $\int e^{i\xi\cdot w}(F\times_\lambda G)(w)\,dw$. We change "w" to "$\eta$", note $\eta\cdot\tilde{\eta} = 0$, and write $i\xi\cdot w - 2i\lambda w\cdot\tilde{\eta} = i(w-\eta)\cdot(\xi-2\lambda\tilde{\eta}) + i\eta\cdot\xi$; this leads at once to what we want.

To go from (6.15) to (6.14), we note that, as an easy consequence of

$$(r\circ_\lambda s)(\xi) = \int e^{-i\xi\cdot\tilde{w}/2\lambda}(F^{-1}r)((\tilde{\xi}-\tilde{w})/2\lambda)s(w)\,dw/(2\lambda)^n,$$

we have $(H|r\circ_\lambda s) = (\bar{r}\circ_\lambda H|s)$ whenever r,s are symbols and $H \in S(\mathbb{R}^{2n})$. Also, $\overline{r\circ_\lambda s} = \bar{s}\circ_\lambda \bar{r}$. Thus, if $f,g \in S(\mathbb{H}^n)$, we have

$$\begin{aligned}(g|f*K_1) = (\tilde{f}*g|K_1) &= (2\pi)^{-2n-1}\int_{\infty}^{\infty}(\overline{F_\lambda f}\circ_\lambda F_\lambda g|F_\lambda K_1)\,d\lambda \\ &= (2\pi)^{-2n-1}\int_{-\infty}^{\infty}(F_\lambda g|F_\lambda f\circ_\lambda F_\lambda K_1)\,d\lambda\end{aligned}$$

so that $F_\lambda(f*K_1) = F_\lambda f\circ_\lambda F_\lambda K_1$. Thus, also,

$$F_\lambda(K_1*f) = \overline{F_\lambda(\tilde{f}*\tilde{K}_1)} = \overline{F_\lambda f}\circ_\lambda\overline{F_\lambda K_1} = F_\lambda K_1\circ_\lambda F_\lambda f.$$

Finally,

$$(g|K_1*K_2) = (\tilde{K}_1*g|K_2) = (2\pi)^{-2n-1}\int_{-\infty}^{\infty}(\overline{F_\lambda K_1}\circ_\lambda F_\lambda g|F_\lambda K_2)$$

$$= (2\pi)^{-2n-1}\int_{-\infty}^{\infty}(F_\lambda g|F_\lambda K_1\circ_\lambda F_\lambda K_2)d\lambda$$

which gives (6.14) and hence part (a) of the theorem.

(b) is, formally, a consequence of (a), if one formally writes

$$f(\xi-\tilde{\eta}/2) = \sum(-\tilde{\eta}/2)^\alpha(\partial^\alpha r)(\xi)/\alpha!.$$

Rather than justify this procedure, we again opt to use Corollary 6.5. Changing our notation, we see that is enough to show this:

Suppose $\mathrm{Re}(k_1+k_2) < -2n-2$. For $\nu = 1,2$, let $j_\nu = -2n-2-k_\nu$, and suppose $K_\nu \in K^{k_\nu}$. Let $K = K_2*K_1$, and $j = j_1+j_2$. For $\nu = 1,2$, say $F_{1/4}K_\nu = r_\nu \sim \sum r_{\nu\ell}$ in S^{j_ν}. Then $F_{1/4}K = r \sim \sum r_\ell$ in S^j, where

$$r_\ell = \sum_{a+b+2|\alpha|=\ell}[(i/2)^{|\alpha|}/\alpha!]\,\partial^\alpha r_{2a}\tilde{\partial}^\alpha r_{1b}.$$

Of course $r_{\nu\ell}, r_\ell = 0$ unless ℓ is even, so it would be better to look at $g_\ell = r_{(2\ell)}$, $g_{\nu\ell} = r_{\nu(2\ell)}$. Also it is more convenient to use F_1 instead of $F_{1/4}$; note that $F_1K \sim \sum 2^\ell r_\ell$, similarly for K_1, K_2. After making these changes, one sees at once that Theorem 6.2(b) is an immediate consequence of Lemma 5.7.

Proof of Theorem 6.3(a) and (b). We work in the analytic category; the arguments in the C^∞ category are the same or simpler. First, we claim that there exist $S' \in \mathrm{Op}(AS^{-j})$ and $P' \in \mathrm{Op}(Z_2^2)$ such that

$$SR = I - P'. \tag{6.17}$$

To see this, say $r \sim \sum r_\ell$. By selecting $R' \in \mathrm{Op}(AS^{-j})$ with symbol $1/r_o$, and by considering $R'R$, we see that we may assume $j = 0$ and $r_o \equiv 1$. The symbol of S' is now constructed just as in the proof of Lemma 5.6. Although Lemma 5.6 only handles the case where $r_\ell = 0$ for all odd ℓ, the construction in the general case is almost identical, and we leave the details to the interested reader.

By (6.17) and Theorem 6.1(c), the null space of $R : S' \to S'$ is contained in Z_2^2. Further, if we write an equation analogous to (6.17) with R^* in place of R, and take adjoints of both sides, we find from these equations that $R : H^o \to H^{-j}$ is Fredholm. Thus its null space is a finite-dimensional subspace of Z_2^2, proving (a). Let P denote the projection onto this null space; by considering the matrix of P with respect to the $\{E_\alpha\}$, as in the proof of Theorem 4.1, we see that $P \in Op(Z_2^2)$. Since $R^* : H^j \to H^o$ is also Fredholm, its range is closed and equals $(I-P)H^o$. Since $R^*S'^* = I - P'^*$ and $R^* = (I-P)R^*$, we have

$$R^*S'^* = I - P - (I-P)P'^* = I - P - Q,$$

say, where $Q \in Op(Z_2^2)$ by Proposition 4.9(a). To complete the proof it suffices to produce $H \in Op(Z_2^2)$ such that $R^*H = Q$. This is done by constructing the matrix columns of H separately, and using Theorem 6.1(a) and (c) exactly as in the proof of Theorem 5.1(d) => (c). Details are left to the interested reader.

Proof of Theorem 6.1(b). We begin by making two reductions. Note first that for any $M,N \in \mathbb{Z}^+$, $Rv = A^N(A^{-N}RA^M)(A^{-M}v)$. Since the Hermite operator 4A is elliptic, $A^{-M}v$ is smooth (resp. analytic) on U. It follows from this formula, with M large and N much larger, that we may assume $v \in H^m$ where $m >> 0$ and $R \in Op(S^j)$ where $j << 0$. Throughout the proof which follows, we shall assume that m and $-j$ are large enough that the arguments are valid. In particular, as a simple consequence of the fact that $R : H^m \to H^{m-j}$ boundedly, we may assume m and $-j$ are large enough that Rv is given by (6.3).

Secondly, fix $\omega \in U$; we need only show Rv is smooth (resp. analytic) near ω. We claim that we may assume $\omega = 0$. To see this, simply define v_ω by $v_\omega(p) = v(p+\omega)$; then

$Rv(p+\omega) = (R_\omega v_\omega)(p)$ for some $R_\omega \in Op(S^j)$ (resp. $Op(AS^j)$), as (6.3) shows.

Now for $w \in \mathbb{R}^n$, put

$$K(p,w) = \int r(p-w/2,q)e^{iw\cdot q}\bar{d}q \tag{6.18}$$

so that

$$Rv(p) = \int K(p,p-p')v(p')dp' \tag{6.19}$$

assuming m and $-j$ are large enough.

Now, if B_1 and B_2 are any open balls about 0 in $\mathbb{R}^n$, it follows easily from (6.18) that K is smooth, with all derivatives bounded, for $(p,w) \in B_1 \times \overline{B}_2^c$. Indeed, for any $\alpha \in (\mathbb{Z}^+)^n$,

$$(-iw)^\alpha K(p,w) = \int \partial_q^\alpha r(p-w/2,q)e^{iw\cdot q}\bar{d}q,$$

so that $\partial_p^\alpha \partial_w^\beta w^\alpha K$ exists and is bounded on $B_1 \times \mathbb{R}^n$ whenever $|\alpha|$ is sufficiently larger than $|\beta|$.

From this, it is easy to prove Theorem 6.1(b) in the smooth case. Indeed, since $R : \mathcal{S} \to \mathcal{S}$, we may assume $v = 0$ in a neighborhood of 0. Theorem 6.1(b) now follows in the smooth case by differentiating under the integral in (6.19).

The analytic case is more subtle. First, say B_1, B_2 are as above; for $\varepsilon > 0$, let $\mathcal{B}_{1\varepsilon} = \{z \in \mathbb{C}^n | \operatorname{Re} z \in B_1, |\operatorname{Im} z| < \varepsilon\}$, $\mathcal{D}_{2\varepsilon} = \{z \in \mathbb{C}^n | \operatorname{Re} z \in \overline{B}_2^c, |\operatorname{Im} z| < \varepsilon\}$. We claim that, if $r \in AS^j$, $-j$ large, then

K has a bounded holomorphic extension to $\mathcal{B}_1 \times \mathcal{D}_{2\varepsilon}$ for some $\varepsilon > 0$. (*)

To see this, suppose first that S is an orthogonal transformation on $\mathbb{R}^n$, let $K_S(p,w) = K(Sp,Sw)$, $r_S(p,q) = r(Sp,Sq)$; then K_S and r_S are related in just the same way K and r are (by a formula exactly analogous to (6.18)). In this manner we easily reduce to showing this. For $\varepsilon,\tau > 0$, let $\mathcal{D}_{2\varepsilon\tau} = \{z \in \mathbb{C}^n \mid \text{Re } z_i > \tau \text{ for all } i, |\text{Im } z| < \varepsilon\}$. Then we need only show that

for all $\tau > 0$, there exists $\varepsilon > 0$ such that K has a bounded holomorphic extension to $\mathcal{B}_{1\varepsilon} \times \mathcal{D}_{2\varepsilon\tau}$. (**)

For this, let us denote the entire extension of r to $\mathbb{C}^{2n}$ also by r. For $t \in \mathbb{R}$, let $\sigma(t)$ be sgn $t = \pm 1$. By deforming contours in (6.18) we may write

$$K(p,w) =$$

$$\int_{\mathbb{R}^n} r(p-w/2,[1+i\eta\sigma(q_1)]q_1,\ldots,[1+i\eta\sigma(q_n)]q_n)e^{iw\cdot([1+i\eta\sigma(q_1)]q_1,\ldots,[1+i\eta\sigma(q_n)]q_n)} \times$$
$$(1+i\eta\sigma(q_1))\ldots(i+i\eta\sigma(q_n))\,đq$$

$$= \int_{\mathbb{R}^n} r(p-w/2,[1+i\eta\sigma(q_1)]q_1,\ldots,[1+i\eta\sigma(q_n)]q_n e^{iw\cdot q} e^{-\eta(w_1|q_1|+\ldots+w_n|q_n|)}(1+i\eta\sigma(q_1)) \times$$
$$\ldots(1+i\eta\sigma(q_n))\,đq$$

for all sufficiently small $\eta > 0$, provided $-j$ is large enough. This formula makes (**) evident, and hence (*) follows.

That Rv is analytic near 0 is not immediately evident from (6.18) and (*), but we show it now. Suppose that

$\{p : |p| < 4a\} \subset U$, where $a > 0$; we shall show Rv is analytic for $|p| < a$. For $|p| < a$, $v(p-p')$ is analytic in p for $|p'| < a$, and for $|p - p'| \leq 3a$; also $K(p,p-p')$ is analytic in p for $|p'| > 3a$. We may write

$$Rv(p) = \int_{|p'|<a} K(p,p')v(p-p')dp' + \int_{\substack{|p'|\geq a \\ |p-p'|\leq 3a}} K(p,p')v(p-p')dp'$$

$$+ \int_{|p'|>3a} K(p,p-p')v(p')dp'.$$

This first integral is then obviously analytic in p for $|p| < a$, and the last is analytic in p for $|p| < a$ by (*). We deal with the middle integral by going to polar coordinates. Write $p' = \rho\omega$ where $\omega \in S^{n-1}$, $\rho > 0$, and let $A = \{p' : |p'| \geq a, |p - p'| \leq 3a\}$. Then, for fixed ω, $\rho\omega \in A$ if $\rho \geq a$ and $|p - \rho\omega|^2 \leq 9a^2$. Write the last condition as $\rho^2 - (2p\cdot\omega)\rho - (9a^2-|p|^2) \leq 0$; we then see $\rho\omega \in A$ iff $a \leq \rho \leq p\cdot\omega + [(p\cdot\omega)^2+(9a^2-|p|^2)]^{1/2} = a + \gamma(p,\omega)$, say. $\gamma(p,\omega)$ is clearly analytic in p for $|p| < a$. We now see that the middle integral equals

$$\int_{S^{n-1}}\int_a^{a+\gamma(p,\omega)} K(p,\rho\omega)v(p-\rho\omega)\rho^{n-1}d\rho\, d\omega$$

$$= \int_{S^{n-1}}\int_0^1 K(p,[a+t\gamma(p,\omega)]\omega)v(p-[a+t\gamma(p,\omega)]\omega)[a+t\gamma(p,\omega)]^{n-1}\gamma(p,\omega)dt\, d\omega$$

Since p can be complexified in the integral (for $|p| < a$), the integral is now easily seen to be analytic in p, as desired.

Proof of Theorem 6.3(c). This is an immediate consequence of Theorem 6.3(b) and Theorem 6.1(a), (b) and (c).

We shall make one final connection with the material of earlier chapters. Let L be a Grušin operator, as in (3.7) (dilation weights (2,1,...,1)). For $\lambda \neq 0$, let $L = L_\lambda$ be as in (3.8); then $L : S(\mathbb{R}^s) \to S(\mathbb{R}^s)$. Write n in place of s. We claim that $L \in Op(AS^k)$ and let L is Hermite-like. Thus (3.10) and Lemma 3.8 follow, for these operators, from Theorems 6.1, 6.2 and 6.3, and we have even more: that these operators are part of a calculus which makes these properties manifest.

Let us see that, in fact, $L \in Op(AS^k)$, and is Hermite-like. We change our notation from that used in and around (3.7) and (3.8), writing p in place of x, and $c_{\alpha\beta m}$ in place of $c''_{\alpha\beta m}$. We write

$$L = \sum_{|\alpha|+m=k} c_{\alpha\beta m} p^\beta (\partial/\partial p)^\alpha (\partial/\partial t)^m + \sum_{|\alpha|+m<k} c_{\alpha\beta m} p^\beta (\partial/\partial p)^\alpha (\partial/\partial t)^m$$

$$= S_1 + S_2, \text{ say,}$$

$$L = \sum_{|\alpha|+m=k} c_{\alpha\beta m} (-i\lambda)^m p^\beta (\partial/\partial p)^\alpha + \sum_{|\alpha|+m<k} c_{\alpha\beta m} (-i\lambda)^m p^\beta (\partial/\partial p)^\alpha .$$

In the sum S_1, since $|\alpha| - |\beta| + 2m = k$ and $m = k - |\alpha|$, we have $|\alpha| + |\beta| = k$; in S_2, $|\alpha| + |\beta| < k$. Since p_j, $\partial/\partial p_j \in Op(AS^1)$ with symbols p_j, iq_j, it is clear from Theorem 6.2(a) that $L \in Op(AS^k)$, and its symbol is

$$\sigma(L)(p,q) = \sum_{|\alpha|+m=k} c_{\alpha\beta m}(-i\lambda)^m p^\beta (iq)^\alpha. \tag{6.20}$$

To see that L is Hermite-like, we must understand why this cannot vanish for $(p,q) \neq (0,0)$. The (usual) symbol at L is also

$$\sigma(\mathrm{L})(p;q,\mu) = \sum_{|\alpha|+m=k} c_{\alpha\beta m}(-i\mu)^m p^\beta (iq)^\alpha$$

if we use $(-q,\mu)$ as dual coordinates to (p,t). L is elliptic for $p \neq 0$, so $\sigma(\mathrm{L})(p;q,\mu)$ does not vanish if $p \neq 0$ and if $(q,\mu) \neq (0,0)$. Since $\lambda \neq 0$, $\sigma(L)(p,q)$ therefore does not vanish if $p \neq 0$. If $p = 0$, $q \neq 0$, we need only concern ourselves with the sum of those terms in (6.20) in which $\beta = 0$. Note, however, that the terms in (6.20) in which $\beta = 0$ are precisely those in which $m = 0$, since $|\alpha| - |\beta| + 2m = |\alpha| + m = k$. Thus $\sigma(L)(0,q) = \sigma(\mathrm{L})(0;q,\lambda) = \sigma(\mathrm{L})(p;q,0)$ for any $p \neq 0$, and this cannot vanish for $q \neq 0$, as desired.

7. Analytic Pseudodifferential Operators on $\mathbb{H}^n$: Basic Properties

Section One - Dependence of Constants in Theorem 2.11 on Degree of Homogeneity

In this section, we shall examine the following situation. Say $k \in \mathbb{C}$, Re $k \geq 0$, and let $\{K^m\}$ be a sequence of distributions on $\mathbb{H}^n$, with $K^m \in AK^{k+m}$ for all m. (m is a superscript, not a power.) Suppose further that there are constants C_1, R_1 such that

$$\begin{aligned} &|\partial^\gamma K^m(u)| < C_1 R_1^{m+\|\gamma\|}\gamma! \text{ whenever } 1 \leq |u| \leq 2, \\ &m \in \mathbb{Z}^+, \ \gamma \in (\mathbb{Z}^+)^{2n+1}. \end{aligned} \tag{7.1}$$

Here $\partial = (\partial/\partial t, \partial/\partial x_1, \ldots, \partial/\partial x_n, \partial/\partial y_1, \ldots, \partial/\partial y_n)$. Under these hypotheses, we want to understand how the constants in Theorem 2.11 depend on m, and to also obtain similar information in a converse direction.

Our motivation is that we shall soon be considering series of the type $\sum_{m=0}^{\infty} K^m$, K^m as above. (7.1) is an extremely natural condition in this context. As we shall easily verify later, it tells us that there is a sector $S = \{\omega \in \mathbb{C}^{2n+1} : |\text{Im}\, \omega| < c_1 |\text{Re}\, \omega|, |\omega| < c_2\}$ in the complexification of $\mathbb{H}^n$, to which all $K^m(w)$ may be holomorphically extended and on which the series converges absolutely.

We shall prove an analogue of the following result of Boutet de Monvel and Kree [9]. Suppose we were on $\mathbb{R}^n$ with dilation weights $\underset{\sim}{a} = (1,\ldots,1)$. Suppose Re $k \geqq 0$, $K^m \in AK^{k+m}$ for all m. Let $J^m = FK^m$. If (7.1) holds (now with $\gamma \in (\mathbb{Z}^+)^n$, not $(\mathbb{Z}^+)^{2n+1}$, then for some $C_2, R_2 > 0$,

$$|\partial^\beta J^m(\xi)| < C_2 R_2^{m+\|\beta\|} m!\beta! \quad \text{for } |\xi| = 1, \text{ all } m,\beta.$$

Conversely, if the J^m satisfy these inequalities, then the K^m satisfy (7.1) for some C_1, R_1--provided $K_\gamma^m = F^{-1}\Lambda_{G^m}$, where $J^m = G^m$ away from 0.

Essentially, the argument in Boutet de Monvel and Kree [9] is as follows. Consider $K_\gamma^m = \partial^\gamma K^m/\gamma!$ for $\|\gamma\| =$ [Re k] + m + 1. The K_γ^m satisfy $|\partial^{\gamma'} K_\gamma^m(u)| < C_3 R_3^{m+\|\gamma'\|}(\gamma')!$ for $1 \leqq |u| \leqq 2$, and all $K_\gamma^m \in AK^{\sigma-1}$ where $\sigma = k -$ [Re k]. Thus, by the first uniformity assertion of Theorem 1.3, we obtain dual estimates

$$|\partial^\beta(-i\xi)^\gamma J^m(\xi)| < C_4 R_4^{m+\|\beta\|}\gamma!\beta! < C_5 R_5^{m+\|\beta\|} m!\beta!$$

for $|\xi| = 1$; this then easily gives the forward implication by use of the Cauchy estimates. For the converse, the same method applied in reverse gives (7.1) for $\gamma \in (\mathbb{Z}^+)^n$, $\|\gamma\| \geqq$ [Re k] + m + 1. To deal with those γ with $\|\gamma\| <$ [Re k] + m + 1 requires an additional analysis. This is as it should be, since the size of K^m depends on more than just the size of J^m away from 0. Since $J^m \in AJ^{-n-k-m}$ and $\mathrm{Re}(-n-k-m)$ is usually very negative, one must know J^m fully, as a distribution.

We shall return to this point presently; for now, we return to $\mathbb{H}^n$ and imitate the analysis we have just presented.

If $\gamma \in (\mathbb{Z}^+)^{2n+1}$, we as usual write $|\gamma| = 2\gamma_1 + \gamma_2 + \ldots + \gamma_{2n+1}$. Let the $\{K^m\}$ be as above. For $|\gamma| = [\mathrm{Re}\ k] + m + 1$ or $[\mathrm{Re}\ k] + m + 2$ ($[\ \ ]$ = greatest integer function), let

$$K^m_\gamma = \partial^\gamma K^m/\gamma!.$$

Let $\sigma = k - [\mathrm{Re}\ k]$. Then for all m, γ, we have $K^m_\gamma \in AK^{\sigma-1}$ or $AK^{\sigma-2}$; note $-(2n+2) < \mathrm{Re}(\sigma-2) < \mathrm{Re}(\sigma-1) < 0$. For all $\gamma' \in (\mathbb{Z}^+)^{2n+1}$, we have

$$|\partial^{\gamma'} K^m_\gamma(u)| < C_1'(R_1')^{m+\|\gamma'\|}(\gamma')! \quad \text{for } 1 \leq |u| \leq 2 \tag{7.2}$$

for some $C_1', R_1' > 0$. Now let

$$J^m = \mathcal{F}K^m,\ J^m_\gamma = \mathcal{F}K^m_\gamma = (i\mu)^\gamma J^m/\gamma!$$

where $\mu = (\lambda, \xi) \in \mathbb{R} \times \mathbb{R}^{2n}$ is dual to $(t,(x,y))$. By (7.2), the fact that all $K^m_\gamma \in AK^{\sigma-1}$ or $AK^{\sigma-2}$, and the first uniformity assertion of Corollary 1.5, we have that (A) and (B) of Corollary 1.5 are satisfied, with constants C,R independent of (m,γ), with the J in (A) and (B) equaling $J^m_\gamma/(R_1')^m$ for any (m,γ). (j in (A) or (B) will equal

$-2n - 2 - \sigma + 1$ or $-2n - 2 - \sigma + 2$, depending on (m,γ); and of course, $p = 2$.) Since (A) and (B) are equivalent to (C) and (D) of Corollary 1.5, we have for some C_2, $R_2 > 0$:

For $|(\lambda,\xi)| = 1$ and all $\lambda, m, a \in \mathbb{Z}^+$, $\beta \in (\mathbb{Z}^+)^{2n}$:

$$\begin{aligned}
(C_1)\quad & |\lambda|^a |\partial_\lambda^a J_\gamma^m(\lambda,\xi)| < C_2 R_2^{a+m} a! \\
& |\xi|^{2a} |\partial_\lambda^a J_\gamma^m(\lambda,\xi)| < C_2 R_2^{a+m} a!^2 \\
(D_1)\quad & |\xi|^{\|\beta\|} |\partial_\xi^\beta J_\gamma^m(\lambda,\xi)| < C_2 R_2^{\|\beta\|+m} \beta! \\
& |\lambda|^{\|\beta\|/2} |\partial_\xi^\beta J_\gamma^m(\lambda,\xi)| < C_2 R_2^{\|\beta\|+m} \beta!^{1/2}
\end{aligned} \tag{7.3}$$

Now let $j = -2n - 2 - k$, and say $J_+^m \sim \sum g_\ell^m$ in $Z^2_{2,j-m}$. Consider (7.3) in the case $\gamma = (r,0,\ldots,0)$, where $2r = [\mathrm{Re}\ k] + m + 1$ or $[\mathrm{Re}\ k] + m + 2$. Then $J_\gamma^m = (-i\lambda)^r J^m/r!$, $J_{\gamma+}^m = (-i)^r J_+^m/r! \sim \sum h_\ell^m$ in $Z^2_{2,j-m+2r} = Z^2_{2,-\sigma-2n-1}$ or $Z^2_{2,-\sigma-2n}$, where

$$h_\ell^m = (-i)^r g_{\ell-r}^m/r! \text{ if } \ell \geqq r,\ h_\ell^m = 0 \text{ if } \ell < r. \tag{7.4}$$

Since the J_γ^m satisfy (7.3), we have, by the uniformity assertion of Theorem 1.7(b):

$$|\partial^\alpha h_\ell^m(\xi)| < C_3'(R_3')^{\|\alpha\|+\ell+m} \|\alpha\|!\,\ell! \quad \text{for } |\xi| = 1 \text{ and all } \alpha \in (\mathbb{Z}^+)^{2n} \tag{7.5}$$

for certain C_3', $R_3' > 0$; so

$$|\partial^\alpha g_\ell^m(\xi)| < C_3'(R_3')^{\|\alpha\|+\ell+m+r}\|\alpha\|!(\ell+r)!r!, \text{ so}$$

$$|\partial^\alpha g_\ell^m(\xi)| < C_3 R_3^{\|\alpha\|+\ell+m}\|\alpha\|!\ell!m!$$

$$\text{for } |\xi| = 1 \text{ and all } \alpha \in (\mathbb{Z}^+)^{2n} \qquad (7.6)$$

for certian $C_3, R_3 > 0$.

Part (a) for the theorem which follows summarizes implications (7.1) => (7.3) => (7.6). Parts (b) and (c) show that the implications have important partial converses. In part (c), it will be helpful to have the following variant of Λ_G (defined in (1.2)), where G is a homogeneous function of degree $j \in \mathbb{C}$, and G is smooth away from 0. Put $\varkappa = -2n - 2 - j$; we assume $N = [\mathrm{Re}\ \varkappa] \geq 0$. We define $\Gamma_G \in S'$ as follows. If $N = 0$, we put $\Gamma_G = \Lambda_G$. If instead $N \geq 1$, we define Γ_G by

$$\Gamma_G(\psi) = \int_{|\mu|\leq N} [\psi(\mu) - \sum_{|\gamma|\leq N} \psi^{(\gamma)}(0)\mu^\gamma/\gamma!]G(\mu)d\mu$$
$$+ \sum_{\substack{|\gamma|\leq N \\ |\gamma|\neq\varkappa}} [N^{|\gamma|-\varkappa}/(|\gamma|-\varkappa)]M(\mu^\gamma G)\psi^{(\gamma)}(0)/\gamma! + \int_{|\mu|>N} \psi(\mu)G(\mu)d\mu$$

for $\psi \in S$.

The main difference with (1.2) is that the integrals are truncated at $|\mu| = N$, not 1. It is easy to compute, using (1.1), that $\Gamma_G \in J^j$. Of course, $\Gamma_G = G$ away from 0.

(b) and (c) taken together yield a sharpening and simplification of Chapter 6 of Peter Heller's 1986 Princeton thesis [42]. He initiated the idea of using Γ_G in place of Λ_G.

Theorem 7.1. Say $k\in\mathbb{C}$, Re $k \geq 0$, $j = -2n - 2 - k$.

(a) Say $K^m \in AK^{k+m}$ for all $m \in \mathbb{Z}^+$ and the $\{K^m\}$ satisfy (7.1). For $\gamma \in (\mathbb{Z}^+)^{2n+1}$, $|\gamma| = [\text{Re } k] + m + 1$ or $[\text{Re } k] + m + 2$, let $K^m_\gamma = \partial^\gamma K^m/\gamma!$, $J^m_\gamma = FK^m_\gamma$. Then for some C_2, R_2, the $\{J^m_\gamma\}$ satisfy (7.3). Further, if $FK^m = J^m$ and $J^m_+ \sim \sum g^m_\ell$ in $Z^2_{2,j-m}$, then the g^m_ℓ satisfy (7.6) for some $C_3, R_3 > 0$.

(b) Suppose, for all $m,\ell \in \mathbb{Z}^+$, g^m_ℓ is homogeneous of degree $j - m - 2\ell$ on $\mathbb{R}^{2n}$ and smooth away from 0. Suppose that (7.6) holds for all m,ℓ.

Then there exist $J^m \in AJ^{n-m}$ so that $J^m_+ \sim \sum g^m_\ell$ in $Z^2_{2,j-m}$ and so that if $J^m_\gamma = (-i\mu)^\gamma J^m/\gamma!$ for $|\gamma| = [\text{Re } k] + m + 1$ or $[\text{Re } k] + m + 2$, then the J^m_γ satisfy (7.3) for some $C_2, R_2 > 0$.

(c) Suppose $J^m \in AJ^{j-m}$ for all $m \in \mathbb{Z}^+$, and $J^m_\gamma = (-i\mu)^\gamma J^m/\gamma!$ for $|\gamma| = [\text{Re } k] + m + 1$ or $[\text{Re } k] + m + 2$. Suppose that the J^m_γ satisfy (7.3).

Then there exist $K^m \in AK^{k+m}$ so that $FK^m = J^m$ away from 0 and so that (7.1) is satisfied for some

C_1,R_1. In fact, if $J^m = G^m \in C^\infty(\mathbb{R}^{2n+1}\setminus\{0\})$ away from 0, we may put $K^m = F^{-1}\Gamma_{G^m}$ or $F^{-1}\Lambda_{G^m}$.

Further, we have the following uniformities:

(1) Suppose $C_1,R_1 > 0$; then there exist $C_2,R_2,C_3,R_3 > 0$ with these properties. If $\{K^m\}$ is any sequence as in (a) satisfying (7.1), and if $\{J^m_\gamma\}$, $\{g^m_\ell\}$ are defined as in (a), then (7.3) and (7.6) hold with those C_2,R_2,C_3,R_3.

(2) Suppose $C_3,R_3 > 0$; then there exist $C_2,R_2 > 0$ with these properties: If $\{g^m_\ell\}$ are as in (b) and satisfy (7.6), then there exist $\{J^m_\gamma\}$ as in (b) satisfying (7.3) with that C_2,R_2. Further, if there is a sector $S = \{|\eta| < c|\xi|\} \subset \mathbb{C}^{2n}$, with the properties that all g^m_ℓ may be extended holomorphically to S and that each $g^m_\ell(\zeta)$ depends holomorphically on a parameter ω for all $\zeta \in S$, and if (7.6) holds for all ω, then we may in addition select the J^m_γ so as to depend holomorphically on ω.

(3) Suppose $C_2,R_2 > 0$; then there exist $C_1,R_1 > 0$ with these properties: If $\{J^m\}$, $\{J^m_\gamma\}$ are as in (c) and satisfy (7.3), $J^m = G^m \in C^\infty(\mathbb{R}^{2n+1}\setminus\{0\})$ away from 0, and if $K^m = F^{-1}\Gamma_{G^m}$ or $F^{-1}\Lambda_{G^m}$, then (7.1) holds with that C_1,R_1.

Proof. (a) has been established above; the first uniformity assertion follows form the proof.

For (b), for any m we choose $r \in Z^+$ with $2r =$ $[\mathrm{Re}\ k] + m + 1$ or $[\mathrm{Re}\ k] + m + 2$. We define $\{h^m_\ell\}$ by (7.4), and observe that they satisfy (7.5) for certain constants C_3', R_3'. (We used the inequalities $(\ell-r)! \leq \ell!/r!$ for $\ell \geq r$ and $m!/r!^2 \leq C\, 2^m$.) Note h^m_ℓ is homogeneous of degree $j - m + 2r - 2\ell = -\sigma - 2n - 1 - 2\ell$ or $-\sigma - 2n - 2\ell$ if $\sigma = k - [\mathrm{Re}\ k]$. We use the uniformity assertion of Theorem 2.12 to construct functions $f^m_+ \sim \sum h^m_\ell$ in $Z^2_{2,j-m+2r}$ ($= Z^2_{2,-\sigma-2n-1}$ or $Z^2_{2,-\sigma-2n}$), $f^m_- \sim \sum (-1)^{\ell+r} h^m_\ell$ in $Z^2_{2,j-m+2r}$, in such a way that for some B, C_4, R_4, c we have $f^m_\pm/(R_3')^m \in Z^2_{2,j-m+2r}$ (B, C_4, R_4, c) for all m. Since $h^m_\ell = 0$ for $\ell < r$, we in fact have $f^m_\pm \in Z^2_{2,j-m}$, and we may select $J^m \in AJ^{j-m}$ such that $J^m_\pm = i^r r!\, f^m_+$. Put $J^m_\gamma = (-i\mu)^\gamma J^m/\gamma!$ for $|\gamma| = [\mathrm{Re}\ k] + m + 1$ or $[\mathrm{Re}\ k] + m + 2$. It suffices to show that there are $B_1, C_5, R_5, c_1, R_6 > 0$ so that

$$J^m_{\gamma+}/(R_6)^m \in Z^2_{2,j-m+|\gamma|}\ (B_1, C_5, R_5, c_1) \text{ for all m.} \qquad (*)$$

Indeed, $j - m + |\gamma| = -\sigma - 2n - 1$ or $-\sigma - 2n$, and we could then use the uniformity assertion of Theorem 1.8(b) to complete the proof of (b).

To prove (*), write $\gamma = (s,\alpha)$ with $s \in \mathbb{Z}^+$, $\alpha \in (\mathbb{Z}^+)^{2n}$, so that $i^{\|\gamma\|} J^m_\gamma = \lambda^s \xi^\alpha J^m/(s!\alpha!)$. Then $i^{\|\gamma\|} J^m_{\gamma\pm} = i^r (r!/s!\alpha!) \xi^\alpha f^m_\pm$.

Note that $2s + \|\alpha\| = |\gamma| = 2r - 1$, $2r$ or $2r + 1$, so that $r!/(s!\alpha!) < R_0^m/\alpha!^{1/2}$ for some absolute constant R_0. It suffices, then to show this:

$$\text{(**)}\quad \begin{array}{l}\text{Suppose } j_1 \in \mathbb{C},\ B, C_4, R_4, c > 0\text{; then there are}\\ B_2, C_6, R_6 > 0 \text{ so that whenever } r \in \mathbb{Z}^+,\ \|\alpha\| \leqq r + 1,\\ f \in Z^2_{2,j_1}(B, C_4, R_4, c) \text{ and } f \sim \sum h_\ell \text{ with } h_\ell = 0 \text{ for}\\ \ell < r, \text{ then } \xi^\alpha f/\alpha!^{1/2} \in Z^2_{2,j_1}(B_2, C_6^{r+1}, R_6, c) \text{ or}\\ Z^2_{2,j_1+1}(B_2, C_6^{r+1}, R_6, c).\end{array}$$

For (**), observe that $\xi^\alpha f \sim \sum H_\ell$ in Z^2_{2,j_1} or Z^2_{2,j_1+1}, where $H_\ell = \xi^\alpha h_{\ell+[\|\alpha\|/2]}$. (**) is now an easy consequence of the definition of Z^2_{2,j_1} and the inequalities $(\xi^\alpha/\alpha!^{1/2})^2 \leqq e^{2n|\xi|^2}$, $(\ell+[\|\alpha\|/2])! \leqq 2^{\ell+\|\alpha\|/2}\ell![\|\alpha\|/2]!$, $[\|\alpha\|/2]!\alpha!^{1/2} \leqq (2n)^{\|\alpha\|/2}$. This proves (b). The second uniformity assertion follows from the proof, and the statement about holomorphic dependence on a parameter follows from the proof and from Corollary 2.14.

For (c), at first we set $K^m = F^{-1}\Gamma_{G^m}$. Evidently, then $K^m \in AK^{k+m}$. We put $\Gamma^m = \Gamma_{G^m}$, $N(m) = [\mathrm{Re}\ k] + m$.

If $\gamma \in (\mathbb{Z}^+)^{2n+1}$, $|\gamma| = N(m) + 1$ or $N(m) + 2$, then

$$F(\partial^\gamma K^m/\gamma!) = (-i\mu)^\gamma \Gamma^m/\gamma! = J^m_\gamma$$

since the homogeneity degree of $\mu^\gamma\Gamma^m/\gamma!$ is greater than

$-2n - 2$. By (7.3) and the second uniformity assertion of Corollary 1.5, we have that there exist $C_1' > 0$, $R_1' > 1$ so that

$$|\partial^{\gamma'} K_\gamma^m(u)| < C_1'(R_1')^{m+\|\gamma'\|}(\gamma')! \text{ for } 1 \leqq |u| \leqq 2,$$
$$m \in \mathbb{Z}^+,\ \gamma' \in (\mathbb{Z}^+)^{2n+1}.$$

Thus

$$|\partial^\gamma K^m(u)| < C_1'(R_1')^{m+\|\gamma\|}\gamma! \text{ for } 1 \leqq |u| \leqq 2,\ m \in \mathbb{Z}^+,$$
$$\gamma \in (\mathbb{Z}^+)^{2n+1},\ |\gamma| \geqq N(m) + 1. \qquad (7.7)$$

Thus we have shown (7.1) for $|\gamma| \geqq N(m) + 1$; it remains to show (7.1) for $|\gamma| \leqq N(m)$. Note that there are at most two K's for which $N(m) < 2$: K^0 and K^1. Since K^0 and K^1 clearly satisfy (7.1) for certain C_1, R_1, we may restrict attention to those K^m with $N(m) \geqq 2$. This restriction will hold for the rest of the proof.

It suffices to show (7.1) for $|\gamma| \leqq N(m) - 1$, for we could then easily obtain the case $|\gamma| = N(m)$ from the cases $|\gamma| = N(m) + \varepsilon, \varepsilon \in \{-2,-1,0,1,2\}$, and the following immediate consequence of Taylor's Theorem:

Suppose $I = (a, a+d) \subset \mathbb{R}$ and f is C^2 in a neighborhood of $\bar{I}$. Then

$$\sup|f'| \leqq 4d^{-1}\sup|f| + (d/2)\sup|f''|,$$

all sups taken over I.

To prove (7.1) in case $|\gamma| \leqq N(m) - 1$, we use (7.3) in the case a = 0, or equivalently $\beta = 0$, namely:

$$\text{For } |\mu| = 1 \text{ and } |\gamma| = N(m) + 1 \text{ or } N(m) + 2, \quad |\mu^\gamma G^m(\mu)| \leqq C_2 R_2^m \gamma!. \tag{7.8}$$

We claim that there are $C_3, R_3 > 0$ so that whenever $|\gamma| \leqq N(m)$, $|\mu| = 1$, we have

$$|\mu^\gamma G^m(\mu)| \leqq C_3 R_3^m \gamma! (N(m) - |\gamma|)!. \tag{7.9}$$

To see this, choose $0 < \varepsilon < 1$ so that the 2n + 1 sets $\{\mu: |\mu_i| > \varepsilon\}$ cover $\{\mu: |\mu| = 1\}$. (In the usual notation $\mu = (\lambda, \xi)$, we are now writing $\mu_1 = \lambda$, $\mu_2 = \xi_1, \ldots, \mu_{2n+1} = \xi_{2n}$.) Write $a_1 = 2$, $a_i = 1$ for $2 \leqq i \leqq 2n + 1$. Fix m; write $N = N(m)$. Then by (7.8), we have that for any μ with $|\mu| = 1$ there is i so that

$$|\mu^\gamma G^m(\mu)| \leqq \varepsilon^{-[(N-|\gamma|)/a_i]-1} |\mu^\gamma \mu_i^{[(N-|\gamma|)/a_i]+1} G^m(\mu)|$$

$$\leqq (2\varepsilon^{-1})^{N+1} C_2 R_2^m \gamma! ([(N-|\gamma|)/a_i]+1)!$$

proving (7.9) at once.

Now write, with $N = N(m)$,

$$(2\pi)^{2n+1} K^m(-u) = [\int_{|\mu| \leqq N} (e^{iu\cdot\mu} - \sum_{|\rho| \leqq N} (iu)^\rho \mu^\rho / \rho!) G^m(\mu)\, d\mu]$$

$$+ [\sum_{\substack{|\rho| \leqq N \\ |\rho| \neq k+m}} (N^{|\rho|-k-m}/(|\rho|-k-m)) M(\mu^\rho G^m/\rho!)(iu)^\rho] + [\int_{|\mu| \geqq N} e^{iu\cdot\mu} G^m(\mu)\, d\mu]$$

$= f_1^m(u) + f_2^m(u) + f_3^m(u)$, say, where the $f_\ell^m (1 \leq \ell \leq 3)$ are the three functions in square parentheses. It suffices to show that for some $C,R > 0$,

$$|\partial^\gamma f_\ell^m(u)| \leq CR^m \gamma! \text{ for } 1 \leq |u| \leq 2, \text{ all } m,\ell \text{ and for } |\gamma| \leq N(m) - 1. \tag{*}$$

In fact, if $|\gamma| \leq N - 1$,

$$|\partial^\gamma f_3^m(u)| \leq \int_{|\mu| \geq N} |\mu^\gamma G^m(\mu)|d\mu = M(|\mu^\gamma G^m|)(k+m-|\gamma|)^{-1}|N^{|\gamma|-k-m}|$$
$$\leq C_3 R_3^m (1/2)\gamma!(N-|\gamma|)!N^{|\gamma|-N} \leq CR^m \gamma!$$

for some $C,R > 0$, by (1.1) and (7.9). This gives (*) if $\ell = 3$.

f_2^m is a polynomial. Note $k + m = N + \sigma$, $\sigma = k - [\text{Re } k]$. Thus the absolute value of the coefficient of u^ρ is no more than

$$|(|\rho|-N-\sigma)^{-1}|N^{-(N+\text{Re }\sigma-|\rho|)}C_3 R_3^m (N-|\rho|)! \leq C_4 R_4^m$$

for some $C_4, R_4 > 0$.

Now, for any $P > 0$, there exist constants $A,Q > 0$ so that if p is any polynomial on $\mathbb{R}^{2n+1}$ the absolute value of whose coefficients do not exceed 1, then

$$|\partial^\gamma p(u)| < AQ^{(\deg p)}\gamma! \quad \text{for } \|u\| \leq P,\ \gamma \in (\mathbb{Z}^+)^{2n+1}.$$

(This is a simple consequence of the Cauchy estimates.)

Applying this to $f_2^m/C_4R_4^m$, we see that (*) holds for $\ell = 2$.

That leaves only f_1^m. We have

$$(\partial^\gamma f_1^m)(u) = \int_{|\mu|\leq N} (e^{iu\cdot\mu} - \sum_{|\rho|\leq N-|\gamma|} (iu)^\rho\mu^\rho/\rho!)(i\mu)^\gamma G^m(\mu)\,d\mu.$$

We claim, however, that

$$\left|e^{iu\cdot\mu} - \sum_{|\rho|\leq N-|\gamma|} (iu)^\rho\mu^\rho/\rho!\right| \leq \sum_{\substack{|\rho|=N-|\gamma|+1 \\ \text{or } N-|\gamma|+2}} |u^\rho\mu^\rho|/\rho! \qquad (**)$$

Accepting (**) for the moment, we see that for $|u| \leq 2$,

$$|(\partial^\gamma f_1^m)(u)| \leq 4^{n+2} \sum_{\substack{|\rho|=N-|\gamma|+1 \\ \text{or } N-|\gamma|+2}} \left(\int_{|\mu|\leq N} |\mu^{\rho+\gamma}G^m(\mu)|\,d\mu\right)/\rho!$$

$$\leq 4^{n+2}C_2R_2^m[\,|N^{1-\sigma}/(1-\sigma)| + |N^{2-\sigma}/(2-\sigma)|\,] \sum_{\substack{|\rho|=N-|\gamma|+1 \\ \text{or } N-|\gamma|+2}} (\rho+\gamma)!\;/\rho!$$

where $\sigma = k - [\mathrm{Re}\, k]$, by (7.8), (1.1) and the fact that $\mu^{\rho+\gamma}G^m$ is homogeneous of degree $(1-\sigma) - (2n+2)$ or $(2-\sigma) - (2n+2)$. There are at most $2(N+2)^{2n+1}$ terms in the last summation, and for some $A,Q > 0$, each term is less than $AQ^N\gamma!$. Thus (*) follows at once.

Thus, to complete the proof of (c), when Γ_{G^m} is used, we need only check (**). This follows at once from the following fact as applied to $f(u) = \cos u\cdot\mu$ or $\sin u\cdot\mu$:

If f is a real-valued C^∞ function on $\mathbb{R}^{2n+1}$, and $\mu = (\lambda, \xi)$, then for any $M \in \mathbb{Z}^+$,

$$|f(\mu) - \sum_{|\rho|\leq M} (\partial^\rho f)(0)\mu^\rho/\rho!| \leq 2 \sum_{\substack{|\rho|=M+1 \\ \text{or } M+2}} |\mu^\rho| \sup_{0\leq r,s\leq 1} |\partial^\rho f(r\lambda, s\xi)|/\rho!.$$

This is a simple replacement for Taylor's Theorem when parabolic homogeneity is used. To see it, observe that we may assume $(\partial^\rho f)(0) = 0$ for all ρ with $|\rho| \leq M$ (otherwise replace $f(u)$ by $(f(u) - \sum_{|\rho|\leq M} (\partial^\rho f)(0)u^\rho/\rho!)$. Then, by Taylor's Theorem,

$$f(\lambda,\xi) = \sum_{\ell=0}^{[M/2]} \lambda^\ell \partial_\lambda^\ell f(0,\xi)/\ell! + \lambda^{[M/2]+1}\partial_\lambda^{[M/2]+1} f(r\lambda,\xi)/([M/2]+1)!$$

$$= \sum_{\ell=0}^{[M/2]} \sum_{\|\alpha\|=M+1-2\ell} \lambda^\ell \xi^\alpha \partial_\lambda^\ell \partial_\xi^\alpha f(0,s_\ell\xi)/\ell!\,\alpha!$$

$$+ \lambda^{[M/2]+1}\partial_\lambda^{[M/2]+1} f(r\lambda,\xi)/([M/2]+1)!$$

for certain $0 \leq r, s_\ell \leq 1$. This proves more than we claimed.

(c) follows, when Γ_{G^m} is used. For the case $K^m = F^{-1}\Lambda_{G^m}$, we need only note

$$(2\pi)^{2n+1}[(F^{-1}\Lambda_{G^m})(-u) - (F^{-1}\Gamma_{G^m})(-u)]$$

$$= \int_{1\leq|\mu|\leq N} \Big(\sum_{|\rho|=k+m} (iu)^\rho \mu^\rho/\rho!)G^m(\mu)\,d\mu\Big)$$

$$= \log N(m) \sum_{|\rho|=k+m} M(\mu^\rho G^m/\rho!)(iu)^\rho.$$

This is again a polynomial, and the absolute value of its coefficients do not exceed $C_5R_5^m$ for some $C_5, R_5 > 0$, by (7.9). Thus this difference can be treated just as we treated f_2^m.

This completes the proof of (c).

The third uniformity assertion follows from the proof. (In case N(0) = 0 or 1, the uniformity in (7.1) when m = 0 or 1 follows from (7.7) for $|\gamma| \geqq 2$. For $|\gamma| = 0$ or 1, uniform estimates follow easily from the methods or Proposition 1.1(a) and from (7.3), which yields uniform estimates for the size of $\partial_\ell^M J^0$, $\partial_\ell^M J^1$ on the unit sphere for $1 \leqq \ell \leqq 2n+1$, $0 \leqq M \leqq M_0$.) This completes the proof of Theorem 7.1.

Had we strictly imitated the method of Boutet de Monvel-Kree [9], then we would have considered K_γ^m, J_γ^m for $|\gamma| = N(m) - 2$ or $N(m) - 1$, not $N(m) + 1$ or $N(m) + 2$. The proof of (c) would then have been much simpler, since $\partial^\beta K^m(0) = 0$ for $|\beta| \leqq N(m) - 1$, and we could then have obtained the estimates on the $\partial^\beta K^m(u)$ (for $1 \leqq |u| \leqq 2$, $|\beta| < N(m) - 2$) simply by integrating the estimates for $|\beta| = N(m) - 2$, $N(m) - 1$. We did not proceed in this manner, because we shall need the following corollary, a Poincaré-type lemma, in the next section of this chapter, and it will be crucial to have (in its notation) $|\gamma| = [\operatorname{Re} \varkappa] + 1$ or $[\operatorname{Re} \varkappa] + 2$.

<u>Corollary 7.2</u>. Suppose $\varkappa \in \mathbb{C}$, $[\operatorname{Re} \varkappa] \geqq 0$. Let $A = \{\gamma \in (\mathbb{Z}^+)^{2n+1} : |\gamma| = [\operatorname{Re} \varkappa] + 1 \text{ or } [\operatorname{Re} \varkappa] + 2\}$. Suppose we are given a set of distributions $\{K_\gamma\}_{\gamma \in A}$ such that $K_\gamma \in \mathcal{A}K^{\varkappa - |\gamma|}$ for all γ, and such that

$$\partial^{\rho}(\gamma!K_{\gamma}) = \partial^{\gamma}(\rho!K_{\rho})$$

for all $\gamma,\rho \in A$. Then there exists $K = F(\{K_{\gamma}\}) \in AK^{\varkappa}$ with

$$\partial^{\gamma}K/\gamma! = K_{\gamma}$$

for all $\gamma \in A$. Further, the map F may be chosen so that the following properties are always satisfied:

Suppose $k \in \mathbb{C}$, $C_4,R_4 > 0$. Then there exists $C_5,R_5 > 0$ as follows.

Say $m \in \mathbb{Z}^+$, $\varkappa = k + m$, $[\mathrm{Re}\,\varkappa] \geqq 0$, and A, $\{K_{\gamma}\}_{\gamma\in A}$ are as above, and in addition we have

$$|\partial^{\gamma'}K_{\gamma}(u)| < C_4R_4^{m+\|\gamma'\|}(\gamma')! \text{ for } 1 \leqq |u| \leqq 2, \quad \gamma' \in (\mathbb{Z}^+)^{2n+1},\ \gamma \in A. \tag{7.10}$$

Then if $K = F(\{K_{\gamma}\})$,

$$|\partial^{\gamma'}K(u)| < C_5R_5^{m+\|\gamma'\|}(\gamma')! \text{ for } 1 \leqq |u| \leqq 2, \quad \gamma' \in (\mathbb{Z}^+)^{2n+1}. \tag{7.11}$$

Further, if the K_{γ} all depend holomorphically on a parameter for $\gamma \in A$, then K also depends holomorphically on this parameter.

<u>Proof</u>. We first describe the map F. Let $J_{\gamma} = FK_{\gamma}$; we have

$$(-i\mu)^{\rho}\gamma!J_{\gamma} = (-i\mu)^{\gamma}\rho!J_{\rho} \text{ for } \gamma,\rho \in A.$$

For any γ, we may define $G_\gamma \in C^\infty(S_\gamma)$ by

$$G_\gamma(\mu) = \gamma! J_\gamma(\mu)/(-i\mu)^\gamma$$

where $S_\gamma = \{\mu \in \mathbb{R}^{2n+1} \setminus \{0\} : \mu_\ell \neq 0 \text{ whenever } \gamma_\ell \neq 0\}$. Since for all $\gamma, \rho \in A$ we have $G_\gamma = G_\rho$ on $S_\gamma \cap S_\rho$, we may define $G \in C^\infty(\mathbb{R}^{2n+1} \setminus \{0\})$ by

$$G = G_\gamma \text{ on } S_\gamma.$$

G is homogeneous of degree $j = -2n - 2 - \varkappa$. We define the desired $K = F(\{K_\gamma\})$ by

$$K = \mathcal{F}^{-1}J, \text{ where } J = \Lambda_G.$$

Let us check that K has the desired properties. In the first place, if $\gamma \in A$,

$$J_\gamma = (-i\mu)^\gamma J/\gamma!$$

so $$K_\gamma = \partial^\gamma K/\gamma!.$$

From this it is clear that there are constants $C, R > 0$ so that

$$|\partial^{\gamma'} K(u)| < CR^{\|\gamma'\|}(\gamma')! \quad \text{for } 1 \leq |u| \leq 2$$

if $|\gamma'| \geq [\text{Re}\,\varkappa] + 1$. Since $K \in \mathcal{K}^\varkappa$, this shows that $K \in A\mathcal{K}^\varkappa$. Hence $J \in A\mathcal{K}^j$.

Let us check that the further properties of F are also satisfied, with F as above. With k, m, C_4, R_4 as in the statement of the corollary, observe that there are constants C_2,R_2 depending only on C_4,R_4, such that (7.3) holds; this follows at once from (7.10) and the first uniformity assertion of Corollary 1.5. By Theorem 7.1(c), K satisfies (7.11) with constants C_5,R_5 depending only on C_4,R_4 as claimed. The statement about holomorphic dependence on a parameter is immediate from the construction. This completes the proof.

Perhaps Corollary 7.2 could be proved directly, without recourse to the Fourier transform. If so, this would lead at once to an alternate proof of the main assertion of Theorem 7.1(c).

In Theorem 7.1(a), if $J_+^m \sim 0$ for all m, then (7.6) has no content. Fortunately, in this case, the situation is very clear-cut.

Proposition 7.3. Say $k \in \mathbb{C}$, Re $k \geqq 0$, $j = -2n - 2 - k$ and $J^m \in AJ^{j-m}$ for all $m \in \mathbb{Z}^+$. Let $J_\gamma^m = (-i\mu)^\gamma J^m/\gamma!$ for $|\gamma| =$ [Re k] + m + 1 or [Re k] + m + 2.

(a) Suppose that the J_γ^m satisfy (7.3). Suppose further that $J_+^m \sim 0$ in $Z^2_{2,j-m}$ for all m.

Then there are constants $B_1,B_2,C_6 > 0$ so that

$$J_+^m, J_-^m \in Z_2^2(B_1,B_2,C_6^m m!^{1/2}) \text{ for all m.} \tag{7.12}$$

(b) Suppose that, for some $B_1,B_2,C_6 > 0$, J_+^m and J_-^m satisfy (7.12). Then for some $C_2,R_2 > 0$, (7.3) holds.

Further, we have the following uniformities (with j fixed):

(1) Say $C_2, R_2 > 0$; then there are $B_1, B_2, C_6 > 0$ as follows. If J^m, J^m_γ are as in (a) and satisfy (7.3) with that C_2, R_2, then (7.12) holds with that B_1, B_2, C_6.

(2) Say $B_1, B_2, C_6 > 0$; then there are $C_2, R_2 > 0$ as follows. If J^m, J^m_γ are as in (b) and satisfy (7.12) with that B_1, B_2, C_6, then (7.3) holds with that C_2, R_2.

Proof. (a) Select $r \in \mathbb{Z}^+$ with $2r = [\text{Re } k] + m + 1$ or $[\text{Re } k] + m + 2$. Let $\gamma = (r,0,\ldots,0)$. Then $J^m_{\gamma\pm} = (-i)^r J_\pm / r! \in Z^2_2$. By the uniformity assertion of Theorem 1.7(b) and Proposition 2.1, there are B_1, B_2, C_6' so that $J^m_{\gamma\pm} \in Z^2_2(B_1, B_2, (C_6')^m)$ for all m. This gives (7.12) at once; the first uniformity assertion follows from the proof.

(b) If $\gamma = (s,\alpha)$, we have

$$i^{|\gamma|} J^m_{\gamma\pm} = \pm(1/s!\,\alpha^{1/2})(\xi^\alpha/\alpha!^{1/2}) J^m_\pm. \qquad (7.13)$$

Note $1/s!\alpha!^{1/2} \leq |\gamma|!^{-1/2} < C^m m!^{-1/2}$. Further, for any $\varepsilon > 0$, $(\xi^\alpha/\alpha!^{1/2})^2 \leq e^{2n\,\varepsilon|\xi|^2}\varepsilon^{-\|\alpha\|}$, so $|\xi^\alpha/\alpha!^{1/2}| \leq e^{n\,\varepsilon|\xi|^2}(\varepsilon^{-1/2})^{\|\alpha\|}$. In (7.13), $\|\alpha\| \leq |\gamma| = [\text{Re } k] + m + 1$ or $[\text{Re } k] + m + 2$.

Choosing $\varepsilon < B_2/n$, and combining the estimates, we see that for some $B_1', B_2', C_6' > 0$,

$$J^m_{\gamma+}, J^m_{\gamma-} \in Z^2_2(B'_1, B'_2, (C'_6)^m)$$

for all m, γ. (7.3) now follows from the first uniformity assertion of Proposition 2.1, and the uniformity assertions of Theorem 2.6 and Theorem 1.8(b). The second uniformity assertion follows from the proof.

Remark. Proposition 7.3(b) is of a stronger nature than Theorem 7.1(b), in the sense that Theorem 7.1(b) only asserts that if the g^m_ℓ satisfy (7.6) then there exists $J^m \sim \sum g^m_\ell$ satisfying (7.3). One could, however, make a stronger statement; for completeness we indicate it, though we shall have no use for it. Our notation is unfortunately badly chosen for this stronger statement; let us change it for this remark only. Returning to the definition of $Z^q_{q,j}(B,C,R,c)$ before Theorem 1.8, we observe that "C" refers to the constant in (i) or (ii) of the definition, and that the same constant C was used in (i) and in (ii) of that definition. For this remark, it is best to allow two different constants C_i, C_{ii} for (i) and (ii), and to define $Z^q_{q,j}(B,C_i,C_{ii},R,c)$ accordingly.

We then assert that if (7.3) holds, there exist $B,C_i,C_{ii},R,C > 0$ so that

$$J^m_{\pm} \in Z^2_{2,j-m}(B, m!^{1/2}C_i, m!C_{ii}, R, c). \tag{7.14}$$

Conversely, if (7.14) holds for a family $\{J^m\}$ with $J^m \in \mathcal{A}J^{j-m}$

then there exist $C_2, R_2 > 0$ so that (7.3) holds. The proof is left to the interested reader.

We shall need only some very rough information about the group Fourier transform of sequences $\{K^m\}$ satisfying (7.1). We look only at the case $k = -2n - 2$. Recall from Proposition 4.8(b) that $\hat{K}^m(\pm 1)$ is a bounded operator on $\mathcal{H}_{\pm 1}$, for all m. In Chapter 8, we shall need an estimate on how the bounds depend on m.

<u>Proposition 7.4</u>. Say $R > 0$; then there exist $C_1, R_1 > 0$ as follows. Suppose $K^m \in \mathcal{A}K^{m-2n-2}$ for all $m \in \mathbb{Z}^+$, and that $|\partial^\gamma K^m(u)| < R^{m+\|\gamma\|}\gamma!$ for $1 \leq |u| \leq 2$, $m \in \mathbb{Z}^+$, $\gamma \in (\mathbb{Z}^+)^{2n+1}$. Say $\hat{K}^m = \underset{\sim}{J}^m, J^m(1) = J_+^m, J^m(-1) = J_-^m$. Then

$$\|J_\pm^m\| < C_1 R_1^m m!^{1/2} \text{ for } m \geq 1.$$

<u>Proof</u>. We begin the proof with a remark about a familiar situation. For $K \in K^{-2n}$, define $c_K \in \mathbb{C}$ as follows. Say $K = G \in C^\infty(\mathbb{H}^n \setminus \{0\})$ away from 0. Define c_K by

$$c_K\delta = \text{P.V.}(TG) - TK. \tag{7.15}$$

By testing both sides of (7.15) on a $\psi \in C_c^\infty$ with $\psi(u) = 1$ for $|u| \leq 1$, we see that there exists $C > 0$ so that

$$|c_K| < C\|K\|_{C^1} \tag{7.16}$$

where $\| \ \|_{C^k}$ denotes C^k norm on the unit sphere $\{u: |u| = 1\}$. An analogous estimate holds for d_K^j, defined for $K \in K^{-2n-1}$ by $d_K^j = \text{P.V.}(Z_jG) - Z_jK$, where $K = G \in C^\infty$ away from 0; similarly with $\overline{Z}_j$ in place of Z_j.

As we observed during the proof of Lemma 4.5, for some $C_2 > 0$, if $K = \text{P.V.}(f) \in AK^{-2n-2}$, then $\|\hat{K}(\pm 1)\| < C_2\|f\|_{C^1}$. By this and (7.16), we see that there exists $C_3 > 0$ so that if $K \in K^{-2n}$ then $\|\hat{TK}(\pm 1)\| < C_3\|K\|_{C^2}$.

Now let us prove the Proposition for $m = 2\ell$ even. Then $J_+^m = i^\ell (T^\ell K^m)\hat{}(1)$; thus

$$\|J_+^m\| < C_3 \|T^{\ell-1}K^m\|_{C^2} < C_3 R^{m+\ell+1}(\ell+1)!$$

assuming, without loss, that $R > 1$. This yields the desired result at once for J_+^m in case m is even; similarly for J_-^m.

Next, let us observe that if $K \in K^{-2n-1}$ then

$$\|\hat{K}(1)\| \leq (1/2) \sum_{j=1}^{n} (\|\widehat{Z_jK}(1)\| + \|\widehat{\overline{Z}_jK}(1)\|). \tag{7.17}$$

Indeed, let $\hat{K}(1) = J$. Then if $v \in H_1^\infty$,

$$\|Jv\| = \|JAA^{-1}v\| = \|(1/2)\sum(JW_jW_j^+ + JW_j^+W_j)A^{-1}v\|$$

$$\leq (1/2)\sum(\|JW_j\| + \|JW_j^+\|)\|v\|$$

as desired, since $\|W_j^+A^{-1}v\|$, $\|W_jA^{-1}v\| \leq \|v\|$ for all j, as is easily seen.

From (7.17) and our earlier observations concerning $Z_jK, \overline{Z}_jK$, we see that for some $C_4 > 0$,

$$\|\hat{K}(\pm 1)\| \leq C_4 \|K\|_{C^2}.$$

Now we can prove the Proposition for $m = 2\ell + 1$ odd. Then $J_+^m = i^\ell (T^\ell K^m)^\wedge(1)$,

$$\|J_+^m\| < C_4 \|T^\ell K^m\|_{C^2} < C_4 R^{m+\ell+1} (\ell+1)!$$

assuming again $R > 1$. Since a similar inequality exists for J_-^m, the proof is complete.

Section Two - Generalized Convolution

Say $\varkappa_1, \varkappa_2 \in \mathbb{C}$, and $K_\nu \in AK^{\varkappa_\nu}$ for $\nu = 1,2$. Let $\varkappa = \varkappa_1 + \varkappa_2 + 2n + 2$. If Re $\varkappa < 0$, Proposition 5.2 explains how to define $K = K_2*K_1$. If, on the other hand, Re $\varkappa \geqq 0$, we seek to define a "generalized convolution" $K = K_2\underline{*}K_1$ by the following rather natural route.

Select $N \in \mathbb{Z}^+$ with $2N >$ Re $\varkappa$. For $\lambda \in \mathbb{R}\setminus\{0\}$, $\xi \in \mathbb{R}^{2n}$, put

$$G(\lambda,\xi) = F(K_2*T^N K_1)(\lambda,\xi)/(-i\lambda)^N. \qquad (7.18)$$

One may then show (for instance, by use of Lemma 5.7 and (2.19)) that G may be extended to a C^∞ function on $\mathbb{R}^{2n+1}\setminus\{0\}$, homogeneous of degree $-2n - 2 - \varkappa$. It is then rather natural to define

$$K_2\underline{*}K_1 = F^{-1}\Lambda_G$$

so that, for instance, $T^N(K_2\underline{*}K_1) = K_2*T^N K_1$. It is less clear that $K_2\underline{*}K_1 \in AK^{\varkappa}$, but this is in fact true. Generalized convolution is then easily seen to be bilinear and associative. It is the purpose of this section to show that this particular choice of $\underline{*}$ leads to very desirable estimates for the size of $K_2\underline{*}K_1$ on the unit sphere, in terms of the sizes of K_2 and K_1 there. This type of estimate is critical for an analytic calculus, for we shall be needing to "convolve" together two series of the type $\sum K^m, K^m$ as in (7.1). Note also that, since such series usually converge only in a neighborhood of 0, such a convolution will only make sense if both sides are first multiplied by characteristic functions of balls centered at 0. Thus it will be crucial to obtain estimates for the size of $\chi_2K_2 * \chi_1K_1 - K_2 \underline{*} K_1$ near 0 if χ_1, χ_2 are characteristic functions of such balls.

Our strategy in this section will be to give an alternative definition for $K_2 \underline{*} K_1$ which will later be seen to be equivalent to (7.19). On the basis of this alternative definition, we will be able to rather readily obtain all the results alluded to in the last paragraph.

Before we do this, it is time to simplify the notation we use to measure the size of elements of the AK^k spaces.

For $C,R > 0$, $-2n - 2 - k \notin \mathbb{Z}^+$, we define

$$AK^k(C,R) = \{K \in AK^k : |\partial^\gamma K(u)| < CR^{\|\gamma\|}\gamma! \text{ whenever } 1 \leqq |u| \leqq 2,$$
$$\gamma \in (\mathbb{Z}^+)^{2n+1}\}.$$

If $-2n - 2 - k \in \mathbb{Z}^+$, we define

$$AK^k(C,R) = \{K = \Lambda_G + \sum_{|\alpha|=-2n-2-k} c_\alpha \partial^\alpha \delta : |\partial^\gamma G(u)| < CR^{\|\gamma\|}\gamma!$$
$$\text{whenever } 1 \leqq |u| \leqq 2,\ \gamma \in (\mathbb{Z}^+)^{2n+1}; \text{ and } |c_\alpha| < C \text{ for all } \alpha\}.$$

Suppose $-2n - 2 - k \in \mathbb{Z}^+$. Say $j_0 \in \mathbb{Z}^+$, $j_0 \geqq -2n - 2 - k$. Note that for every $C,R > 0$ there exists $C_1 > 0$ so that if $K \in AK^k, K = G \in C^\infty(\mathbb{H}^n \setminus \{0\})$ away from 0, and if:

1. $|\partial^\gamma G(u)| < CR^{\|\gamma\|}\gamma!$ whenever $1 \leqq |u| \leqq 2$,
$\gamma \in (\mathbb{Z}^+)^{2n+1}$ (7.20)

2. $|K(\psi)| < C_1$ whenever $\psi \in C_c^\infty(H^n)$,

supp $\psi \subset \{u : |u| \leqq 1\}$, $|(\partial^\gamma\psi)(u)| \leqq 1$ (7.21)

whenever $|\gamma| \leqq j_0$

then $K \in AK^k(C,R)$. (This is clear once we pick $\psi_0 \in C_c^\infty$ with supp $\psi_0 \subset \{|u| \leqq 1\}$, $\psi_0 = 1$ near 0, and examine $K(u^\gamma\psi_0)$ for $|\gamma| \leqq j_0$.) Conversely, for any k with $-2n - 2 - k \in \mathbb{Z}^+$, and any $c,R > 0$, there exists $j_0 > -2n - 2 - k$ and $C_1 > 0$ so that (7.21) holds for all $K \in AK^k(C,R)$. (This is clear from the definition of Λ_G.)

Here now is the main theorem of this section.

<u>Theorem 7.5</u>. For any $\varkappa_1,\varkappa_2 \in \mathbb{C}$ there is a bilinear map $\underline{*}$ from $AK^{\varkappa_2} \times AK^{\varkappa_1}$ to $AK^\varkappa$ where $\varkappa = \varkappa_1 + \varkappa_2 + 2n + 2$, with these properties:

(a) Say $K_\nu \in AK^{\varkappa_\nu}$, $\nu = 1,2$.

If Re $\varkappa < 0$, then $K_2\underline{*}K_1$ is a "replacement" for $K_2 * K_1$. Indeed, we have the following.

If Re $\varkappa \geqq 0$, $N \in \mathbb{Z}^+$, $2N >$ Re $\varkappa$, and if G is defined by (7.18) for $\lambda \in \mathbb{R}\setminus\{0\}$, $\xi \in \mathbb{R}^{2n}$, then G may be extended to a C^∞ function on $\mathbb{R}^{2n+1}\setminus\{0\}$, homogeneous of degree $-2n - 2 - \varkappa$, and (7.19) holds.

If L (resp. R) is homogeneous and left (resp. right) invariant, with homogeneity degree ℓ(resp. r), and if $\ell + r > \mathrm{Re}\,\varkappa$, then

$$RLK = RK_2 * LK_1. \tag{7.22}$$

If $\hat{K}_\nu = \underset{\sim}{J}_\nu$, $\hat{K} = \underset{\sim}{J}$, all in the $\mathcal{Q}$ sense, then

$$J_+ = J_{2+}J_{1+},\ J_- = J_{2-}J_{1-} \quad \text{on } \mathcal{H}^\infty. \tag{7.23}$$

Further:

If $\mathcal{F}K_\nu = F_\nu$, $\mathcal{F}K = F$, $F_{\nu+} \sim \sum g_{\nu\ell}$ in Z^2_{2,j_ν}, $F \sim \sum g_\ell$ in $Z^2_{2,j}$ then

$$g_\ell = \sum_{a+b+|\alpha|=\ell} [(2i)^{|\alpha|}/\alpha!]\partial^\alpha g_{2a}\tilde{\partial}^\alpha g_{1b} \quad . \tag{7.24}$$

with $\tilde{\partial}$ as before (5.7). Here $j_\nu = -2n - 2 - \varkappa_\nu$, $j = -2n - 2 - \varkappa$.

In addition, $\underline{*}$ has these properties.

For $\nu = 1,2$, suppose $k_\nu \in \mathbb{C}$, and $R_\nu > 0$. Say $0 < r_2 < r_1$. Then there exist $A,R,A_0,R_0,r > 0$ as follows.

Suppose, for $\nu = 1,2$, $C_\nu > 0$, and $K_\nu \in \mathcal{AK}^{k_\nu+m_\nu}(C_\nu R_\nu^{m_\nu},R_\nu)$ for some $m_\nu \in \mathbb{Z}^+$. (7.25)

Let $k = k_1 + k_2 + 2n + 2$, $m = m_1 + m_2$. (7.26)

Let $K = K_2 \underline{*} K_1$. Then we have (b) and (c):

(b) $K \in \mathcal{AK}^k(AC_1C_2R^m,R)$ (7.27)

(c) For $\nu = 1,2$, let χ_ν be the characteristic function of $\{u \in \mathbb{H}^n : |u| < r_\nu\}$. Then there exists Q analytic on $\{u : |u| < r\}$ such that

$$\chi_2 K_2 * \chi_1 K_1 = K + Q \text{ on } \{u : |u| < r\} \tag{7.28}$$

and

$$|\partial^\gamma Q(u)| < A_0 C_1 C_2 R_0^{m+\|\gamma\|} \gamma! \text{ for all } \gamma \in (\mathbb{Z}^+)^{2n+1}, \text{ if } |u| < r. \tag{7.29}$$

(d) Further, say $K_1 = K_{1\omega}$, $K_2 = K_{2\omega}$ depend holomorphically on a parameter $\omega \in \Omega$, and satisfy (7.25) for all $\omega \in \Omega$. Then $K_2 \underline{*} K_1$ depends holomorphically on ω; further, for any fixed u with $|u| < r$, Q(u) will be a holomorphic function of ω.

To prove Theorem 7.5, we shall need several lemmas. First, however, a remark about conditions like (7.27).

Say $K \in AK^{k+m}(A,R)$, so that K satisfies

$$|\partial^\gamma K(u)| < AR^{\|\gamma\|}\gamma! \text{ for } \gamma \in (\mathbb{Z}^+)^{2n+1},\ 1 \leq |u| \leq 2. \tag{7.30}$$

If K is homogeneous, we of course have for $u \neq 0$,

$$|\partial^\gamma K(u)| < A|u|^{\mathrm{Re}\ k+m}(R^{\|\gamma\|}|u|^{-|\gamma|})\gamma!. \tag{7.31}$$

If K is not homogeneous, then $k + m \in \mathbb{Z}^+$ and $K = K' + p(u)\log|u|$ for some homogeneous K',p with p a polynomial, and $\partial^\gamma K = K'_\gamma + (\partial^\gamma p)(u)\log|u|$ for some homogeneous K'_γ. Examining (7.30) for $|u| = 1$, we see that for $u \neq 0$

$$|K'_\gamma(u)| < A|u|^{k+m}(R^{\|\gamma\|}|u|^{-|\gamma|})\gamma!. \tag{7.32}$$

Next, using (7.30) and (7.32) for $|u| = 2$, we see that for $|u| = 2$,

$$|(\partial^\gamma p)(u)\log 2| < A(1+2^{k+m})R^{\|\gamma\|}\gamma!. \tag{7.33}$$

From (7.32) and (7.33), we see that if $u \neq 0$,

$$|\partial^\gamma K(u)| < A[1+(1+2^{k+m})(|\log|u||/\log 2]|u|^{k+m}(R^{\|\gamma\|}|u|^{-|\gamma|})\gamma!. \tag{7.34}$$

A bit of additional notation. We fix $S_0 > 0$ so that

$$|uv| \leqq S_0(|u|+|v|) \text{ for all } u,v \in \mathbb{H}^n.$$

(This is the $\mathbb{H}^n$ triangle inequality; see [20] page 449.)

If $N \in \mathbb{Z}^+$, $g \in \mathcal{E}'(\mathbb{H}^n)$, we as usual say that g is of order N if for some $C > 0$

$$|g(\psi)| < C|\psi|_{C^N} \quad \text{for all } \psi \in C_c^\infty.$$

We define $\|g\|_{(N)}$ to be the least such C.

If $S_1, S_2 \subset \mathbb{H}^n$, we let $S_2S_1 = \{vu : v \in S_2,\ u \in S_1\}$.

If $f \in \mathcal{D}'(\mathbb{H}^n)$, we as usual let sing $\text{supp}_a f$ denote the analytic singular support of f, that is, the complement of the largest open set on which f is an analytic function.

We shall need three lemmas which will enable us to control analyticity of convolutions.

Part (a) of the first lemma is standard. We include a proof (adapted from [9]) for completeness.

Lemma 7.6(a). If $f,g \in \mathcal{E}'(\mathbb{H}^n)$, then

$$\text{sing supp}_a(g*f) \subset (\text{sing supp}_a g)(\text{sing supp}_a f).$$

Quantitatively, suppose that for $j = 1,2$, B_j, S_j are compact subsets of $\mathbb{H}^n$, and U_j are bounded open subsets of $\mathbb{H}^n$, with $B_j \subset U_j$, $B_j \subset S_j$. Suppose P is compact and disjoint from $\overline{U}_2\overline{U}_1$. Suppose also $R_1, R_2 > 0$, $M,N \in \mathbb{Z}^+$.

Then there exist $C,R > 0$ as follows:

Say $f,g \in \mathcal{E}'(\mathbb{H}^n)$, f of order M, g of order N. Say supp $f \subset S_1$, $\text{sing supp}_a f \subset B_1$, supp $g \subset S_2$, $\text{sing supp}_a g \subset B_2$. Say

$$|\partial^\gamma f(u)| < C_1 R_1^{\|\gamma\|}\gamma! \quad \text{for all } \gamma \in (\mathbb{Z}^+)^{2n+1},\ u \notin U_1,$$

$$|\partial^\gamma g(u)| < C_2 R_2^{\|\gamma\|}\gamma! \quad \text{for all } \gamma \in (\mathbb{Z}^+)^{2n+1},\ u \notin U_2.$$

$$\text{Then } |\partial^\gamma (g*f)(u)| < C(C_2+\|g\|_N)(C_1+\|f\|_M)R^{\|\gamma\|}\gamma! \quad \text{for } u \in P \text{ and all } \gamma. \tag{7.35}$$

(b) Say $k \in \mathbb{C}$, $\varepsilon > 0$. Say $\overline{V} \subset U$, $\overline{U} \subset W \subset \mathbb{H}^n$, U,V,W bounded open sets. Say $N \in \mathbb{Z}^+$, $R > 0$. Then there exist $C, R_1 > 0$ as follows.

Suppose $K \in \mathcal{AK}^{k+m}$ for some $m \in \mathbb{Z}^+$, and that

$$|\partial^\gamma K(u)| < R^{\|\gamma\|}\gamma! \quad \text{for } \gamma \in (\mathbb{Z}^+)^{2n+1},\ 1 \le |u| \le 2.$$

Suppose f is a distribution which is zero on U, and that either:

1. $f \in \mathcal{E}'$, supp $f \subset W$, f of order N, or

2. f is a function, and $|f(u)| < A(1+|u|)^{-2n-2-\mathrm{Re}\,k-m-\varepsilon}$ for all $u \in \mathbb{H}^n$. Then $K*f$ is analytic on U. Further, in Case 1,

$$|\partial^\gamma (K*f)(v)| < C\|f\|_{(N)} R_1^{m+\|\gamma\|}\gamma! \text{ for all } \gamma \in (\mathbb{Z}^+)^{2n+1},\ v \in V.$$

In Case 2,

$$|\partial^\gamma (K*f)(v)| < CA(4S_0)^m R_1^{\|\gamma\|}\gamma! \text{ for all } \gamma \in (\mathbb{Z}^+)^{2n+1},\ v \in V.$$

Proof. (a) It suffices to prove analyticity of $g*f$, and that (7.35) holds, in a neighborhood of each point $u \in P$. By translating if necessary, we may assume $u = 0$. We think of the underlying manifold of $\mathbb{H}^n$ as $\mathbb{R}^{2n+1}$ (not $\mathbb{R} \times \mathbb{C}^n$). We write points in $\mathbb{H}^n$ as $v = (t,x,y) = (t,w)$ where $w = (x,y) \in \mathbb{R}^{2n}$. For $w = (x,y)$, $w' = (x',y') \in \mathbb{R}^{2n}$, write $\langle w,w'\rangle = y\cdot x' - x\cdot y'$.

Fix $s \in \mathbb{R}^{2n+1}$, with all coordinates s_j irrational. For $N \in \mathbb{Z}$, let $\mathcal{R}_N$ denote the dyadic grid of closed rectangles of sides 2^N and let $\mathcal{R}_N + s$ denote this grid of rectangles translated by s. By enlarging U_1, U_2 if necessary, we may assume that for some N, $\overline{U}_1$ is a finite union of cubes in $\mathcal{R}_N$, $\overline{U}_2$ is a finite union of cubes in $\mathcal{R}_N + s$, and $U_1 = \mathrm{int}\,\overline{U}_1, U_2 = \mathrm{int}\,\overline{U}_2$. Then U_1^c is also a union of cubes in $\mathcal{R}_N$; let $\{Q_i\}$ denote those which intersect $S_1 \cap U_1^c$. These then cover $S_1 \cap U_1^c$. Also U_2^c is a union of cubes in $\mathcal{R}_N + s$; let $\{Q_i' + s\}$ denote those which intersect $S_2 \cap U_2^c$. Let $A_1 \subset \mathbb{R}$ denote the set of coordinates of the vertices of the Q_i, and let $A_2 \subset \mathbb{R}$ denote the set of coordinates of the vertices of the $Q_i' + s$. Since all s_j are irrational, the finite sets A_1 and $-A_2$ are a positive distance apart.

Let χ_j be the characteristic function of U_j $(j=1,2)$, $f_1 = \chi_1 f$, $g_1 = \chi_2 g$, $f_2 = f - f_1$, $g_2 = g = g_1$. Then for u close enough to 0,

$$\begin{aligned}(g*f)(u) &= (g_2*f_1)(u) + (g_1*f_2)(u) + (g_2*f_2)(u) \\ &= f_1(g_{2u}) + g_1(f_{2u}) + g_2*f_2(u),\end{aligned}$$

where $g_{2u}(v) = g_2(uv^{-1})$, $f_{2u}(v) = f_2(v^{-1}u)$. By complexifying u in the first two terms, one sees at once that they are analytic near 0. As for g_2*f_2, note that $g_2*f_2(0)$ is a finite sum of terms of the form

$$\int_{w_1' \varepsilon I_1} \cdots \int_{w_{2n}' \varepsilon I_{2n}} \int_{t' \varepsilon I_{2n+1}} g(-t',-w')f(t',w')dt'dw'.$$

Here each I_j is an interval, one of whose endpoints is in A_1 and the other of whose endpoints is in $-A_2$. By using the fact that A_1 and $-A_2$ are a positive distance apart, one sees that for $u = (t,x,y) = (t,w)$ close enough to 0, $(g_2*f_2)(u)$ is a finite sum of terms of the form

$$\int_{w_1'\in I_1(w)} \cdots \int_{w_{2n}'\in I_{2n}(w)} \int_{t'\in I_{2n+1}(w,w')} g(t-t'-2\langle w,w'\rangle, w-w')f(t',w')dt'dw'. \quad (7.36)$$

Here, for $1 \leq j \leq 2n$, each $I_j(w)$ is an interval with one endpoint in A_1 and the other endpoint of the form $w_j - b_j$ $(b_j \varepsilon A_2)$. Also $I_{2n+1}(w,w')$ is an interval with one endpoint in A_1 and the other endpoint of the form $t - 2\langle w,w'\rangle - c$ $(c\varepsilon A_2)$.

In (7.36), we make a linear substitution for t', and then for w', to convert all integrals to integrals from 0 to 1. The integral then becomes

$$\int_0^1 \cdots \int_0^1 p_3(r,u)g(P_2(r,u))f(P_1(r,u))dr_1 \ldots dr_{2n+1}$$

where p_3, and the components of P_2, P_1 are polynomials, and $g(P_2(r,u))$, $f(P_3(r,u))$ are analytic in u for u near 0, $r \in [0,1]\times\ldots\times[0,1]$. We can complexify u in this integral. This proves g*f is analytic near 0, and the estimate (7.35) follows from the proof.

(b) We shall complexify v in the expression

$$(K*f)(v) = \int K(vw^{-1})f(w)dw.$$

It is easy to complexify v in Case 1 of (b), and to prove Case 1 of (b) by use of (7.31) and (7.34).

One can also complexify v in Case 2 of (b), as we shall show. In this case, by compactness one easily reduces to the case where V and U are balls of the form $V = \{u : |uv_0^{-1}| < r_1\}$, $U = \{u : |uv_0^{-1}| < r_2\}$, where $v_0 \in \mathbb{H}^n$, $0 < r_1 < r_2 < 1$. By replacing f by a translate we may also assume $v_0 = 0$. There exists $a > 0$ so that $|vw^{-1}| \geqq a$ for all $v \in V$, $w \notin U$. Further, by (7.31) and (7.34), it is clear that there exists $c > 0$ such that $K(u)$ has an extension to a function $K(u+i\mu)$, holomorphic in the region $\mathcal{U} = \{u + i\mu \in \mathbb{C}^{2n+1} : |u| \geqq a, |\mu| < c\,|u|\}$ (here $|\ |$ denotes $\mathbb{H}^n$ norm), and that for $u + i\mu \in \mathcal{U}$,

$$|K(u+i\mu)| < A_1 2^m (1+|u|)^{\mathrm{Re}\, k+m+\varepsilon/2}$$

for some constant A_1; c and A_1 depend only on U, V, k, ε and R.

Next, temporarily write $\circ$ for group multiplication, and further write

$$(t,x,y)\circ(t',x',y') = (t+t'+2(x'\cdot y-x\cdot y'),x+x',y+y')$$

even for $(t,x,y) \in \mathbb{C} \times \mathbb{C}^n \times \mathbb{C}^n$. Next, if $v \in V$, $w = (t,x,y) \in U$, $\mu = (\tau,\xi,\eta) \in \mathbb{R}^{2n}$, note

$$(v+i\mu)\circ w^{-1} = v\circ w^{-1} + i\mu_w$$

where $\mu_w = \mu + (2(\xi\cdot y-\eta\cdot x),0,0)$.

We claim that:

There exists $\delta > 0$, depending only on r_1 and r_2, so that if $|\mu| < \delta$ then

$$|\mu_w| < c|vw^{-1}| \text{ for all } v \in V,\ w \notin U. \tag{7.37}$$

Indeed, $|\mu_w| \leq S_0(|\mu|+|2(\xi\cdot y-\eta\cdot x)|^{1/2})$
$\leq S_0(|\mu|+2^{1/2}|\xi|^{1/2}|y|^{1/2}+|\eta|^{1/2}|x|^{1/2}) \leq S_0(|\mu|+b|\mu|^{1/2}|w|^{1/2})$ where $b = 2^{3/2}$. We choose δ of the form $\varepsilon^2 r_2$, $\varepsilon < 1$ to be chosen presently. Then $|\mu| < \delta \Rightarrow |\mu_w| < s'\varepsilon\,|w|$ for some $s' > 0$. Now ε need only be chosen so that $v \in V$, $w \notin U \Rightarrow |w| \leq (c/s'\varepsilon)\,|vw^{-1}|$; this is easy to do, so (7.37) is established. Thus $v \in V$, $w \notin U$, $|\mu| < \delta \Rightarrow (v+i\mu)\circ w^{-1} \in \mathcal{U}$.

Finally, then for $|\mu| < \delta$, $v \in V$, $(K*f)(v)$ has a holomorphic extension to

$$(K*f)(v+i\mu) = \int K((v+i\mu)\circ w^{-1})f(w)dw$$

which satisfies

$$|(K*f)(v+i\mu)| \leq A_1A2^m\int(1+|vw^{-1}|)^{\mathrm{Re}\,k+m+\varepsilon/2}(1+|w|)^{-2n-2-\mathrm{Re}\,k-m-\varepsilon}dw.$$

Since $(1+|vw^{-1}|)^{\ell} \leq S_0^{\ell}(1+|v|)^{|\ell|}(1+|w|)^{\ell}$ for any $\ell \in \mathbb{R}$, and since $1 + |v| \leq 2$ for $v \in V$, Case 2 of (b) follows at once.

<u>Proposition 7.7</u>. Suppose $\overline{V} \subset U$, $\overline{U} \subset W \subset \mathbb{H}^n$, U,V,W bounded open sets. Say $k \in \mathbb{C}$, $R_1,R_2 > 0$. Then there exist $C,R_3 > 0$ as follows:

Suppose $K \in AK^{k+m}(1,R_1)$ for some $m \in \mathbb{Z}^+$.

Suppose $f \in L^1$, supp $f \subset W$, f is analytic on U, and that $|\partial^\gamma f(u)| < A_0 R_2^{\|\gamma\|}\gamma!$ for all $u \in U$, $\gamma \in (\mathbb{Z}^+)^{2n+1}$.

Then $K*f$ is analytic on U. Further

$$|\partial^\gamma (K*f)(v)| < (A_0+\|f\|_1)R_3^{m+\|\gamma\|}\gamma! \text{ for all } v \in V,\ \gamma \in (\mathbb{Z}^+)^{2n+1}.$$

Proof. This is immediate from Lemma 7.6(a), (7.31) and (7.34).

Lemma 7.8. Suppose $r_1, r_2 > 0$, $r_1 \neq r_2$. Select $s > 0$ so that $|v| \leq s$, $|u| \leq \min(r_1,r_2) \Rightarrow |vu| < \max(r_1,r_2)$. Select s_1 with $0 < s_1 < s$. For $\nu = 1,2$, let χ_ν denote the characteristic function of $B_\nu = \{u : |u| < r_\nu\}$.

Suppose $k_\nu \in \mathbb{C}$, and suppose $R_\nu > 0$, for $\nu = 1,2$. Then there exist $A,R > 0$ with the following properties.

Suppose, for $\nu = 1,2$, $K_\nu \in AK^{k_\nu+m_\nu}$ for some $m_\nu \in \mathbb{Z}^+$. Let $k = k_1 + k_2 + 2n + 2$, $m = m_1 + m_2$. Then:

(a) Suppose $|\partial^\gamma K_\nu(u)| < R_\nu^{\|\gamma\|}\gamma!$ for all $\gamma \in (\mathbb{Z}^+)^{2n+1}$, $1 \leq |u| \leq 2$. If $\operatorname{Re} k + m < 0$, then $K_2*(1-\chi_1)K_1$ is analytic for $|u| < r_1$, and

$$|\partial^\gamma[K_2*(1-\chi_1)K_1](u)| < AR^{m+\|\gamma\|}\gamma! \text{ for } \gamma \in (\mathbb{Z}^+)^{2n+1},\ |u| \leq r_1/2.$$

(b) Suppose $K_\nu \in AK^{k_\nu+m_\nu}(1,R_\nu)$ for $\nu = 1,2$. Say $\operatorname{Re} k + m < 0$. Then $K_2*K_1 \in AK^{k+m}(AR^m,R)$. If $K_1 = K_{1\omega}, K_2 = K_{2\omega}$ depend holomorphically on a parameter ω, then so does $K_{2\omega}*K_{1\omega}$.

(c) Suppose $K_\nu \in AK^{k_\nu+m_\nu}(1,R_\nu)$ for $\nu = 1,2$. Then $\chi_2K_2*\chi_1K_1$ is analytic for $0 < |u| < s$. Further,

$$|\partial^\gamma(\chi_2K_2*\chi_1K_1)(v)| < AR^{m+\|\gamma\|}\gamma! \quad \text{for } |v| = s_1. \qquad (7.38)$$

If $K_1 = K_{1\omega}$, $K_2 = K_{2\omega}$ depend holomorphically on a parameter ω, then $(\chi_2K_{2\omega}*\chi_1K_{1\omega})(u)$ is holomorphic in ω for $0 < |u| < s$.

Proof. Clearly we may assume, without loss of generality, that $r_2 > r_1$.

(a) is immediate from Lemma 7.6(b), Case 2. As for (b), note first that there are only finitely many values of m_1,m_2 for which the hypothesis Re $k + m < 0$ could hold. This fact, together with the usual decomposition (5.1) for K_2*K_1, makes it clear that there exist $A' > 0$, $j_0 \in \mathbb{Z}^+$, both depending only on R_1,R_2, so that $|(K_2*K_1)(\psi)| < A'$ whenever $\psi \in C_c^\infty(\mathbb{H}^n)$, supp $\psi \subset \{u : |u| \leq 1\}$, $|(\partial^\gamma\psi)(u)| \leq 1$ whenever $|\gamma| \leq j_0$. This is what we need to know for the condition (7.21). For (7.20) (with $G = K_2*K_1$ away from 0, $C = AR^m$), note that for $|u| < s$ we may write

$$K_2*K_1 = K_2*(1-\chi_1)K_1 + K_2*\chi_1K_1 = K_2*(1-\chi_1)K_1 + \chi_2K_2*\chi_1K_1. \qquad (7.39)$$

Because of (a), it is now clear that in order to prove (b) we need only show (c), with $s_1 = \min(r_1/2,s/2)$. We would then obtain relations like (7.20) for $|u| = s_1$, not for $1 \leqq |u| \leqq 2$, but the information is easily carried over to $1 \leqq |u| \leqq 2$ by use of the homogeneity of $K_2 * K_1$. However (7.38), the main assertion of (c), is immediate from Lemma 7.6(a), (7.31) and (7.34). The statement about holomorphic dependence on a parameter in (b) is easily proved through use of (5.1) and the map Q of Proposition 5.2(a). Similarly, breaking up $\chi_1 K_1$ and $\chi_2 K_2$ into two parts, one supported very close to 0, enables one to prove the statement about holomorphic dependence on a parameter in (c).

Remark. Note that Fact 5.10 does indeed follow from Lemma 7.8(b), as we claimed when we presented that Fact. Indeed, note that (in the notation of Fact 5.10), on any set Ω' of the type described there, there must exist $C_1', C_2', R_1', R_2' > 0$ so that $K_{\nu\omega} \in AK^{k_\nu}(C_\nu, R_\nu)$ for all $\omega \in \Omega'$, $\nu = 1,2$.

Proof of Theorem 7.5. Our first task is to define $K = K_2 \underline{*} K_1$. If Re $\kappa < 0$, we set $K_2 \underline{*} K_1 = K_2 * K_1$. Say then Re $\kappa \geqq 0$.

Let us motivate our procedure. We do not use the definitions (7.18), (7.19); rather we give a different definition which we later prove to be equivalent to (7.18), (7.19). Our strategy will be to first specify what $K_\gamma = \partial^\gamma K/\gamma!$ is for $|\gamma| = [\text{Re } \kappa] + 1$ or $[\text{Re } \kappa] + 2$ and then to invoke Corollary 7.2 to produce the needed K. If * were Euclidean convolution, we could define $\gamma! K_\gamma$ by

$$\gamma!K_\gamma = \partial^{\gamma_2}K_2 * \partial^{\gamma_1}K_1, \tag{7.40}$$

with γ_1, γ_2 chosen so that $\gamma_1 + \gamma_2 = \gamma$; this makes sense since $\mathrm{Re}\, \varkappa_1 + \mathrm{Re}\, \varkappa_2 - |\gamma| = \mathrm{Re}\, \varkappa - |\gamma| - 2n - 2 < -2n - 2$. We wish to avoid differentiating kernels the real parts of whose homogeneity degrees are less than or equal to $-2n - 2$, if at all possible. For this reason, we would distinguish three cases:

$$\mathrm{Re}\, \varkappa_1 > -2n - 2, \ \mathrm{Re}\, \varkappa_2 > -2n - 2 \tag{7.41}$$

$$\mathrm{Re}\, \varkappa_1 > -2n - 2, \ \mathrm{Re}\, \varkappa_2 \leqq -2n - 2 \tag{7.42}$$

$$\mathrm{Re}\, \varkappa_1 \leqq -2n - 2, \ \mathrm{Re}\, \varkappa_2 > -2n - 2. \tag{7.43}$$

If we have (7.41), we would be sure to choose $\gamma_\nu (\nu=1,2)$ in (7.40) so that $\mathrm{Re}\, \varkappa_\nu - |\gamma_\nu| > -2n - 2$ for $\nu = 1,2$, and $\gamma_1 + \gamma_2 = \gamma$; it is easy to see that this is always possible, at least if $n > 1$. (For this, for $\nu = 1,2$, write $a_\nu = \mathrm{Re}\, \varkappa_\nu + 2n + 2$. Then $a_1 + a_2 = \mathrm{Re}\, \varkappa + 2n + 2 \geqq 2n + |\gamma|$. It is then easy to write $\gamma = \gamma_1 + \gamma_2$ where $a_\nu - |\gamma_\nu| > 0$--at least if $n > 1$.) If we have (7.42) we would choose $\gamma_1 = \gamma, \gamma_2 = 0$, so that at least $\mathrm{Re}\, \varkappa_1 - |\gamma_1| > -2n - 2$. (This follows from $\mathrm{Re}\, \varkappa_1 - |\gamma_1| \geqq \mathrm{Re}\, \varkappa - |\gamma| \geqq -2$.) Similarly, in case (7.43) we would choose $\gamma_1 = \gamma$, $\gamma_2 = 0$. In this way we would be able to take full advantage of (7.27) as well as Lemma 7.8(b) (to estimate $\gamma!K_\gamma$ for $|u| = 1$) and Corollary

7.2. The trouble is that $*$ is not Euclidean convolution, so we need a replacement for the identity $\partial^{\gamma}(f*g) = \partial^{\gamma_1}f*\partial^{\gamma_2}g$; this unfortunately forces us to contend with more technicalities. There will also be a slight additional technicality in the case $n = 1$.

We are writing $\partial_1 = T, \partial_{j+1} = \partial/\partial x_j, \partial_{j+n+1} = \partial/\partial y_j$ $(1 \leqq j \leqq n)$. We intend to avail ourselves of these widely-applicable identities:

$$\begin{aligned}
\partial_1(f*g) &= f*Tg \\
&= Tf*g \\
\partial_{j+1}(f*g) &= f*X_jg - (2y_jf)*Tg - f*(2y_jTg) \\
&= X_j^Rf*g + (2y_jTf)*g + Tf*2y_jg \quad \text{if } 1 \leqq j \leqq n \\
\partial_{j+n+1}(f*g) &= f*Y_jg + (2x_jf)*Tg + f*(2x_jTg) \\
&= Y_j^Rf*g - (2x_jTf)*g - Tf*2x_jg \quad \text{if } 1 \leqq j \leqq n
\end{aligned}$$

The third identity, for instance, follows from $\partial_j = X_j - 2y_jT$, and $y_j(F*G) = y_jF*G + F*y_jG$.

Now let $\mathcal{U}$ denote the set of finite sets of ordered pairs of distribution on $\mathbb{H}^n$; thus $u \in \mathcal{U}$ if for some $N \in \mathbb{N}$, $u = \{(f_\ell, g_\ell) : 1 \leqq \ell \leqq N, f_\ell, g_\ell \in \mathcal{D}'(\mathbb{H}^n)\}$. We abuse notation by writing such a set of ordered pairs as a formal sum $u = \sum_{\ell=1}^{N} (f_\ell, g_\ell)$. (The formal sum of two ordered pairs is <u>not</u> another ordered pair.) If $u = (f,g) \in \mathcal{U}$, $1 \leqq m \leqq 2n + 1$, we define elements $P_m u, Q_m u \in \mathcal{U}$ as follows:

$$P_1 u = (f, Tg)$$

$$Q_1 u = (Tf, g)$$

and if $1 \leqq j \leqq n$, $P_{j+1}u = (f, X_j g) + (-2y_j f, Tg) + (f, -2y_j Tg)$

$$Q_{j+1}u = (X_j^R f, g) + (2y_j Tf, g) + (Tf, 2y_j g)$$

$$P_{j+n+1}u = (f, Y_j g) + (2x_j f, Tg) + (f, 2x_j Tg) \tag{7.44}$$

$$Q_{j+n+1}u = (Y_j^R f, g) + (-2x_j Tf, g) + (Tf, -2x_j g)$$

Then $f, g \in \mathcal{E}'$, $u = (f, g)$ and $\sum(f_\ell, g_\ell) = P_j u$ or $Q_j u$, $1 \leqq j \leqq 2n+1$, we have

$$\partial_j (f * g) = \sum f_\ell * g_\ell. \tag{7.45}$$

If now $u = \sum_{\ell=1}^{N} (F_\ell, G_\ell) \in \mathcal{U}$, let us put, for $1 \leqq j \leqq 2n+1$,

$$P_j u = \sum_{\ell=1}^{N} P_j (F_\ell, G_\ell), \quad Q_j u = \sum_{\ell=1}^{N} Q_j (F_\ell, G_\ell).$$

Now, suppose $K^\nu \in \mathcal{AK}^{k^\nu}$ for $\nu = 1,2$. Suppose that $(\mathrm{Re}\, k^1 + 2n+2) + (\mathrm{Re}\, k^2 + 2n+2) > 4$, so that $\mathrm{Re}\, k^\nu + 2n + 2 > 2$ for $\nu = 1$ or 2. For $u = (K^2, K^1)$ of this form, and $1 \leqq j \leqq 2n+1$, let us define $S_j u \in \mathcal{U}$ as follows:

$$S_j u = P_j u \text{ if } \mathrm{Re}\, k^1 + 2n + 2 > 2$$

$$= Q_j u \text{ if } \mathrm{Re}\, k^1 + 2n + 2 \leqq 2 \text{ (in which case } \mathrm{Re}\, k^2 + 2n + 2 > 2).$$

Finally, let

$V = \{u = \sum(K^{\ell 2}, K^{\ell 1}) : K^{\ell\nu} \in \mathcal{A}K^{k^{\ell\nu}}$ for all ℓ, ν, where

$h(u) = (\mathrm{Re}\ k^{\ell 1}+2n+2) + (\mathrm{Re}\ k^{\ell 2}+2n+2)$ is independent of $\ell\}$.

If $u = \sum(K^{\ell 2}, K^{\ell 1}) \in V$, and if $h(u) > 4$, we define

$S_j u = \sum S_j (K^{\ell 2}, K^{\ell 1})$. Then $S_j u \in V$ and

$h(S_j u) = h(u) - 2$ if $j = 1$, $h(S_j u) = h(u) - 1$ if $2 \leqq j \leqq 2n+1$.

Thus, if $u \in V$, $\gamma \in (\mathbb{Z}^+)^{2n+1}$ and $h(u) > 3 + |\gamma|$, we can form

$S^\gamma u = S_1^{\gamma_1} \ldots S_{2n+1}^{\gamma_{2n+1}} u$, and $h(S^\gamma u) = h(u) - |\gamma|$.

Now, returning to the notation of the statement of the theorem, let us write

$$u = (K_2, K_1).$$

If $|\gamma| = [\mathrm{Re}\ \varkappa] + 1$ or $[\mathrm{Re}\ \varkappa] + 2$, then

$h(u) = (\mathrm{Re}\ \varkappa_1+2n+2) + (\mathrm{Re}\ \varkappa_2+2n+2) \geqq 2n + |\gamma| > 3 + |\gamma|$ if $n > 1$.

Let us assume $n > 1$ for now, then; we explain the necessary modifications to handle the case $n = 1$ later. Then we can form

$$S^\gamma u = \sum_{\ell=1}^{N} (K_{\ell 2}, K_{\ell 1}),$$

say. Note $h(S^\gamma u) = h(u) - |\gamma| = (\mathrm{Re}\ \varkappa - [\mathrm{Re}\ \varkappa]) + 2n + 1$ or $(\mathrm{Re}\ \varkappa - [\mathrm{Re}\ \varkappa]) + 2n$; then $2n \leqq h(S^\gamma u) < 2n + 2$. Thus for all $\ell, \nu, K_{\ell\nu} \in \mathcal{A}K^{k_{\ell\nu}}$ where $\mathrm{Re}\ k_{\ell 1} + \mathrm{Re}\ k_{\ell 2} = h(S^\gamma u) - 4n - 4 < -2n - 2$, and we can form

$$K_\gamma = \sum_{\ell=1}^{N} (K_{\ell 2} * K_{\ell 1})/\gamma!. \tag{7.47}$$

(Because of (7.45), we have every heuristic right to think of this as a replacement for $\partial^\gamma(K_2 * K_1)/\gamma!$.) We shall now define $K = K_2 * K_1$ in such a way that $K_\gamma = \partial^\gamma K/\gamma!$ for all γ. We shall be using Corollary 7.2, so we shall need to show that, if $|\gamma|$, $|\rho| \in \{[\mathrm{Re}\ \varkappa] + 1, [\mathrm{Re}\ \varkappa] + 2\}$, then

$$\partial^\rho(\gamma! K_\gamma) = \partial^\gamma(\rho! K_\rho). \tag{*}$$

For this, note that $S^\gamma = U_1 \ldots U_{\|\gamma\|}$ where each $U_i = P_j$ or Q_m for some j or m. In the notation of (7.47) we may write

$$K_{\ell\nu} = D_{\ell\nu} K_\nu \tag{7.48}$$

for $\nu = 1,2$, where $D_{\ell\nu}$ is a homogeneous differential operator. Thus $S^\gamma u = \sum (D_{\ell 2} K_2, D_{\ell 1} K_1)$. Now choose $\phi \in C_c^\infty(\mathbb{H}^n)$ with $\phi(t,z) = 1$ for $|(t,z)| < 1$, and let $\phi_\varepsilon(t,z) = \phi(\varepsilon^2 t, \varepsilon z)$, for $\varepsilon > 0$. Put $K_{1\varepsilon} = \phi_\varepsilon K_1$. Put $u_\varepsilon = (K_{2\varepsilon}, K_{1\varepsilon})$, and note

$$U_1 \ldots U_{\|\gamma\|} u_\varepsilon = \sum (D_{\ell 2} K_{2\varepsilon}, D_{\ell 1} K_{1\varepsilon}).$$

Thus, by (7.45),

$$\partial^\gamma(K_{2\varepsilon} * K_{1\varepsilon}) = \sum D_{\ell 2} K_{2\varepsilon} * D_{\ell 1} K_{1\varepsilon}.$$

By (7.47) and Proposition 5.2(h), then, if $|\gamma| = [\mathrm{Re}\ \varkappa] + 1$ or $[\mathrm{Re}\ \varkappa] + 2$, then

$$\gamma!K_\gamma = \lim_{\varepsilon\to 0} \partial^\gamma(K_{2\varepsilon}*K_{1\varepsilon}) \tag{7.49}$$

where the limit is taken in the sense of distributions. (*) is an immediate consequence of (7.49).

We may therefore put $K = F(\{K_\gamma\})$ as in Corollary 7.2, and we say $K = K_2\underline{*}K_1$. It is easy to see, from the above construction, as well as the definition of the map F (see the proof of Corollary 7.2) that $\underline{*}$ is bilinear from $AK^{\varkappa_2} \times AK^{\varkappa_1}$ to $AK^{\varkappa}$.

Let us now prove (a) and (b), still assuming $n > 1$, then indicate the necessary modifications for $n = 1$. (a) is immediate if $\mathrm{Re}\,\varkappa < 0$, by Lemma 5.7 and Proposition 5.5(b). If $\mathrm{Re}\,\varkappa \geq 0$, let L and R be as in (7.22). Write

$$RL = \sum p_\rho(t,x,y)\,\partial^\rho$$

where the p_ρ are homogeneous polynomials. A homogeneity check reveals $|\rho| \geq r + \ell$ for all ρ; hence $|\rho| \geq [\mathrm{Re}\,\varkappa] + 1$ for all ρ. We may therefore write $\rho = \rho_1 + \rho_2$ where $|\rho_2| = [\mathrm{Re}\,\varkappa] + 1$ or $[\mathrm{Re}\,\varkappa] + 2$. By the definition of K and (7.49),

$$RLK = \sum p_\rho \partial^{\rho_1}(\rho_2!K_{\rho_2}) = \lim_{\varepsilon\to 0} \sum p_\rho \partial^{\rho_1}\partial^{\rho_2}K_{2\varepsilon}*K_{1\varepsilon} = \lim_{\varepsilon\to 0} RL(K_{2\varepsilon}*K_{1\varepsilon})$$

$$= \lim_{\varepsilon\to 0} RK_{2\varepsilon}*LK_{1\varepsilon} = RK_2*LK_1$$

where the limits are in the distribution sense; we used Proposition 5.2(h).

This proves (7.22). In particular

$$T^N K = K_2 * T^N K_1 \tag{7.50}$$

for $2N > \operatorname{Re} \varkappa$. Since $K = F(\{K_\gamma\})$ as in Corollary 7.2, it follows at once from the construction of Corollary 7.2 that if $FK = G \in C^\infty(\mathbb{R}^{2n+1} \setminus \{0\})$ away from 0 then $K = F^{-1}\Gamma_G$. By (7.50), however, G is given by (7.18), so that K is given by (7.19). The remaining statements in (a) follow at once from (7.50), Lemma 5.7 and Proposition 5.5(b).

Let us now prove (b), still assuming $n > 1$. For $\nu = 1,2$, suppose $k_\nu \in \mathbb{C}$, $R_\nu > 0$, $K_\nu \in AK^{k_\nu + m_\nu}$ for some $m_\nu \in \mathbb{Z}^+$, and that (7.25), (7.26) hold. We put $K = K_2 * K_1$, and if $|\gamma| \in \{[\operatorname{Re} k] + m + 1, [\operatorname{Re} k] + m + 2\}$, we put $K_\gamma = \partial^\gamma K/\gamma!$. To prove (b), we must estimate the "size" of K_γ in the sense made clear by (7.10) of Corollary 7.2, with constants depending only on k_ν and R_ν. Writing

$$K_\gamma = \sum_{\ell=1}^{N} (D_{\ell 2} K_2 * D_{\ell 1} K_1)/\gamma! \tag{7.51}$$

in the notation of (7.47) and (7.48), and noting by (7.44) that $N \leq 3^{\|\gamma\|} \leq 3^{|\gamma|} \leq 3^{[\operatorname{Re} k]+m+2}$, we see that it suffices to consider each term separately.

We need to use another property of the map S^γ $(\gamma \in (\mathbb{Z}^+)^{2n+1})$. With notation as directly after (7.46), say $u = (K^2, K^1) \in V$ where $K^\nu \in AK^{k^\nu}$, $h(u) > 3 + |\gamma|$, and $S^\gamma u = \sum (K_\gamma^{\ell 2}, K_\gamma^{\ell 1})$ where

$K_\gamma^{\ell\nu} \in AK^{k^{\ell\nu\gamma}}$. Suppose that for some ℓ,ν we have

$$\operatorname{Re} k^{\ell\nu\gamma} + 2n + 2 \leqq 0. \qquad (7.52)$$

Then it must be the case that

$$\operatorname{Re} k^{\nu} + 2n + 2 \leqq 0 \qquad (7.53)$$

and for some $\alpha,\beta \in (\mathbb{Z}^+)^n$ we have

$$K_\gamma^{\ell\nu} = \pm(2x)^\alpha(2y)^\beta K^\nu. \qquad (7.54)$$

This is easily seen by induction on $|\gamma|$ and by use of (7.44) and (7.46). (Loosely speaking, here is why this holds: In the process of applying S_i's repeatedly, we never differentiate a distribution if the real part of its homogeneity degree is $\leqq -2n - 2$; at worst we multiply it by $\pm 2x_j$ or $\pm 2y_j$. Further, in the process, a distribution the real part of whose homogeneity degree is $> -2n - 2$ will never be differentiated to the point where the real part of the new homogeneity degree is $\leqq -2n - 2$. This, then, is why we used both the P_j and Q_j to define S_j; as we said at the outset of the proof, we wished to avoid differentiating any distribution if the real part of its homogeneity degree is $\leqq -2n - 2$.)

Now let us return to the notation of the theorem and of (7.51). In (7.51), we may write any $D_{\ell\nu}$ in the form $W_1^{\ell\nu}\ldots W_N^{\ell\nu}$ where each $W_i^{\ell\nu} \in A = \{T, X_j, Y_j, X_j^R, Y_j^R : 1 \leqq j \leqq n\}$

or $B = \{\pm 2x_j, \pm 2y_j\}$. Letting $p_{\ell\nu} = \#\{i : W_i^{\ell\nu} \in A\}$, it is clear from repeated applications of (7.44) that $p_{\ell 1} + p_{\ell 2} = \|\gamma\|$ for all ℓ. Further, $\#\{i : W_i^{\ell\nu} \in B\} \leq \|\gamma\|$. It now follows easily from Lemma 2.5 and (7.25) that

$$\sup_{1 \leq |u| \leq 2} |\partial^{\rho}(D_{\ell\nu}K_{\nu})| < A'_{\nu}C_{\nu}(R'_{\nu})^{m_{\nu}+\|\rho\|}(\|\rho\|+p_{\ell\nu})!$$

$$< A''_{\nu}C_{\nu}(R''_{\nu})^{m_{\nu}+\|\rho\|}\|\rho\|!p_{\ell\nu}!$$

$$\text{for all } \rho \in (\mathbb{Z}^{+})^{2n+1},$$

for certain $A'_{\nu}, R'_{\nu}, A''_{\nu}, R''_{\nu} > 0$ depending only on the k_{ν} and R_{ν}.

Say that $D_{\ell\nu}K_{\nu} \in AK^{k_{\ell\nu}}$ for all ℓ,ν. Note, by the discussion surrounding (7.52)-(7.54), that if some ℓ,ν we have $\text{Re } k_{\ell\nu} + 2n + 2 \leq 0$, then $\text{Re } k_{\nu} + 2n + 2 \leq 0$ and all $W_i^{\ell\nu} \in B$. Then if $j_0 \in \mathbb{Z}^{+}$, there exists $C_0 > 0$, depending only on j_0, k_{ν} and R_{ν}, so that for such ℓ,ν we have

$|D_{\ell\nu}K_{\nu}(\psi)| < C_0C_{\nu}$ whenever $\psi \in C_c^{\infty}(\mathbb{H}^n)$, supp $\psi \subset$

$\{(t,z) : |(t,z)| \leq 1\}$

and $|(\partial^{\rho}\psi)(t,z)| \leq 1$ whenever $|\rho| \leq j_0$.

Thus, for $\nu = 1,2$, there exist $A'''_{\nu}, R'''_{\nu} > 0$, depending only on the k_{ν} and R_{ν}, so that for all ℓ,ν,

$$D_{\ell\nu}K_{\nu} \in AK^{k_{\ell\nu}}(A'''_{\nu}C_{\nu}(R'''_{\nu})^{m_{\nu}}p_{\ell\nu}!, R'''_{\nu}).$$

Since $p_{\ell 1}!p_{\ell 2}!/\|\gamma\|! \leq 1$ for all ℓ,ν, Lemma 7.8(b) and (7.9) together imply that for some $C_4, R_4 > 0$, depending only on the k_ν and R_ν, we have $K_\gamma \in AK^{k+m-|\gamma|}(C_4C_1C_2R_4^m, R_4)$ for all γ. By (*) and Corollary 7.2, $K = K_2 \underline{*} K_1 \in AK^{k+m}(AC_1C_2R^m, R)$ for some A,R depending only on the k_ν and R_ν. This proves (b) for $n > 1$.

If $n = 1$, $u = (K_2, K_1)$, $h(u) = (\mathrm{Re}\,\varkappa_1+4) + (\mathrm{Re}\,\varkappa_2+4)$, then if $|\gamma| = [\mathrm{Re}\,\varkappa] + 1 \geq 1$ we have $h(u) \geq 3 + |\gamma|$; while if $|\gamma| = [\mathrm{Re}\,\varkappa] + 2 \geq 2$ we have $h(u) \geq 2 + |\gamma|$. In either case we may either write

(1) $S^\gamma = S_i S^{\gamma'}$ or (2) $S^\gamma = S_i S_j S^{\gamma'}$,

where $i,j = 1,2$ or 3 and where $h(u) \geq 4 + |\gamma'|$. We may not be able to apply S^γ to u, but we can at least form

$$S^{\gamma'}u = \sum_{\ell=1}^{M} (K'_{\ell 2}, K'_{\ell 1}),$$

say. Then we define

$P_i S^{\gamma'} u = \sum_{\ell=1}^{N} (K_{\ell 2}, K_{\ell 1})$ in Case (1); $P_i P_j S^{\gamma'} u = \sum_{\ell=1}^{N} (K_{\ell 2}, K_{\ell 1})$ in Case 2.

We now define K_γ as in (7.47), again verify (7.49) and (*), put $K = F(\{K_\gamma\})$, and verify (a). The slight problem arises in (b), since by using the P_i, P_j we may have now differentiated distributions the real parts of whose homogeneities were ≤ -4. However, say that $K'_{\ell\nu} \in AK^{k'_{\ell\nu}}$.

The arguments for (b) in the case $n > 1$ show clearly that for $1 \leqq \ell \leqq M$ there exist $p'_{\ell 1}, p'_{\ell 2}$ with $p'_{\ell 1} + p'_{\ell 2} = \|\gamma'\|$, and A'_1, A'_2, R'_1, R'_2 depending only on the k_ν and R_ν, so that for all ℓ, ν, $K'_{\ell\nu} \in AK^{k'_{\ell\nu}}(A'_\nu C_\nu (R'_\nu)^{m_\nu} p'_{\ell\nu}!, R'_\nu)$. Note also $\text{Re } k'_{\ell\nu} \geqq \min(-4, \text{Re } k'_\nu)$. Since we need now to apply the P's at most twice to get the $K_{\ell\nu}$, it is now easy to see that for $1 \leqq \ell \leqq N$ there exists $p_{\ell 1}, p_{\ell 2}$ with $p_{\ell 1} + p_{\ell 2} = \|\gamma\|$, and $A''_1, A''_2, R''_1, R''_2$ depending only on the k_ν and the R_ν, so that for all $\ell, \nu, K_{\ell\nu} \in AK^{k_{\ell\nu}}(A''_\nu C_\nu (R''_\nu)^{m_\nu} p_{\ell\nu}!, R''_\nu)$. (b) now follows at once as in the case $n > 1$.

We now tackle (c), one of the main points of the entire discussion.

With Q_j as in (7.44), let us write, for $(f,g) \in \mathcal{U}$,

$$Q^\gamma(f,g) = \sum (D^\gamma_{\ell 2} f, p^\gamma_{\ell 1} g). \tag{7.56}$$

Here (dropping γ superscripts) $D_{\ell 2}$ is a homogeneous differential operator of homogeneous degree $d_{\ell 2}$, $p_{\ell 1}$ is a homogeneous polynomial of degree $d_{\ell 1}$, and

$$d_{\ell 2} - d_{\ell 1} = |\gamma|. \tag{7.57}$$

If at least one of f,g has compact support, $\partial^\gamma(f*g) = \sum D_{\ell 2} f * p_{\ell 1} g$. Thus, by (7.49), Proposition 5.2(h) and (7.57), if $|\gamma| \geqq [\text{Re } k] + m + 1$ then

$$\partial^\gamma K = \sum D_{\ell 2} K_2 * p_{\ell 1} K_1.$$

Writing $K_\nu = \chi_\nu K + (1-\chi_\nu)K_\nu$ in this expression, we find, for $|\gamma| \geqq [\mathrm{Re}\ k] + m + 1$,

$$\partial^\gamma K = \partial^\gamma (\chi_2 K_2 * \chi_1 K_1) + \partial^\gamma G + \partial^\gamma H + F_\gamma \tag{7.58}$$

where
$$G = (1-\chi_2)K_2 * \chi_1 K_1$$
$$H = \chi_2 K_2 * (1-\chi_1)K_1$$
$$F_\gamma = \sum D^\gamma_{\ell 2}(1-\chi_2)K_2 * p^\gamma_{\ell 1}(1-\chi_1)K_1 .$$

For some $s_1 > 0$, depending only on r_1 and r_2, and for some $A_3, R_3 > 0$, depending only on $k_1, k_2, R_1, R_2, r_1, r_2$ and s_1, we have

$$|\partial^\rho G(u)| < A_3 C_1 C_2 R_3^{m+\|\rho\|} \rho! \quad \text{for all } \rho \in (\mathbb{Z}^+)^{2n+1}, \quad \text{if } |u| < s_1. \tag{7.59}$$

This follows at once from Lemma 7.6(a) and (7.31) and (7.34). Further, for some s_2 depending only on r_1 and r_2,

$$H(u) = 0 \quad \text{if } |u| < s_2 \tag{7.60}$$

and
$$F_\gamma(u) = [\sum D^\gamma_{\ell 2} K_2 * (1-\chi_1) p^\gamma_{\ell 1} K_1](u) \text{ if } |u| < s_2. \tag{7.61}$$

The number of terms in the summation for F_γ does not exceed $3^{\|\gamma\|}$. Further, we may write $D_{\ell 2} = W_1 \ldots W_N$, $p_{\ell 1} = \xi_1 \ldots \xi_M$, where each $W_i \in A = \{T, X_j, Y_j, X_j^R, Y_j^R : 1 \leqq j \leqq n\}$ or $B = \{\pm 2x_j, \pm 2y_j : 1 \leqq j \leqq n\}$, and each $\xi_i \in B$. Now, from (7.44), $\#\{i : W_i \in A\} = \|\gamma\|$, while $\#\{i : W_i \in B\} + \#\{i : \xi_i \in B\} \leqq \|\gamma\|$. From Lemma 2.5,

$$\sup_{1\leq|u|\leq 2} |\partial^\rho(D_{\ell 2}K_2)| < A_4C_2R^{m_2+\|\rho\|+\|\gamma\|}(\|\rho\|)!(\|\gamma\|)!$$

$$\sup_{1\leq|u|\leq 2} |\partial^\rho(p_{\ell 1}K_1)| < A_5C_1R_5^{m_1+\|\rho\|+\|\gamma\|}(\|\rho\|)!$$

for all $\rho \in (\mathbb{Z}^+)^{2n+1}$, where A_4, R_4, A_5, R_5 depend only on k_1, k_2, R_1 and R_2. From Lemma 7.8(a) and (7.61), we see that F_γ is analytic for $|u| < r_1$, and

$$|(\partial^\rho F_\gamma)(u)| < A_6C_1C_2R_6^{m+\|\rho\|+\|\gamma\|}\rho!\gamma! \text{ if } |u| < s_2/2$$

where A_6, R_6 depend only on k_1, k_2, R_1, R_2 and r_1. In particular,

$$|F_\gamma(0)| < A_6C_1C_2R_6^{m+\|\gamma\|}\gamma!. \tag{7.62}$$

Now, for $|u|$ sufficiently small, let us form

$$F(u) = \sum_{|\rho|\geq[\mathrm{Re}\ k]+m+1} F_\rho(0)u^\rho/\rho!.$$

By (7.62) this is analytic near 0. We have

$$|\partial^\gamma F(u)| < A_7C_1C_2R_7^{m+\|\gamma\|}\gamma! \text{ if } |u| < s_3, \gamma \in (\mathbb{Z}^+)^{2n+1}. \tag{7.63}$$

Here s_3, A_7, R_7 are constants depending only on k_1, k_2, R_1, R_2 and r_1, and s_3 is chosen with $s_3 < r_1$. We claim that, if $|\gamma| \geq [\mathrm{Re}\ k] + m + 1$, then

$$\partial^\gamma F(u) = F_\gamma(u) \quad \text{for } |u| < s_3.$$

Indeed, $\partial^\gamma F(u) = \sum F_{\gamma+\rho}(0)u^\rho/\rho!$ (sum over $\rho \in (\mathbb{Z}^+)^{2n+1}$), so it suffices to observe that $F_{\gamma+\rho} = \partial^\rho F_\gamma$. This is evident from (7.58).

With $r = \min(s_1, s_2, s_3)$, $Q' = -F - G$, we have from (7.58), (7.59), (7.60) and (7.63):

$$\partial^{\gamma}(\chi_2 K_2 * \chi_1 K_1 - K - Q') = 0 \text{ for } |u| < r, \ |\gamma| \geq [\mathrm{Re}\, k] + m + 1 \qquad (7.64)$$

$$|(\partial^{\gamma} Q')(u)| < A_8 C_1 C_2 R_8^{m+\|\gamma\|} \gamma! \text{ for } |u| < r, \gamma \in (\mathbb{Z}^+)^{2n+1}. \qquad (7.65)$$

Here A_8, R_8 depend only on k_1, k_2, r_1, r_2 and r. This is almost all that we need, except that (7.64) tells us only that

$$\chi_2 K_2 * \chi_1 K_1 - K - Q' = p \text{ on } \{u : |u| < r\} \qquad (7.66)$$

where p is a polynomial all of whose terms have homogeneity degree less than $[\mathrm{Re}\, k] + m + 1$. To complete the proof, it suffices to show that

$$|(\partial^{\gamma} p)(u)| < A_9 C_1 C_2 R_9^{m+\|\gamma\|} \gamma! \text{ if } |u| < r, \gamma \in (\mathbb{Z}^+)^{2n+1} \qquad (7.67)$$

where A_9 and R_9 depend only on $k_1, k_2, R_1, R_1, r_1, r_2$ and r. However, by (7.66), Lemma 7.8(c), part (b) of this Theorem and (7.65), we clearly have (7.67) for $|u| = r/2$. But p is a polynomial of degree at most $[\mathrm{Re}\, k] + m$, so, using Taylor's Theorem on p about any point u_0 with $|u_0| = r/2$, we see by enlarging A_9 and R_9 if necessary, (7.67) holds for $|u| < r$. Setting $Q = Q' + p$, we have completed the proof of (c).

As for (d), the fact that $K_2 * K_1$ depends holomorphically on the parameter is immediate from Lemma 7.8(b) if $\mathrm{Re}\, k + m < 0$.

If Re $k + m \geqq 0$, we define G as in (7.18). Note (by (2.20)) that $G(\lambda,\xi)$ is a holomorphic function of the parameter for $\lambda \neq 0$. Since, in (7.18), $F(t^N(K_2 * T^N K_1))(0,\xi)$ is clearly a non-zero multiple of $G(0,\xi)$ for $\xi \neq 0$, $G(0,\xi)$ is holomorphic in the parameter also for $\xi \neq 0$. Hence Γ_G depends holomorphically on the parameter, and therefore $K_2 * K_1$ does also. Possibly decreasing r, we now see from Lemma 7.8(c) that $Q(u) = (\chi_2 K_2 * \chi_1 K_1)(u) - K(u)$ is holomorphic in the parameter for $0 < |u| < r$. Since (7.29) is satisfied for all $\omega \in \Omega$, there is a point u_0 with $0 < |u_0| < r$ so that $Q(0)$ may be obtained by means of a power series expansion about u_0 for any $\omega \in \Omega$. An argument involving uniform convergence of holomorphic functions on compact subsets of Ω now shows $Q(0)$ is holomorphic in ω as well. This completes the proof of Theorem 7.5.

Although our arguments have been lengthy and careful, in one important way they have been crude. In (c) of the Theorem, say r_1 and r_2 are small. If m is large, then Q should be small on $\{u : |u| < r\}$, with size decreasing geometrically in m, since $\chi_2 K_2 * \chi_1 K_1$ and K are in some sense small there. The point is, we have not kept careful track of the dependence of R_0 in (c) on r_1 and r_2. Fortunately, dilations provide us an extremely easy way to recover this information, and to prove the following lemma, which is the backbone of all that follows.

Lemma 7.9 (Localization Lemma). For $\nu = 1,2$, suppose $k_\nu \in \mathbb{C}$, and $R_\nu > 0$. Suppose $0 < r_2 < r_1$. Then there exist $A,R,r > 0$ with these properties.

Suppose, for $\nu = 1,2, K_\nu \in AK^{k_\nu+m_\nu}(C_\nu R_\nu^{m_\nu}, R_\nu)$

for some $m_\nu \in \mathbb{Z}^+$, $C_\nu > 0$. (7.68)

Let $k = k_1 + k_2 + 2n + 2$, $m = m_1 + m_2$, $K = K_2 * K_1$. For $c > 0$, $\nu = 1,2$, let $\chi_{\nu c}$ be the characteristic function of $\{u \in \mathbb{H}^n : |u| < cr_\nu\}$.

Then there exists Q_c analytic on $\{u : |u| < cr\}$ such that

$$\chi_{2c}K_2 * K_1 = \chi_{2c}K_2 * \chi_{1c}K_1 = K + Q_c \text{ on } \{u : |u| < cr\} \quad (7.69)$$

and

$$|\partial^\gamma Q_c(u)| < AC_1C_2c^{\text{Re } k+m-|\gamma|}(1+|\log c|)^2 R_0^{m+\|\gamma\|}\gamma! \quad (7.70)$$

for $\gamma \in (\mathbb{Z}^+)^{2n+1}$, $|u| < cr$.

Further if K_2, K_1 depend holomorphically on a parameter $\omega \in \Omega$ and satisfy (7.68) for all ω, then $Q_c(u)$ is also a holomorphic function of ω for $0 < |u| < cr$.

Proof. If K_1, K_2 and K are homogeneous, this is a simple consequence of Theorem 7.5(c), in the case $c = 1$, together with use of dilations. In the general case, let us write, for $\nu = 1,2$, $K_\nu = K_\nu' + p_\nu(u)\log|u|$, $K = K' + p(u)\log|u|$, with the K'_ν, K', p_ν, p homogeneous, and with the p_ν, p polynomials. (If $k_\nu + m_\nu \notin \mathbb{Z}^+$, then of course, $p_\nu = 0$.) Now for $\varepsilon > 0$, recall $D_\varepsilon(t,z) = (\varepsilon^2 t, \varepsilon z)$, and write $D_\varepsilon f = f \circ D_\varepsilon$ for f a function. For $f,g \in \mathcal{E}'$, we have $D_\varepsilon(f*g) = \varepsilon^{2n+2} D_\varepsilon f * D_\varepsilon g$. Now choose r,Q as in Theorem 7.5(c). We may assume r is sufficiently small that if $f,g \in \mathcal{D}'$, supp $f \subset \{u : |u| \leq r_2\}$, supp $g \subset \{u : |u| \geq r_1\}$, then $f*g = 0$ on $\{u : |u| \leq r\}$. With $\chi_{\nu 1} = \chi_\nu$, we have on $\{u : |u| < cr\}$,

$$\begin{aligned}
\chi_{2c}K_2 * K_1 &= \chi_{2c}K_2 * \chi_{1c}K_1 = D_{1/c}D_c(\chi_{2c}K_2 * \chi_{1c}K_1) \\
&= c^{2n+2} D_{1/c}(\chi_2 D_c K_2 * \chi_1 D_c K_1) \\
&= c^{k+m} D_{1/c}[\chi_2 K_2 * \chi_1 K_1 + \log c(\chi_2 K_2 * \chi_1 p_1 + \chi_2 p_2 * \chi_1 K_1) \\
&\qquad + (\log c)^2 \chi_2 p_2 * \chi_1 p_1] \\
&= K + Q_c, \text{ where}
\end{aligned}$$

$$\begin{aligned}
Q_c = c^{k+m} D_{1/c}[Q - p \log c &+ (\log c)\chi_2 K_2 * \chi_1 p_1 \\
&+ (\log c)\chi_2 p_2 * \chi_1 K_1 + (\log c)^2 \chi_2 p_2 * \chi_1 p_1].
\end{aligned} \tag{7.71}$$

We obtain the needed estimate for each term separately. The term involving Q is estimated by (7.29); the term involving p log c is handled through (7.27) and (7.33). The next term is handled through Lemma 7.6(a), (7.31), (7.34) and (7.33). The last two terms are handled through (7.33) and Proposition 7.7. This establishes (7.69) and (7.70).

For the statement about holomorphic dependence, note that since K_1, K_2, K depend holomorphically on ω, $p_1(u)$, $p_2(u)$ and $p(u)$ will also be holomorphic functions of ω for any u. Indeed, this is easily verified if one first examines $K_1(u)$, $K_2(u)$, $K(u)$ in the case $|u| = 1$, then in the case $|u| = 2$, and then uses homogeneity. The case $u = 0$ is trivial since the polynomials are homogeneous, so that they are 0 at the origin if they are not constant. The statement about holomorphic dependence of $Q_c(u)$ on ω is now immediate from (7.71) and Theorem 7.5(c). This completes the proof.

We also have an analogue of Proposition 7.7 with a "smallness" conclusion if f has small support.

<u>Proposition 7.10</u>. Suppose $u_0 \in U$, an open subset of $\mathbb{H}^n$. Further, suppose $c_0 > 0$, and $\{u : |u^{-1}u_0| < c_0\} \subset U$. Say $k \in \mathbb{C}$, $R_1, R_2 > 0$, $0 < r < 1$. Then there exist $C, R_3 > 0$ as follows: Suppose $K \in AK^{k+m}(1,R_1)$ for some $m \in \mathbb{Z}^+$. Suppose $f \in L^1$, f is analytic on U, supp $f \subset \overline{U}$, and that

$$|\partial^\gamma f(u)| < R_2^{\|\gamma\|}\gamma! \quad \text{for all } u \in U,\ \gamma \in (\mathbb{Z}^+)^{2n+1}.$$

For any $c > 0$, let χ_c be the characteristic function of $\{u : |u^{-1}u_0| < c\}$. Then if $0 < c < c_0$,

$$|\partial^\gamma (K * \chi_c f)(u)| < Cc^{m-|\gamma|}(1+|\log c|)R_3^{m+\|\gamma\|}\gamma!$$

for $|u^{-1}u_0| < rc$, $\gamma \in (\mathbb{Z}^+)^{2n+1}$.

<u>Proof</u>. By means of a translation, we may assume $u_0 = 0$. With notation as in the proof of Lemma 7.9, we write

$$K*\chi_c f = D_{1/c} D_c (K*\chi_c f) = c^{2n+2} D_{1/c}(D_c K*\chi_1 D_c f)$$
$$= c^{k+m+2n+2} D_{1/c}[(K*\chi_1 D_c f) + \log c(p*\chi_1 D_c f)] \qquad (7.72)$$

if $K = K' + p \log|u|$, K', p homogeneous, p a polynomial. Now $\chi_1 D_c f$ is analytic for $|u| < 1$, and satisfies

$$|\partial^\gamma(\chi_1 D_c f)(u)| < c_0^{|\gamma|} R_2^{\|\gamma\|} \gamma! \text{ if } |u| < 1, \ \gamma \in (\mathbb{Z}^+)^{2n+1}.$$

Note that the right side of this inequality does not depend on c. Thus, by Proposition 7.7 and (7.70), there exist $C_1, R_4 > 0$, depending only on k, R_1, R_2, r so that

$$|\partial^\gamma(K*\chi_1 D_c f)| < C_1 R_4^{m+\|\gamma\|}\gamma! \text{ for } |u| < r, \gamma \in (\mathbb{Z}^+)^{2n+1}$$

and $$|\partial^\gamma(p*\chi_1 D_c f)| < C_1 R_4^{m+\|\gamma\|}\gamma! \text{ for } |u| < r, \gamma \in (\mathbb{Z}^+)^{2n+1}.$$

The Proposition follows at once from these relations and (7.72).

The localization lemma easily gives us precise information on the composition of right-invariant operators whose kernels are series of truncated elements of the AK spaces. Suppose $C_1, C_2, R_1, R_2 > 0$, $\{K_1^m\}$, $\{K_2^m\}$ are sequences of distributions, and $K_\nu^m \in AK^{k_\nu}(C_\nu R_\nu^m, R_\nu)$ for $\nu = 1,2$. Then for some $r_\nu > 0$, $\sum K_\nu^m$ converges to an analytic function on $B_\nu \setminus \{0\}$, where $B_\nu = \{u : |u| < r_\nu\}$. (We used (7.31) and (7.34).)

With $\chi_{\nu c}$ the characteristic funciton of $\{u : |u| < cr_\nu\}$, for sufficiently small c the localization lemma easily shows

$$(\chi_{2c}\sum K_2^m) * (\chi_{1c}\sum K_1^m) = \sum K^m + Q$$

on a sufficiently small neighborhood of 0, where for some $C,R > 0$, $K^m \in AK^{k+m}(CR^m,R)$, $k = k_1 + k_2 + 2n + 2$. (For good convergence in the series for Q, we want c to be small enough that $cR_0 < 1$, where R_0 is the constant arising as in (7.70).)

But, far more interestingly, the localization lemma also allows us to deal with non-invariant operators. In the next section, we shall obtain a composition rule for these, analogous to the Kohn-Nirenberg product formula for pseudodifferential operators.

Section Three - Analytic Pseudodifferential Operators on $\mathbb{H}^n$, Products and Adjoints

We are now ready to define the study analytic pseudodifferential operators on $\mathbb{H}^n$. As we explained in the introduction to this book, then, we will be looking at operators of the form $\mathcal{K}$, where for $f \in C_c^\infty$,

$$(\mathcal{K}f)(u) = (K_u * f)(u),$$

where u ranges over an open subset U of $\mathbb{H}^n$. For any u, K_u will always be a distribution on $\mathbb{H}^n$ which agrees with a function away from 0; and we will systematically write

$$K(u,w) = K_u(w)$$

for $w \neq 0$. We call K the core of $\mathcal{K}$, and write

$$K = \varkappa(\mathcal{K}),\ \mathcal{K} = \mathcal{O}(K).$$

The class of cores which we shall be considering is $C^k(U)$ $(k \in \mathbb{C})$. This class was defined in (0.16)-(0.22) of the Introduction to this book--at least if U is relatively compact. (Note that $|\,|$ in (0.16)-(0.22) is to be interpreted as the usual $\mathbb{H}^n$ length.) If U is not relatively compact, a slightly different definition is needed. (0.22) would imply that there exists $s > 0$ so that for any $u \in U$, $1 \leq |w| \leq 2$, one can define a function $K^m(v+u,w)$, which is holomorphic in v in a polydisc in $\mathbb{C}^{2n+1}$, centered at 0 and with all polyradii s, and which agrees with $K^m(v+u,w)$ for $v \in \mathbb{R}^{2n+1}$. If U is not relatively compact, that is not quite what we want. For (t,x,y), $(t,x',y') \in \mathbb{C} \times \mathbb{C}^n \times \mathbb{C}^n$, we shall write

$$(t,x,y) \circ (t',x',y') = (t+t'+2(x'\cdot y-x\cdot y'), x+x', y+y').$$

Of course, $\circ$ agrees with $\mathbb{H}^n$ multiplication if the points are in $\mathbb{R} \times \mathbb{R}^n \times \mathbb{R}^n$, and of course $\circ$ is associative. We would be happier if we instead knew we could define $K^m(v\circ u,w)$ as a holomorphic function of v in a uniform polydisc centered at $0 \in \mathbb{C}^{2n+1}$ (independent of u,w). We would like to have a notation for those cores in $C^k(U)$ for which particular constants "work." For these reasons, we make the definition which follows.

For $s > 0$, we let $D_s = \{v \in \mathbb{C}^{2n+1} : |v_1| < s^2, |v_j| < s$ for $2 \leq j \leq 2n+1\}$, a polydisc, and let $\overline{D}_s$ be its closure.

Definition. For $k \in \mathbb{C}$, U an open subset of $\mathbb{H}^n$, $r,s,R,R' > 0$, we let $C^k(U;r,s,R,R')$ denote the space of all cores of the form

$$K_u(w) = \chi(w)\left[\sum_{m=0}^{\infty} K_u^m(w) + Q(u,w)\right] \tag{7.73}$$

where χ is the characteristic function of $B_r = \{w : |w| < r\}$;

for each m, $K_u^m \in AK^{k+m}$ for all $u \in U$, and depends analytically on u; in fact, for $u \in U$, there is a family $K_{v\circ u}^m$ which is holomorphic for v in a neighborhood of $\bar{D}_s$ and which agrees with K_{vu}^m for $v \in \bar{D}_s \cap \mathbb{H}^n$; (7.75)

Q is analytic in $(u,w) \in U \times B_r$; in fact, for $u \in U$, $w \in B_r$, there is a funciton of v which is holomorphic in a neighborhood of $\bar{D}_s$, denoted $Q(v\circ u,w)$, and which agrees with $Q(vu,w)$ for $v \in \bar{D}_s \cap \mathbb{H}^n$; (7.75)

If $K^m(u,w) = K_u^m(w)$ for $w \neq 0$, then:

For v in a neighborhood of $\bar{D}_s$, $u \in U$, $0 < |w| \leqq r$, $\sum K^m(v\circ u,w)$ converges absolutely, and is holomorphic in v, real analytic in w; (7.76)

and for any compact set $E \subset U$, there exists $C_E > 0$ so that for $v \in \bar{D}_s$, $u \in E$:

$$K_{v\circ u}^m \in AK^{k+m}(C_E R^m, R); \tag{7.77}$$

$$|\partial_w^\rho[\Sigma K^m(v\circ u,w)]| < C_E (R')^{\|\rho\|}\rho! \text{ for } 1/3R \leqq |w| \leqq r,\ \rho \in (\mathbb{Z}^+)^{2n+1} \tag{7.77'}$$

$$|\partial_w^\gamma Q(v\circ u,w)| < C_E R^{\|\rho\|}\rho! \qquad \text{for } w \in B_r,\ \rho \in (\mathbb{Z}^+)^{2n+1}. \tag{7.78}$$

We also let $C^k(U) = \bigcup_{r,s,R,R'>0} C^k(U;r,s,R,R')$ and

$$C^k(U;r,s,R) = \bigcup_{R'>0} C^k(U;r,s,R,R').$$

As we indicated in the Introduction to this book, the conditions (7.74), (7.77) imply:

Let $V = \{v \circ u : v \in \bar{D}_s, u \in U\}$. Then there is a truncated sector $S = \{\omega \in \mathbb{C}^{2n+1} : |\text{Im } \omega| < c|\text{Re } \omega|, |\omega| < b\}$, such that we may extend $K^m(u,w)$ to a holomorphic function $K^m(\zeta,\omega)$ in a neighborhood of $V \times S$, (*) and such that $\sum K^m(\zeta,\omega)$ converges absolutely and uniformly on compact subsets of $V \times S$ to a holomorphic function of (ζ,ω).

Indeed, by (7.74), (7.77) and (7.31), (7.34), for some $b' > 0$ we easily obtain estimates of the form

$$|\partial_w^\rho K^m(\zeta,w)| < C_1 R_1^{\|\rho\|} 2^{-m} \rho!$$

for (ζ,w) in compact subsets of $V \times \{0 < |w| < b'\}$. (*) then follows at once. Conversely, if we have (7.74) and (*) for some $s,c,b > 0$, then for any compact $E \subset U$, we have an estimate

$$|\partial_w^\rho K^m(v\circ u,w)| < C_E' R^{\|\rho\|} \rho! \quad \text{for } b/4 \leq |w| \leq b/2,$$
$$\rho \in (\mathbb{Z}^+)^{2n+1}, \; v \in \bar{D}_s.$$

By arguments analogous to those used to prove (7.31) and (7.34), we obtain an estimate of the form

$$|\partial_w^\rho K^m(v\circ u,w)| < C_E R^{\|\rho\|+m} \rho! \text{ for } 1 \leq |w| \leq 2,$$
$$\rho \in (\mathbb{Z}^+)^{2n+1}, \; v \in \bar{D}_s, \; u \in E$$

which is (7.77), at least if $\text{Re}\, k + m + 2n + 2 \notin \mathbb{Z}^-$. Thus, (7.77) could virtually be replaced by (*), and it is for this reason that we consider $C^k(U)$ to be an extremely natural space of cores.

We have the following obvious relationships:

If $U \subset V$, U,V, open, then $C^k(V) \subset C^k(U)$;

If $K \in C^k(U)$, $r_1 > 0$ and χ_1 is the characteristic function of $\{w : |w| < r_1\}$, then $\chi_1(w)K(u,w) \in C^k(U)$.

These simple facts will enable us, in essence, to reduce the study of cores $K \in C^k$ to the study of their germs (for u near a fixed point u_0 and for w near 0), though we shall not explicitly use such terminology.

We shall use the notation

$$K \sim \sum K^m$$

if we have (7.73). The K^m are not uniquely determined by K; if k is an integer, we may add various homogeneous polynomials to the K^m. We say K^o is the principal core of K.

There is, of course, some redundancy built into conditions (7.73)-(7.78). If w is restricted to lie in a compact subset of B_r, (7.75) of course automatically implies (7.78) for some C,R. Also (7.77) implies (7.76) if w is additionally restricted to satisfy $0 < |w| < 1/2R$. Indeed, by (7.77), (7.31) and (7.34), for any $v \in \overline{D}_s$, any compact

$E \in U$, and any $u \in E$, we have that

$$|\partial_w^\rho K^m(v\circ u,w)| < BC_E(2R|w|)^m(1+|\log|w||)(2R)^{\|\rho\|}|w|^{\mathrm{Re}\ k-|\rho|}\rho! \quad (7.79)$$

with B a constant which depends only on R. (7.76) with $0 < |w| < 1/2R$ follows at once. This is why we chose the region $1/3R \leqq |w| \leqq r$ for the normalization (7.77)'.

Note also, for future use, that for any compact $E \subset U$ there exists $s' > 0$ so that for any $u \in E$ the polydisc $\{v \in \mathbb{C}^{2n+1} : |v_j - u_j| < s'$ for $1 \leqq j \leqq 2n+1\}$ is contained in $\{v\circ u : v \in \overline{D}_s\}$. This, together with (7.79) and the Cauchy estimates, shows that there exists $R_{1E} > 0$ so that for all $u \in E$,

$$|\partial_u^\gamma \partial_w^\rho K^m(u,w)| < BC_E(2R|w|)^m(1+|\log|w||)R_{1E}^{\|\gamma\|}(2R)^{\|\rho\|}|w|^{\mathrm{Re}\ k-|\rho|}\rho!\gamma! \ . \quad (7.80)$$

We systematically write

$$K^+(u,w) = \chi(w)[\sum_{\mathrm{Re}\ k+m>-2n-2} K^m(u,w) + Q(u,w)]$$

$$K^- = K - K^+.$$

For $K \in C^k(U)$, $f \in C_c^\infty$, $u \in U$, we define

$$(\mathcal{K}f)(u) = (K_u * f)(u). \quad (7.81)$$

Then

$$\mathcal{K} : C_c^\infty \to C^\infty(U) \quad (7.82)$$

where $C^\infty(U)$ is the space of functions which are C^∞ on U.

(This is justified as follows. As is easily seen, we need only show that we can differentiate under the integral in

$$\int K^+(u,w)\, f(w^{-1}u)\, dw.$$

This is justified through (7.76) for $|w| \geq 1/3R$ and through (7.80), with $\rho = 0$, for $0 < |w| \leq 1/3R$.)

If $V, W \subset \mathbb{H}^n$ are nonempty, we put

$\text{dist}(V,W) = \inf|vw^{-1}|$ for $v \in V$, $w \in W$.

If $V = \emptyset$ or $W = \emptyset$, we put $\text{dist}(V,W) = \infty$.

If $V \subset \mathbb{H}^n$, let V^c be the complement of V.

The same reasoning as was used to show (7.82) shows that if $U_1 \subset U$, $r < \text{dist}(U_1, U^c)$, $K \in C^k(U;r,s,R)$ and if $\mathcal{K} = \mathcal{O}(K)$, then

$$\mathcal{K} : C^\infty(U) \to C^\infty(U_1).$$

If $\Omega \subset \mathbb{C}^M$ for some M, Ω open, and if for each $\omega \in \Omega$ we are given $K^\omega \in C^k(U;r,s,R)$, we say that K^ω depends holomorphically on ω if the following conditions hold.

Say that in fact

$$K_u^\omega(w) = \chi(w)\,[\textstyle\sum K_u^{\omega m}(w) + Q^\omega(u,w)]$$

as in (7.73). We then require, for each $u \in U$, that for

each m, $K^{\omega m}_{v\circ u}$ be holomorphic in (v,ω) for v in a neighborhood of $\overline{D}_s$ and ω in Ω; that $Q^{\omega}(v\circ u,w)$ be holomorphic in (v,ω,w) for v in a neighborhood of $\overline{D}_s$ and $(\omega,w) \in \Omega \times B_r$; and that there be constants, corresponding to C_E in (7.77), (7.77)', (7.78) which are independent of ω, for ω in any compact subset of Ω.

As in the introduction to the book, we write $u = (t,x,y) \in \mathbb{H}^n$, $\partial = (\partial/\partial t, \partial/\partial x_1,\ldots,\partial/\partial x_n, \partial/\partial y_1,\ldots,\partial/\partial y_n)$, $\mathcal{U} = (T,X_1^R,\ldots,X_n^R,Y_1^R,\ldots,Y_n^R)$. For $\gamma \in (\mathbb{Z}^+)^{2n+1}$, we write $\sigma_\gamma(\mathcal{U})$ to denote the symmetrization of $\mathcal{U}^\gamma$, that is, the coefficient of τ^γ in the multinomial expansion of $(\gamma!/\|\gamma\|!)(\tau\cdot\mathcal{U})^{\|\gamma\|}$. (Here $\tau \in \mathbb{R}^{2n+1}$.)

Finally, if f is a function on $\mathbb{H}^n$, $\gamma \in (\mathbb{Z}^+)^{2n+1}$, we define the function $M^\gamma f$ on $\mathbb{H}^n$ by $(M^\gamma f)(w) = w^\gamma f(w)$; similarly we define $M^\gamma f$ if f is a distribution.

Now we turn to the fundamental results on composition, adjoints, and analytic regularity.

<u>Theorem 7.11</u> (a) (Adjoints). Suppose $K \in \mathcal{C}^k(U;r,s,R)$; let $\mathcal{K} = \mathcal{O}(K)$. Then there exists

$$K^* : C_c^\infty(U) \to C^\infty(U)$$

so that

$$(g,\mathcal{K}f) = (\mathcal{K}^*g,f)$$

for all $f,g \in C_c^\infty(U)$. Accordingly, $\mathcal{K}$ extends to an operator

$$K : \mathcal{E}'(U) \to \mathcal{D}'(U). \tag{7.83}$$

Here $\mathcal{E}'(U)$ is the space of distributions of compact support within U, and $\mathcal{D}'(U)$ is the dual of $C_c^\infty(U)$. Further:

$$\text{If } g \in C_c^\infty(U),\ u \in U \text{ and } \operatorname{dist}(u, \operatorname{supp} g) > r, \text{ then } (K^*g)(u) = 0. \tag{7.84}$$

Thus, if $U_1 \subset U$ and $\operatorname{dist}(U_1, U^c) > r$, then

$$K^* : C_c^\infty(U_1) \to C_c^\infty(U) \tag{7.85}$$

and

$$K : \mathcal{E}'(U_1) \to \mathcal{E}'(U). \tag{7.86}$$

Further, if $r < A = \min(s, 1/2R)$, then there exists $K^* \in C^{\overline{k}}(U)$ so that $K^* = \mathcal{O}(K^*)$. In fact, if K is as in (7.73), then

$$K_u^*(w) = \chi(w)\,[\textstyle\sum (K^*)_u^m(w) + Q^*(u,w)] \tag{7.87}$$

where

$$(K^*)_u^m = \sum_{a+|\gamma|=m} \overline{M^\gamma \sigma_\gamma(U) K_u^a/\gamma!} \tag{7.88}$$

and $\qquad Q^*(u,w) = \overline{Q(w^{-1}u, w^{-1})}.$

We also have these uniformities: if $r < A$, then there exist $s^*, R^* > 0$ so that $K' \in C^k(U; r,s,R) \Rightarrow (K')^* \in C^{\overline{k}}(u,r,s^*,R^*)$; further if K' depends holomorphically on a parameter, so does $\overline{(K')}^*$.

(b) (Pseudolocality) Suppose $K \in \mathcal{C}^k(U;r,s,R)$. Suppose $V \subset W \subset U$, V,W open, that $r > \operatorname{diam} W = \sup_{u,v\in W} |uv^{-1}|$, that $f \in \mathcal{E}'(W)$, and that f is analytic on V.

Let $\mathcal{K} = \mathcal{O}(K)$.

Then $\mathcal{K}f$ is analytic on V.

(c) (Composition) Suppose $u_0 \in U \subset \mathbb{H}^n$, U open, and for $\nu = 1,2, k_\nu \in \mathbb{C}$, $r_\nu, s_\nu, R_\nu, R'_\nu > 0$, and $r_1 \neq r_2$. Say also $r_2 < s_1$ or $r_1 < s_1$. (7.89)

Suppose also $r_1 < \operatorname{dist}(u_0, U^c)$. (7.90)

Then there exist $r,s,R,R' > 0$ and open sets $U_0, U_1 \subset U$ with $u_0 \in U_0 \cap U_1$, as follows.

Suppose $K_\nu \in \mathcal{C}^{k_\nu}(U;r_\nu,s_\nu,R_\nu,R'_\nu)$ for $\nu = 1,2$ and that $K_\nu \sim \sum K_\nu^m$. Let $\mathcal{K}_\nu = \mathcal{O}(K_\nu)$. (7.91)

Let $k = k_1 + k_2 + 2n + 2$. (7.92)

Then there exists $K \in \mathcal{C}^k(U;r,s,R,R')$ with the following properties: (7.93)

Let $\mathcal{K} = \mathcal{O}(K)$. Suppose $f \in \mathcal{E}'(U_1)$.

Then $\mathcal{K}_1 f \in \mathcal{E}'(U)$ and we may form $\mathcal{K}_2\mathcal{K}_1 f$. We have

$$\mathcal{K}f = \mathcal{K}_2\mathcal{K}_1 f \text{ on } U_0. \tag{7.94}$$

Also

$$K \sim \sum K^m \tag{7.95}$$

where

$$K_u^m = \sum_{|\gamma|+a+b=m} [(-M)^\gamma K_{2u}^b] \underline{*} [\sigma_\gamma(\mathcal{U}) K_{1u}^a]/\gamma! \qquad (7.96)$$

Further, if K_1, K_2 depend holomorphically on a parameter, we may choose K so as also to depend holomorphically on this parameter. (7.97)

(d) (Analytic hypoellipticity). Suppose $u_o \in U \subset \mathbb{H}^n$, U open, and $k_1, k_2 \in \mathbb{C}$. Suppose we have (7.91). Suppose also $k = k_1 + k_2 + 2n + 2 = -2n - 2$. Define K_u^m by (7.96), and suppose that for all $u \in U$,

$$K_u^o = K_{2u}^o * K_{1u}^o = \delta(w). \qquad (7.98)$$

$$K_u^m = \sum_{|\gamma|+a+b=m} [(-M)^\gamma K_{2u}^b] \underline{*} [\sigma_\gamma(\mathcal{U}) K_{1u}^a]/\gamma! = 0 \text{ for } m > 0. \qquad (7.99)$$

Suppose $V \subset W \subset U$, V,W open, and that $r_1 > \text{diam } W$. Suppose $g \in \mathcal{E}'(W)$ and that $\mathcal{K}_1 g$ is analytic on V. Then g is analytic on V. (7.100)

Suppose we have in addition (7.89) and (7.90). Then we of course have (7.93), (7.97), for some $r, s, R, R' > 0$ and $U_0, U_1 \subset U$ with $u_0 \in U_0 \cap U_1$. Thus

$$\mathcal{K} \sim \delta; \qquad (7.101)$$

in other words, $\mathcal{K} = I + \mathcal{Q}$ where the core of $\mathcal{Q}$ has the form

$\chi(w)Q(u,w)$ with χ the characteristic function of $\{w : |w| < r\}$ and Q satisfying relations like (7.75), (7.78) (with different constants).

$$\mathcal{Q} \text{ is analytic regularizing in the sense that if } f \in \mathcal{E}'(U_1) \text{ then } \mathcal{Q}f \text{ is analytic on } U_0. \tag{7.102}$$

Then if $f \in \mathcal{E}'(U_1)$,

$$K_2K_1f = f + \mathcal{Q}f = f + h, \text{ say, on } U_0, \tag{7.103}$$

where h is analytic on U_0.

Further, if K_1, K_2 depend holomorphically on a parameter, we may choose K so as also to depend holomorphically on this parameter.

As we have indicated in the Introduction, formulas (7.88) and (7.96) will be derived as consequences of a version of Taylor's Theorem for $\mathbb{H}^n$, which we state and prove now.

<u>Proposition 7.12</u> (Taylor's Theorem for $\mathbb{H}^n$). Suppose $u \in \mathbb{H}^n$, and f is analytic near u. We may then extend f to a holomorphic function, also denoted f, in some connected complexified neighborhood of u in $\mathbb{C}^{2n+1}$. Likewise, for any $\gamma \in (\mathbb{Z}^+)^{2n+1}$, we may extend $\sigma_\gamma(\mathcal{U})f$ to a holomorphic function, also denoted $\sigma_\gamma(\mathcal{U})f$, in the same neighborhood of u in $\mathbb{C}^{2n+1}$. Thus, we may form $f(v \circ u)$, and more generally $(\sigma_\gamma(\mathcal{U})f)(v \circ u)$, for v in a neighborhood of some closed polydisc $\overline{\mathcal{D}}_r$, where

$\mathcal{D}_r = \{v \in \mathbb{C}^{2n+1} : |v_j| < r_j \text{ for all } j\}$. Here $r = (r_1,\ldots,r_{2n+1})$, all $r_j > 0$. Then

$$f(v \circ u) = \sum_\gamma v^\gamma [\sigma_\gamma(\mathcal{U})f](u)/\gamma! \qquad \text{for } v \in \mathcal{D}_r. \tag{7.104}$$

Further,
$$|[\sigma_\gamma(\mathcal{U})f](u)| \leqq \gamma! \sup_{v\in\mathcal{D}_r} |f(v\circ u)|/r^\gamma. \tag{7.105}$$

More generally, say $s \in \mathbb{R}^{2n+1}$, $0 < s_j < r_j$ for all j, and $\rho > 0$ is sufficiently small that whenever $v \in \mathbb{C}^{2n+1}$, $|v_j| \leqq \rho$ for all j, then $v\circ w \in \mathcal{D}_r$ for all $w \in \overline{\mathcal{D}}_s$. Then, whenever $u' = w\circ u$ for some $w \in \mathcal{D}_s$, we have a Cauchy integral formula:

$$[\sigma_\gamma(\mathcal{U})f](u') = (2\pi i)^{-2n-1}\gamma! \int_{S_\rho} [f(v\circ u')/v^\gamma]d\Sigma(v) \tag{7.106}$$

where $S_\rho = \{v \in \mathbb{C}^{2n+1} : |v_j| = \rho \text{ for all } j\}$, and $d\Sigma(v)$ is the usual surface measure on S_ρ. In particular, then,

$$|[\sigma_\gamma(\mathcal{U})f](w\circ u)| \leqq \gamma!\rho^{-\|\gamma\|}\sup_{v\in\mathcal{D}_r} |f(v\circ u)| \text{ for all } w \in \overline{\mathcal{D}}_s. \tag{7.107}$$

Proof. This might be called "Taylor's Theorem in exponential coordinates;" formally

$$f(v\circ u) = [\exp(v\cdot\mathcal{U})f](u) = \sum [(v\cdot\mathcal{U})^m f](u)/m!$$

formally yielding (7.104).

It is easy to give an elementary proof. Observe first that we may assume $u = 0$; otherwise, we replace f by a right translate and use the right invariance of $\mathcal{U}$. For

(7.104) and (7.105), then, it suffices to check that

$$\sigma_\gamma(\mathcal{U}) = \partial^\gamma \text{ at } 0, \text{ for all } \gamma. \tag{7.108}$$

To see this, observe that for $u \in \mathbb{H}^n$, $s \in \mathbb{R}$, $F \in C^\infty(\mathbb{H}^n)$ we have

$$[(u\cdot\partial)F](su) = \partial/\partial s[F(su)] = [(u\cdot\mathcal{U})F](su) \tag{7.109}$$

since, if $G(v) = F(v\circ su)$,

$$\partial/\partial s[F(su)] = \lim_{h\to 0}[F(hu\circ su) - F(su)]/h = [\partial G(hu)/\partial h]\big|_{h=0}$$
$$= [(u\cdot\mathcal{U})G](0) = [(u\cdot\mathcal{U})F](su).$$

(We used the fact that lines through 0 are one-parameter subgroups.) From (7.109), repeated differentiation of $F(su)$ with respect to s yields:

$$[(u\cdot\partial)^m F](su) = (\partial^m/\partial s^m)[F(su)] = [(u\cdot\mathcal{U})^m F](su) \text{ for all } m \in \mathbb{Z}^+.$$

In particular,

$$[(u\cdot\partial)^m F](0) = [(u\cdot\mathcal{U})^m F](0)$$

for all m and all $F \in C^\infty$. (7.108) follows at once from an examination of the coefficient of u^γ in this identity in the case $m = \|\gamma\|$. This proves (7.104) and (7.105).

For (7.106), let $V = \{w\circ u : w \in \mathcal{D}_s\}$. Note that (7.106) reduces to the ordinary Cauchy integral formula if $u' = 0 \in \mathbb{H}^n$. It then follows for $u' \in V \cap \mathbb{H}^n$ by the right invariance of $\sigma_\gamma(\mathcal{U})$. Since both sides of (7.106) are clearly holomorphic in $u' \in V$, and agree for $u' \in V \cap \mathbb{H}^n$, they are clearly equal for all $u' \in V$. (7.106) follows, and (7.107) is an immediate consequence of (7.106) and the associativity of $\circ$. This completes the proof of Proposition 7.12.

Proof of Theorem 7.11 (a) Let us first observe that (7.85) and (7.86) follow easily once we have (7.83) and (7.84). Indeed, (7.85) is obvious. For (7.86), say we have $h \in \mathcal{E}'(U_1)$, and let $P = \{u \in \mathbb{H}^n : \text{dist}(u, \text{supp } h) \leq r\}$. It suffices to show that

$$\text{supp } \mathcal{K}h \subset P \tag{7.110}$$

since P is a compact subset of U. We need only show then that $(\mathcal{K}^*g, h) = 0$ for all $g \in C_c^\infty(U \setminus P)$, and for this we need only show $(\mathcal{K}^*g)(u_1) = 0$ for all $u_1 \in \text{supp } h$. This is evident from (7.84), since dist(supp h, supp g) > r.

Now, for the remaining assertions of (a), note that we may assume $r < A$. Otherwise, let χ' be the characteristic function of $\{u : |u| < A/2\}$, and let $K'(u,w) = [1 - \chi'(w)]K(u,w)$, which, for any compact $E \subset U$, is bounded and analytic in $E \times (B_r \setminus B_{A/2})$. Let $\mathcal{K}' = \mathcal{O}(K')$. $\mathcal{K}'$ clearly has an adjoint $(\mathcal{K}')^*$, where

$$[(\mathcal{K}')^* f](u) = \int \overline{K'(v, vu^{-1})} f(v)\,dv = \int \overline{K'(vu, v)} f(vu)\,dv$$

which maps $C_c^\infty(U)$ to $C^\infty(U)$ and $C_c^\infty(U_1)$ to $C_c^\infty(U)$. K' then clearly lets us reduce to the case $r < A$.

Now say K is as in (7.73)-(7.78). The adjoint of the operator with core $\chi(w)Q(u,w)$ is the operator with core $\chi(w)Q^*(u,w)$, where $Q^*(u,w) = \overline{Q(w^{-1}u,w^{-1})}$. Q^* is analytic in (u,w) for $(u,w) \in U \times B_r$, by (7.78). (Note $B_r \subset D_s$.) Also, by (7.78), there exists $q > 0$ so that for any $u \in U$, $w \in B_r$, there is a holomorphic funciton $Q(v\circ u,\omega+w)$ of $(v,\omega) \in D_s \times D_q$, agreeing with $Q(vu,\omega+w)$ for $v \in \mathbb{H}^n$, $vu \in U$ and $\omega = w \in \mathbb{H}^n$, and which for any compact set $E \subset U$, satisfies $|Q(v\circ u,\omega+w)| < C_E$ for $u \in E$, $(v,\omega) \in D_s \times D_q$, $w \in B_r$. Selecting $s' > 0$, $0 < q' < q$ so that $(D_{q'}+B_r)\circ D_{s'} \subset D_s$, we see $Q^*(v\circ u,\omega+w)$ has an analogous extension for $(v,\omega) \in D_{s'} \times D_{q'}$, and thus an analogue of (7.78) holds for Q^*. In short, we may assume $Q = 0$.

If $K = K^+$, it is also easy to proceed. For then we may as well assume $\operatorname{Re} k > -2n - 2$. We have $K \in L^1$, and the adjoint of K is the operator with core $\chi(w)J^*(u,w)$ where, for $0 < |w| < r < s$,

$$J^*(u,w) = \sum_m \overline{K^m(w^{-1}u,w^{-1})} = \sum_m\sum_\gamma \overline{(-w)^\gamma \sigma_\gamma(U)K^m(u,w^{-1})}/\gamma!.$$

The inner series converges absolutely, by (7.105) and the fact that $|w| < r < s \Rightarrow w \in D_s$. By (7.105) and (7.80) (with $\rho = 0$), the sum of the inner series is

$O((2Rr)^m(1+|\log|w||)|w|^{\operatorname{Re} k})$ and hence, since $r < A$, the double sum converges absolutely for $u \in U$, $0 < |w| < r$. Rearranging the series, we write

$$J^*(u,w) = \sum (K^*)^m(u,w)$$

with $(K^*)^m$ as in (7.88). We only have to check that we have an analogue of (7.77) for the $(K^*)^m$. Now $(K^*)^m$, as expressed in (7.88), is a sum of $\binom{m+2n+1}{m}$ terms (the number of ways of writing m as a sum of 2n + 2 nonnegative integers). This quantity is less than $(m+2n+1)^{2n+1}$, so we may consider each term separately. By (7.77), there exists $q > 0$ so that for any m, any compact $E \subset U$, and $u \in E$ and any w with $1 \leqq |w| \leqq 2$, there is a holomorphic function $K^m(v\circ u,-\omega-w)$ of $(v,\omega) \in D_s \times D_q$, agreeing with $K^m(vu,-\omega-w)$ for $v \in \mathbb{H}^n$, $vu \in U$ and $\omega \in \mathbb{H}^n$, and satisfying $|K^m(v\circ u,-\omega-w)| < C_E R^m$ on $D_s \times D_q$. For $v \in \overline{D_{s/2}}$, we have an estimate of the form

$$|\sigma_\gamma(\mathcal{U})K^a(v\circ u,-\omega-w)| < C_E R_1^{a+\|\gamma\|}\gamma! = C_1 R_1^m \gamma! \tag{7.111}$$

by (7.107) of Proposition 7.12; here R_1 is a constant depending only on R and s. (We are denoting the holomorphic extension of $\sigma_\gamma(\mathcal{U})K^a$ also by $\sigma_\gamma(\mathcal{U})K^a$.) This then gives an estimate

$$|(\omega+w)^\gamma \sigma_\gamma(\mathcal{U})K^a(v\circ u,-\omega-w)| < C_E R_2^m \gamma! \tag{7.112}$$

with R_2 depending only on R and s, for $(v,\omega) \in \overline{D_{s/2}} \times D_q$. This completes the proof of (a), if $K = K^+$. ((7.77)' for $\sum (K^*)^m$ may be dealt with just as we dealt with Q, since $\sum (K^*)^m(u,w) = \sum K^m(w^{-1}u,w)$ for $0 < |w| \leq r$.)

Thus, to finish the proof of (a), we may assume $K = \chi(w)K_u^m$, where $\mathrm{Re}\, k + m \leq -2n - 2$. Changing k if necessary, we may as well assume $m = 0$. Write $J_u = K_u^o$. If $J_u = \sum c_\alpha(u)\partial_w^\alpha \delta$ is supported at 0, the result is easily established. Thus we may assume $J_u = \Lambda_{G_u}$ for a suitable function $G_u(w)$.

Observe, in passing, that if $G \in C^\infty(\mathbb{H}^n\setminus\{0\})$, G is homogeneous of degree j and

$$\int_{1<|u|\leq 2} u^\gamma G(u)\,du = 0 \quad \text{whenever } |\gamma| = -j - 2n - 2, \tag{7.113}$$

then

$$\Lambda_G = \Lambda_{\tilde{G}}. \tag{7.114}$$

This is clear from (1.2) and the definition

$$\tilde{\Lambda}_G(f) = \overline{\Lambda_G(\tilde{f})} \tag{7.115}$$

Now, in our situation, let

$$H_u(w) = \chi(w)\widetilde{\sum_\gamma M^\gamma \sigma_\gamma(U)J_u/\gamma!} \tag{7.116}$$

where, by definition, $\sigma_\gamma(U)J_u = \Lambda_{\sigma_\gamma(U)G_u}$. Away from $w = 0$, $H_u(w)$ clearly equals $\chi(w)G(w^{-1}u,w^{-1})$, with the sum converging absolutely, by (7.105). It is then easy to see, as in the analysis leading to (7.112), that $H \in C^{\overline{k}}(U)$. Let $\mathcal{H}$ be the operator with core H. To complete the proof of (a), it will suffice to show that

$$(g, \mathcal{K} f) = (\mathcal{H} g, f)$$

for all $f, g \in C_c^\infty$. For this let $D = \{\zeta \in \mathbb{C} : \zeta + k + 2n + 2 \notin \mathbb{Z}^-\} \cup \{0\}$. For $\zeta \in D$, let

$$G_u^\zeta(w) = G_u(w)|w|^\zeta, \quad J_u^\zeta = \Lambda_{G_u^\zeta}, \quad \mathcal{K}^\zeta = \mathcal{O}(J^\zeta).$$

Also define $H_u^{\overline{\zeta}} \in C^{\overline{k+}}$ by (7.116) with J_u replaced by J_u^ζ, and let $\mathcal{H}^{\overline{\zeta}} = \mathcal{O}(H^{\overline{\zeta}})$. By the part of the proof dealing with K^+, we have

$$(g, \mathcal{K}^\zeta f) = (\mathcal{H}^{\overline{\zeta}} g, f) \tag{7.117}$$

if $\mathrm{Re}(\zeta + k) + 2n + 2 > 0$. Thus it suffices to show that both sides are holomorphic in $\zeta \in D$. This is clear, however, since at any point of D, the sum for H_u^ζ consists of finitely many easily-controlled terms, plus a series which converges to a continuous function of u,w, with compact support in w, and whose values vary holomorphically in ζ. (We are using (7.80).) This completes the proof of (a), since the uniformities in (a) follow from the proof.

For (b), let $P = \{v \circ u : v \in D_s, u \in U\} \subset \mathbb{C}^{2n+1}$. For $p \in P$, $|w| < r$, we may form $K_p(w) = \chi(w)[\sum K_p^m(w) + Q(p,w)] \in \mathcal{E}'$.

For $u \in V$, set $g(p,u) = (K_p * f)(u)$. Since $r > \text{diam } W$, by Lemma 7.6(a) we have that for any $p \in \mathcal{P}$, $g(p,u)$ is analytic on V. Also, by (7.35), for any compact subsets E of V and F of $\mathcal{P}$, there exist $C_{EF}, R_{EF} > 0$ so that

$$|\partial_u^\gamma g(p,u)| < C_{EF} R_{EF}^{|\gamma|} \gamma! \qquad \text{for } u \in E,\ p \in F;$$

this is so because, by (7.77), (7.77)', (7.78) and (7.79), K_p satisfies appropriate estimates uniformly for $p \in F$. Also, for fixed $u \in V$, $g(p,u)$ is holomorphic in $p \in \mathcal{P}$, as one sees easily by considering K^+, K^- separately. From these facts, it is clear that $g(p,u)$ is real analytic for $(p,u) \in \mathcal{P} \times U$. But $(\mathcal{K}f)(u) = g(u,u)$ for $u \in V$, so (b) follows.

For (c), we assume to begin that $r_2 < r_1$; this restriction will be removed at the end of the proof of (c). Further, for $0 < c < 1$, we let $\chi_{\nu c}$ be the characteristic function of $\{w : |w| < cr_\nu\}$ $(\nu=1,2)$. We let

$$K_\nu^c(u,w) = \chi_{\nu c}(w) K_\nu(u,w), \quad \mathcal{K}_\nu^c = \mathcal{O}(K_\nu^c).$$

We begin by proving that there exists $c_0 > 0$, depending only on the quantities mentioned in (7.89), so that (c) holds when $K_\nu, \mathcal{K}_\nu$ are replaced by $K_\nu^c, \mathcal{K}_\nu^c$, provided $c \leq c_0$. At the end of the proof of (c) we deal with the given K_1, K_2.

To prove the conclusions of (c), we shall take our motivation from (0.8) and (0.23) of the Introduction. Actually, before attempting to make (0.8) and (0.23) rigorous, we make a formal change of variables in those formal equations: if $\varkappa(\mathcal{K}_1) = K_1$, $\varkappa(\mathcal{K}_2) = K_2$, $\mathcal{K} = \mathcal{K}_2\mathcal{K}_1$, $\varkappa(\mathcal{K}) = K$, we have formally

$$(\mathcal{K}f)(u) = \int K_{2u}(w)[\int K_1(w^{-1}u, w^{-1}uv^{-1})f(v)dv]dw.$$

Formally writing

$$K_1(w^{-1}u, w^{-1}uv^{-1}) = \sum (w^{-1})^\gamma [\sigma_\gamma(\mathcal{U})K_{1u}](w^{-1}uv^{-1})/\gamma!$$

we find formally

$$(\mathcal{K}f)(u) = (K_u * f)(u),\ K_u = \sum_\gamma (-M)^\gamma K_{2u} * \sigma_\gamma(\mathcal{U})K_{1u}/\gamma!\ .$$

Proceeding now rigorously, we fix $u \in U$, $f \in C_c^\infty$, and $c > 0$ with $cr_1 < 1/2R_1$. We examine

$$g(v,\xi) = ([\chi_{1c}K_{1(v\circ u)}]*f)(\xi) \qquad \text{for } v \in D_{s_1},\ \zeta \in \mathbb{H}^n.$$

Later we will set $v = w^{-1}$, $\xi = w^{-1}u$. Of course, by definition,

$$\chi_{1c}K_{1(v\circ u)} = \chi_{1c}[\sum K^m_{1(v\circ u)} + Q_{1(v\circ u)}]$$

which exists by (7.79) with $\rho = 0$, since $cr_1 < 1/2R_1$. It is further evident from (7.79) with $\rho = 0$ that g is smooth in ξ for all v, and that $\partial_\xi^\gamma g(v,\xi)$ is holomorphic in $v \in D_{s_1}$ for all $\gamma \in (\mathbb{Z}^+)^{2n+1}$. We may therefore expand

$$g(v,\xi) = \sum v^\gamma [\partial_v^\gamma g](0,\xi)/\gamma!$$

for $v \in D_{s_1}$. The Cauchy integral formula shows that this expression converges in the topology of $C^\infty(D_{s_1} \times \mathbb{H}^n)$. For $f \in C_c^\infty$, $u \in U$, $w \in D_{s_1} \cap \mathbb{H}^n$, we have by (7.108) and the right invariance of $\mathcal{U}$,

$$(K_1^c f)(w^{-1}u) = g(w^{-1}, w^{-1}u) = \sum (-w)^\gamma [(\sigma_\gamma(\mathcal{U})[\chi_{1c}K_{1u}*f])(w^{-1}u)]/\gamma!.$$

To be very clear, $\chi_{1c}K_{1u}*f(\xi)$ is a function of u and ξ, and $\sigma_\gamma(\mathcal{U})$ is applied in the u variable. By our analysis of g, we see that for fixed u, the series for $(K_1^c f)(w^{-1}u)$ converges in the topology of $C^\infty(D_{s_1} \cap \mathbb{H}^n)$. Observe further that $\sigma_\gamma(\mathcal{U})[\chi_{1c}K_{1u}*f] = [\chi_{1c}\sigma_\gamma(\mathcal{U})K_{1u}]*f$, as is easily seen from (7.80) (with $\rho = 0$, Re $k + m > -2n - 2$).

Now say $U_1 \subset U$ is open, $\text{dist}(U_1, U^c) > cr_1$, supp $f \subset U_1$. Then $K_1^c f \in C_c^\infty(U)$ by (a). Say now c also satisfies $0 < cr_2 < 1/2R_2$. We may look at $(K_2^c K_1^c f)(u)$ for $u \in U$; if $h(w) = (K_1^c f)(w^{-1}u)$, this is

$$(\chi_{2c}K_{2u}*K_1^c f)(u) = (\chi_{2c}K_{2u})(h)$$

(the distribution $\chi_{2c}K_{2u}$ applied to h). Because of the convergence of the series for h in $C^\infty(D_{s_1} \cap \mathbb{H}^n)$, we see

$$(K_2^c K_1^c f)(u) = \sum_\gamma ([\chi_{2c}(-M)^\gamma K_{2u}]*[\chi_{1c}\sigma_\gamma(\mathcal{U})K_{1u}]*f)(u)/\gamma! .$$

Of course, convolution of distributions of compact support is associative, so we should now examine

$[\chi_{2c}(-M)^{\gamma}K_{2u}]*[\chi_{1c}\sigma_{\gamma}(U)K_{1u}]$. We may write

$$\chi_{2c}(-M)^{\gamma}K_{2u} = \sum_b \chi_{2c}(-M)^{\gamma}K^b_{2u} + \chi_{2c}(-M)^{\gamma}Q_{2u} \tag{7.118}$$

with convergence in E', by (7.79) with $\rho = v = 0$. We may therefore convolve both sides of the right with $\chi_{1c}\sigma_{\gamma}(U)K_{1u}\varepsilon E'$ and bring the convolution through the summation, and still have convergence in E'. Therefore one can convolve both sides on the right with f, and bring this convolution through the summation. Expanding $\chi_{1c}\sigma_{\gamma}(U)K_{1u}$ in a like manner to (7.118), and using similar arguments, we now see

$$\begin{aligned}(K^c_2K^c_1f)(u) &= \sum_{\gamma}\sum_{b}\sum_{a}(\chi_{2c}(-M)^{\gamma}K^b_{2u}*\chi_{1c}\sigma_{\gamma}(U)K^a_{1u}*f)(u)/\gamma! \\ &+ \sum_{\gamma}\sum_{b}(Q^{\gamma b1}_{cu}*f)(u) + \sum_{\gamma}\sum_{a}(Q^{\gamma a2}_{cu}*f)(u) + \sum_{\gamma}(Q^{\gamma 3}_{cu}*f)(u)\end{aligned} \tag{7.119}$$

where

$$Q^{\gamma b1}_{cu} = [\chi_{2c}(-M)^{\gamma}K^b_{2u}]*[\chi_{1c}\sigma_{\gamma}(U)Q_{1u}/\gamma!]$$

$$Q^{\gamma a2}_{cu} = [\chi_{2c}(-M)^{\gamma}Q_{2u}]*[\chi_{1c}\sigma_{\gamma}(U)K^a_{1u}/\gamma!]$$

$$Q^{\gamma 3}_{cu} = [\chi_{2c}(-M)^{\gamma}Q_{2u}]*[\chi_{1c}\sigma_{\gamma}(U)Q_{1u}/\gamma!]$$

Now put $K^{\gamma ab}_{v\circ u} = (-M)^{\gamma}K^b_{2(v\circ u)} \underline{*}\, \sigma_{\gamma}(U)K^a_{1(v\circ u)}/\gamma! \;\varepsilon\; AK^{k+m}$, where $m = a + b + |\gamma|$. Let $s = \min(s_1,s_2)/2$, and select $\tau > 0$ so that

$$\text{if } v \in \overline{D}_s, \text{ and } v' \in \mathbb{C}^{2n+1} \text{ with } |v'_j| \leq \tau \text{ for all } j \text{ then } v'\circ v \in D_{2s}. \tag{7.120}$$

It is now time to estimate all the relevant quantities. In the rest of the proof of (c), $u \in U$ will be held fixed. Any constant called "C", with one or more subscripts, depends on u, but may be chosen to be the same for u ranging over any fixed compact subset of U. All other constants may be chosen independently of $u \in U$.

Now, by definition, there are $C_1, R_1 > 0$ so that, for all a,

$$K^a_{1v\circ u} \in AK^{k_1+a}(C_0R_0^a, R_0) \quad \text{for } v \in D_{2s}. \tag{7.121}$$

Thus, by (7.120), (7.107) and the definition of $AK^k(C,R)$ (before (7.20)), for all a, γ,

$$\sigma_\gamma(U)K^a_{1(v\circ u)}/\gamma! \in AK^{k_1+a}(C_1\tau^{-\|\gamma\|}R_1^a, R_1) \quad \text{for } v \in \bar{D}_s. \tag{7.122}$$

Also, arguments such as those leading to (7.111) and (7.112), involving passing to polydiscs in the w variable, together with the definition of $AK^k(C,R)$, show that there exist $C_2, R_2 > 0$ so that

$$M^\gamma K^b_{2v\circ u} \in AK^{k_2+b}(C_2R_2^{b+|\gamma|}, R_2) \quad \text{for } v \in D_{2s}. \tag{7.123}$$

Applying Theorem 7.5(b), together with (7.121), we see that there are $C_3, R_3 > 0$ so that, if $m = a + b + |\gamma|$,

$$K^{\gamma ab}_{v\circ u} \in AK^{k+m}(C_3R_3^m, R_3) \quad \text{for } v \in \bar{D}_s. \tag{7.124}$$

In addition, by Lemma 7.9, there are $C_4, R_4 > 0$ and r_3 with $0 < r_3 < r_1$ so that

$$\chi_{2c}(-M)^{\gamma}K^{b}_{2(v\circ u)} * \chi_{1c}\sigma_{\gamma}(U)K^{a}_{1(v\circ u)}/\gamma! = K^{\gamma ab}_{v\circ u} + Q^{\gamma[a,b]}_{c(v\circ u)} \quad \text{on } \{w : |w| < cr_3\} \tag{7.125}$$

where

$$|\partial^{\rho}Q^{\gamma[a,b]}_{c(v\circ u)}(w)| < C_4 c^{\operatorname{Re} k+m-|\rho|}(1+|\log c|)^2 R_4^{m+\|\rho\|}\rho! \quad (|w| < cr_3) \tag{7.126}$$

for $v \in D_s$, $\rho \in (\mathbb{Z}^+)^{2n+1}$. By Lemma 7.9, $K^{\gamma ab}_{v\circ u}$ and $Q^{\gamma[a,b]}_{c(v\circ u)}$ are holomorphic in $v \in D_s$.

Define K^m_u by (7.96), so that

$$K^m_u = \sum_{|\gamma|+a+b=m} K^{\gamma ab}_u .$$

The number of terms in the summation is at most $\binom{m+2n+2}{m} \leq (m+2n+2)^{2n+2}$. Thus c is sufficiently small that $cr_3 < 1/2R_3$, it follows from (7.79) that $\sum_{\gamma}\sum_{b}\sum_{a} K^{\gamma ab}_{v\circ u} = \sum K^m_{v\circ u}$ converges in the sense of distributions on $\{w : |w| < cr_3\}$; in fact, if finitely many terms are omitted, the series converges absolutely in L^1 on that domain. Further, by (7.124), $\chi_{3c}(w)\sum K^m_u(w) \in C^k(U)$ if χ_{3c} is the characteristic function of $\{w : |w| < cr_3\}$, $cr_3 < 1/2R_3$. If also $c < 1/2R_4$, then by (7.126) $\sum_{\gamma}\sum_{b}\sum_{a} Q^{\gamma[a,b]}_{c(v\circ u)}(w)$ is analytic for $|w| < cr_3$. Thus, if c is sufficiently small,

$$J_u^c(w) = \chi_{3c}(w)[\sum K_u^m(w) + Q_c^{\gamma[a,b]}(w)] \in C^k(U).$$

Let $\mathcal{J}^c = \mathcal{O}(J^c)$. There is then clearly an open set U_2 with $u_o \in U_2 \subset U$ so that if c is sufficiently small that all previous restrictions on it hold and also small enough that $\operatorname{dist}(\{u : |uu_o^{-1}| < c_1r_3\}, U^c) > cr_1$, then if $c_1 = c/2$, $f \in C_c^\infty$, $\operatorname{supp} f \subset \{u : |uu_o^{-1}| < c_1r_3\}$, $u \in U_2$, then the first (triple) summation on the right side of (7.119) is equal to $(\mathcal{J}^cf)(u)$.

Thus, at least if $f \in C_c^\infty$, we need only deal with the remaining three summations in (7.119). By (7.106) and (7.78), we have that for some $C_5, R_5 > 0$,

$$|\partial_w^\rho[\sigma_\gamma(\mathcal{U})Q_1(v\circ u,w)]/\gamma!| < C_5R_5^{|\gamma|+\|\rho\|}\rho! \quad \text{for } |w| < r_1 \tag{7.127}$$

for $v \in D_s$, $\rho \in (\mathbb{Z}^+)^{2n+1}$. In addition, for any $\varepsilon > 0$ there exist $\delta, C_6, R_6 > 0$ so that

$$|\partial_w^\rho[M^\gamma Q_2(v\circ u,w)]| < C_6\varepsilon^{|\gamma|}R_6^{\|\rho\|}\rho! \quad \text{for } |w| < \delta \tag{7.128}$$

if $v \in D_{2s}$, $\rho \in (\mathbb{Z}^+)^{2n+1}$. (7.128) is clear from the Cauchy estimates and (7.78), when a sufficiently small polydisc in w is used. We fix $\varepsilon = \min(1/2R_5, \tau/2, 1)$, ($\tau$ as in (7.120)), and choose and fix δ, C_6, R_6 accordingly. Also, select $c_2 > 0$ so that if $c \leq c_2$, then c satisfies all earlier restrictions.

For the second summation in (7.119), let χ denote the characteristic function of $\{u : |u| < c_2r_1\}$. Observe that we may choose $r_4 > 0$ depending only on r_1 and r_2, so that $c \leq c_2$,

$$Q_{cu}^{\gamma bl} = [\chi_{2c}(-M)^{\gamma} K_{2u}^{b}] * [\chi\sigma_{\gamma}(U) Q_{1u}/\gamma!] \quad \text{for } |w| < cr_4.$$

For some c_3 with $0 < c_3 \leqq c_2$, if also $c \leqq c_3$, then we may assume r_4 is small enough that also:

1. $Q_{cu}^{\gamma bl}(w)$ is actually a function of w for $|w| < cr_4$ (not just a distribution).

2. For fixed such w, it is a function of u which may be extended to $Q_c^{\gamma bl}(v \circ u, w)$ a holomorphic function of $v \in D_s$.

3. $Q_c^{\gamma bl}(v \circ u, w)$ is, for fixed $v \in D_s$, analytic in w for $|w| < cr_4$.

(For assertion 3, we used Lemma 7.6(a).) In addition, by (7.123), for some $N, L, C_7, R_7 > 0$ independent of c, γ, b,

$$\|\chi_{2c}(-M)^{\gamma} K_{2(v\circ u)}^{b}\|_{(N)} \leqq C_7 R_7^{|\gamma|+b} c^{|\gamma|+b-L}$$

in the notation of Lemma 7.6(a). By (7.127) and Lemma 7.6(a), if $c \leqq c_3$ and $c \leqq 1/2R_5R_7$, we may assume r_4 is small enough that

4. $|\partial_u^{\rho} Q_c^{\gamma bl}(v \circ u, w)| < c^{-L} C_8 R_8^{\|\rho\|} 2^{-\|\gamma\|-b} \rho!$

for $v \in D_s$, $|w| < cr_4$, $\rho \in (\mathbb{Z}^+)^{2n+1}$.

Clearly, then $\sum_{\gamma}\sum_{b} Q_{cu}^{\gamma bl}(w)$ is analytic in w for $|w| < cr_4$; moreover, $\chi_{4c}(w)\sum_{\gamma}\sum_{b} Q_{cu}^{\gamma bl}(w) \in C^k(U)$ if χ_{4c} is the characteristic function of $\{w : |w| < cr_4\}$. Thus, as is easily seen, for $f \in C_c^\infty$, we need only now concern ourselves with the third and fourth summations in (7.119).

We handle $Q_c^{\gamma a2}$ in a rather analogous manner.

There exists $r_5 > 0$, depending only on r_1 and r_2, so that

$$Q_{cu}^{\gamma a2} = \chi_{2c}(-M)^\gamma Q_{2u} * \sigma_\gamma(\mathcal{U}) K_{1u}^a/\gamma! \quad \text{for } |w| < cr_5.$$

By Proposition 7.7, $Q_c^{\gamma a2}$ satisfies analogues of properties 1,2,3 above for $Q_c^{\gamma b1}$. (In the analogues, c can now be arbitrary with $0 < c < 1$, and we may assume r_5 is small enough that the properties hold for $|w| < cr_5$.) Let us also assume c is small enough that $cr_2 < \delta$ (see (7.128)). Then we may assume that r_5 is small enough that there exist $L, C_9, R_9 > 0$, independent of c, γ, a, so that for all $v \in D_s$,

$$|\partial_w^\rho Q_c^{\gamma a2}(v \circ u, w)| < C_9 R_9^{a+\|\rho\|} c^{a-L-|\rho|_2-\|\gamma\|} \rho!$$

for $|w| < cr_5$, $\rho \in (\mathbb{Z}^+)^{2n+1}$.

We used (7.122), (7.128), the fact that $\varepsilon, \varepsilon^2 \leq \tau/2$, and Proposition 7.10. Clearly, then, for $c < 1/2R_9$, $\sum_\gamma \sum_a Q_{cu}^{\gamma a2}(w)$ is analytic in w for $|w| < cr_5$; moreover, $\chi_{5c}(w) \sum_\gamma \sum_a Q_{cu}^{\gamma a2}(w) \in C^k(U)$ if χ_{5c} is the characteristic function of $\{w : |w| < cr_5\}$.

That leaves us with the last summation in (7.119). There exists $r_6 > 0$, depending only on r_1 and r_2, so that if c is small enough that $cr_2 < \delta$, $0 < c < 1$, then $Q_c^{\gamma 3}$ satisfies analogues of properties 1,2,3 above for $Q_c^{\gamma b1}$ for $|w| < cr_6$; and further, there exist $C_{10}, R_{10} > 0$ possibly depending on c but independent of γ, so that

$$|\partial_w^\rho Q_c^{\gamma 3}(v\circ u,w)| < C_{10}R_{10}^{\|\rho\|}\rho!2^{-|\gamma|} \text{ for } |w| < cr_6,\ \rho \in (\mathbb{Z}^+)^{2n+1}.$$

We used (7.128), (7.127), the fact that $\varepsilon \leqq 1/2R_5$, and Lemma 7.6(a). This estimate then easily enables us to dispense with the last summation of (7.119). This establishes (c) with the restriction $f \in C_c^\infty$, $r_2 < r_1$, K_ν^c in place of K_ν, $c \leqq c_o$. Indeed, (7.97) and the uniformity (the assertion that r,c,R,R',U_o,U_1 can be chosen to depend only on the data in (7.89)) follow from the proof.

We retain the restriction $f \in C_c^\infty$, $r_2 < r_1$, and show how to deal with the original K_1,K_2. Select $c \leqq c_o$, and produce $U_1,U_o,\mathcal{K}^c$ as before. We may assume U_1 is sufficiently small that $\text{dist}(U_1,U^c) > r_1$, and so that if $f \in C_c^\infty(U_1)$ then $\text{dist}(u_o,\text{supp}[\mathcal{K}_1 f-\mathcal{K}_1^c f]) > cr_2$. Then, if $f \in C_c^\infty(U_1)$, then $\mathcal{K}_1 f$, $\mathcal{K}_1^c f \in C_c^\infty(U)$, and $\mathcal{K}_2^c\mathcal{K}_1 f = \mathcal{K}_2^c\mathcal{K}_1^c f$ near u_o. Accordingly it suffices to show that for some $r,s,R,R' > 0$, and U_1' open with $u_o \in U_1'$, there exists $Q \in \mathcal{C}^k(U;r,s,R,R')$, $Q \sim 0$, so that if $\mathcal{Q} = \mathcal{O}(Q)$, $f \in C_c^\infty(U_1')$, then

$$(\mathcal{K}_2-\mathcal{K}_2^c)\mathcal{K}_1 f = \mathcal{Q}f \quad \text{near } u_o. \qquad (*)$$

Select $d > 0$ with $0 < dr_1 < cr_2$; then if U_2 is sufficiently small, $f \in C_c^\infty(U_2)$, we have $(K_2-K_2^c)K_1 f = (K_2-K_2^c)(K_1-K_1^d)f$. Now write $G_1 = K_1 - K_1^d$, $G_2 = K_2 - K_2^c$; then for $f \in C_c^\infty(U_1')$,

$$[(K_2-K_2^c)K_1 f](u) = \iint G_2(u,v)G_1(v^{-1}u, v^{-1}uw^{-1})f(w)\,dwdv$$

so

$$[(K_2-K_2^c)K_1 f](u) = (Q_u * f)(u) \tag{7.129}$$

where

$$Q(u,w) = \int_{cr_2<|v|<r_2} G_2(u,v)G_1(v^{-1}u, v^{-1}w)\,dv. \tag{7.130}$$

In the integral $|v| < r_2 < s_1$ and $|v| > cr_2 > dr_1$. Thus it is easy to complexify u and w and to show that Q satisfies conditions analogous to (7.75), (7.78) for w sufficiently small. This verifies (*), and hence (c) for $f \in C_c^\infty$, $r_2 < r_1$, since the uniformities follow from the proof.

Still assume $f \in C_c^\infty$; we show how to deal with the case $r_2 > r_1$. Select c with $cr_2 < r_1$; it will suffice again to verify an analogue of (*). Again select d with $0 < dr_1 < cr_2$ and put $G_1 = K_1 - K_1^d$, $G_2 = K_2 - K_2^c$; then we have (7.129), (7.130) for $f \in C_c^\infty(U_2)$, $u_0 \in U_2$, U_2 sufficiently small. We may extend G_1 analytically to

$A = \{vu : v \in D_{s_1} \cap \mathbb{H}^n, u \in U\} \times \{w : dr_1 < |w| < r_1\}$. Call the extended function G_1, and define $H_p(w) = G_1(wp,w)$ for $(wp,w) \in A$. Note $r_1 < s_1$, and now select $\rho > 0$ so that

$|w| < \rho$, $|v| < r_1 \Rightarrow |vw| < s_1$. Then if p is such that $wp \in U$ for some w with $|w| < \rho$, then $H_p(v)$ is analytic for $dr_1 < |v| < r_1$. Put

$$F(u,p,w) = (G_{2u}*H_p)(w) \qquad (7.131)$$

for $u, wp \in U$, $|w| < \rho$. Then

$$Q(u,w) = F(u,w^{-1}u,w) \qquad (7.132)$$

if $|w| < \rho$.

In (7.131), $H_p(v)$ is analytic for $dr_1 < |v| < r_1$, and 0 otherwise, while $G_{2u}(v)$ is analytic for $cr_2 < |v| < r_2$, 0 otherwise. Since $dr_1 < cr_2 < r_1 < r_2$, Lemma 7.6(a) implies that F is analytic in w for w sufficiently small. Since the analyticity of F in (u,p) is apparent, these considerations and (7.132) give (*) without difficulty.

Thus we have shown (c) for $f \in C_c^\infty$. Finally we deal with the general case $f \in \mathcal{E}'$. First observe that, if $\text{dist}(U_1,U^c) > r_1$, then we may extend K_1^* to map $C^\infty(U)$ to $C^\infty(U_1)$, as follows. Take V open, $\overline{V} \subset U$, $\text{dist}(U_1,V^c) > r_1$; choose $\psi \in C^\infty$, $\psi = 1$ on V, $\psi = 0$ on U^c. For $f \in C^\infty(U)$, define $K_1^* f = K^*(\psi f) \in C^\infty(U_1)$; this is independent of the choices made, by (7.84). Also, if $G \in C^\infty(U)$, $f \in \mathcal{E}'(U_1)$, we have $(f, K_1^* G) = (K_1 f, G)$, by (7.86) with V in place of U.

We may thus consider $K_1^* K_2^* g \in C^\infty(U_1)$ for any $g \in C_c^\infty(U_0)$.

Since

$$(K_1^* K_2^* g, f) = (K^* g, f)$$

for all $f \in C_c^\infty(U_1)$, it follows that $K_1^* K_2^* g = K^* g$ on U_1 for any $g \in C_c^\infty(U_0)$. Thus $K_2 K_1 f = Kf$ on U_0 for all $f \in E'(U_1)$. This concludes the proof of (c).

For (d), observe first that, since the core of Q has the form stated in (7.101), Q is clearly analytic regularizing in the sense of (7.102).

Thus we need only prove (7.100). Say $u_0 \in V$; we need only show g is analytic near u_0. Select $r_2 < s_1$. Select $c > 0$ and V_1 a neighborhood of u_0 sufficiently small that

(i) $cr_1 < \operatorname{dist}(u_0, U^c)$; $cr_1 < r_2$

(ii) For all $f \in E'(V_1)$, $\operatorname{diam}(\operatorname{supp} K_1^c f) < r_2$.

Using K_1^c in place of K_1, select r, K, U_0, U_1 as in (7.101)-(7.103). Then if $f \in E'(U_1)$,

$$K_2 K_1^c f = Kf = f + h \quad \text{on} \quad U_0 \tag{7.138}$$

where h is analytic on U_0.

Next select an open neighborhood of u_0, $V' \subset V_1 \cap U_1$, so small that $K_1 g' = K_1^c g'$ on V' for all $g' \in E'(V')$.

Now let χ be the characteristic function of V'. By (b), $K_1[(1-\chi)g]$ is analytic on V', so $K_1(\chi g) = K_1^c(\chi g)$ is analytic on V'. Further, by (ii), diam(supp $K_1^c(\chi g)) < r_2$, so by (b) $K_2 K_1^c \chi g$ is analytic on V'. By (7.135), g is analytic on $V' \cap U_0$, as desired.

Of course, Theorem 7.11(d) is not a useful criterion for analytic hypoellipticity of K_1, since it requires that one know the existence of $\{K_2^m\}$ satisfying (7.98), (7.99), and estimates akin to (7.77). Our main result, presented in the next chapter, is that, on relatively compact open subsets of U, knowing the existence of K_2^o satisfying (7.98) Is already sufficeint; one can construct the other K_2^m from it. Thus, if the principal core of K_1 is invertible in the sense of (7.98), K_1 will be analytic hypoelliptic. Further, the necessary and sufficient conditions for (7.98) to be satisfied for some K_2^o were already discussed, in Theorems 5.1 and 5.9.

8. Analytic Parametrices

In any good analytic calculus there must be a simple criterion for inversion of operators, modulo operators with analytic kernels. What we are going to show now is that if $K_1 \sim \sum K_1^m \in C^k(U)$; if U' is open, $\overline{U}'$ compact and $\overline{U}' \subset U$; and if there exists $K_{2u}^o \in AK^{k_2}$ satisfying (7.98) for $u \in U'$, then there exists $K_2 \sim \sum K_2^m \in C^{k_2}(U')$ $(k_1+k_2+2n+2=-2n-2)$ so that (7.99) is also satisfied on U'. Thus, $\mathcal{K}_1 = \mathcal{O}(K_1)$ is analytic hypoelliptic in the sense of Theorem 7.11(d). Thus for operators whose "principal cores" K_1^o satisfy (7.98) for some K_2^o, we have analytic hypoellipticity. Further, the necessary and sufficient conditions for (7.98) to be satisfies for some K_2^o were already discussed, in Theorems 5.1 and 5.9.

Precisely, we have the following theorem.

Theorem 8.1. Suppose $K_1 \in C^{k_1}(U;r_1,s_1,R_1)$ and that $K_1 \sim \sum K_1^m$. Let $\mathcal{K}_1 = \mathcal{O}(K_1)$.

Suppose that for all $u \in U$, K_{1u}^o satisfies the equivalent conditions of Theorem 5.1. In particular, say $\hat{K}_{1u}^o = \underset{\sim}{J}_{1u}^o$ in the $\mathcal{Q}$ sense. We assume only:

$$(\mathcal{F}K_{1u}^o)(0,\xi) \neq 0 \text{ for } \xi \neq 0; \; J_{1u+}^o v = 0, \; v \in H^\omega \Rightarrow v = 0;$$

and $J_{1u-}^o v = 0, \; v \in H^\omega \Rightarrow v = 0.$

Then K_1 is analytic hypoelliptic on U, in the following sense:

Suppose $V \subset W \subset U$, V,W open, and that $r_1 > \text{diam } W$. Suppose $g \in E'(W)$ and that $K_1 g$ is analytic on V. Then g is analytic on V.

Further, say $u_o \in U$ and $r_1 < \text{dist}(u_o, U^c)$. Suppose $u_o \in P \subset U$, P compact. Then there exists an open set U' with $P \subset U' \subset U$, and there exists $K_2 \in C^{k_2}(U')$ (where $k_1 + k_2 + 2n + 2 = -2n - 2$), $Q \in C^{-2n-2}(U')$, and open sets $U_1, U_o \subset U'$ with $u_o \in U_1 \cap U_o$, so that if $K_2 = \mathcal{O}(K_2)$, $\mathcal{Q} = \mathcal{O}(Q)$, we have:

$$\begin{aligned} &\text{If } f \in E'(U_1) \text{ then } K_1 f \in E'(U'); \\ &K_2 K_1 f = f + \mathcal{Q} f \quad \text{on} \quad U_o; \end{aligned} \tag{8.1}$$

and $\mathcal{Q}f$ is analytic on U_o. (8.2)

The core of $\mathcal{Q}$ has the form $\chi(w)Q(u,w)$ with χ the characteristic function of $\{u : |u| < r\}$, and Q as in (7.75), (7.78), for appropriate r,s,R, but for $u \in U'$ (not U).

Further, suppose K_1 in addition depends holomorphically on a parameter $\omega \in \Omega \subset \mathbb{C}^M$. Suppose E is compact and $\emptyset \neq E \subset \Omega \cap \mathbb{R}^M$. Then there exists an open set Ω' with $E \subset \Omega'$ as follows:

We may choose U', U_1, U_o as above independent of $\omega \in \Omega'$, and $K_2 \in C^{k_2}(U')$ depending holomorphically on $\omega \in \Omega'$, so that (8.1) and (8.2) hold for all $\omega \in \Omega'$. The core of $\mathcal{Q}$ has the form $\chi(w)Q_\omega(u,w)$ with χ the characteristic function of $\{u : |u| < r\}$, and Q_ω as in (7.75), (7.78), for appropriate r,s,R independent of $\omega \in \Omega'$, for $u \in U'$; further, $Q_\omega(u,w)$ depends holomorphically on $\omega \in \Omega'$ for $u \in U'$, $|w| < r$.

Theorem 8.1 is an immediate consequence of Theorem 7.11(d) and the following assertion:

Lemma 8.2. Suppose K_1 is as in Theorem 8.1. Say U' is open, $\bar{U}' \subset U$, $\bar{U}'$ compact. Then there exists $K_2 \sim \sum K_2^m \in C^{k_2}(U')$ with (7.98) and (7.99) being satisfied for $u \in U'$.

Further, suppose K_1 in addition depends holomorphically on a parameter $\omega \in \Omega \subset \mathbb{C}^M$. Then K_2 may be chosen to depend holomorphically on $\omega \in \Omega'$.

Indeed, let us use this to prove Theorem 8.1. To prove the analyticity of g, one need only use (7.100) with U replaced by a U' as in Lemma 8.2, which also contains W. To prove the statements concerning the existence of K_2,Q, we simply select U' as in Lemma 8.2 so that U' also contains P and satisfies $\text{dist}(u_0,(U')^c) > r_1$. Then we apply (7.103) with U' in place of U.

Thus we need only prove Lemma 8.2. K_2^o is, as we said, immediately produced through Theorems 5.1 and 5.9. The remaining K_2^m may then be produced algebraically from (7.79) but it is quite difficult to prove that they satisfy the needed estimates.

We first simplify our work by reducing to the case $k_1 = -2n - 2$, $K^o_{1u}(w) = \delta(w)$ for all u. Heuristically, we should be able to do this by selecting $J \in C^{k_2}(U)$ with $J \sim K^o_2$. Then, by (7.96), if $\mathcal{J} = \mathcal{O}(J)$, then the principal core of $\mathcal{J}\mathcal{K}_1$ is $\delta(w)$. If we can find $\mathcal{J}'$ with core in $C^{-2n-2}(U)$ to invert $\mathcal{J}\mathcal{K}_1$ on the left modulo an analytic regularizing operator, we could set $\mathcal{K}_2 = \mathcal{J}'\mathcal{J}$.

One must be careful in executing this plan, since Theorem 7.11(c) and (d) require that f have small support and that $\mathcal{K}f$ be looked at only near u_o. We shall explain exactly how to proceed in a moment. First, however, we introduce some useful notation.

If there is a $K \in C^k(U)$ with $K \sim \sum K^m$, let us say that $\{K^m\}_{m\in\mathbb{Z}^+}$ is in $FC^k(U)$, <u>formal</u> $C^k(U)$. It is suggestive to abuse notation and write $\sum K^m$ for $\{K^m\}$. We say K^o is the principal core of $\sum K^m$; all the other K^m we call "lower" cores. If $K^m_\nu \in FC^{k_\nu}(U)$ for $\nu = 1,2$, let us define $\sum K^m_2 \# \sum K^m_1 = \sum K^m \in FC^k(U)$ $(k=k_1+k_2+2n+2)$ by

$$K^m_u = \sum_{|\gamma|+a+b=m} [(-M)^\gamma K^b_{2u}] \underline{*} [\sigma_\gamma(\mathcal{U}) K^a_{1u}]/\gamma! \qquad (8.3)$$

as in (7.95).

We shall often use the following fact:

Suppose $\sum K^m$ is a linear combination of expressions of the form $\sum K_2^m \# \sum K_1^m$. Say that for some m, $K_u^m(w)$ is a polynomial in w for each u. Then $K^m \equiv 0$. (8.4)

To see this, fix any u. Note that $\mathcal{F}K_u^m$ is supported at 0 for all u; but by (8.3) and (7.19), $\mathcal{F}K_u^m = \Lambda_G$ for some G. Thus $G \equiv 0$, so $K_u^m \equiv 0$.

Clearly, under the conditions of Lemma 8.2, we are seeking to find $\sum K_2^m \in \mathcal{F}C^{k_2}(U')$ so that $\sum K_2^m \# \sum K_1^m = \delta$. It will be easy to reduce to the case $K_{1u}^0(w) = \delta(w)$, once we prove that # is associative.

Proposition 8.3. # is associative. That is, if $\sum K_\nu^m \in FC^{k_\nu}(U)$ for $\nu = 1,2,3$, then

$$[\sum K_3^m \# \sum K_2^m] \# \sum K_1^m = \sum K_3^m \# [\sum K_2^m \# \sum K_1^m]. \tag{8.5}$$

Proof. Fix $u_0 \in U$. It suffices to prove (8.5) in a neighborhood of u_0.

For $\nu=1,2,3$, select $K_\nu \in C^{k_\nu}(U;r_\nu,s_\nu,R_\nu)$ with $K_\nu \sim \sum K^m$. Without loss we may assume

$$r_1 < r_2 < r_3 \tag{8.6}$$

$$\max(r_1,r_2,r_3) < \min(s_1,s_2,s_3,\mathrm{dist}(u_0,U^c)). \tag{8.7}$$

Using $\mathcal{K}_2,\mathcal{K}_1$ we may produce, as in Theorem 7.11(c), a $\mathcal{K}$ (corresponding to their "product"), a U_0 and a U_1.

However, we write $\mathcal{K}_{21}$, U_0^{21} and U_1^{21} instead of $\mathcal{K}$, U_0,U_1. Say $K_{21} = \varkappa(\mathcal{K}_{21}) \in C^{k_{21}}(U;r_{21},s_{21},R_{21})$. Similarly we form $\mathcal{K}_{32}, K_{32} \in C^{k_{32}}(U;r_{32},s_{32},R_{32})$, and U_0^{32},U_1^{32}, using $\mathcal{K}_3,\mathcal{K}_2$. Without loss we may assume

$$\max(r_1,r_2,r_3,r_{21},r_{32}) < \min(s_1,s_2,s_3,s_{21},s_{32},\text{dist}(u_0,U^c)) \tag{8.8}$$

since to achieve this we need only multiply the various K's by the characteristic functions $\chi(w)$ of small balls, and shrink the U_0's and U_1's. We impose no further restriction on r_3; but, without loss, we reduce r_1 and r_2 in order that we may assume: u_0 has an open neighborhood V so that if

$P_1 = \{u : \text{dist}(u,\bar{V}) \leqq r_1\}$ and $P_2 = \{u : \text{dist}(u,P_1) \leqq r_2\}$ then

$$\text{diam } P_2 < r_3. \tag{8.9}$$

We impose no further restrictions on r_2,r_{32}, so $\mathcal{K}_2,\mathcal{K}_{32}$, U_0^{32},U_1^{32}, may rest without being further altered. Without loss, we reduce r_1 and shrink V in order that we may assume:

$$r_1 < r_{32} \tag{8.10}$$

$$P_1 \subset U_0^{32}. \tag{8.11}$$

We impose no further restrictions on r_1. Without loss, we reduce r_{21} in order that we may assume:

$$r_{21} < r_1. \tag{8.12}$$

Now (using (8.8), (8.10), (8.12)) form $K_{3(21)}$ corresponding to the "product" of K_3, K_{21}, and $K_{(32)1}$ corresponding to the "product" of K_{32}, K_1. We shall show:

For some open U_0, U_1, U with $u_0 \in U_0 \cap U_1$, if $f \in E'(U_0)$ then

$$K_{3(21)}f - K_{(32)1}f \text{ is analytic on } U_1, \text{ for all } f \in E'(U_0). \quad (8.13)$$

To see this, note that by (7.110), if $f \in E'(V)$ then $\text{supp } K_1 f \subset P_1$, $\text{supp } K_2 K_1 f \subset P_2$. For the same reasons, and by (8.12), if $f \in E'(V)$ then $\text{supp } K_{21} f \subset P_1 \subset P_2$. Also, by Theorem 7.11(c), if $f \in E'(U_0^{21})$ then

$$K_2 K_1 f = K_{21} f + g$$

where g is zero on U_1^{21}. Now say $\text{supp } f \subset V \cap U_0^{21}$. Then $\text{supp } g \subset P_2$, so by (8.9) and Theorem 7.11(b), $K_3 g$ is analytic on U_1^{21}. Since $K_{21}f, K_2 K_1 f \in E'(U)$, we may form $K_3 K_2 K_1$ and $K_3 K_{21} f$, and their difference is analytic in int P_2. Thus for V_0, V_1 sufficiently small neighborhoods of u_0,

$$K_3 K_2 K_1 f - K_{3(21)} f \text{ is analytic on } V_1 \text{ for all } f \in E'(V_0). \quad (8.14)$$

On the other hand, by (8.11), if $\text{supp } f \subset V$ then $K_3 K_2 K_1 f = K_{32} K_1 f$ on U_1^{32}. Then for W_0, W_1 sufficiently small neighborhoods of u_0,

$$K_3 K_2 K_1 f = K_{(32)1} f \text{ on } W_1 \text{ for all } f \in E'(W_0). \qquad (8.15)$$

(8.13) now follows at once from (8.14), (8.15).

Let $K = K_{3(21)} - K_{(32)1}$, $K = \varkappa(K)$. We now know $K : E'(U_0) \to C^\infty(U_1)$, and a review of our arguments (especially those proving Theorem 7.11(b)), shows that this map is continuous if $E'(U_0)$ is given the weak topology. The easy Theorem 5.2.6 of Volume I of [45] now tells us that the kernel of K msut be C^∞ on $U_0 \times U_1$. Thus $K(u,w)$ must be smooth in (u,w) for $u \in U_0$, w sufficiently small. Say $K \sim \sum K^m$. Then $\sum \partial_w^\alpha K^m(u,w)$ is smooth in w for small w, for all α. Since this series can never blow up as $w \to 0$, all $K^m(u,w)$ must be polynomials in w. If we now use (8.4), the proof is clearly complete.

It is easy to now reduce Lemma 8.2 to the case $K_{1u}^0 \equiv \delta(w)$. In this case, we claim something stronger than Lemma 8.2. Namely, we assert:

<u>Lemma 8.4</u>. Suppose $K_1 \sim \sum K_1^m \in C^{-2n-2}(U)$ and that the principal core of K_1 is $\delta(w)$. Then there exists $K_2 \sim \sum K_2^m \in C^{-2n-2}(U)$ so that (7.98) and (7.99) hold. Further, if K_1 in addition depends holomorphically on a parameter, K_2 may be chosen to depend holmorphically on this parameter.

Indeed, suppose Lemma 8.4 were known. For Lemma 8.2, we need to find $\sum K_2 \in FC^{k_2}(U')$ so that

$$\sum K_2^m \# \sum K_1^m = \delta . \tag{8.16}$$

For this, we simply use Theorems 5.1 and 5.9 to produce $K_{2u}^o(w) \in AK^{k_2}$ for all $u \in U'$, depending analytically on u, and satisfying

$$K_{2u}^o * K_{1u}^o = \delta(w) \tag{8.17}$$

for all $u \in U'$. Moreover, we may extend K_{2u}^o to $K_{2v\circ u}^o$ for $v \in D_s$, $u \in U'$; this depends holomorphically on v, and $K_{2v\circ u}^o * K_{1v\circ u}^o = \delta(w)$ for all $v \in D_s$, $u \in U'$. Moreover, for some $C_2, R_2 > 0$ we have for all $u \in U'$, $v \in D_s$:

$$|\partial_w^\rho K_2^o(v\circ u, w)| < C_2 R_2^{\|\rho\|} \rho! \quad \text{for } 1 \leqq |w| \leqq 2, \quad \rho \in (\mathbb{Z}^+)^{2n+1} \tag{8.18}$$

by (7.77) and Theorem 5.9.

Accordingly, $K_2^o \in FC^{k_2}(U')$. We then have $K_2^o \# \sum K_1^m \in FC^{-2n-2}(U')$ with principal core $\delta(w)$. Thus, by Lemma 8.4, there exists $K_3 \sim \sum K_3^m \in FC^{-2n-2}(U')$ with $\sum K_3^m \# (K_2^o \# \sum K_1^m) = \delta$. By Proposition 8.3, then, $(\sum K_3^m \# K_2^o) \# \sum K_1^m = \delta$ also. This establishes Lemma 8.2, given Lemma 8.4. The statement about holomorphic dependence on a parameter follows from the argument just given.

In proving Lemma 8.4, there is not much mystery about what the K_2^m must be. We must have

$$K_{2u}^{o} = \delta(w) \tag{8.19}$$

and we must then have, for $m \geq 1$,

$$K_{2u}^{m} = \sum_{\substack{|\gamma|+a+b=m \\ b\neq m}} [(-M)^{\gamma} K_{2u}^{b}] \underline{*} [\sigma_{\gamma}(U) K_{1u}^{a}]/\gamma! + p_{mu} \tag{8.20}$$

for certain homogeneous polynomials p_{mu}, of homogeneous degree $m - 2n - 2$. (We are using two facts here:

(i) if $K \in AK^{k}$, then $K\underline{*}\delta = 0$ iff K is a polynomial;

(ii) if $K_{\nu} \in AK^{k_{\nu}}$ $(\nu=1,2)$, then $K_2\underline{*}K_1\underline{*}\delta = K_2\underline{*}K_1$.

For (i), note that by (7.18), (7.19), (7.22), we have $K\underline{*}\delta = 0$ iff $T^{N}K = 0$ for some N; this is true iff FK is supported at 0, since FK is smooth away from 0. For (ii), let $K = K_2\underline{*}K_1$, $K' = K\underline{*}\delta$. By (7.18), (7.19), $FK' = FK$ away from 0; but then by (7.19), FK' and FK both equal Λ_G for some G. Hence $K' - K = 0$.)

(8.19), (8.20) recursively define K_2^m, up to a knowledge of the p_m. The difficulty, then, is in proving that $\sum K_2^m \in FC^{-2n-2}(U)$ for appropriate p_{mu}.

We have now reached a very important crossroads. For, in Chapter 7 and up to now in this chapter, all the results would have evident analogues if we were using Euclidean convolution in place of $\mathbb{H}^n$ convolution, with proofs the same or easier. But at this point, it is <u>crucial</u> that $\mathbb{H}^n$

convolution be used. If K_2^m is as in (8.19), (8.20), where now $\underline{*}$ and $\sigma_\gamma(U)$ are replaced by those analogous operations which arise from the Euclidean convolution structure, then in general, $\sum K_2^m$ is <u>not</u> an element of FC^{-2n-2}, regardless of what the p_{mu} are.

This is clear from the following simple example. Say $U = \mathbb{H}^n$, that $K_{1u}^1(w) = K_1^1(w) = -K(w)$ is independent of u, and that $K_{1u}^m(w) = 0$ for $m > 2$. Thus

$$\sum K_1^m = \delta - K.$$

From (8.20) we see

$$K_{2u}^m = K\underline{*}K\ldots\underline{*}K + p_{mu} \tag{8.21}$$

where $K\underline{*}\ldots\underline{*}K$ is an m-fold "convolution," and $\underline{*}$ arises from Euclidean convolution. (Of course, this is just the assertion that the formal inverse of $\delta - K$ is given by the Neumann series.) Fix any u_o, and put $K^m = K_{2u_o}^m$. If $\sum K^m$ were in $FC^{-2n-2}(\mathbb{H}^n)$, then (7.1) would certainly hold for some C_1, R_1. Define J_γ^m as after (7.2) for $|\gamma| = m - 2n - 1$ or $m - 2n$. The second inequality of (D_1) in (7.3), for $\lambda = 1$, shows that there exists a $B > 0$ as follows. For every m there exists $C_1(m) > 0$ so that the entire extension in ξ of $J_\gamma^m(1,\xi)$ to $\mathbb{C}^{2n}$ satisfies:

$$|J_\gamma^m(1,\zeta)| < C_1(m)e^{B|\zeta|^2}.$$

Accordingly, for some $C(m) > 0$, if $J^m = \mathcal{F}K^m$,

$$|J^m(1,\zeta)| < C(m)e^{B|\zeta|^2}. \tag{8.22}$$

But by (7.21),

$$J^m(1,\zeta) = (\mathcal{F}K^m)(1,\zeta) = [(\mathcal{F}K)(1,\zeta)]^m. \tag{8.23}$$

In general, $(\mathcal{F}K)(1,\zeta)$ (the entire extension of $(\mathcal{F}K)(1,\xi)$ to $\mathbb{C}^{2n}$) will grow like a Gaussian on $\mathbb{C}^{2n}$, say on the order of $e^{A|\zeta|^2}$. By (8.23), $J^m(1,\zeta)$ should grow like $e^{mA|\zeta|^2}$, which flagrantly contradicts (8.22).

Probably, then, the operator $f \to (\delta-K)*f$ (Euclidean convolution) is <u>not</u> in general analytic hypoelliptic, although we have made no effort to prove this.

Be that as it may, this simple example tells us two things. First, we are again faced with a subtle problem, since we are claiming that Lemma 8.4 holds for $\mathbb{H}^n$ convolution, although its analogue fails for Euclidean convolution. Second, since we lack any effective means to exploit subtle differences between $\mathbb{H}^n$ and Euclidean convolution by just working on the group side, we shall have to use the Fourier transforms.

Perhaps one could prove that the type of trouble manifested in (8.23) cannot arise for $\mathbb{H}^n$ convolution by using $\mathcal{F}$ to convert $\mathbb{H}^n$ convolution into, not a product as in (8.23), but an oscillatory integral as in (6.6). This would then require a detailed study of such oscillatory integrals.

However, we do not need to develop such techniques. We shall prove Lemma 8.4 by adapting the plan of chapters four and five. That is, we shall use the two Fourier transforms in turn.

Specifically, suppose that, with K_1^m, K_2^m as in (8.19), (8.20), we have $FK_{\nu u}^m(1,\xi) \sim \sum g_{\nu u\ell}^m$ in $Z^2_{2,-m}$. It is not difficult, using (8.13), to determine the $\{g_{2u\ell}^m\}$ in terms of the $\{g_{1u\ell}^m\}$. Estimating the $g_{2u\ell}^m$ through Theorem 7.1(a) and invoking Theorem 7.1(b) and (c), we shall be able to produce a "first guess" $\sum K_{II}^m \in FC^{-2n-2}(U)$ so that $FK_{IIu}^m(1,\xi) \sim \sum g_{2u\ell}^m$ also. For this step, Euclidean convolution would work as well as $\mathbb{H}^n$ convolution. Finally--and what has no analogue for Euclidean convolution—we shall use the group Fourier transform and Proposition 7.3 to control the "error" $K_2^m - K_{II}^m$.

Now that we have indicated what we plan to do, let's do it.

<u>Proof of Theorem 8.1</u>. As we have said above, we have only to prove Lemma 8.4. As usual, we treat only the case where K_1 does not depend on a parameter; the case of holomorphic dependence on a parameter will follow from our proof.

We define K_2^m by (8.19), (8.20) with $p_{mu} \equiv 0$. That is:

$$K_{2u}^0 = \delta(w) \tag{8.24}$$

$$K_{2u}^{m} = \sum_{\substack{|\gamma|+a+b=m \\ b\neq m}} [(-M)^{\gamma} K_{2u}^{b}] \underline{*} [\sigma_{\gamma}(U) K_{1u}^{a}]/\gamma! \quad \text{for } m \geq 1. \qquad (8.25)$$

The proof has two parts. In the first, we use the Euclidean Fourier transform; in the second, the group Fourier transform.

Part I. Say, in fact, that $K_1 \in C^{-2n-2}(U;r_1,s,R_1)$. We think of $u \in U$, $v \in D_{s/2}$ as being fixed until the end of Part I of the proof. Any constant called "C", with one or more subscripts, depends on u, but may be chosen to be the same for u ranging over any fixed compact subset of U. All other constants may be chosen independent of $u \in U$. All constants may be chosen independent of $v \in D_{s/2}$.

We adopt the notation defined in the paragraph before Lemma 5.7. We let $\sqrt{-1}$ be the square root of -1 which we have previously called "i," and instead use the letter "i" as an index. We also define $M_0, M_i : S'(\mathbb{H}^n) \rightarrow S'(\mathbb{H}^n)$ $(1 \leq i \leq 2n)$ as follows, for $1 \leq j \leq n$:

$$\begin{aligned}(M_0 f)(t,x,y) &= tf(t,x,y);\ (M_j f)(t,x,y) \\ &= x_j f(t,x,y);\ (M_{j+n} f)(t,x,y) \qquad (8.26)\\ &= y_j f(t,x,y)\end{aligned}$$

Say $K \in AK^k$, $j = -2n-2-k$, and $(FK)(1,\xi) \sim \sum g_\ell(\xi)$ in $Z_{2,j}^2$. Then, by the formula for g_ℓ in Theorem 2.11,

$$F(M_i K)(1,\xi) \sim \sum (-\sqrt{-1})(\partial_i g_\ell)(\xi) \text{ in } Z^2_{2,j-1} \tag{8.27}$$

if $1 \leq i \leq 2n$, where $\partial_i = \partial/\partial\xi_i$.

$$F(M_o^i K)(1,\xi) \sim \sum (-\sqrt{-1})^i [(\ell+i)!/\ell!] g_{\ell+i}(\xi) \text{ in } Z^2_{2,j-2i}, \tag{8.28}$$

if $i \in \mathbb{Z}^+$.

Returning to the situation at hand, let us say

$$F[\sigma_\gamma(U) K^m_{1v\circ u}/\gamma!](1,\xi) \sim \sum g^m_{1\gamma\ell}(\xi) \text{ in } Z^2_{2,-m} \tag{8.29}$$

and $$(FK^m_{2v\circ u})(1,\xi) \sim \sum g^m_{2,\ell}(\xi) \text{ in } Z^2_{2,-m}. \tag{8.30}$$

Also, let us put

$$g^m_{1\gamma\ell;\alpha} = (2\sqrt{-1})^{|\alpha|} \tilde{\partial}^\alpha g_{1\gamma\ell}/\alpha! \text{ for } \alpha \in (\mathbb{Z}^+)^{2n}. \tag{8.31}$$

(As usual, since $\alpha \in (\mathbb{Z}^+)^{2n}$, $|\alpha| = \sum_{i=1}^{2n} \alpha_i$. At times we have written $\|\alpha\|$ for this; both mean the same thing.)

In (8.25), write $\gamma = (i,\beta) \in \mathbb{Z}^+ \times (\mathbb{Z}^+)^{2n}$,

and write $$g^m_{1i\beta\ell;\alpha} = g^m_{1\gamma\ell;\alpha}. \tag{8.32}$$

From (7.24), we find

$$g^m_{2,\ell} = -\sum_{\substack{2i+|\beta|+a+b=m \\ b<m}} \sum_{c+d+|\alpha|=\ell} (\sqrt{-1})^{|\beta|+i} [(d+i)!/d!] g^a_{1i\beta c;\alpha} \partial^\alpha \partial^\beta g^b_{2,d+i}. \tag{8.33}$$

In the summation, $c,d \in \mathbb{Z}^+$. Observe that, since $K^o_{2u} \equiv \delta(w)$, there is only one nonzero term in the summation with $b = 0$. In that term, $\alpha = \beta = 0$, $d = i = 0$, $a = m$, $c = \ell$; the term is just $g^m_{100\ell;0}$. We now iterate (8.33) by replacing $g^b_{2,d+i}$ by its expression as a summation. We continue this process of replacing all $g^{b'}_{2,d'+i'}$ by summations until it is exhausted by virtue of only terms with b' = 0 being reached. The result is

$$g^m_{2,\ell} = \sum_{I\in \mathrm{I}} A_I g^{a_0}_{1i_1\beta^1 c_0;\alpha^1} \partial^{\alpha^1+\beta^1} g^{a_1}_{1i_2\beta^2 c_1;\alpha^2} \partial^{\alpha^2+\beta^2} \cdots \partial^{\alpha^N+\beta^N} g^{a_N}_{100c_N;0}. \tag{8.34}$$

Each $\partial^{\alpha^\nu+\beta^\nu}$ is applied to everything on its right. I here is shorthand for

$$(N, a_0, \ldots, a_N, c_0, \ldots, c_N, i_1, \ldots, i_N, \alpha^1, \ldots, \alpha^N, \beta^1, \ldots, \beta^N)$$

and I is the set of those I satisfying (i), (ii) and (iii) below, with

$$a = \sum_{\nu=0}^{N} a_\nu,\ c = \sum_{\nu=0}^{N} c_\nu,\ i = \sum_{\nu=1}^{N} i_\nu,\ \alpha = \sum_{\nu=1}^{N} \alpha^\nu,\ \beta = \sum_{\nu=1}^{N} \beta^\nu:$$

(i) $N \in \mathbb{Z}^+$, $N + 1 \leqq m$

(ii) $a_0, \ldots, a_N, i_1, \ldots, i_N \in \mathbb{Z}^+$, $\beta^1, \ldots, \beta^N \in (\mathbb{Z}^+)^{2n}$, and

$$a + 2i + |\beta| = m \tag{8.35}$$

and $a_N > 0$, $a_{\nu-1} + 2i_\nu + |\beta^\nu| > 0$ for $1 \leqq \nu \leqq N$;

(iii) $c_0, \ldots, c_N \in \mathbb{Z}^+$, $\alpha^1, \ldots, \alpha^N \in (\mathbb{Z}^+)^{2n}$, and

$$c + |\alpha| = \ell + i, \text{ and} \tag{8.36}$$

$$d_N = c_N - i_N \geqq 0,\ d_{N-1} = c_{N-1} + c_N + |\alpha^N| - (i_{N-1}+i_N) \geqq 0, \ldots$$
$$d_1 = c_1+\ldots+c_N + |\alpha^2| +\ldots+ |\alpha^N| - (i_1+\ldots+i_N) \geqq 0 \tag{8.37}$$

Further A_I is a constant; explicitly,

$$A_I = [(d_N+i_N)!/d_N!]\ldots[(d_1+i_1)!/d_1!] \tag{8.38}$$

We need to estimate the sum in (8.34). First some simple observations. If $m = 0$, $g^m_{2,\ell} = 1$ if $\ell = 0$, 0 otherwise. Let us estimate $\#I$ if $m > 0$. The number of N satisfying (i) is m. For fixed N satisfying (i), the number of $a_o,\ldots,a_N$, $i_1,\ldots,i_N$, $\beta^1,\ldots,\beta^N$ satisfying (8.35) is no more than the number of ways of writing m as a sum of $N + 1 + 2N + 2nN$ elements of $\mathbb{Z}^+$. This number is

$$\binom{m+n+1+2N+2nN-1}{m} \leqq 2^{m+3N+2nN} < 2^{m+3m+2nm} = 2^{(2n+4)m},$$

for any N. For fixed $i = i_1+\ldots+i_N < m$ as in (8.35), the number of $c_o,\ldots,c_N,\alpha^1,\ldots,\alpha^N$ satisfying (8.36) is the number of ways of writing $\ell + i$ as a sum of $N + 1 + 2nN$ elements of $\mathbb{Z}^+$. This number is

$$\binom{\ell+i+N+1+2nN-1}{\ell+i} \leqq 2^{\ell+i+N+2nN} < 2^{\ell+m+(2n+1)m} = 2^{\ell+(2n+3)m}$$

for any i. All together, then

$$\# I < m2^{\ell+(4n+7)m}, \text{ if } m > 0. \tag{8.39}$$

It is also easy to estimate $|A_I|$. By (8.37), (8.36) and the fact that $i = i_1+\ldots+i_N < m$, it follows at once that $d_\nu < \ell + m$ for $\nu = 1,\ldots,N$. Write each factor $(d_\nu+i_\nu)!/d_\nu!$ as a product $(d_\nu+1)\ldots(d_\nu+i_\nu)$. This leads to a total of i factors in (8.38). By inspection,

$$|A_I| < (\ell+m+1)\ldots(\ell+m+i) = (\ell+m+i)!/(\ell+m)! \leqq 2^{\ell+m+i}i!.$$

Thus

$$|A_I| < 2^{\ell+2m}i! \tag{8.40}$$

It remains to estimate the various g's. For some $C_2,R_2 > 0$ we have an estimate

$$|\partial_w^\rho \sigma_\gamma(U)K^m_{1\nu\circ u}(w)/\gamma!| < C_2R_2^{\|\rho\|+m+\|\gamma\|}\rho! \quad \text{for } 1 \leqq |w| \leqq 2,\ \rho \in (\mathbb{Z}^+)^{2n+1} \tag{8.41}$$

by (7.77) and (7.107). This implies, by Theorem 7.1(a), that there exist $C_3,R_3 > 0$ so that

$$|\partial^\beta g^m_{1\gamma\ell}(\xi)| < C_3R_3^{|\beta|+\|\gamma\|+\ell+m}|\beta|!\ell!m! \quad \text{for } |\xi| = 1,\ \beta \in (\mathbb{Z}^+)^{2n}.$$

Thus by (8.31), there are $C_4,R_4 > 0$ so that for any $\alpha \in (\mathbb{Z}^+)^{2n}, m,\gamma,\ell$, we have

$$|\partial^\beta g^m_{1\gamma\ell;\alpha}(\xi)| < C_4R_4^{|\alpha|+|\beta|+\|\gamma\|+\ell+m}|\beta|!\ell!m! \text{ for } |\xi| = 1,$$
$$\beta \in (\mathbb{Z}^+)^{2n}.$$

Thus there is a polydisc $D = \{\zeta \in \mathbb{C}^{2n} : |\zeta_i| < q_i\}$ so that all $g^m_{1\gamma\ell;\alpha}$ have holomorphic extensions to $\mathcal{D} = \{\zeta \in \mathbb{C}^{2n} :$ $\zeta - \xi \in D$ for some $\xi \in \mathbb{R}^{2n}$ with $|\xi| = 1\}$, and so that, for some $C_5, R_4 > 0$,

$$|g^m_{1\gamma\ell;\alpha}(\zeta)| < C_5 R_5^{|\alpha|+\|\gamma\|+\ell+m} \ell!m! \quad \text{for } \zeta \in D. \tag{8.42}$$

Select $0 < q_i' < q_i$ $(1 \le i \le 2n)$, put $D' = \{\zeta : |\zeta_i| < q_i'\}$, $\mathcal{D}' = \{\zeta : \zeta - \xi \in D'$ for some ξ with $|\xi| = 1\}$. We may of course extend $g^m_{2,\ell}$ holomorphically to $\mathcal{D}$, by (8.34). By Lemma 2.5 and (8.34), there are $C_6, C_7, C_8, C_9, R_6, R_7, R_8, R_9 > 0$ so that if $\zeta \in \mathcal{D}'$,

$$\begin{aligned} |g^m_{2,\ell}(\zeta)| &< \sum_{I \in \mathrm{I}} |A_I| C_6 R_6^{|\alpha|+|\beta|} (|\alpha|+|\beta|)! C_5^m R_5^{|\alpha|+|\beta|+i+c+a} c_0! \ldots c_N! a_0! \ldots a_N! \\ &< C_7 R_7^{\ell+m} \sum_{I \in \mathrm{I}} i!|\alpha|!|\beta|!c!a! < C_7 R_7^{\ell+m} \sum_{I \in \mathrm{I}} (\ell+i)!i!a!|\beta|! \\ &< C_8 R_8^{\ell+m} \sum_{I \in \mathrm{I}} \ell!m! < C_9 R_9^{\ell+m} \ell!m!. \end{aligned}$$

For the first inequality, we used (8.42); for the second, (8.35), (8.36) and (8.40); for the third, (8.36); and for the last, (8.39).

Thus there are $C_{10}, R_{10} > 0$ so that for $|\xi| = 1$,

$$|\partial^\alpha g^m_{2,\ell}(\xi)| < C_{10} R_{10}^{|\alpha|+\ell+m} |\alpha|!\ell!m!. \tag{8.43}$$

By Theorem 7.1(b) and (c) we may find $K^m_{\mathrm{II}v\circ u} \in AK^{-2n-2+m}$ so that for some $C_{11}, R_{11} > 0$,

$$|\partial^\gamma_w K^m_{\mathrm{II}v\circ u}(w)| < C_{11} R_{11}^{m+\|\gamma\|} \gamma! \text{ for all } m \in \mathbb{Z}^+, \gamma \in (\mathbb{Z}^+)^{2n+1} \tag{8.44}$$

and so that

$$\mathcal{F}K^m_{IIv\circ u}(1,\xi) \sim \sum g^m_{2,\ell}(\xi) \text{ in } Z^2_{2,-m}, \tag{8.45}$$

and so that $\mathcal{F}K^m_{IIv\circ u}$ is of the form Λ_G, for any m,v,u.

For m = 0, we simply choose

$$K^0_{IIv\circ u} \equiv \delta(w). \tag{8.46}$$

We now allow u,v to vary over the set $U \times D_{s/2}$. Then $v\circ u \in P = \{v\circ u : v \in D_{s/2}, u \in U\}$. By (2.23), the $g^m_{1\gamma\ell;\alpha}$ depend holomorphically on $v\circ u \in P$, so that the $g^m_{2,\ell}$ do likewise by (8.34). In (8.43), C_{10} and R_{10} are independent of v, C_{10} is uniform in u on compact subsets of U, and R_{10} is independent of $u \in U$. By the second and third uniformity assertions of Theorem 7.1 (to deal with K^m_{II}, $m \geqq 2n + 2$) and by Corollary 2.15 (to deal with K^m_{II}, $m < 2n + 2$), we may choose the $K^m_{IIv\circ u}$ in (8.44) to depend holomorphically on $v\circ u \in P$, and so that C_{11}, R_{11} are independent of v, C_{11} is uniform in u on compact subsets of U, and R_{11} is independent of $u \in U$. Thus $\sum K^m_{II} \in \mathcal{F}C^{-2n-2}(U)$.

For $m \in \mathbb{Z}^+$, $u \in U$, $v \in D_{s/2}$, we put

$$K^m_{Ev\circ u} = K^m_{2v\circ u} - K^m_{IIv\circ u}. \tag{8.47}$$

E stands for "error." In the next part of the proof, we will show that $\sum K^m_E \in \mathcal{F}C^{-2n-2}(U)$. Since we know

$\sum K^m_{II} \in FC^{-2n-2}(U)$, this will serve to complete the proof.

By (8.47), (8.30) and (8.45),

$$\mathcal{F}K^m_{Ev\circ u}(1,\xi) \in Z^2_2(\mathbb{R}^{2n}) \text{ for all m.} \tag{8.48}$$

Also, by (8.47), (8.25) and (7.19), $\mathcal{F}K^m_{Ev\circ u}$ is also of the form Λ_G, for any m,v,u.

Note also that for any $\gamma \in (\mathbb{Z}^+)^{2n+1}$, $\mathcal{F}[(-M)^\gamma K^m_{2v\circ u}](1,\xi)$ has the same asymptotics in $Z^2_{2,-m-|\gamma|}$ as $\mathcal{F}[(-M)^\gamma K^m_{IIv\circ u}](1,\xi)$ does, by (8.30), (8.45), (8.27) and (8.28). Thus, from (7.24), if $k \in \mathbb{C}$, $K \in AK^k$, $\mathcal{F}[(-M)^\gamma K^m_{2v\circ u} \underline{*} K](1,\xi)$ has the same asymptotics in $Z^2_{2,-m-|\gamma|-k-2n-2}$ as $\mathcal{F}[(-M)^\gamma K^m_{IIv\circ u} \underline{*} K](1,\xi)$ does. By (8.25) it now follows that if

$$\sum K^m_{II} \# \sum K^m_1 = -\sum K^m_3 \tag{8.49}$$

so that

$$-\sum K^m_3 \in FC^{-2n-2} \tag{8.50}$$

then

$$\mathcal{F}K^m_{3v\circ u}(1,\xi) \in Z^2_2(\mathbb{R}^{2n}) \text{ for all m,v,u.} \tag{8.51}$$

Further, if we substract the equations implicit in (8.49) from (8.25), and use (8.24), (8.46), we find

$$K^0_E \equiv 0 \tag{8.52}$$

$$K^m_{Ev\circ u} = \sum_{\substack{|\gamma|+a+b=m \\ b<m}} [(-M)^\gamma K^b_{Ev\circ u}] \underline{*} [\sigma_\gamma(U) K^a_{1v\circ u}]/\gamma! + K^m_{3v\circ u}. \tag{8.53}$$

Part II. From (8.50), (8.51), (8.52) and (8.53), and Proposition 7.3, we shall obtain the needed estimates on the K_E^m. Part I could have been carried out equally well, and even more easily, if Euclidean convolution had been used instead of $\mathbb{H}^n$ convolution. But Part II will have no analogue for Euclidean convolution. We shall use the group Fourier transform and Proposition 7.3. Part II will be formally similar in many ways to Part I, though trickier. A major difference is that, because of Proposition 7.3, victory will be measured in terms of square roots of factorials instead of factorials themselves.

As in Part I, we think of $u \in U$, $v \in D_{s/4}$ as being fixed until the end of the proof. We start afresh in naming our constants; the constants are not necessarily the same as those in Part I. Again, any constant called "C," with one or more subscripts, depends on u, but may be chosen to be the same for u ranging over any fixed compact subset of U. All other constants may be chosen independent of $u \in U$. All constants may be chosen independent of $v \in D_{s/4}$.

To complete the proof, it will be enough to show that for some $B_1, B_2, C_2', R_2' > 0$,

$$\mathcal{F}K^m_{Ev\circ u}(1,\xi),\ \mathcal{F}K^m_{Ev\circ u}(-1,\xi) \in Z_2^2(B_1,B_2,C_2'(R_2')^m m!^{1/2}) \tag{8.54}$$

for all m.

Indeed, if this were known, we could then use Proposition 7.3(b), Theorem 7.1(c) and the fact that $FK^m_{Ev\circ u}$ is of the Λ_G for any m,v,u, to conclude: For some $C_2'', R_2'' > 0$,

$$|\partial^\gamma_w K^m_{Ev\circ u}(w)| < C_2''(R_2'')^{m+\|\gamma\|}\gamma! \text{ for } 1 \leq |w| \leq 2,\ \gamma \in (\mathbb{Z}^+)^{2n+1} \tag{8.55}$$

with the above-stated conventions on the dependence of the constants on v,u. For $u \in U$, $v \in D_{s/4}$, $K^m_{Ev\circ u}$ depends holomorphically on v, by (8.47). Thus $\sum K^m_E \in FC^{-2n-2}(U)$, as desired.

Thus we need only prove (8.54). As we said, we shall need to use the group Fourier transform for this. With M_o, M_i $(1 \leq i \leq 2n)$ as in (8.26), we need now to understand what happens to the M_i when the group Fourier transform is applied. We will state the result in equation (8.61) below.

In the notation of Chapters 4 and 5, let us put

$$\underset{\sim}{P}_i = \sqrt{-1}(\underset{\sim}{W}_i^+ - \underset{\sim}{W}_i),\quad \underset{\sim}{Q}_i = \underset{\sim}{P}_{i+n} = -(\underset{\sim}{W}_i + \underset{\sim}{W}_i^+) \quad (1 \leq i \leq n) \tag{8.56}$$

so that if $f \in S(\mathbb{H}^n)$,

$$(X_i f)^\wedge = \hat{f}(-\sqrt{-1}\underset{\sim}{P}_i),\quad (Y_i f)^\wedge = \hat{f}(-\sqrt{-1}\underset{\sim}{P}_{i+n}). \tag{8.57}$$

Further, if $\underset{\approx}{R}$ is an operator family with $R(\lambda) : H^\infty \to H^\infty$ for all λ, define new operator families $\underline{d}_i\underset{\sim}{R}$ $(1 \leq i \leq 2n)$ as follows: $(\underline{d}_i\underset{\sim}{R}) = (d_{i\lambda}(R(\lambda)))$ where

$$d_{i\lambda}(R(\lambda)) = (4\sqrt{-1}\lambda)^{-1}[R(\lambda),Q_{i\lambda}],\ d_{(i+n)\lambda}(R(\lambda)) = -(4\sqrt{-1}\lambda)^{-1}[R(\lambda),P_{i\lambda}] \quad (1 \leqq i \leqq n) \tag{8.58}$$

as operators on $\mathcal{H}^\infty$.

We also define a map $\underline{d}_o$, on operator families $\underset{\sim}{J}$ which are homogeneous of degree j for some j (see the definition before Proposition 4.7), and which satisfy $J(\lambda) : \mathcal{H}^\infty \to \mathcal{H}^\infty$ for all λ. Our definition of $\underline{d}_o\underset{\sim}{J}$ is:

$$\underline{d}_o\underset{\sim}{J} = ((j/2\lambda)J(\lambda) - d'_{o\lambda}(J(\lambda))) \tag{8.59}$$

where

$$d'_{o\lambda}(J(\lambda)) = (4\lambda)^{-1}\sum_{i=1}^{2n}(P_{i\lambda}d_{i\lambda}(J(\lambda)) + d_{i\lambda}(J(\lambda))P_{i\lambda}) \tag{8.60}$$

as operators on $\mathcal{H}^\infty$. $\underline{d}_o\underset{\sim}{J}$ is homogeneous of degree $j - 2$.

What we claim is this. Say $K \in \mathcal{AK}^k$; then

$$(M_iK)^\wedge = -\sqrt{-1}\,\underline{d}_i\hat{K} \quad \text{for } 0 \leqq i \leqq 2n \tag{8.61}$$

in the $\mathcal{Q}$ sense. This is to be thought of as very analogous to the Euclidean Fourier transform taking multiplication to $-\sqrt{-1}$ times differentiation. Note that, from (8.58), the $d_{i\lambda}$ are derivations if $1 \leqq i \leqq 2n$.

To prove (8.61) when $1 \leqq i \leqq n$, note that $4TM_i = Y_i^R - Y_i$. This shows $-4\sqrt{-1}\lambda(M_iK)^\wedge(\lambda) = -\sqrt{-1}(Q_{i\lambda}\hat{K}(\lambda) - \hat{K}(\lambda)Q_{i\lambda})$, yielding (8.61) when $1 \leqq i \leqq n$. The case $n + 1 \leqq i \leqq 2n$ is analogous. For the case $i = 0$, we write $K = K' + p(x)\log|x|$ with K',p

homogeneous of degree k, p a polynomial and recall from (1.3) that, for $r > 0$,

$$(D^r K - r^k K - r^k (\log r) p \mid f) = 0$$

for all $f \in S$. We look at the case $\hat{f} \in \mathcal{Q}$. By the first line of the proof of Proposition 4.8(b), we have $(D^r K - r^k K \mid f) = 0$. Now we may obtain an analogue of Euler's identity by differentiating with respect to r and setting $r = 1$:

$$((2t\partial/\partial t + \sum x_i \partial/\partial x_i + \sum y_i \partial/\partial y_i - k)K \mid f) = 0 \quad \text{if } \hat{f} \in \mathcal{Q}.$$

Thus $(2\partial_t M_o K + \sum \partial_{x_i} M_i K + \sum \partial_{y_i} M_{i+n} K + jK \mid f) = 0$ if $\hat{f} \in \mathcal{Q}$ where $\partial_t = \partial/\partial t$, etc., $j = -2n - 2 - k$, and all sums are from $i = 1$ to n. In the relation, we write $\partial_{x_i} = (X_i + X_i^R)/2$, $\partial_{y_i} = (Y_i + Y_i^R)/2$. We invoke (8.57), (8.61) for $1 \leqq i \leqq 2n$, and $(\partial_t M_o K)\hat{}(\lambda) = -\sqrt{-1}\lambda (M_o K)\hat{}(\lambda)$. The result is (8.61) for $i = 0$. This proves (8.61). By the way, it is clear from (8.61) that the $\underline{d}_i$ commute when applied to operator families of the form $\hat{K}$, $K \in \mathcal{A}\mathcal{K}^k$ for some k.

Returning now to (8.52), (8.53), it is convenient for us to first take $\tilde{\ }$. We obtain

$$\tilde{K}_E^o = 0 \tag{8.62}$$

$$\tilde{K}^m_{Ev\circ u} = -\sum_{\substack{|\gamma|+a+b=m \\ b<m}} (\widetilde{[\sigma_\gamma(U) K^a_{1v\circ u}]/\gamma!} \,\underline{*}\, M^\gamma \tilde{K}^b_{Ev\circ u} + \tilde{K}^m_{3v\circ u}. \tag{8.63}$$

(We have used the fact that $\widetilde{K'\underline{*}K''} = \tilde{K}''\underline{*}\tilde{K}'$ for K',K" in the AK^k spaces; this is a simple consequence of (7.18), (7.19).) Now put

$$\underset{\sim}{J}_E^m = (\tilde{K}^m_{Ev\circ u})^\wedge, \quad \underset{\sim}{J}_3^m = (\tilde{K}^m_{3v\circ u})^\wedge, \quad \underset{\sim}{J}^m_{1\gamma} = (\widetilde{[\sigma_\gamma(U)K^m_{1v\circ u}]/\gamma!})^\wedge$$

for $m \in \mathbb{Z}^+$, $\gamma \in (\mathbb{Z}^+)^{2n+1}$. From (8.62), (8.63) we have

$$\underset{\sim}{J}_E^o = 0,$$

$$\underset{\sim}{J}_E^m = \underset{\sim}{J}_3^m - \sum_{\substack{|\gamma|+a+b=m\\ b<m}} (-\sqrt{-1})^{\|\gamma\|} \underset{\sim}{J}^a_{1\gamma}\underline{d}^\gamma \underset{\sim}{J}_E^b \tag{8.64}$$

where $\underline{d} = (\underline{d}_o,\ldots,\underline{d}_{2n})$. In the summation, of course, there is no non-zero term with b = 0. We now iterate (8.64) by replacing $\underset{\sim}{J}_E^b$ by its expression as $\underset{\sim}{J}_3^b$ minus a summation. We continue the process of replacing all $\underset{\sim}{J}_E^{b'}$ by $\underset{\sim}{J}_3^{b'}$ minus summatinos until it is exhausted by virtue of only terms with b' = 0 being reached. The result is

$$\underset{\sim}{J}_E^m = \sum_{I\in \mathrm{I}} A_I \underset{\sim}{J}^{a_o}_{1\gamma^1}\underline{d}^{\gamma^1} \underset{\sim}{J}^{a_1}_{1\gamma^2}\underline{d}^{\gamma^2} \cdots \underset{\sim}{J}^{a_{N-1}}_{1\gamma^N}\underline{d}^{\gamma^N} \underset{\sim}{J}_3^{a_N}. \tag{8.65}$$

Each $\underline{d}^{\gamma^\nu}$ is applied to everything to its right. *I* here is shorthand for $(N,a_o,\ldots,a_N,\gamma^1,\ldots,\gamma^N)$, and I is the set of those *I* satisfying:

$$a_o,\ldots,a_N \in \mathbb{Z}^+, \quad \gamma^1,\ldots,\gamma^N \in (\mathbb{Z}^+)^{2n+1},$$
$$a_o+\ldots+a_N + |\gamma^1|+\ldots+|\gamma^N| = m \tag{8.66}$$

and $a_N > 0$, $a_{\nu-1} + |\gamma^\nu| > 0$ for $1 \leqq \nu \leqq N$.

Further, A_I is a constant; in fact

$$A_I = 1 \quad \text{for all } I. \tag{8.67}$$

By reasoning entirely analogous to that in Part I, we see that

$$\# I \leqq m2^{(2n+4)m}. \tag{8.68}$$

We have to estimate each term of (8.65).

In (8.65), we could write each $\gamma^\nu = (i^\nu,\beta^\nu) \in \mathbb{Z}^+ \times (\mathbb{Z}^+)^{2n}$, $\underline{d}^{\gamma^\nu} = \underline{d}_0^{i^\nu} \underline{d}_1^{\beta_1^\nu} \cdots \underline{d}_{2n}^{\beta_{2n}^\nu}$. Each $\underline{d}_0$ in this product is to be applied, in (8.65), to an operator family which is homogeneous of degree $-\sum_{\ell=\nu+1}^{N} (a_\ell+|\gamma^\ell|) - a_\nu - q$ where $|\beta^\nu| \leqq q < |\gamma^\nu|$. All we need to know about this degree is that it is between 0 and m, which is clear from (8.66).

We need compact notation. If $\underset{\sim}{J}$ is an operator family such that $J(\lambda) : H^\infty \to H^\infty$ for all λ, we define $\underline{d}_0'\underset{\sim}{J}$ by

$$\underline{d}_0'\underset{\sim}{J} = (d_{0\lambda}'(J(\lambda))),$$

$d_{0\lambda}'$ as in (8.60). Further, it will be notationally efficient to write

$$d_{i\lambda}' = d_{i\lambda} \quad \text{for } 1 \leqq i \leqq 2n,$$

for then we can manipulate $d_\lambda' = (d_{0\lambda}', d_{1\lambda}', \ldots, d_{2n\lambda}')$. We shall use our usual notation that $J_+ = J(1)$, $J_- = J(-1)$, for

$\underset{\sim}{J} \in (AK^k)^\wedge$ for any k. We want to use (8.65) to examine $J^m_{E\pm}$. We put $d'_{i+} = d'_{i1}$, $d'_+ = d'_1$, the "1" subscript meaning $\lambda = 1$; similarly with $-$. We write (8.65) in the form

$$\underset{\sim}{J}^m_E = \sum_{I\in I} \underset{\sim}{J}^m_{EI} \tag{8.69}$$

where $\underset{\sim}{J}^m_{EI} = A_I \underset{\sim}{J}^{a_o}_{1\gamma^1} \underline{d}^{\gamma^1} \underset{\sim}{J}^{a_1}_{1\gamma^2} \underline{d}^{\gamma^2} \cdots \underset{\sim}{J}^{a_{N-1}}_{1\gamma^N} \underline{d}^{\gamma^N} \underset{\sim}{J}^{a_N}_{3}$.

We now fix I and examine J^m_{EI+}. We use the observations of the previous paragraph, together with (8.59). If $\gamma^\nu = (i^\nu, \beta^\nu)$, then $i^1+\ldots+i^N \leq m/2$. So we see:

(*) J^m_{EI+} is the sum of at most $2^{m/2}$ terms of the form

$$A'_L J^{a_o}_{1\gamma^1+} (d'_+)^{\rho^1} J^{a_1}_{1\gamma^2+} (d'_+)^{\rho^2} \cdots J^{a_{N-1}}_{1\gamma^N+} (d')^{\rho^N} J^{a_N}_{3+}. \tag{8.70}$$

Here L is shorthand for $(\rho^1,\ldots,\rho^N)$. The ρ's satisfy:

$$\rho^1,\ldots,\rho^N \in (\mathbb{Z}^+)^{2n+1}, \text{ and}$$

$$b_L = m - (a_o+\ldots+a_N + |\rho^1|+\ldots+|\rho^N|) \geq 0. \tag{8.71}$$

Also A'_L is a constant, satisfying

$$|A'_L| \leq (m/2)^{b_L/2} \tag{8.72}$$

for all L, by the observaiton of the previous paragraph.

As it turns out, there is no advantage to be gained at this point by not expanding out all the commutators implicit in (8.70). Perhaps this is a little surprising, but it has to do with $J_{3+}^{a_N}$ being in $Z_2^2(H_1)$. In any event, we fix L and we do use (8.58) and (8.60) to expand out all the commutators implicit in (8.70), simply obtaining from any expression of the form $[R,P_i]$ the two separate terms RP_i and $-P_iR$. Viewed in this way, (8.58) is a sum of two terms, while (8.60) is a sum of 8n terms. Let P = {products of the form $p = P_{i_{1+}} \dots P_{i_{\ell+}}$ where $\ell \in \mathbb{Z}^+$.} If $p \in P$ is of this form, we say $\#p = \ell$. The element $p \in P$ with $\#p = 0$ is defined to be I, the identity operator.

Expanding out the commutators, and using the fact that $\|\rho^1\|+\dots+\|\rho^N\| \leqq m$, we see that:

(**) (8.70) is the sum of at most $(8n)^m$ terms of the form

$$A\, p_0 J_{1\gamma^1_+}^{a_0} p_1 J_{1\gamma^2_+}^{a_1} p_2 \dots J_{1\gamma^N_+}^{a_{N-1}} p_N J_{3+}^{a_N} p_{N+1} \tag{8.73}$$

where $p_0,\dots,p_{N+1} \in P$, and

$\#p_0+\dots+\#p_{N-1} = |\rho^1|+\dots+|\rho^N|$. Thus, setting $b = b_L$, we see

$$b = m - (a_0+\dots+a_N+\#p_0+\dots+\#p_{N+1}) \geqq 0. \tag{8.75}$$

Also A is a constant, satisfying

$$|A| \leqq m^{b/2}. \tag{8.76}$$

(We have thrown away the 4's arising from (8.58) and (8.60)

and the 2's from (8.72).)

Using (8.69), (8.68), (*) and (**), we may summarize all the information we need as follows: J^m_{E+} is the sum of at most $m[2^{(2n+5)}\cdot 8n]^m$ terms of the form (8.73), where $a_o,\ldots,a_N \in \mathbb{Z}^+$, $\gamma^1,\ldots,\gamma^N \in (\mathbb{Z}^+)^{2n+1}$,

$$|\gamma^1|+\ldots+|\gamma^N| \leqq m \tag{8.77}$$

and (8.74), (8.75), (8.76) hold. Similarly for J^m_{E-}.

So far, in Part II of the proof, we have just been doing algebra and crude counting arguments. We must now obtain estimates for expressions of the form (8.73). For this, we shall ultimately establish an analogue of Lemma 2.5.

We are trying to establish that for some $B_1, B_2, C_2', R_2' > 0$, $G^m_+(\xi) = \mathcal{F}K^m_{Ev\circ u}(1,\xi), G^m_-(\xi) = \mathcal{F}K^m_{Ev\circ u}(-1,\xi) \in Z^2_2(B_1,B_2,C_2'(R_2')^m m!^{1/2})$ for all m. By reasoning similar to that of the proof of Proposition 5.5(d),

$$J^m_{E+} = \mathcal{W}_1 G^m_+,\ J_- = \mathcal{W}_{-1} G^m_-,$$

if $\mathcal{F}'_{\mathbb{C}^n}$ is identified with $\mathcal{F}_{\mathbb{R}^{2n}}$. By the uniformities in Theorem 2.3(a) $\Leftrightarrow$(d), and by the Remark following the proof of Proposition 4.2, we see that it is equivalent to show that for some $C_4, R_4 > 0$ we have

$$\|p'J^m_{E\pm}p''\|_2 < C_4R_4^{M+m}(M+m)!^{1/2} \tag{8.78}$$

for all $m,M \in \mathbb{Z}^+$ and all $p',p'' \in P$ with $\#p' + \#p'' = M$. Here $\| \ \|_2$ denotes Hilbert-Schmidt norm. We now see that it will be enough to show that:

(***) For some $C_5,R_5 > 0$

$$m^{b/2}\|p_oJ^{a_o}_{1\gamma^1+}p_1J^{a_1}_{1\gamma^2+}p_2\cdots J^{a_{N-1}}_{1\gamma^N+}p_NJ^{a_N}_{3+}p_{N+1}\|_2 < C_5R_5^m m!^{1/2} \tag{8.79}$$

for all $m \in \mathbb{Z}^+$, $N \in \mathbb{Z}^+$, $p_0,\ldots,p_{N+1} \in P$, $a_0,\ldots,a_N \in \mathbb{Z}^+$, $\gamma^1,\ldots,\gamma^N \in (\mathbb{Z}^+)^{2n+1}$ so that

$$b = m - (a_0+\ldots+a_N+\#p_0+\ldots+\#p_{N+1}) \geq 0,$$

and $$|\gamma^1|+\ldots+|\gamma^N| \leq m.$$

Indeed, this would lead directly to (8.78) in case $M = 0$. The general case follows from replacing m by $m + M$ in (***) and also replacing p_0 by $p'p_0$, p_{N+1} by $p_{N+1}p''$.

As we said, we have to find an analogue of Lemma 2.5, and we have yet to indicate what properties of the $J^m_{1\gamma+}$, J^m_{3+} that are analogous to analyticity and that we plan to exploit. The key to our estimate is the operator D which has already been so useful in Proposition 4.4 and Lemma 4.5. Recall $\mathcal{R}_0 = \mathcal{L}_0 + iTD$. Thus if $K \in AK^k$ for some k, $\hat{K} = \underset{\sim}{J}$, then $(DK)^{\wedge} = \underset{\sim\sim}{DJ} = (D_\lambda(J(\lambda))$, where if $S : H^\infty_\lambda \to H^\infty_\lambda$ we define

$$D_\lambda S = \lambda^{-1}[A_\lambda, S]. \tag{8.80}$$

(When we fix λ and leave off the λ subscript in this last identity, as is our custom, we shall have another operator called "D" which acts on operators on H_λ; this can hardly be confused with the other "D" which acts on functions on $\mathbb{H}^n$.) Now J^m_{3+} satisfies an estimate akin to (8.78), since $\sum K^m_3 \in FC^{-2n-2}$ by (8.50). Since $A = (\sum_{i=1}^{2n} P_i^2)/4$, there exist $C_6, R_6 > 0$ so that for any $L, M, N \in \mathbb{Z}^+$, if $p', p'' \in P$ and $\#p' + \#p'' = L$, then

$$\| A^M p' J^m_{3+} p'' A^N \|_2 < [C_6 R_6^{L+M+N+m} (L+M)!^{1/2}](M+N)!. \tag{8.81}$$

As a simple consequence of this and (8.80), if $L, M \in \mathbb{Z}^+$, $p', p'' \in P$ and $\#p' + \#p'' = L$, then

$$\| D^M (p' J^m_{3+} p'') \|_2 < [C_6 2^M R_6^{L+M+m} (L+m)!^{1/2}] M!. \tag{8.82}$$

This is the analogue of analyticity for J^m_{3+} which we plan to exploit. As for the J^m_{1+}, we intend to use Proposition 7.4 in the following form, which is an immediate consequence of that proposition:

Suppose $R > 0$; then there exist $C', R' > 0$ as follows. Suppose $K^m \in AK^{m-2-2}$ for all $m \in \mathbb{Z}^+$, and

$$|\partial^\gamma K^m(w)| < R^{M+\|\gamma\|}\gamma! \text{ for } 1 \leqq |w| \leqq 2,\ m \in \mathbb{Z}^+,\ \gamma \in (\mathbb{Z}^+)^{2n+1} \tag{8.83}$$

Suppose $\hat{K}^m = \underset{\sim}{J}^m$; then for all $M \in \mathbb{Z}^+$,

$$\|D^M J^m_{\pm}\| < C'(R')^{m+M} m!^{1/2} M! \quad \text{if } m \geqq 1.$$

This follows at once from Proposition 7.4, since for some $C'', R'' > 0$ we have an estimate

$$|\partial^\gamma D^M K^m(w)| < C''(R'')^{m+M+\|\gamma\|} M!\gamma!$$

$$\text{for } 1 \leqq |w| \leqq 2,\ m, M \in \mathbb{Z}^+,\ \gamma \in (\mathbb{Z}^+)^{2n+1},$$

as an elementary application of Lemma 2.5 shows.

(8.83) might be applied, for instance, in the case $K^m = K^m_{1v\circ u}$. But the estimate (***) does not follow so directly from (8.82) and (8.83); we shall need to proceed by a rather circuitous route. In fact, (***) will follow from the following key lemma, which is stated with fresh notation.

<u>Lemma 8.5</u>. Suppose $R > 0$; then there exist $C', R' > 0$ as follows. Suppose $L \in \mathbb{Z}^+$, and $K_\nu \in AK^{a_\nu - 2n-2}$ for $1 \leqq \nu \leqq L$, where $a_\nu \in \mathbb{Z}^+$. Suppose also that:

$$\text{For any } \nu, \text{ if } a_\nu = 0, \text{ then } K_\nu = \delta. \tag{8.84}$$

Suppose that

$$|\partial^{\nu} K_{\nu}(w)| < R^{m_{\nu}+\|\gamma\|}\gamma! \text{ for } 1 \leqq |w| \leqq 2,\ \gamma \in (\mathbb{Z}^{+})^{2n+1},\quad 1 \leqq \nu \leqq L. \tag{8.85}$$

Let $\hat{K}_{\nu} = \underset{\sim}{J}_{\nu}$, $a = \sum_{\nu=1}^{L} a_{\nu}$.

Also suppose that $S \in Z_2^2(H_1)$, and that

$$\|A^M S A^N\|_2 < R^{M+N}(M+N)! \quad \text{for } M,N \in \mathbb{Z}^+. \tag{8.86}$$

Let $T_1 = \{P_{1+},\ldots,P_{2n+}\}$ (the usual operators on H_1.) and let $T = T_1 \cup \{I\}$, I being the identity operator on H_1. Say $Q_1,\ldots,Q_L \in T$, and that $\#\{Q_{\nu} : Q_{\nu} \in T_1\} = L_1$. Then

$$\|Q_1 J_{1+} Q_2 J_{2+} \cdots Q_L J_{L+} S\|_2 < C_1 R_1^{a+L}(a+L_1)!^{1/2}. \tag{8.87}$$

Similarly with $J_{\nu-}$, H_{-1} in place of $J_{\nu+}$, H_1.

We shall prove the lemma in a moment. First let us see how (***) follows from it. Firstly, note that

$$J^{0}_{1\gamma+} = \delta \text{ if } \gamma = 0,\ 0 \text{ otherwise},$$

since $K^{0}_{1v'\circ u'}(w) = \delta(w)$ for all $u' \in U'$, $v' \in D_{s/4}$. Next note that in (***) there is no loss in assuming

$$\#p_{\nu} = 0 \text{ or } 1 \text{ for } 0 \leqq \nu \leqq N-1$$

for we can easily reduce to this case as follows: If $p_{\nu} = P_{i_{1}+}\cdots P_{i_{\ell}+}$, say, with $\ell > 1$, we only need rewrite

$p_\nu = P_{i_1+}J^o_{10+}P_{i_2+}J^o_{10+}\cdots P_{i_\ell+}$. Also observe that we may assume $\#p_\nu + a_\nu \geqq 1$ for $0 \leqq \nu \leqq N-1$; otherwise $p_\nu = I$, $J^{a_\nu}_{1\gamma^\nu+} = I$, and the product of the two may be discarded. Also we may assume $a_N \geqq 1$; otherwise $J^{a_N}_{3+} = 0$. Since $b \geqq 0$, these relations tell us

$$m \geqq N + 1.$$

We now recall (8.41) and use (8.81) with $p' = p_N$, $p'' = p_{N+1}$, and invoke Lemma 8.5 with $L = N - 1$, $J_{\nu+}$ an appropriate multiple of $J^{a_o}_{1\gamma^\nu+}$ for $1 \leqq \nu \leqq N-1$, and S an appropriate multiple of $p_N J^{a_N}_{3+} p_{N+1}$. With $a' = a_o+\ldots+a_{N-1}$, $\Gamma = \|\gamma^1\|+\ldots+\|\gamma^{N-1}\|$, we see that for some $C_7, C_8, R_7, R_8 > 0$, the left side of (8.79) is bounded by

$$m^{b/2}[C_7R_7^{a'+N-1}(a'+\#p_o+\ldots+\#p_{N-1})!^{1/2}][C_2^{N-1}R_2^{\Gamma}] \times$$
$$[C_6R_6^{\#p_N+\#p_{N+1}}(\#p_N+\#p_{N+1}+a_N)!^{1/2}]$$
$$\leqq m^{b/2}C_8R_8^m(a_o+\ldots+a_N+\#p_o+\ldots+\#p_{N+1})!^{1/2}$$
$$= C_8R_8^m m^{b/2}(m-b)!^{1/2} \leqq C_8R_8^m m^{b/2} m^{(m-b)/2}$$

which yields (8.70) at once. (Here C_2, R_2 are as in (8.41) and C_6, R_6 are as in (8.81).)

Thus we need only estalbish Lemma 8.5. For this, we shall first need to prove two sublemmas, the first of which really looks very much like Lemma 2.5. Indeed, the method of proof we shall give could also have been used to establish a variant of Lemma 2.5.

Sublemma 8.6. Suppose $R > 0$; then there exist $C_1, R_1 > 0$ as follows. Suppose $L \in \mathbb{Z}^+$, $S_1, \ldots, S_L \in \mathcal{B}(\mathcal{H}_1)$, $S_\nu : \mathcal{H}^\infty \to \mathcal{H}^\infty$ for all ν, and $D^M S_\nu$, initially defined on $\mathcal{H}^\infty$, has a bounded extension to $\mathcal{H}$ for all M, ν. Suppose

$$\|D^M S_\nu\| < R^M M! \quad \text{for all } M, \nu . \tag{8.88}$$

Suppose also $S \in Z_2^2(\mathcal{H}_1)$, and

$$\|D^M S\|_2 < R^M M! \quad \text{for all } M. \tag{8.89}$$

Suppose $D_1, \ldots, D_L \in \{D, I\}$, and $\#\{\nu : D_\nu = D\} = N$; then

$$\|D_1 S_1 D_2 S_2 \ldots D_L S_L S\|_2 < C_1 R_1^L N! .$$

(Here each D_ν is applied to everything to its right.)

Proof. D is, of course, a derivation. Let us not yet adopt the notation of the statement of the sublemma. We shall need to prove a simple variant of Leibniz's rule, the latter which of course states:

$$D^N (S_1 \ldots S_\ell) = \sum_{N_1 + \ldots + N_\ell = N} [N!/(N_1! \ldots N_\ell !)] (D^{N_1} S_1) \ldots (D^{N_\ell} S_\ell)$$

if $N, \ell \in \mathbb{Z}^+$, and $S_1, \ldots, S_\ell : \mathcal{H}^\infty \to \mathcal{H}^\infty$. The variant we shall need is:

Suppose $S_1,\ldots,S_\ell : \mathcal{H}^\infty \to \mathcal{H}^\infty$, $D_1,\ldots,D_\ell \in \{D,I\}$, and $\#\{\nu : D_\nu = D\} = N$. Then

$$D_1S_1D_2S_2\ldots D_\ell S_\ell = \sum_{I\in\mathcal{J}} c_I (D^{N_1}S_1)\ldots(D^{N_\ell}S_\ell) \tag{8.90}$$

where I is shorthand for $(N_1,\ldots,N_\ell)$, $\mathcal{J}$ is the set of those I with $N_1+\ldots+N_\ell = N$, and the c_I are certain constants with

$$0 \leqq c_I \leqq N!/(N_1!\ldots N_\ell!) \tag{8.91}$$

This follows from a simple induction on ℓ, the case $\ell = \ell_0$ following from the case $\ell = \ell_0 - 1$ trivially if $D_1 = I$, and by use of the following identities if $D_1 = D$: If $N = N_1+\ldots+N_\ell$, then

$$\begin{aligned} N!/(N_1!\ldots N_\ell!) &\geqq (N-1)!/(N_2!\ldots N_\ell!) && \text{if } N_1 = 1 \\ &= \sum_{\nu=2}^{\ell} (N-1)!N_\nu/(N_2!\ldots N_\ell!) && \text{if } N_1 = 0. \end{aligned}$$

Note also in (8.90) that

$$\#\mathcal{J} = \binom{N+\ell-1}{N} \leqq 2^{N+\ell-1}. \tag{8.92}$$

If we now adopt the notation of the statement of the sublemma, we see that the sublemma is an immediate consequence of (8.90) with $\ell = L + 1$, $D_{L+1} = I$, $S_{L+1} = S$. Indeed, by (8.92), (8.90) expresses $D_1S_1D_2S_2\ldots D_LS_LS$ as a sum of at most 2^{N+L} terms, each of whose Hilbert-Shcmidt norms does not exceed $R^N N!$ by (8.88), (8.89) and the general inequality $\|S_1' \ldots S_{L+1}'\|_2 \leq \|S_1'\| \ldots \|S_L'\| \, \|S_{L+1}'\|_2$ for operators $S_1', \ldots, S_{L+1}'$. This proves the sublemma.

Sublemma 8.7. Suppose $R > 0$; then there exist $C_1, R_1 > 0$ as follows. Suppose $L \in \mathbb{Z}^+$, $S_1, \ldots, S_L \in \mathcal{B}(\mathcal{H}_1)$, $S_\nu : \mathcal{H}^\infty \to \mathcal{H}^\infty$ for all ν, and $D^m S_\nu$, initially defined on $\mathcal{H}^\infty$, has a bounded extension to $\mathcal{H}$ for all M, ν. Suppose

$$\|D^M S_\nu\| < R^M M! \qquad \text{for all } M, \nu. \tag{8.93}$$

Suppose also $S \in Z_2^2(\mathcal{H}_1)$, and

$$\|A^M S A^N\|_2 < R^{M+N}(M+N)! \qquad \text{for all } M, N. \tag{8.94}$$

Suppose $A_1, \ldots, A_L \in \{A, I\}$, and $\#\{\nu : A_\nu = A\} = r$. Then

$$\|A_1 S_1 \ldots A_L S_L S\|_2 < C_1 R_1^L r!. \tag{8.95}$$

Proof. This is again based on a simple identity. We do not yet adopt the notation of the statement of the sublemma. Say $S_1, \ldots, S_\ell : \mathcal{H}^\infty \to \mathcal{H}^\infty$, $A_1, \ldots, A_\ell \in \{A, I\}$ and $\#\{\nu : A_\nu = A\} = r$. Then:

(*) $A_1 S_1 \ldots A_\ell S_\ell$ is the sum of 2^r terms of the form

$$D_1 S_1 D_2 S_2 \ldots D_{\ell-1} S_{\ell-1} D_\ell S_\ell A^m \tag{8.96}$$

where $D_1, \ldots, D_\ell \in \{D, I\}$, and where $\#\{\nu : D_\nu = D\} + m = r$.

(*) is easily proved by induction on ℓ, the case $\ell = \ell_0$

following from the case $\ell = \ell_o - 1$ trivially if $A_1 = I$, and, if $A_1 = D$, from the identity

$$AS' = DS' + S'A$$

for operators S' on H^∞. In applying this identity, however, one must be careful to observe that

$$[S_1D_2S_2..D_{\ell-1}S_{\ell-1}D_\ell S_\ell A^{m-1}]A = S_1D_2S_2...D_{\ell-1}S_{\ell-1}D_\ell S_\ell A^m \quad (8.97)$$

if $D_2,\ldots,D_\ell \in \{D,I\}$, $m - 1 \in \mathbb{Z}^+$. (Remember our convention that each D_ν acts on <u>everything</u> to its right.) (8.97) follows from repeated use of the identity

$$[DS']A = D(S'A)$$

for operators S' on H^∞. (This follows from the derivation law and DA = 0.)

If we now adopt the notation of the sublemma, we see that the sublemma follows easily from (*) in the case $\ell = L + 1$, $A_{L+1} = I$, $S_{L+1} = S$. Indeed, by (8.94), and since $(M+m)! \leqq 2^{M+m}m!M!$, we have

$$||D^M SA^m||_2 < 2^M R^{M+m} 2^{M+m} m!M!,$$

$$||D^M(DSA^m)||_2 < 2^{m+1}R^{M+m+1}2^{M+m+1}(m+1)!M!$$

for all M,m. Thus, by Sublemma 8.6, for some $C_2,R_2 > 0$ the Hilbert-Schmidt norm of the typical term (8.96) does not exceed $(C_2R_2^L N!)(R_2^m m!)$ if, in (8.96), $\#\{\nu : D_\nu = D\} = N$. By hyothesis, $N!m! \leqq (N+m)! = r!$. Since $A_1S_1 \ldots A_LS_LS$ is the sum of 2^r terms of type (8.96), the proof is complete.

With the aid of Sublemma 8.7, we are finally in position to attack the key Lemma 8.5. First some motivation. Notice that Sublemma 8.7 has the appearance of a variant of Lemma 8.5. Indeed, by (8.83), the operators $J_{\nu+}$ of Lemma 8.5 satisfy estimates

$$\|D^M J_{\nu+}\| < C'(R')^{a_\nu+M} a_\nu!^{1/2} M! \tag{8.98}$$

for certain $C', R' > 0$. If we replaced each S_ν by $J_{\nu+}$ in (8.95), we would then have an estimate

$$\|A_1J_{1+} \ldots A_LJ_{L+}S\| < C_2R_2^{a+L} a!^{1/2} r!.$$

This is rather like (8.87), since if we expanded out all $A = \sum P_i{}^2/4$, there would be $2r$ P's, and $a!^{1/2} r! \leqq (a+2r)!^{1/2}$. Of course, this does not prove (8.87), but does help to motivate what we are about to do, which is to reduce Lemma 8.5 to Sublemma 8.7.

Before we do this, however, let us now explain briefly why we could not work directly with the derivations d'_+ in (8.70), and had to bring in the derivation D. Stated very briefly, the trouble is this: D is the commutator with A, so victory for the D's is measured in terms of factorials. d_{i+} is essentially the commutator with a P_ν, so victory for the d's is measured in terms of square roots of factorials. But the crucial facts (8.90), (8.91), involve factorials, not square roots of factorials.

That, in a very few words, is why we have had to pursue the roundabout route of proving Sublemma 8.7 and seeking to derive Lemma 8.5 from it. We shall do so by means of a device, which is analytical rather than combinatorial in nature. The idea behind it is easy to explain. What is needed are analogues of the Riesz transforms. Suppose for the moment that we knew that there exists $\Phi \in \mathcal{A}K^{-2n-1}$ with $\hat{\Phi} = \underset{\sim}{A}^{-1/2}$, so that convolution on the right with Φ would be like applying $L_0^{-1/2}$. Suppose also, in Lemma 8.5, that all $Q_\nu \in \mathcal{T}$, so that $L_1 = L$. Say, for instance, $Q_1 = P_{1+}$, $Q_2 = P_{2+}$. Then we could rewrite

$$Q_1 J_{1+} Q_2 J_{2+} = (P_1 J_{1+} A^{-1/2}) A (A^{-1/2} P_{2+} J_{2+}). \tag{8.99}$$

Now, for $1 \leqq i \leqq 2n$, $K \in \mathcal{A}K^k$ for some k, we can define

$$\mathcal{R}_i K = X_i^R (K \underline{*} \Phi),\ \mathcal{R}^i K = X_i \Phi \underline{*} K$$

where we are writing $X_{i+n} = Y_i$, $X_{i+n}^R = Y_i^R$ for $1 \leqq i \leqq n$. The $\mathcal{R}_i, \mathcal{R}^i$ are analogues of the Riesz transforms. Note $\mathcal{R}_i K, \mathcal{R}^i K \in \mathcal{A}K^k$. Most importantly, it is clear from the results of Theorem 7.5 that for some $C > 0$ the $\mathcal{R}_i K_\nu / C$, $\mathcal{R}^i K_\nu / C$ satisfy estimates akin to (8.86), with some new R;

C and the new R depend only on the old R. Thus if we define $M_{i}\underset{\sim}{J}_{\nu} = (R_{i}K_{\nu})^{\wedge}$, $M^{i}\underset{\sim}{J}_{\nu} = (R^{i}K_{\nu})^{\wedge}$, then the $(M_{i}J_{\nu})_{+}$, $(M^{i}J_{\nu})_{+}$ would satisfy estimates akin to (8.98), with a new C' and R' depending only on R. However, from (8.99) above,

$$Q_1J_{1+}Q_2J_{2+} = -(M_1J_1)_{+}A(M^2J_2)_{+}.$$

Applying the same reasoning to each group of terms $Q_{\ell}J_{\ell+}Q_{\ell+1}J_{(\ell+1)+}$, for ℓ odd, we see that if L is even we have only to estimate

$$\|J'_{1+}AJ'_{2+}J'_{3+}AJ'_{4+}J'_{5+}A\ldots AJ'_{L+}S\|_2,$$

for certain $J'_{\nu+}$ much like the J_{ν} of the statement of the Lemma. Such an expression can be estimated in the manner that we indicated after (8.98), and the estimate needed on (8.87) would be obtained, since the number of A's is L/2 = $L_1/2$ under our assumptions.

There are several minor problems with this plan - L_1 might not be L, L might not be even - and one major problem - we don't know that there exists $\Phi \in AK^{-2n-1}$ with $\hat{\Phi} = \underset{\sim}{A}^{-1/2}$. There is such a Φ in K^{-2n-1}, however - see [18]. Although we believe this Φ is, in fact, in AK^{-2n-1}, we have made no effort to prove it. Instead we opt to use some exact formulae of [28] to prove replacements for Φ.

We define $\Phi_1, \Phi_2 \in AK^{-2n-1}$ as follows:

$$\Phi_1(t,z) = 2^{n-3/2}\pi^{-n-1}[\Gamma(\tfrac{n}{2})\Gamma(\tfrac{n+1}{2})/\Gamma(\tfrac{1}{2})] \times$$

$$(|z|^2+it)^{-(n+1)/2}(|z|^2-it)^{-n/2}, \tag{8.100}$$

$$\Phi_2 = \overline{\Phi}_1.$$

Then by Proposition 7.1 of [28], we have:

$$\hat{\Phi}_1 = \underset{\sim}{B}_1, \ \hat{\Phi}_2 = \underset{\sim}{B}_2 \tag{8.102}$$

where the operator families $\underset{\sim}{B}_1, \underset{\sim}{B}_2$ are defined as follows:

$$\begin{aligned} B_{1\lambda}E_{\alpha,\lambda} &= (2|\lambda|)^{-1/2}[\Gamma(|\alpha|+n/2)/\Gamma(|\alpha|+(n+1)/2)]E_{\alpha,\lambda} \text{ if } \lambda > 0 \\ &= (2|\lambda|)^{-1/2}[\Gamma(|\alpha|+(n+1)/2)\Gamma(|\alpha|+(n/2)+1)]E_{\alpha,\lambda} \text{ if } \lambda < 0 \end{aligned} \tag{8.103}$$

while $\underset{\sim}{B}_2$ does just the opposite:

$$\begin{aligned} B_{2\lambda}E_{\alpha,\lambda} &= (2|\lambda|)^{-1/2}[\Gamma(|\alpha|+(n+1)/2)\Gamma(|\alpha|+(n/2)+1)]E_{\alpha,\lambda} \text{ if } \lambda > 0 \\ &= (2|\lambda|)^{-1/2}[\Gamma(|\alpha|+n/2)/\Gamma(|\alpha|+(n+1)/2)]E_{\alpha,\lambda} \text{ if } \lambda < 0 \end{aligned} \tag{8.104}$$

(In the notation of Proposition 7.1 of [28], these are the cases $k = n+1/2$, $j = 1/2$, $\gamma = n/2$ or $(n+1)/2$.) The forms of $\underset{\sim}{B}_1, \underset{\sim}{B}_2$ show that they are "approximate" square roots of $\underset{\sim}{A}^{-1}$. Moreover,

$$\underset{\sim}{B}_1\underset{\sim}{B}_2 = \underset{\sim}{B}_2\underset{\sim}{B}_1 = \underset{\sim}{A}^{-1}. \tag{8.105}$$

Thus $\Phi_1 * \Phi_2 = \Phi_2 * \Phi_1$ is the fundamental solution of L_o.

Using Φ_1, Φ_2 in place of Φ, we can now give a proof of Lemma 8.5 which solidifies all of our heuristics.

Proof of Lemma 8.5. We define $\Phi_1, \Phi_2, \underset{\sim}{B}_1, \underset{\sim}{B}_2$ by (8.100), (8.101), (8.103), (8.104), and note (8.102) and (8.105).

For $1 \leqq i \leqq 2n$, $K \in AK^k$ for some k, we define

$$R_i K = X_i^R (K \underline{*} \Phi_1), \; R^i K = X_i \Phi_2 \underline{*} K$$

where we are writing $X_{i+n} = Y_i$, $X_{i+n}^R = Y_i^R$ for $1 \leqq i \leqq n$. The R_i, R^i are analogues of the Riesz transforms. We shall also need to define, for $K \in AK^k$,

$$NK = L_o(\Phi_2 \underline{*} K \underline{*} \Phi_1).$$

(N would do nothing if convolution were commutative.) Clearly $R_i, R^i, N : AK^k \to AK^k$. We also write, if $\hat{K} = \underset{\sim}{J}$, $1 \leqq i \leqq 2n$,

$$(R_i K)^\wedge = M_i \underset{\sim}{J} = (M_{i\lambda} J(\lambda)), \; (R^i K)^\wedge = M^i \underset{\sim}{J} = (M^{i\lambda} J(\lambda)),$$

$$(NK)^\wedge = \mathcal{A} \underset{\sim}{J} = (\mathcal{A}_\lambda J(\lambda)).$$

The M_i, M^i are analogues of multipliers, and $\mathcal{A}$ would be the identity if everything commuted. Explicitly,

$$M_i \underset{\sim}{J} = \sqrt{-1}\, \underset{\sim}{P}_i \underset{\sim}{J} \underset{\sim}{B}_1, \; M^i \underset{\sim}{J} = \sqrt{-1}\, \underset{\sim}{B}_2 P_i J, \; \mathcal{A} \underset{\sim}{J} = \underset{\sim}{B}_2 \underset{\sim}{J} \underset{\sim}{B}_1 \underset{\sim}{A}.$$

Now, for some $C_2, R_2 > 0$, depending only on R, we have an estimate

$$|\partial^\gamma K'_\nu(w)| < C_2 R_2^{m_\nu + \|\gamma\|} \gamma! \text{ for } 1 \leqq |w| \leqq 2,$$
$$\gamma \in (\mathbb{Z}^+)^{2n+1},\ 1 \leqq \nu \leqq L, \tag{8.106}$$

if $K'_\nu = K_\nu$, $\mathcal{N}K_\nu$, $\mathcal{R}_i K_\nu$ or $\mathcal{R}^i K_\nu$ for any i. This is an easy consequence of (8.85) and Theorem 7.5. (If $K'_\nu = \mathcal{R}_i K_\nu$ or $\mathcal{N}K_\nu$, an easy application of Lemma 2.5 is needed to handle the X_i^R or the $\mathcal{L}_o$.) By (8.83), then, there are $C_3, R_3 > 0$, depending only on R_2, so that if

$$S_\nu = J_{\nu+},\ \mathcal{d}_1 J_\nu(1),\ \mathcal{M}_{i1} J_\nu(1) \text{ or } \mathcal{M}^{i1} J_\nu(1), \tag{8.107}$$

then

$$\|D^M S_\nu\| < C_3 (R_3)^{a_\nu + M} a_\nu!^{1/2} M! \quad \text{for all } M \in \mathbb{Z}^+, \nu. \tag{8.108}$$

(The case $a_\nu = 0$ is not dealt with in (8.83). But then K'_ν could only be one of the fixed finite collection $\{\delta, X_1^R\Phi_1, \ldots, X_{2n}^R\Phi_1, X_1\Phi_2, \ldots, X_{2n}\Phi_2\} \subset$ APV. As we remarked during the proof of Lemma 4.5, if $K = \text{P.V.}(f) \in$ APV, $D^M K = \text{P.V.}(D^M f)$ for all $M \in \mathbb{Z}^+$. Also, as is clear from (4.6), $D\delta = 0$. Thus, in our case, if $M \geqq 1$, $D^M K'_\nu$ must equal $\text{P.V.}(D^M f)$, where f is an element of a fixed finite collection of real analytic functions. (8.108) now follows if $a_\nu = 0$ by the usual fact that, for some $C_o > 0$, if $K = \text{P.V.}(f)$, $\hat{K} = \underset{\sim}{J}$, then $\|J_+\| < C_o \|f\|_{C^1}$.)

Now for (8.87), let us first assume L_1 is even, and let $\mathcal{L} = \{\ell : Q \in \mathcal{T}\}$. Say $\mathcal{L} = \{\ell_i : 1 \leqq i \leqq L_1\}$, where

where $\ell_1 < \ell_2 < \cdots < \ell_{L_1}$. Say i is <u>odd</u>, $Q_{\ell_i} = P_{j+}$, $Q_{\ell_{i+1}} = P_{k+}$. Then $Q_\nu = I$ if $\ell_i < \nu < \ell_{i+1}$. We write the string

$$Q_{\ell_i} J_{\ell_i +} \cdots Q_{\ell_{i+1}} J_{\ell_{i+1}+} = -(M_{j1} J_{\ell_i +}) A [\mathcal{A}_1 J_{\ell_i +1}(1))] \times$$

$$[\mathcal{A}_1 J_{\ell_i +2}(1)] \cdots [\mathcal{A}_1 J_{\ell_{i+1}-1}(1)][M^{k1} J_{\ell_{i+1}+}]$$

so

$$Q_{\ell_i} J_{\ell_i +} \cdots Q_{\ell_{i+1}} J_{\ell_{i+1}+} = -S_{\ell_i} A S_{\ell_i +1} \cdots S_{\ell_{i+1}} \tag{8.109}$$

for certain $S_\nu (\ell_i \leq \nu \leq \ell_{i+1})$ as in (8.107).

In the expression $Q_1 J_{1+} \cdots Q_L J_{L+} S$ we now replace each string of the form of the left side of (8.109), for ℓ_i odd, by the corresponding string on the right side. We leave along any other factors in the expression. In this manner we have now rewritten the expression in the form $\pm A_1 S_1 A_2 S_2 \cdots A_L S_L S$, where $A_i \in \{A, I\}$ for all i, where $\#\{A_i : A_i = A\} = L_1/2$. By (8.108) and Sublemma 8.7, for some C_4, R_4 depending only on R we have an estimate

$$\|Q_1 J_{1+} \cdots Q_L J_{L+} S\|_2 = \|A_1 S_1 \cdots A_L S_L S\|_2 < C_4 R_4^{a+L} a_1!^{1/2} \cdots a_L!^{1/2} (L_1/2)!$$

$$< C_4 R_4^{a+L} (a+L_1)!^{1/2}$$ as desired, if L_1 is even.

If L_1 is odd, we put $J_{0+} = I$, and rewrite

$$\|Q_1 J_{1+} \cdots Q_L J_{L+} S\|_2 = 4 \| \sum_{i=1}^{2n} A^{-1} P_i (P_i J_{0+} Q_1 J_{1+} \cdots Q_L J_{L+} S) \|_2$$

$$\leq 8n \sum_{i=1}^{2n} \|P_i J_{0+} Q_1 J_{1+} \cdots Q_L J_{L+} S\|_2 < 8nC' C_4 R_4^{a+L+1} (a+L_1+1)!^{1/2}$$

$< C_1 R_1^{a+K} (a+L)!^{1/2}$ for an absolute constant C' and for certain $C_4, R_4, C_1, R_1 > 0$ depending only on R. We have used the case where L_1 is even, as well as the fact that each $\underset{\sim}{A}^{-1} \underset{\sim}{P}_i$, being the Fourier transform of an element of AK^{-2n-1}, is a family of bounded operators. This completes the proof of lemma 8.5, and with it, the proof of Theorem 8.1.

9. Applying the Calculus

In this chapter we derive a number of other results about the calculus, which are useful in applications. The chapter ends with a proof that the Kohn Laplacian, and a parametrix for it, lie in the systems analogue of the calculus after a contact transformation, under natural hypotheses.

1. When using the calculus, it is frequently simplest to work with formal sums. To make this easier to do, we add some elementary facts to (8.4) and Proposition 8.3, for later reference.

(i) Suppose $K \in \mathcal{C}^k(U)$, $\mathcal{K} = \mathcal{O}(K)$, $U_o \subset U$, U_o open, and $\mathcal{K}f = 0$ on U_o for all $f \in C_c^\infty(U_o)$. Then $K_u \equiv 0$ for all $u \in U_o$. Indeed, the kernel of $\mathcal{K} : C_c^\infty(U_o) \to C^\infty(U_o)$ must be zero. Thus, for any $u \in U_o$, the distribution $K_u(uw^{-1})$ must be zero for $w \in U_o$. By analyticity, $K_u \equiv 0$.

(ii) For $k \in \mathbb{C}$, let $F\mathcal{P}^k(U) = \{\sum K^m \in F\mathcal{C}^k(U)$: for all $u \in U$, and all m, $K_u^m(w)$ is a polynomial in w$\}$. Clearly $F\mathcal{P}^k = \{0\}$ unless $k \in \mathbb{Z}$.

Suppose $\sum K_\nu^m \in F\mathcal{C}^{k_\nu}(U)$ for $\nu = 1,2$, and either $\sum K_2^m \in F\mathcal{P}^{k_2}(U)$ or $\sum K_1^m \in F\mathcal{P}^{k_1}(U)$. Then we claim $\sum K_2^m \# \sum K_1^m = 0$. Indeed, by (8.4), it suffices to show that if K^m is given by (8.3), then K_u^m is always a polynomial in w. Since $\mathcal{F}K_u^m$ is smooth away from 0, it suffices to show $T^N K_u^M = 0$ for N sufficiently large. This is immediate from (8.3) and (7.22).

(iii) Suppose $K_2 \in \mathcal{C}^k(U)$, $K_2 \sim \sum K_2^m$, and $\sum K_2^m \in F\mathcal{P}^k(U)$. Suppose $V \subset\subset U$, V open. Then $K_2 \sim 0$ also as an element of $\mathcal{C}^k(V)$. This may be shown directly, but also follows at once from Theorem 7.11(c), with $K_1 = I$. Indeed, this result tells us that each $u_0 \in U$ has neighborhoods U_0, U_1 so that for some $K \in \mathcal{C}^k(U)$, if $K = \mathcal{O}(K)$, then $Kf = K_2 f$ on U_1 for all $f \in \mathcal{E}'(U_0)$; further, $K \sim \sum K_2^m \# \delta = 0$ (by (ii)). By (i), $K_{2u} = K_u$ for $u \in U_0 \cap U_1$, so $K_2 \sim 0$ as an element of $\mathcal{C}^k(U_0 \cap U_1)$. The result we wanted follows from the compactness of $\overline{V}$.

A slight refinement of this argument shows that in fact $K_2 \sim 0$ as an element of $\mathcal{C}^k(U)$, but we shall not need this.

(iv) Suppose $K \in \mathcal{C}^k(U)$, $K \sim \sum K^m$, and also $K \sim \sum K_1^m$. Then $\sum(K^m - K_1^m) \in F\mathcal{P}^k(U)$. Indeed, for each u, $\sum[K_u^m(w) - K_{1u}^m(w)]$ must be smooth in w, for w sufficiently small. Then $\sum \partial_w^\alpha [K_u^m(w) - K_{1u}^m(w)]$ is smooth in w for small w, for all α. Since this series can never blow up as $w \to 0$, we must have $\sum(K^m - K_1^m) \in F\mathcal{P}^k$.

Whenever we have $\sum(K^m - K_1^m) \in F\mathcal{P}^k(U)$, we say $\sum K^m = \sum K_1^m \bmod F\mathcal{P}$.

(v) Suppose $\sum K^m \in FC^k(U)$. Then $\sum K^m = \delta \# \sum K^m$ mod FP, and $(\sum K^m)\#\delta = \delta\#\sum K^m$. For the first, say $\delta\#\sum K^m = \sum K_1^m$. Then by (8.3), $K_{1u}^m = \delta \underline{*} K_u^m$ for all u. By (7.18), (7.19), $FK_{1u}^m = FK_u^m$ away from 0, so $K_{1u}^m - K_u^m$ is a polynomial, as desired. For the second fact, note by (7.18), (7.19) that $\delta \underline{*} K_u^m = K_u^m \underline{*} \delta$.

(vi) For $\sum K^m \in FC^k(U)$, define $(K^*)_u^m$ by (7.88).

Say $\sum K_\nu^m \in FC^{k_\nu}(U)$ for $\nu = 1,2$. We claim that if $\sum K_2^m \# \sum K_1^m = \sum K^m$, then $\sum (K_1^*)^m \# \sum (K_2^*)^m = \sum (K^*)^m$ mod FP.

Indeed, this need only be proved in a neighborhood of a point $u_o \in U$. For $\nu = 1,2$, for r_ν sufficiently small, we may form $K_\nu \in C^{k_\nu}(U;r_\nu,s_\nu,R_\nu)$ where $K_{\nu u}(w) = \chi_\nu(w)\sum K_{\nu u}^m(w)$, and χ_ν is the characteristic function of $\{|w| < r_\nu\}$. Let $\mathcal{K}_\nu = \mathcal{O}(K_\nu)$. We call r_ν the <u>support radius</u> of $\mathcal{K}_\nu$. By Theorem 7.11(a), if the r_ν are sufficiently small, then $\mathcal{K}_\nu^* = \mathcal{O}(K_\nu^*)$ where $K_{\nu u}^*(w) = \chi_\nu(w)\sum (K_\nu^*)_u^m(w)$, and $K_\nu^* \in C^{\bar{k}_\nu}(U;r_\nu,s_\nu^*,R_\nu^*)$. Further, as is apparent, the s_ν^*, R_ν^* are unaffected if we further decrease the r_ν. We further decrease the r_ν so that $\max(r_1,r_2) < \min(s_1,s_2,s_1^*,s_2^*,\mathrm{dist}(u_o,U^c))$. Using $\mathcal{K}_2$, $\mathcal{K}_1$, we may produce, as in Theorem 7.11(c) a $\mathcal{K}$ corresponding to their "product," a U_o and a U_1. We also produce a $\mathcal{K}'$, corresponding to the "product" of $\mathcal{K}_1^*$ and $\mathcal{K}_2^*$. By shrinking U_o and U_1, we may assume the support radius of $\mathcal{K}$ is small enough that $\mathcal{K}^*$ also has a core in the calculus. If V_o is a sufficiently small neighborhood of u_o, then for $f,g \in C_c^\infty(V_o)$, $(f,K^*g) = (Kf,g) = (K_2K_1f,g) = (f,K_1^*K_2^*g) = (f,K'g)$. If we now use (i) and (iv), we find what we want.

(vii) If $\sum K^m \in FC^k(U)$, then $\sum (K^{**})^m = \sum K^m \bmod FP$. Indeed, if $\varkappa(K) = \chi \sum K^m$ and K has sufficiently small support radius, this can be read off $(Kf,g) = (f,K^*g) = (K^{**}f,g)$ for $f,g \in C_c^\infty(U)$.

(viii) In the situation of Theorem 8.1, note that the arguments given after Lemma 8.4 show that we may arrange $\sum K_2^m \# \sum K_1^m = \delta$ on all of U'.

(ix) Suppose $K \in C^k(U;r,s,R)$, $K = O(K)$, and K is self-adjoint as an operator on $C_c^\infty(U)$. Then $\sum K^m = \sum (K^*)^m \bmod FP$. Indeed, we need only prove this on some neighborhood of a fixed $u_o \in U$. For $0 < c < 1$, let χ_c be the characteristic function of $\{w : |w| < cr\}$; let $K^c(u,w) = \chi_c(w)K(u,w)$; let $K^c = O(K^c)$. Choose c sufficiently small that Theorem 7.11(a) guarantees $(K^c)^*$ has a core $(K^c)^*$ in the calculus, with $(K^c)^* \sim \sum (K^*)^m$. For f,g in some neighborhood V of u_o, $Kf = K^cf$ on V for all $f \in C_c^\infty(V)$. Thus for $f,g \in C_c^\infty(V)$, $(f,Kg) = (Kf,g) = (K^cf,g) = (f,(K^c)^*g)$, and the desired result follows by (i).

More generally, suppose Φ is an analytic function in a neighborhood of $\overline{U}$, $\Phi(u) \geqq c \geqq 0$ for $u \in \overline{U}$, and for some $\varepsilon > 0$, Φ has an extension so that $\Phi(v \circ u)$ is defined for

$u \in \overline{U}$, $v \in \overline{D}_s$ and is holomorphic in v for $v \in D_s$. (Notation as before (7.73).)

Let M_Φ denote the operator of multiplication by Φ, acting on $\mathcal{E}'(U)$; then M_Φ is in the calculus, and $\varkappa(M_\Phi) = \Phi(u)\delta(w) \in C^{-2n-2}(U)$. We let $L^2_\Phi(u)$ denote L^2 (as a vector space) with $(\ ,\ )_\Phi$ as inner product. Suppose now $K \in C^k(U;r,s,R)$, $\mathcal{K} = \mathcal{O}(K)$, and $\mathcal{K}$ is self-adjoint relative to $(\ ,\ )_\Phi$ as an operator on $C_c^\infty(U)$. Then $\mathcal{K}^* = M_\Phi \mathcal{K} M_\Phi^{-1}$, and we have $\sum (K^*)^m = \Phi(u)\delta(w) \# \sum K^m \# \sum \Phi(u)^{-1}\delta(w)$ mod $F\mathcal{P}$. Indeed, $M_\Phi \mathcal{K}$ is self-adjoint, and we can apply the previous result together with the associative law for #.

(x) Conversely, suppose Φ is as in (ix), $\sum K_1^m \in FC^k(U)$ and $\sum (K_1^*)^m = \Phi(u)\delta(u) \# \sum K^m \# \Phi(u)^{-1}\delta(w)$ mod $F\mathcal{P}$. Then we claim that there exists $K \in C^k(U)$ so that $\mathcal{K} = \mathcal{O}(K)$ is self-adjoint relative to $(\ ,\ )_\Phi$ as an operator on $C_c^\infty(U)$, and so that if $K \sim \sum K^m$, then $\sum K^m = \sum K_1^m$ mod $F\mathcal{P}$. Indeed we may select any $\mathcal{K}_1$ with $\varkappa(\mathcal{K}_1) \sim \sum K_1^m$ and with small enough support radius that Theorem 7.11(a) guarantees that $\mathcal{K}_1^*$ and $(M_\Phi \mathcal{K}_1)^*$ have cores in the calculus. The support radius of $(M_\Phi \mathcal{K}_1)^*$ equals that of $\mathcal{K}_1$, so the support radius of $M_\Phi^{-1}\mathcal{K}_1^* M_\Phi = M_{\Phi^{-1}}(M_\Phi \mathcal{K}_1)^*$ equals that of $\mathcal{K}_1$. Then we need only set $\mathcal{K} = (\mathcal{K}_1 + M_{\Phi^{-1}}\mathcal{K}_1^* M_\Phi)/2$.

As the arguments show, there exist $r_0, s, R > 0$ so that if $r < r_0$, we can choose $\mathcal{K}$ so that we also have $\varkappa(\mathcal{K}) \in C^k(U;r,s,R)$.

2. Theorem 8.1 provides us with left parametrices for suitable operators in the calculus. It is easy, however, to use our results to obtain right parametrices under natural conditions.

<u>Proposition 9.1</u>. Suppose that $K_2 \in C^{k_2}(U;r_2,s_2,R_2)$ and that $K_2 \sim \sum K_2^m$. Let $\mathcal{K}_2 = \mathcal{O}(K_2)$. Say $(\tilde{K}_{2u}^o)^\wedge = \underset{\sim}{J}_{2u}^o$ in the $\mathcal{Q}$ sense. We assume for all $u \in U$, that

$(FK_{2u}^o)(0,\xi) \neq 0$ for $\xi \neq 0$; $J_{2u}^{o*}v = 0$, $v \in \mathcal{H}^\omega \Rightarrow v = 0$;

and $J_{2u-}^{o*}v = 0$, $v \in \mathcal{H}^\omega \Rightarrow v = 0$.

Suppose $u_0 \in P \subset U$, P compact. Then there exists an open U' with $P \subset U' \subset U$ and there exist $K_1 \in C^{k_1}(U')$ (where $k_1 + k_2 + 2n + 2 = -2n - 2$), $Q \in C^{-2n-2}(U')$ and open sets $U_1, U_0 \subset U'$ with $u_0 \in U_1 \cap U_0$ so that if $\mathcal{K}_1 = \mathcal{O}(K_1)$, $\mathcal{Q} = \mathcal{Q}(Q)$ we have:

If $f \in \mathcal{E}'(U_1)$ then $\mathcal{K}_1 f \in \mathcal{E}'(U')$;

$$\mathcal{K}_2\mathcal{K}_1 f = f + \mathcal{Q}f \quad \text{on } U_0; \tag{9.1}$$

and $\mathcal{Q}f$ is analytic on U_0. The core of $\mathcal{Q}$ is as after (8.2).

Further, we may arrange that $\sum K_2^m \# \sum K_1^m = \delta$ on all of U'.

(There is also a statement for holomorphic dependence on a parameter, as in Theorem 8.1.)

Proof. Using Theorem 8.1 and (viii) of #1, we may select $\sum K_o^m \in FC^{k_1}(U')$ so that $\sum K_o^m \# \sum (K_2^*)^m = \delta$ on U'. By (vi), (vii) and (ii) of #1, and (8.4), $\sum K_2^m \# \sum (K_o^*)^m = \delta$ on U' also. The result follows by putting $K_1^m = (K_o^*)^m$ and using Theorem 7.11(c).

Accordingly, $u = \mathcal{K}_1 f$ gives a local solution of the equation $\mathcal{K}_2 u = f$ --modulo an analytic function. If $\mathcal{K}_2$ is a differential oeprator, we can rid ourselves of the analytic function error by using the Cauchy-Kowalewski Theorem. One can then derive a variety of regularity properties for the solution by using the methods of [20].

Suppose now $K_1 \in C^{k_1}(U)$, $K_1 \sim \sum K_1^m$, and $U' \subset U$, U' open. Define k_2 by $k_1 + k_2 + 2n + 2 = -2n - 2$. Suppose that there exist $K_2 \in C^{k_2}(U')$ and $K_3 \in C^{k_2}(U')$ so that $\sum K_2^m \# \sum K_1^m = \delta$ on U' and $\sum K_1^m \# \sum K_3^m = \delta$ on U'. Suppose $u_o \in U'$, and the support radius r_1 of $\mathcal{K}_1 = \mathcal{O}(K_1)$ is less than $\mathrm{dist}(u_o, (U')^c)$. Then, on U', $\mathcal{K}_1$ has a left analytic parametrix near u_o (in the sense of Theorem 8.1) and a right analytic parametrix near u_o (in the sense of Proposition 9.1). In fact, on U', $\mathcal{K}_1$ has a two-sided analytic parametrix near u_o. For we must have $\sum K_2^m = \sum K_3^m \bmod F\mathcal{P}$ by the associative law applied to $\sum K_2^m \# \sum K_1^m \# \sum K_3^m$, together with (v) of #1. By (ii) of #1, we then have $\sum K_1^m \# \sum K_2^m = \sum K_2^m \# \sum K_1^m = \delta$. Thus, by Theorem 7.11(c) we may find $\mathcal{K}_2 \in C^{k_2}(U')$ so that $\varkappa(\mathcal{K}_2) = K_2 \sim \sum K_2^m$, and so that

K_2 is both a left and right analytic parametrix for K_1 near u_o, in the sense of Theorem 8.1 and Proposition 9.1.

In particular, say K_1, K_2, U, U', u_o are as in Theorem 8.1 and (viii) of #1, Φ is as in (ix) of #1, and that, in addition, K_1 is self-adjoint relative to $(\ , \ )_\Phi$ as an operator on $C_c^\infty(U)$. Then $\sum K_2^m \# \sum K_1^m = \delta$ on U', and by (vi), (ii) and (ix) of #1, and (8.4), we have

$\Phi(u)\delta(w)\#\sum K_1^m \#\Phi(u)^{-1}\delta(w)\# \ (K_2^*)^m = \delta$ on U' as well. Then $\sum K_1^m \#\Phi(u)^{-1}\delta(w)\#\sum(K_2^*)^m\#\Phi(u)\delta(w) = \delta$ on U' as well, and thus $\sum K_2^m = \Phi(u)^{-1}\delta(w)\#\sum(K_2^*)^m\#\Phi(u)\delta(w)$ mod FP. Thus by (x) of #1, on U' we may find a two-sided analytic parametrix K_2 for K_1 near u_o, so that K_2 is self-adjoint relative to $(\ , \ )_\Phi$ on $C_c^\infty(U')$.

3. The operators in our calculus act by convolution on the left. It is easy, however, to derive corresponding results for convolution on the right, by use of the simple identity

$$(Kf)^\sim = \tilde{K}\tilde{f}$$

where, if $\varkappa(K) = K \in C^k(U)$, then $(\tilde{K}g)(v) = (g*\tilde{K}_{v^{-1}})(v)$ for $g \in C_c^\infty$.

4. The conditions on the principal core in Theorem 8.1,

(i) $(FK_{1u}^o)(0,\xi) \neq 0$ for $\xi \neq 0$;

(ii) $J_{1u\pm}^o$ are injective on H^ω

were assumed to hold for all $u \in U$. It is of interest to note that if the conditions hold at a single $u_o \in U$, they must hold in a whole neighborhood of u_o, so that K will be analytic hypoelliptic near u_o. This is of course evident for condition (i), since $\mathcal{F}K^o_{lu}$ is homogeneous and depends analytically on u. As for condition (ii), let

$R_u = J^o_{lu+}$ or J^o_{lu-}, $R = R_{u_o}$, $j = -2n - 2 - k$.

As we have seen in the proof of Theorem 5.1(d) $\Rightarrow$ (c), R^* is Fredholm from H^j to H^o, so R is Fredholm from H^o to H^{-j}. Further, as we have seen, $[R^*H^j]^{\perp} = \{v \in H^{\omega} : Rv = 0\}$, so the kernel of $R : H^o \to H^{-j}$ is $\{0\}$. Since R has closed range, there therefore exists $C > 0$ so that $\|Rv\|_{-j} \geq C\|v\|_o$ for all $v \in H^o$. But then Proposition 5.5(e) shows that this type of estimate must hold for R_u as well if u is close to u_o. (Proposition 5.5(e) is to be applied to $K = \tilde{K}^o_{lu} - \tilde{K}^o_{lu_o}$, and the identity $\|R^*_u - R^*\|_{j,0} = \|R_u - R\|_{0,-j}$ is to be used.) Thus (ii) follows for u near u_o.

5. It is more elaborate to check that our main results remain true for systems, but this is in fact so. An $N \times N$ system is simply a matrix K of cores so that for some k,U, all entries are in $C^k(U)$. The operator $(Kf)(u) = (K_u * f)(u)$ makes sense for column vectors f of elements of $\mathcal{E}'(U)$. (If one wants K to act on the right, as in #2 above, one would

use row vectors f.) We are asserting, in particular:

Proposition 9.2. The analogues of Theorem 3.2(b) $\Leftrightarrow$ (c), Theorem 3.5, Theorem 4.1, Theorem 5.1, Theorem 5.9, and Theorem 8.1 and Proposition 8.8 all hold for $N \times N$ systems. In these theorems, the analogue of a statement such as "$(FK_1)(0,\xi) \neq 0$ for $\xi \neq 0$" is "$(FK_1)(0,\xi)$ is an invertible matrix for $\xi \neq 0$." The analogue of a statement such as "J_{1+} is injective on H^{ω}" is "J_{1+} is injective on $\oplus H^{\omega}$" (the direct sum of N copies of H^{ω}.)

Almost all our other results remain valid for systems as well, but we have not listed them in Proposition 9.2 since the proofs of these results are immediate consequences of the "scalar" case. The results indicated in Proposition 9.2 follow from making very simple changes in the proofs of the scalar cases of these results. This task, though tedious, is not nearly as irritating as one might think. Indeed, as we said, the analogues of the results preparing for those theorems—such as the results of Chapters 1 and 2, and Theorem 7.11--are immediate consequences of the scalar case.

The scalar case also immediately implies analogues of the adjoint and product rules, Theorems 7.11(a) and (c), for $M \times N$ systems, even for $M \neq N$.

For the results which follow, we look at the calculus of operators which act by convolution on the right, as in #3 above.

6. When one attempts to apply the pseudolocality result Theorem 7.11(b), the restriction that f has small support is at times not met. This situation will arise in the next chapter. In such situations, it is no longer a good idea to cut off cores by characteristic functions χ as in (7.73). Instead, we shall need to use special cutoff functions for the only time in this book. (We shall not, however, need any further intricate combinatorics.)

Say then that $K \in C^k(U;r,s,R)$, and that $0 < a < b < r$. We may ([78]) select a sequence $\{\phi_N\} \subset C_c^\infty(\mathbb{H}^n)$ with $\phi_N(w) = 1$ for $|w| \leq a$, $\phi_N(w) = 0$ for $|w| \geq b$, and so that for some $C_1, R_1 > 0$,

$$|\partial^\alpha \phi_N(w)| < C_1 R_1^{\|\alpha\|} N^{\|\alpha\|} \text{ for } \|\alpha\| \leq N,\ w \in \mathbb{H}^n. \tag{9.2}$$

(As usual $\|\alpha\| = \alpha_1 + \ldots + \alpha_n$.) It is easy to see by Leibniz's rule that if V is open and g is analytic in a neighborhood of $\overline{V}$, then for certain $C_2, R_2 > 0$, $|\partial^\alpha(\phi_N g)(w)| < C_2 R_2^{\|\alpha\|} N^{\|\alpha\|}$ for $\|\alpha\| \leq N$, $w \in V$. By making simple modifications in the proof of Sublemma 8.6, we see that there exist $C_3, R_3 > 0$ so that for all N, all $M \leq N$, all $w \in V$ and all choices of $D_1, \ldots, D_M \in \{T, X_1, \ldots, X_n, Y_1, \ldots, Y_n, X_1^R, \ldots, X_n^R, Y_1^R, \ldots, Y_n^R\}$, we have

$$|D_1 \ldots D_M (\phi_N g)(w)| < C_3 R_3^M N^M . \tag{9.3}$$

We define

$$K_{Nu}(w) = \phi_N(w)K_u(w) \quad \text{for } u \in U,\ w \in \mathbb{H}^n, \tag{9.4}$$

and for $f \in C_c^\infty(U)$ we define

$$(\mathcal{K}_N f)(u) = f * K_{Nu}(u). \tag{9.5}$$

Of course $\mathcal{K}_N$ depends on the choice of a,b and ϕ_N; when we (rarely) need to indicate the dependency we write

$$\mathcal{K}_N = \mathcal{K}_{N,a,b,\phi_N}. \tag{9.6}$$

Say $u \in U$, $f \in C_c^\infty(U)$. If $\zeta \in C_c^\infty$ equals 1 near u, and $\zeta(w) = 0$ for $|u^{-1}w| > a/2$, then for v near u,

$$(\mathcal{K}_N f)(v) = (\zeta f * K_v)(v) + [(1-\zeta)f * K_{Nv}](v) \tag{9.7}$$

so

$$(\mathcal{K}_N f)(v) = (\mathcal{K}(\zeta f))(v) + [(1-\zeta)f](G_{Nv}) \tag{9.8}$$

where $G_{Nv}(w) = K_N(v, w^{-1}v)$. We may use (9.8) to define $\mathcal{K}_N f \in \mathcal{D}'(U)$ for $f \in \mathcal{E}'(U)$. The second term on the right side of (9.8) is clearly smooth in v near u for any $f \in \mathcal{E}'(U)$. The point is that it can be estimated. Letting $S = \{T, X_1, \ldots, X_n, Y_1, \ldots, Y_n\}$, we have this:

Proposition 9.3(a). Suppose $K \in \mathcal{C}^k(U; r, s, R)$. Suppose $u \in U$, $f \in \mathcal{E}'(U)$, f has order L, and f is analytic near u. Then there are $C_4, R_4 > 0$ and a neighborhood V of u so that for all N,

$$|D_1 \ldots D_M(K_N f)(v)| < C_4 R_4^M N^{M+L} \text{ whenever } M \leqq N - L,\ v \in V,$$

and $D_1, \ldots, D_M \in S$.

(b) Suppose $W \subset \mathbb{H}^n$ is open, g is a smooth function on W, and for some $C, R, L' > 0$,

$$|D_1 \ldots D_M g(w)| < CR^M N^{M+L'} \text{ whenever } M \leqq N - L',\ w \in W \tag{9.9}$$

and $D_1, \ldots, D_M \in S$.

Then g is analytic on W.

Proof. (a) Clearly it is enough to show that for any $c > 0$, there exist C_4, R_4, V so that for all N

$$|D_1' \ldots D_I' D_1 \ldots D_M G_{Nv}(w)| < C_4 R_4^M N^{M+L} \text{ where } M \leqq N - L,$$
$$I \leqq L,\ v \in V,\ |u^{-1}w| > c \tag{9.10}$$

and $D_1', \ldots, D_I'$, $D_1, \ldots, D_M \in S$. Here the D_j', D_k are differentiations with respect to w and v respectively. But (9.10) is clear from (9.3) and from iterating relations such as $X_1[F(v,v)] = F_1(v,v) + F_2(v,v)$ where F_1 denotes the result of applying X_1 to F in the first variable, similarly F_2.

(b) We may assume $w = 0$. Put $\mathcal{U}_L = (T, X_1, \ldots, X_n, Y_1, \ldots, Y_n)$. Note that (9.9) also holds with $\sigma_\alpha(\mathcal{U}_L)$ in place of $D_1 \ldots D_M$, if $\|\alpha\| = M$. Indeed, $\sigma_\alpha(\mathcal{U}_L)$ is simply an average of terms of the form $D_1 \ldots D_M$. Since $\sigma_\alpha(\mathcal{U}_L) = \partial^\alpha$ at 0 (see (7.108)), this completes the proof.

The utility of Proposition 9.3 was indicated on page 12 of the Introduction. We will use it in the next chapter.

7. Of course, the calculus we have presented can be transplanted onto any real analytic odd-dimensional manifold M, via an analytic diffeomorphism. This is most natural if M is a <u>contact</u> manifold, namely a 2n + 1-dimensional manifold on which there exists a one-form σ with $\sigma\Lambda d\sigma\Lambda\ldots d\sigma \neq 0$ everywhere. (Here there are n factors of $d\sigma$.) σ is called a contact form. An example is $M = \mathbb{H}^n$, $\sigma = \tau = dt + 2\sum(x_j dy_j - y_j dx_j)$. Note that τ is left invariant, since it satisfies $\tau(T) = 1$, $\tau(X_j) = \tau(Y_j) = 0$ for all j. More generally, we may let M be any CR manifold with non-degenerate Levi form, with σ being a 1-form annihilating $T_{1,0} \oplus T_{0,1}$ at each point. (We use the notation and terminology of [19].)

A diffeomorphism between open subsets of two contact manifolds is called a contact transformation if it preserves the contact form at each point, up to a constant multiple (depending on the point). Darboux's theorem implies that if M is an analytic contact manifold with contact form σ, then (M,σ) is locally diffeomorphic to $(\mathbb{H}^n, \tau)$ by an analytic contact transformation. This observation will enable us, in #9 below, to identify the Kohn Laplacian and a parametrix for it as being within our calculus, on a non-degenerate analytic CR manifold, under Kohn's hypotheses.

Since we do want to view our pseudodifferential operators as living on contact manifolds, it is only natural at this point to ask whether our classes of operators are invariant under analytic contact transformations. (Then we would be able to define our calculus, on an analytic contact manifold M, as that calculus which reduces to the one we have already seen, under <u>any</u> analytic contact transformation taking an open subset of M to $\mathbb{H}^n$. Invariance under general diffeomorphisms is clearly out of the question because we are using a non-isotropic notion of homogeneity.) Invariance under an analytic contact transformation could certainly not be true without some qualification, since in (7.73) χ will in general be mapped to the characteristic function of an open set which is not a ball. However, we have the following result (which we will not use, so we only sketch the proof):

<u>Proposition 9.4</u>. Suppose $U \subset \mathbb{H}^n$ is open, $u_0 \in U$, $r,s,R,R_1 > 0$. Suppose $\psi^{-1} : U \to U'$ is an analytic contact transformation. Then there exist $r',s',R',R_1' > 0$, $U_0 \subset U$ a neighborhood of u_0, as follows. Let $U_0' = \psi^{-1}(U_0)$.

Say $K \in C^k(U_0;r,s,R,R_1)$. Then there exists $K' \in C^k(U_0';r',s',R',R_1')$ so that

$$(\mathcal{K}g)' = \mathcal{K}'g' \quad \text{for all } g \in \mathcal{E}'(U_0).$$

Here $g' = g\circ\psi$, $\mathcal{K} = \mathcal{O}(K)$, $\mathcal{K}' = \mathcal{O}(K')$, $g' = g\circ\psi$, $(\mathcal{K}g)' = (\mathcal{K}g)\circ\psi$.

Proof (sketch): If $g \in C_c^\infty(U')$, we may write symbolically

$$(\mathcal{K}g)'(u) = \int g(v)K(\psi(u), v^{-1}\psi(u))dv$$

$$= \int g'(v)K(\psi(u), \psi(v)^{-1}\psi(u))\,|\det(D\psi(v))|\,dv. \qquad (9.11)$$

The crux of the matter is this lemma, which we shall eventually use with $w = v^{-1}u$:

Lemma 9.4. With ψ as in Propsition 9.4, $v \in U'$ fixed, put

$$\Psi(w) = \psi(v)^{-1}\psi(vw)$$

for $vw \in U'$. Write $\Psi(w) = (\Psi_0(w), \Psi_1(w), \ldots, \Psi_{2n}(w))$ in $\mathbb{H}^n$ coordinates, $w = (t,x,y)$. Then $L\Psi_0(0) = 0$ if L is any differential operator of degree less than or equal to 2 which involves only the x_j's and the y_j's.

Proof. Since left translations are clearly contact transformations, Ψ is also a contact transformation, which takes 0 to 0. The condition that Ψ preserves τ at each point w, up to a constant multiple $f(w)$, is equivalent to the following relations for all w near 0, and $1 \leq k, \ell \leq n$, where all functions are to be evaluated at $w = (t,x,y)$, and the sums are from j = 1 to n:

$$\partial\Psi_o/\partial t + 2\sum(\Psi_j\partial\Psi_{j+n}/\partial t - \Psi_{j+n}\partial\Psi_j/\partial t) = f \tag{9.12}$$

$$\partial\Psi_o/\partial x_k + 2\sum(\Psi_j\partial\Psi_{j+n}/\partial x_k - \Psi_{j+n}\partial\Psi_j/\partial x_k) = -2y_k f \tag{9.13}$$

$$\partial\Psi_o/\partial y_\ell + 2\sum(\Psi_j\partial\Psi_{j+n}/\partial y_\ell - \Psi_{j+n}\partial\Psi_j/\partial y_\ell) = 2x_\ell f. \tag{9.14}$$

If deg $L \leqq 1$, the lemma follows at once from (9.13), (9.14) and $\Psi(0) = 0$. We also find $\partial^2\Psi_o/\partial x_\ell\partial x_k = 0$ at $w = 0$ by differentiating (9.13) and $\partial^2\Psi_o/\partial y_\ell\partial y_k = 0$ at $w = 0$ by differentiating (9.14). Finally, $\partial^2\Psi_o/\partial x_k\partial y_\ell = 0$ at $w = 0$, as we see by differentiating (9.13) with respect to y_ℓ and (9.14) with respect to x_k and adding the two resulting equations. This proves the lemma.

Keeping the notation of the lemma, but writing $w' = (x,y)$, we now see from the lemma and the fact that Ψ is a diffeomorphism near 0:

For some $S \in GL_{2n}(\mathbb{R})$, $c \in \mathbb{R}\setminus\{0\}$,

$D\Psi(w) = (ct,Sw') + o(|w|)$.

In fact the error is bounded in $|\,|$ by $C|t|$. From this and the lemma, we see:

$$\Psi(w) = (ct,Sw') + o(|w|). \tag{9.15}$$

In fact the error is bounded by $C(|w'|^{3/2}+|t|)$. Now we allow v to vary in Lemma 9.4. In (9.15) we write

$\Psi = \Psi_v$, $c = c(v)$, $S = S_v$; these depend analytically on v. Putting $w = v^{-1}u$, writing $c(v) = c(uu^{-1}v)$, $S_v = S_{uu^{-1}v}$, and using a power series expansion as in the right-multiplication analogue of Proposition 7.12, we now see

$$\Psi(v)^{-1}\psi(u) = (c(u)t, S_u w') + o(|v^{-1}u|) \tag{9.16}$$

if $v^{-1}u = (t,w')$ and v is sufficiently close to u. For $u \in U$, $w = (t,w') \in \mathbb{H}^n$, write $\rho_u(w) = (c(u)t, S_u w')$. The point is that $\rho_u(D_r w) = D_r(\rho_u(w))$ if D_r is a dilation. If K is as in (7.73), it is then clear that

$$\chi_1(w)[\sum K^m_{\psi(u)}(\rho_u(w)) + Q(\psi(u), \rho_u(w))] \tag{9.17}$$

is in $C^k(U_1')$ for any relatively compact open subset U_1' of U' and χ_1 the characteristic function of a sufficiently small ball. (One must note $\log|w| - \log|\rho_u(w)|$ is homogeneous of degree zero in w.)

This observation, (9.11) and (9.16) come pretty close to proving Proposition 9.4, but we must still explain what we are going to do with the $o(|v^{-1}u|)$ in (9.16) and the $|\det D\psi(v)|$ in (9.11). The tern $o(|v^{-1}u|)$ is in fact an analytic function of u and $v^{-1}u$, say $F(u, v^{-1}u)$. Note also $|\rho_u(w)| \sim |w|$ and that the radius of convergence of the power series for $K^m_{\psi(u)}$ at $\rho_u(w)$ is also on the order of $|w|$, by (7.79). For w small, then it is valid to expand

$K^m_{\psi(u)}(\rho_u(w)+W)$ in power series about $W = 0$ and to put $W = F(u,w)$. After one does this, it is then valid to expand each $F(u,w)^\alpha$ as a power series in w $(\alpha \varepsilon (\mathbb{Z}^+)^{2n+1})$. The terms of this power series will all be $o(|w|^{|\alpha|})$. In this way one readily sees that one obtains an element K'' of $C^k(U''_o)$ by replacing $\rho_u(w)$ by $\rho_u(w) + F(u,w)$ in (9.17), if U''_o is a sufficiently small neighborhood of $\psi^{-1}(u_o)$. (One may need to reduce the support of χ_1 to a smaller ball.) If U'_o is sufficiently small, we obtain an element K' of $C^k(U'_o)$ by multiplying $K''(u,w)$ by $|\det D\psi(uw^{-1})|$, since one can also expand this function in power series about $w = 0$ as in the right-multiplication analogue of Proposition 7.12. (Again one may need to reduce supp χ_1.) This K' is as desired. Further details are left to the interested reader.

Note that the principal cores are related by

$$(K')^0_u(w) = |\det D\psi(u)| K^o_{\psi(u)}(\rho_u(w)).$$

8. In the applications which follow, we shall frequently suppose ourselves in the following situation. We use the notation and terminology of [19] or of [20], Sections 2 and 13.

Let M be a $2n + 1$-dimensional C^∞ CR manifold. Suppose $p \varepsilon M$, and that M is an analytic CR manifold near p, and that the Levi form is nondegenerate near p. Thus M is an analytic manifold near p, and $T_{1,0}$, $T_{0,1}$ are analytic subbundles of $\mathbb{C}TM$ near p. Suppose we are given a smooth Hermitian metric $<,>$ on $\mathbb{C}TM$ which is analytic in a neighborhood of p, and which is compatible with the CR structure. (That is, $T_{1,0} \perp T_{0,1}$ and $\overline{<Z,W>} = <\bar{Z},\bar{W}>$ for $Z,W \in T_{1,0}$. In particular, if M is the boundary of a domain D in $\mathbb{C}^{n+1}$, with $M = \partial D$ smooth, and analytic near p, we could use the restriction of the ambient metric to M.) We call such a metric a CR metric. Then we may find $W_1,\ldots,W_n$ analytic sections of $T_{1,0}$ near p, which are orthonormal in the metric at each point. Conversely, if we were instead to begin by choosing analytic sections $W_1,\ldots,W_n$ of $T_{1,0}$ near p, which span $T_{1,0}$ at each point near p, we could then choose a Hermitian metric $<,>$ on $\mathbb{C}TM$, analytic near p, and compatible with the CR structure, such that $W_1,\ldots,W_n$ are orthonormal at each point near p. We let $\mathcal{U}$ be an open neighborhood of p such that $\mathcal{U}$ is an analytic CR manifold, $<,>$ is analytic on $\mathcal{U}$, and $W_1,\ldots,W_n$ are orthonormal at each point of $\mathcal{U}$, and such that there exists an analytic contact transformation $\psi : \mathcal{U}' \to \mathcal{U}$, $\mathcal{U}' \subset \mathbb{H}^n$, $\psi(\mathcal{U}') = \mathcal{U}$.

We shall let $\Lambda^{0,q}$, $L_2^{0,q}$, $(\mathcal{E}')^{0,q}$, $(\mathcal{D}')^{0,q}$ denote (0,q) forms which near any point have coefficients which are respectively smooth, in L^2, in $\mathcal{E}'$ or in $\mathcal{D}'$, with respect to a local basis of smooth (0,q) forms. Say now $\overline{\eta}_1,\ldots,\overline{\eta}_n$ is a basis for $\Lambda^{0,1}$ on $\mathcal{U}$, and all are real analytic. We may expand $f \in (\mathcal{D}')^{0,q}(\mathcal{U})$ in the form $f = \sum f_I \overline{\eta}^I$ where the I are multi-indices. We then write f_ψ to denote the row vector $(f_I \circ \psi)_{|I|=q}$, all of whose entries are in $(\mathcal{D}')^{0,q}(\mathcal{U}')$.

We abbreviate $\mathcal{O}^k = \mathcal{O}(C^k)$, the space of operators with cores as in (7.77)-(7.78).

In the applications which follow, we shall frequently be seeking to establish that an operator K has the following properties: for an open neighborhood $U \subset \mathcal{U}$ of p,

(i) $K : (\mathcal{E}')^{0,q}(U) \to (\mathcal{D}')^{0,q'}(U)$ and

(ii) K <u>is in $\mathcal{O}^k(\psi^{-1}(U))$ after the contact transformation ψ</u>. Explicitly, this last condition means that there exists a K_ψ in the systems analogue of $\mathcal{O}^k(\psi^{-1}(U))$ (acting by convolution on the right) so that

$$(Kf)_\psi = K_\psi f_\psi$$

on $\psi^{-1}(U)$, for $f \in (\mathcal{E}')^{0,q}(U)$. We sometimes refer to K_ψ as the <u>pullback</u> of K to $\mathbb{H}^n$ (through ψ). Frequently K will also satisfy this: for an open neighborhood $U_o \subset\subset U$,

(iii) $K : (\mathcal{E}')^{0,q}(U_o) \to (\mathcal{E}')^{0,q'}(U)$.

Also, for $U' \subset \mathbb{H}^n$ open, we let $\mathcal{Q}(U') = \{K' = \mathcal{O}(K') \in \mathcal{O}^k(U')$ for some $k : K' \sim 0\}$, and we say K is <u>in $\mathcal{Q}(\psi^{-1}(U))$ after</u>

a contact transformation ψ if $K_\psi \in \mathcal{Q}(\psi^{-1}(U))$.

On the "germ" level, the definitions of the underlined terms in the last paragraph do not depend on the choice of $\bar{\eta}_1,\ldots,\bar{\eta}_n$. Thus, suppose we are given K, U satisfying (i), (ii) with $U \subset\subset \mathcal{U}$. Say we are given another basis $\eta_1',\ldots,\eta_n'$ for $\Lambda^{0,1}$ on $\mathcal{U}$. Then there exists an open neighborhood U_0' of p and an operator K' with $K' : (\mathcal{E}')^{0,q}(U_0') \to (\mathcal{E}')^{0,q'}(U)$, $K' : (\mathcal{E}')^{0,q}(U) \to (\mathcal{D}')^{0,q}(U)$, and so that K' is in $\mathcal{O}^k(\psi^{-1}(U))$ after the contact transformation ψ—where the basis $\bar{\eta}_1',\ldots,\bar{\eta}'$ is used in pulling back to $\mathbb{H}^n$—and so that $Kf = K'f$ on U_0' for $f \in (\mathcal{E}')^{0,q}(U_0')$. This is easy to see from the product Theorem 7.11(c), since the operator of multiplication by a real analytic function g on $\psi^{-1}(\mathcal{U})$ is in $\mathcal{O}^{-2n-2}(\psi^{-1}(U))$, with core $g(u)\delta(w)$. (Since we are only interested in behavior near p, we can always reduce the support radius of the pullback K_ψ of K so that the hypotheses of Theorem 7.11(c) apply).

Say $\mathcal{U}_0$ is an open neighborhood of p, $\mathcal{U}_0 \subset \mathcal{U}$, and $R : (\mathcal{E}')^{0,q}(\mathcal{U}_0) \to (\mathcal{E}')^{0,q'}(M)$, or even just $(\mathcal{E}')^{0,q}(\mathcal{U}_0) \to (\mathcal{D}')^{0,q'}(\mathcal{U}_0)$. Then we say R is locally in $\mathcal{O}^k$ near p, after ψ, if there exist K, U_0, U satisfying (i), (ii), (iii) with $U \subset \mathcal{U}_0$, so that $Rf = Kf$ on U_0 for all $f \in (\mathcal{E}')^{0,q}(U_0)$. We say R is locally in $\mathcal{O}^k$ near p, after ψ, up to an analytic regularizing error, if there exist K, U_0, U satisfying (i), (ii), (iii), with $U \subset \mathcal{U}_0$, so that $Rf - Kf$ is analytic on U_0 for all $f \in (\mathcal{E}')^{0,q}(U_0)$.

Note that, in these definitions, one can always assume that

U_0 and the support radius of the pullback of K to $\mathbb{H}^n$ are as small as one likes. Also, when proving that an operator R has one of these properties, by producing a K, U_0, U_1, it is not necessary to show that condition (iii) holds. For one can always shrink U_0 and the support radius of K to force it to hold.

If $R : (E')^{0,q}(M) \to (E')^{0,q'}(M)$, we say R is <u>analytic pseudolocal on $\mathcal{U}$</u> if whenever $f \in (E')^{0,q}(M)$, and f is analytic on $U \subset \mathcal{U}$, we then have Rf is analytic on U.

Say now $R_1 : (E')^{0,q_1}(M) \to (E')^{0,q_2}(M)$, $R_2 : (E')^{0,q_2}(M) \to (E')^{0,q_3}(M)$ are both analytic pseudolocal on $\mathcal{U}$. Suppose also each R_ν is locally in C^{k_ν} near p, after ψ, up to an analytic regularizing error ($\nu = 1,2$). Let $k = k_1 + k_2 + 2n + 2$. Then R_2R_1 is also analytic pseudolocal on $\mathcal{U}$ and locally in $\mathcal{O}^k$ near p, after ψ, up to an analytic regularizing error. Indeed, choose the appropriate K_ν, U_0^ν, U^ν for R_ν. By reducing support radii and the size of the U_0^ν, and by passing to $U^1 \cap U^2$, we may assume $U^1 = U^2 = U$. Say that the pullback of K_ν is in $\mathcal{O}^{k_1}(\psi^{-1}(U);r_\nu,s_\nu,R_\nu)$. By reducing r_1 and U_0^1 further, we may assume $r_1 < s_1$, $r_1 < \text{dist}(\psi^{-1}(p),\psi^{-1}(U)^c)$, $r_1 < r_2$, and that $f \in (E')^{0,q_1}(U_0^1) \Rightarrow K_1 f \in (E')^{0,q_2}(U_0^2)$. Then if $f \in (E')^{0,q_1}(U_0^1)$, $R_2R_1f - K_2K_1f$ is analytic on $U_0^1 \cap U_0^2$, and Theorem 7.11(c) applies to complete the proof.

9. Next, we show that the Kohn Laplacian, and a parametrix for it, lie in our calculus after a contact transformation, under natural hypotheses.

Let $M, \mathcal{U}$ be as in the second paragraph of #8; suppose that the Levi form is non-degenerate everywhere on M. For the time being, all considerations will be local, so we may assume $M = \mathcal{U}$.

Let the one-form σ be an analytic section of $(T_{1,0} \oplus T_{0,1})^{\perp}$ near p, satisfying $\langle\sigma,\sigma\rangle = 1$ in the induced metric on $\mathbb{C}T^*M$. As we have said in #5, σ is a contact form. Let $b_{jk} = \langle W_j, W_k\rangle_L = \frac{1}{2}i\langle\sigma, [W_j, \overline{W}_k]\rangle$. ($\langle,\rangle_L$ denotes Levi form.) Then $B = (b_{jk}(p))$ is a self-adjoint matrix. Choose an orthogonal transformation $R = (r_{jk})$ with RBR^{-1} diagonal. Then if $W_i' = \sum r_{ij} W_j$, the $\{W_i'\}$ are also orthonormal at each point near p, and if $B' = (b_{jk}')$, $b_{jk}' = \langle W_j', W_k'\rangle_L$, then B' is diagonal at p. Thus: we may assume B is diagonal at p. Let $\lambda_j = b_{jj}(p)$, the eigenvalues of the Levi form. We may assume that for some $k \geqq 0$, $\lambda_j > 0$ for $1 \leqq j \leqq k$, $\lambda_j < 0$ for $k + 1 \leqq j \leqq n$, so that M is k-strongly pseudoconvex near p.

Suppose now S is the vector field dual to σ, so that $W_1, \ldots, W_n, \overline{W}_1, \ldots, \overline{W}_n, S$ is an orthonormal basis for $\mathbb{C}TM$. Denote the dual frame for $\mathbb{C}T^*M$ by $\omega_1, \ldots, \omega_n, \overline{\omega}_1, \ldots, \overline{\omega}_n, \sigma$.

As in [20], we use the notation "$\mathcal{E}$" for error terms. If $\sum_J \phi_J \overline{\omega}^J$ is a C^∞ form (the J are multi-indices), then

$\mathcal{E}(\phi)$ will denote an expression of the form $\sum_{JK} a_{JK}\phi_J\bar{\omega}^k$ with a_{JK} analytic. $\mathcal{E}(W\phi)$ will denote an expression of the form $\sum_{JK\ell} a_{JK\ell}(W_\ell\phi_J)\bar{\omega}^k$ with $a_{JK\ell}$ analytic; similarly $\mathcal{E}(\bar{W}\phi)$, and we abbreviate $\mathcal{E}(A) + \mathcal{E}(B)$ as $\mathcal{E}(A,B)$. We then have

$$[W_j,\bar{W}_\ell]f = -2ib_{j\ell}Sf + \mathcal{E}(Wf,\bar{W}f) \tag{9.18}$$

for C^∞ functions f. Following the computation of [20], Sections 13 and 5, but using (9.18) instead of (13.3) of [20], we see

$$\Box_b\phi = \sum_I(-\frac{1}{2}\sum_1^n(W_m\bar{W}_m+\bar{W}_mW_m)+ig_IS)\phi_I)\bar{\omega}^I+\mathcal{E}(W\phi,\bar{W}\phi,\phi)$$

if $\phi = \sum_I \phi_I\bar{\omega}^I \in C^\infty(\Lambda^{0,q})$. Here g_I is an analytic function with

$$g_I(p) = \alpha_I = \sum_{m\notin I}\lambda_m - \sum_{m\in I}\lambda_m. \tag{9.19}$$

Now let $a_j = |\lambda_j|^{1/2}$, $1 \leq j \leq n$. Select an analytic contact transformation ψ taking (M,σ) (near p) to $(\mathbb{H}^n,\tau)$. Under ψ, say $\psi(p) = p'$, and let us say $W_1,\ldots,W_k$, $W_{k+1},\ldots,W_n$ are mapped to vector fields $a_1Z''_1,\ldots,a_kZ''_k$, $a_{k+1}\bar{Z}''_{k+1},\ldots,a_n\bar{Z}''_n$. These vector fields, being annihilated by τ, are in the span of $Z_1,\ldots,Z_n$, $\bar{Z}_1,\ldots,\bar{Z}_n$ at each point, with analytic coefficients. Thus, after ψ, $\bar{\partial}_b,\mathcal{D}_b$ and $\Box_b$ are in $\mathcal{O}^{-2n-3}$, $\mathcal{O}^{-2n-3}$ and $\mathcal{O}^{-2n-4}$ respectively, on any neighborhood of p whose closure is contained in $\mathcal{U}$.

Of course ψ carries Sf to $Tf + \mathcal{E}(Zf,\bar{Z}f)$; so at p',

$$[Z_j'', \overline{Z}_\ell'']f = -2i\delta_{j\ell}Tf + E(Zf, Zf), \quad f \in C^\infty.$$

Suppose, in fact, at u, $Z_j'' = \sum_\ell (r_{j\ell}(u)Z_\ell + s_{j\ell}(u)\overline{Z}_\ell)$. Define $Z_j' = \sum_\ell (r_{j\ell}(p')Z_\ell + s_{j\ell}(p')\overline{Z}_\ell)$. Then, at p',

$$[Z_j'', \overline{Z}_\ell'']f = [Z_j', \overline{Z}_\ell']f + E(Zf, \overline{Z}f)$$

for $f \in C^\infty$. But clearly $[Z_j', \overline{Z}_\ell'] = c_{j\ell}T$ for some constant $c_{j\ell}$; so

$$[Z_j', \overline{Z}_\ell'] = -2i\delta_{j\ell}T. \tag{9.20}$$

Now let

$$L_I' = -\frac{1}{2}\sum_1^n \lambda_m(Z_m'\overline{Z}_m' + \overline{Z}_m'Z_m') + i\alpha_I T,$$

a self-adjoint operator. Then the system of cores which is a diagonal matrix and has the core $L_I'\delta$ in the (I,I) slot is clearly, at u = p', the principal core of the pullback of $\Box_b$. By Theorem 8.1 and its analogue for systems, and #4 above, the pullback of $\Box_b$ will have a parametrix in our calculus provided the L_I' satisfy the conditions of Theorem 5.1 for all I--that is, if L_I' is hypoelliptic for all I. Let $Z_j' = X_j' + iY_j'$. By (9.20) the map

$$(X_j', Y_j', T) \to (X_j, Y_j, T)$$

is a Lie algebra automorphism of the Lie algebra of $\mathbb{H}^n$. Exponentiating it, we obtain a Lie group automorphism of

$\mathbb{H}^n$, which carries L_I' onto

$$L_I = -\frac{1}{2}\sum_1^n |\lambda_m| (Z_m\bar{Z}_m + \bar{Z}_m Z_m) + i\alpha_I T.$$

L_I' is hypoelliptic if and only if L_I is. Let $K_I = L_I\delta$, $\underset{\sim}{J}_I = \hat{K}_I$. We check condition (e) of Theorem 5.1. We find $K_I(0,\xi) = (\sum|\lambda_m||\xi_m|^2)/4 \neq 0$ for $\xi \neq 0$. Also

$$J_{I+}E_\beta = (-\sum(2\beta_j+1)|\lambda_m|+\alpha_I)E_\beta$$

$$J_{I-}E_\beta = (-\sum(2\beta_j+1)|\lambda_m|-\alpha_I)E_\beta.$$

Since $|\alpha_I| \leqq \sum|\lambda_m|$, J_{I+}, J_{I-} are injective on $\mathcal{H}^\omega$ as long as they do not annihilate E_o. This can happen only if $|\alpha_I| = \sum|\lambda_m|$, which by (9.19) can occur only if $I = \{1,\ldots,k\}$ or $I = \{k+1,\ldots,n\}$. Thus if $q \neq k$ or $n - k$, Theorem 5.1(e) will be satisfied for all the $L_I\delta$.

In particular, by Theorem 8.1, on $\mathcal{U}$, $\Box_b$ must be analytic hypoelliptic on q-forms for $q \neq k$ or $n - k$. This was originally shown in [81], [77], and [78]. But Theorem 8.1, Proposition 9.1 and #2 above also give us an analytic parametrix for $\Box_b$ of a very precise form:

<u>Theorem 9.6</u>. Let M be a nondegenerate smooth CR manifold of dimension $2n + 1$, which is k-strongly pseudoconvex and analytic on an open neighborhood $\mathcal{U}$ of a point p. Suppose we have a CR metric on M which is analytic on $\mathcal{U}$. Suppose

$0 \leq q \leq n$, $q \neq k$ or $n - k$. Then p has open neighborhoods $U_o \subset U \subset \mathcal{U}$ as follows.

There exist operators $K, \mathcal{Q}_1, \mathcal{Q}_2 : (E')^{0,q}(U_o) \to (E')^{0,q}(U)$, and $(E')^{0,q}(U) \to (\mathcal{D}')^{0,q}(U)$, which are in $\mathcal{O}^{-2n}(\psi^{-1}(U))$, $\mathcal{Q}(\psi^{-1}(U))$, $\mathcal{Q}(\psi^{-1}(U))$ respectively after an analytic contact transformation ψ from $\mathbb{H}^n$, and which have the following properties:

(a) $K\Box_b f = f + \mathcal{Q}_1 f$, $\Box_b K f = f + \mathcal{Q}_2 f$ on U_o, for $f \in (E')^{0,q}(U_o)$;

(b) $K, \mathcal{Q}_1, \mathcal{Q}_2 : (C_c^\infty)^{0,q}(U_o) \to (C_c^\infty)^{0,q}(U)$; $K = K^*$ and $\mathcal{Q}_1^* = \mathcal{Q}_2$ on $(C_c^\infty)^{0,q}(U_o)$;

(c) Say $W \subset U_o$. Say $g \in (E')^{0,q}(U_o)$ and g is analytic on W. Then Kg is analytic on W.

(d) If $f \in (E')^{0,q}(U_o)$, then $\mathcal{Q}_1 f$, $\mathcal{Q}_2 f$ are analytic on U_o.

10. Now let us suppose that M, $\mathcal{U}$ are as in Theorem 9.6, and additionally that M is compact. Let $\Box_b^q$ denote $\Box_b$ on (0,q) forms. If $q \neq k$ or $n - k$, then $\Box_b$ is hypoelliptic, and analytic hypoelliptic near p. As is well-known [19], [20], there is a "Hodge theory" for $\Box_b^q$. Thus range $\Box_b^q$ is closed, and ker $\Box_b^q$ is finite-dimensional. Of course $L_2^{0,q}(M) = \text{ran } \Box_b^q \oplus \ker \Box_b^q$, and the functions in ker $\Box_b^q$ are real analytic on $\mathcal{U}$. Let P_q be the projection in $L_2^{0,q}(M)$ onto ker $\Box_b^q$ and let N_q be the self-adjoint operator

annihilating ker $\square_b^q$ and inverting $\square_b^q$ on ran $\square_b^q$ so that $N_q\square_b^q = \square_b^q N_q = I - P_q$, $P_qN_q = 0$ and $\vartheta_b N_q$, $\bar\partial_b N_q$, $\bar\partial_b\vartheta_b N_q$, $\vartheta_b\bar\partial_b N_q : L_2^{0,q} \to L_2^{0,q'}$ (q' suitable). Clearly $N_q : \Lambda^{0,q} \to \Lambda^{0,q}$ since $\square_b^q$ is hypoelliptic, so we may extend N_q : $(\mathcal{E}')^{0,q} \to (\mathcal{E}')^{0,q}$; also we may extend $P_q : (\mathcal{E}')^{0,q} \to (\mathcal{E}')^{0,q}$. Since $\square_b^q$ is analytic hypoellilptic on $\mathcal{U}$, N_q is analytic pseudolocal on $\mathcal{U}$.

Let S_q, T_q denote the projection onto ker $\bar\partial_b$, ker ϑ_b (respectively) in $L_2^{0,q}$. We then have the following corollary, of which part (b) is well-known.

Corollary 9.7. In the situation of Theorem 9.6, assume M is also compact. Again assume $q \neq k$ or $n - k$. Then:

(a) N_q, S_q, T_q are locally in $\mathcal{O}^{-2n}$, $\mathcal{O}^{-2n-2}$, $\mathcal{O}^{-2n-2}$ (respectively) near p after ψ, up to an analytic regularizing error.

(b) (i) $S_q, T_q : (\mathcal{E}')^{0,q}(M) \to (\mathcal{E}')^{0,q}(M)$ are analytic pseudo- on $\mathcal{U}$ (in the sense explained in #8 above.)

(ii) Suppose $f \in (\mathcal{E}')^{0,q}(M)$, and $\bar\partial_b f$ is analytic on $U \subset \mathcal{U}$. Then $(I-S_q)f$ is analytic on U. If instead $\vartheta_b f$ is analytic on $U \subset \mathcal{U}$, then $(I-T_q)f$ is analytic on U.

Proof. Select K, $\mathcal{Q}_1$, $\mathcal{Q}_2$, U_0, U as in Theorem 8.11. Then we claim that for $f \in (\mathcal{E}')^{0,q}(U_0)$ we have

$$N_q f = \mathcal{K} f + g \tag{9.21}$$

where g is analytic on U_o. This gives (a) for N_q. To see (9.21), note that for $f \in (\mathcal{E}')^{0,q}(U_o)$, we may write $\mathcal{K}f = N_q \Box_b^q \mathcal{K} f + P\mathcal{K}f = N_q(f+\mathcal{Q}f) + P\mathcal{K}f$ as desired, by the analytic pseudolocality of N on $\mathcal{U}$. For S_q, we need only note

$$I - S_q = \vartheta_b \bar{\partial}_b N_q \tag{9.22}$$

on $L_2^{0,q}$. Indeed, if $f \in L_2^{0,q}$, and $g = \vartheta_b \bar{\partial}_b N_q f$, then $f - g = \bar{\partial}_b \vartheta_b N_q f + P_q f \in \ker \bar{\partial}_b$, $g \perp \ker \bar{\partial}_b$. By (9.22), we may extend $S_q : (\mathcal{E}')^{0,q} \to (\mathcal{E}')^{0,q}$. This gives (a) and (b)(i) at once for S_q; similarly for T_q. By self-adjointness of S_q, we also have $I - S_q = N_q \vartheta_b \bar{\partial}_b$ on $(\mathcal{E}')^{0,q}$, giving (b)(ii) for S_q; similarly for T_q.

10. Analytic Pseudolocality of the Szegö Projection and Local Solvability

Let M be a smooth compact CR manifold of dimension 2n+1. Suppose $\mathcal{U} \subset M$ is open, real analytic, and strictly pseudoconvex. Under a further "global" assumption on M, which is automatically satisfied if M is the boundary of a bounded smooth pseudoconvex domain in $\mathbb{C}^{n+1}$, or more generally, if the range of $\bar{\partial}_b : C^\infty(M) \to \Lambda^{0,1}(M)$ is closed in the C^∞ topology, we shall show that the Szegö projection S on M is "analytic pseudolocal" on $\mathcal{U}$. That is:

(i) If $V \subset \mathcal{U}$ is open, and f is real analytic on V, then Sf is also real analytic on V.

Among other consequences, this will lead to an analogue of the Greiner-Kohn-Stein Theorem for the Lewy equation on $\mathbb{H}^n$ ([35]; that is,

(ii) If $p \in \mathcal{U}$, then $\partial_b u = f$ is locally solvable near p if and only if Sf is real analytic near p.

Here S is the orthogonal projection onto ker $\bar{\partial}_b$ in L^2, where we are taking $\bar{\partial}_b : L^2(M) \to (L^2)^{0,1}(M)$ to be the closure of $\bar{\partial}_b : C^\infty(M) \to \Lambda^{0,1}(M)$.

We must be careful about the <u>class</u> of f under consideration in (i) and (ii), since our global hypothesis will be too weak to imply $S : \mathcal{E}'(M) \to \mathcal{E}'(M)$. (i) and (ii) will always hold for $f \in L^2$. <u>If</u>, in fact, $S : C^\infty(M) \to C^\infty(M)$ continuously, then of course $S : \mathcal{E}' \to \mathcal{E}'$, and, under the global hypothesis, we shall show that (i) and (ii) hold for $f \in \mathcal{E}'$.

(Cases in which it is known that $S : C^\infty(M) \to C^\infty(M)$ continuously include: M strictly pseudoconvex, $n > 1$ (see equation (10.12) below); M the boundary of a bounded, smooth strictly pseudoconvex domain in $\mathbb{C}^2$ ([52]); M the boundary of a bounded smooth pseudoconvex domain in $\mathbb{C}^2$, if each point is of finite type ([57]); M the boundary of a bounded smooth pseudoconvex domain in $\mathbb{C}^n$ $(n>2)$, if each point is of finite ideal type [57]. We do not need to assume M has any of these forms; but our theorem covers all these cases, since the range of $\bar{\partial}_b : C^\infty(M) \to \Lambda^{0,1}(M)$ is closed in the C^∞ topology in all these cases, so the global hypothesis holds, as we shall explain presently.)

<u>Part</u> of our global assumption is:

For all $f \in C^\infty(M)$ there exists $u \in (\mathcal{E}')^{0,1}$ so that

$$\vartheta_b u = (I-S)f. \tag{10.1}$$

This is a very weak version of a "closed range" hypothesis for ϑ_b, since it heuristically states that the range of ϑ_b is the orthocomplement of the kernel of $\bar{\partial}_b$. We will discuss conditions under which (10.1) holds, in a moment.

Under condition (10.1), we shall show that the map $f \to Sf|_{\mathcal{U}}$ is continuous from $C^\infty(M)$ to $C^\infty(\mathcal{U})$. Let $\mathcal{V} \supset \mathcal{U}$ be any fixed open subset of M such that the map $f \to Sf|_{\mathcal{V}}$ is continuous from $C^\infty(M)$ to $C^\infty(\mathcal{V})$. Then of course we may extend $S : \mathcal{E}'(\mathcal{V}) \to \mathcal{E}'$. Under the <u>full</u> global assumption, stated below, we shall show that (i) and (ii) above hold for $f \in \mathcal{E}'(\mathcal{V}) + L^2(M)$. Define $\mathcal{D}(S)$, the domain of S, to be $\mathcal{E}'(\mathcal{V}) + L^2(M)$.

The <u>full</u> global hypothesis on M is the following:

For all $f \in \mathcal{D}(S)$ there exists $u \in (\mathcal{E}')^{0,1}$ so that

$$\bar{\partial}_b u = (I-S)f. \tag{10.2}$$

(10.2) follows if one knows that the range of $\bar{\partial}_b$: $C^\infty(M) \to \Lambda^{0,1}(M)$ is closed in the C^∞ topology, or more generally, from the following criterion:

<u>Proposition 10.1</u>. Let M be a smooth, compact CR manifold.

Say $t \geqq 0$. Suppose that there exist $s \in \mathbb{R}$, $C > 0$ so that:

for all $u \in C^\infty(M)$ there exists $v \in H^t(M)$ so that for some sequence $\{v_m\} \in C^\infty(M)$, $v_m \to v$ in H^t, $\bar{\partial}_b v_m \to \bar{\partial}_b u$ in H^s, and $\|v\|_t \leqq C\|\bar{\partial}_b u\|_s$. (10.3)

Then for all $f \in \mathcal{D}(S)$ with $(I-S)f \quad H^{-t}$, there exists $\omega \in (\mathcal{E}')^{0,1}$ so that $\bar{\partial}_b \omega = (I-S)f$. (10.4)

(Here H^s, H^t denote Sobolev spaces.)

<u>Proof</u>. Let $\bar{\partial}_b|_s^t$ denote the closure of $\bar{\partial}_b : C^\infty \to \Lambda^{0,1}$, considered as an operator from H^t to H^s. In this proof, we

simply write $\bar{\partial}_b = \bar{\partial}_b|_s^t$. Then, in (10.3), we may simply write $\bar{\partial}_b v = \bar{\partial}_b u$. If P is the projection onto ker $\bar{\partial}_b$ in H^t, we may assume $v = (I-P)v$, since $\|(I-P)v\|_t \leq \|v\|_t$. But then clearly $v = (I-P)u$. Thus we may assume $v = (I-P)u$, and consequently the range of $\bar{\partial}_b$ is closed in H^s. Then $\bar{\partial}_b^* : H^{-s} \to H^{-t}$ also has closed range. Indeed, suppose $g \in H^{-s}$, g orthogonal to ker $\bar{\partial}_b^*$, and $g \in$ dom $\bar{\partial}_b^*$. Select $g_1 \in H^s$ so that $\langle g,h\rangle_{-s} = \langle g_1,h\rangle$ for all $h \in H^{-s}$. Then $g_1 = \bar{\partial}_b w$ for some $w \in H^t$ with $\|w\|_t \leq C\|g_1\|_s$, and we have

$$\langle g_1,g_1\rangle_s = \langle g_1,g\rangle = \langle\bar{\partial}_b w,g\rangle = \langle w,\bar{\partial}_b^* g\rangle \leq \|w\|_t \|\bar{\partial}_b^* g\|_{-t}$$

$$\leq C\|g_1\|_s \|\bar{\partial}_b^* g\|_{-t};$$

hence $\|g\|_{-s} = \|g_1\|_s \leq C\|\bar{\partial}_b^* g\|_{-t}$, proving that the range of $\bar{\partial}_b^*$ is closed. Thus ran $\bar{\partial}_b^* = \{u \in H^{-t} : \langle u,v\rangle = 0$ for all $v \in$ ker $\bar{\partial}_b\} = \{u \in H^{-t} : \langle u,v\rangle = 0$ for all $v \in$ ker $\bar{\partial}_b \cap C^\infty\}$, and the Proposition follows at once.

Evidently, then, (10.4) would hold if one knew any of the following:

The range of $\bar{\partial}_b : C^\infty(M) \to \Lambda^{0,1}(M)$ is closed in the C^∞ topology; (10.5)

The range of the closure of $\bar{\partial}_b|_{C^\infty}$, as an operator on H^t, is closed; (10.6)

For some $s \geqq t$, $C > 0$, we have: For all $u \in C^{\infty}(M)$ there exists $v \in H^{s+1}(M)$ so that $\bar{\partial}_b v = \bar{\partial}_b u$ and $\|v\|_t \leqq C\|\bar{\partial}_b u\|_s$. (10.7)

Now we describe situations under which (10.5), (10.6) or (10.7) hold for all $t \geqq 0$. Hence, our theorem will hold in any of these situations.

(10.5) holds if M is strictly pseudoconvex, $n > 1$ ([19], page 88). If M is the boundary of a bounded smooth pseudoconvex domain in $\mathbb{C}^{n+1}$, we have (10.6), for all $t \geqq 0$ ([58]). In the latter setting, (10.7) can be obtained more simply than (10.6) just by combining the main theorem of [55] with the method of Kohn-Rossi ([58], pages 540-1). In fact, (10.7) holds for all $t \geqq 0$ if M is the boundary of a smooth pseudoconvex complex manifold $D \subset\subset D'$, another complex manifold, provided there exists a nonnegative function λ on D' which is strongly plurisubharmonic in a neighborhood of M. (For this, again combine [55] with [58], pages 540-1. The latter argument gives, in this setting, $\bar{\partial}_b u = \bar{\partial}_b v + \Theta_b$ where Θ is a weighted harmonic form, and where Θ_b, its restriction to $\mathbb{C}TM$, is plainly in the range of $\bar{\partial}_b$. It is then easy to express $\Theta_b = \bar{\partial}_b w$ where w satisfies good estimates, simply by using the fact that any two norms on a finite-dimensional spaces are equivalent.) In fact, in any of these settings, we do have (10.5), as an argument of Hörmander [56], coupled with the arguments already cited, show.

Here, then, is the theorem we shall prove.

Theorem 10.2. Let M be a smooth compact CR manifold of dimension 2n+1. Suppose $\mathcal{U} \subset M$ is open, and is a real analytic, strictly pseudoconvex CR manifold. Suppose <,> is a smooth Hermitian metric on $\mathbb{C}TM$, analytic on $\mathcal{U}$, which is compatible with the CR structure.

Let S be the Szegö projection (onto ker $\bar{\partial}_b$ in L^2). Assume (partial global assumption):

$$\text{For all } f \in C^\infty(M) \text{ there exists } u \in (\mathcal{E}')^{0,1} \text{ so that } \vartheta_b u = (I-S)f. \tag{10.8}$$

Then:

$$\text{The map } f \to Sf|_{\mathcal{U}} \text{ is continuous from } C^\infty(M) \text{ to } C^\infty(\mathcal{U}). \tag{10.9}$$

Fix an open subset $\mathcal{V} \supset \mathcal{U}$ of M so that the map $f \to Sf|_{\mathcal{V}}$ is continuous from $C^\infty(M)$ to $C^\infty(\mathcal{V})$, and let $\mathcal{D}(S) = \mathcal{E}'(\mathcal{V}) + L^2(M)$. Assume (full global assumption):

$$\text{For all } f \in \mathcal{D}(S) \text{ there exists } u \in (\mathcal{E}')^{0,1} \text{ so that } \vartheta_b u = (I-S)f. \tag{10.10}$$

Then:

(a) $S : \mathcal{D}(S) \to \mathcal{E}'$ is analytic pseudolocal on $\mathcal{U}$. That is, if $f \in \mathcal{D}(S)$ is analytic on an open subset V of $\mathcal{U}$, then Sf is analytic on V.

(b) If $g \in \mathcal{D}(S)$, $p \in \mathcal{U}$, the following are equivalent:

(i) $\square_b f = g$ is locally solvable near p (for $f \in \mathcal{E}'$);

(ii) $\bar{\partial}_b \omega = g$ is locally solvable near p (for $\omega \in (\mathcal{E}')^{0,1}$);

(iii) Sg is real analytic near p.

(c) Let V be an open subset of $\mathcal{U}$. If $f \in \mathcal{D}(S)$ and $\bar{\partial}_b f$, or even just $\square_b f$, is analytic on V, then (I-S)f is analytic on V.

In (d), (e) we assume $\mathcal{U}$ sufficiently small that there is an analytic contact transformation $\psi : \mathcal{U}' \to \mathcal{U}$, $\mathcal{U}' \subset \mathbb{H}^n$. Let $p \in \mathcal{U}$.

(d) S is locally in $\mathcal{O}^{-2n-2}$ near p, after ψ, up to an analytic regularizing error.

(e) For certain open neighborhoods $U_0 \subset\subset U \subset\subset \mathcal{U}$ of p, there exists an operator K so that $K : \mathcal{E}'(U_0) \to \mathcal{E}'(U)$ and $\mathcal{E}'(U) \to \mathcal{D}'(U)$, so that K is locally in $\mathcal{O}^{-2n}(\psi^{-1}(U))$ after ψ, and so that whenever $f \in \mathcal{E}'(U_0)$,

$$\square_b^0 K f = (I-S)f + g_1, \quad K\square_b^0 f = (I-S)f + g_2 \qquad (10.11)$$

on U_0, where g_1, g_2 are analytic on U_0. Also, $K : C_c^\infty(U_0) \to C^\infty(U)$, and $K = K^*$ on $C_c^\infty(U_0)$.

(f) Say $p \in \mathcal{U}$. Then there exist $g \in \mathcal{E}'(M)$ so that $\bar{\partial}_b \omega = g$ is not locally solvable near p. (This part is well-known.)

Corollary 10.3. Let $D \subset \mathbb{C}^{n+1}$ be a bounded pseudoconvex domain with smooth boundary M. Suppose $\mathcal{U} \subset M$ is open, real analytic, and strictly pseudoconvex. Suppose <,> is a smooth Hermitian metric on $\mathbb{C}TM$, analytic on $\mathcal{U}$, which is compatible with the CR structure. Then (10.9) holds. Fix an open subset $\mathcal{V} \supset \mathcal{U}$ of M so that the map $f \to Sf|_{\mathcal{V}}$ is continuous from $C^\infty(M)$ to $C^\infty(\mathcal{V})$, and let $\mathcal{D}(S) = \mathcal{E}'(\mathcal{V}) + L^2(M)$.

Then Theorem 10.2 (a)-(f) hold. Further, (b)(i)-(iii) are equivalent to:

(b)(iv) Sg may be extended to be a holomorphic function in a neighborhood of p in $\mathbb{C}^{n+1}$.

Proof of Theorem 10.2 and Corollary 10.3. Corollary 10.3 will follow at once from Theorem 10.2 and the remarks before it. For (b)(iii) =>(b)(iv), we note that Sg is real analytic near p, and that $\bar{\partial}_b Sg = 0$ near p. (b)(iv) then follows from a theorem of Tomassini [82]. (b)(iv) =>(b)(iii) is trivial.

We now show how Theorem 10.2(b) and (c) follow immediately once (10.9), (a), (d) and (e) are shown. For now, assume (10.9), (a), (d) and (e).

For (c), assume $\Box_b f$ is analytic on V, $p \in V$. To show (I-S)f is analytic near p, note that we may assume $\mathcal{U}$ is small enough that (e) applies. Select U_0 as in (e). By (a), and the fact that multiplication by a C_c^∞ function preserves $\mathcal{D}(S)$, we may assume supp f is contained in any neighborhood V_0 of p that we please ($V_0 \subset U_0$). We choose V_0 so that (e) and Theorem 7.11(b) give us (c).

For (b), (i) => (ii) is trivial. For (ii) => (iii), we may assume $w \in \mathcal{E}'(\mathcal{U})$; then (ii) => (iii) is clear since $Sg = S(g-\bar{\partial}_b\omega)$ is analytic near p by analytic pseudolocality of S. For (iii) => (i), write $g = Sg + (I-S)g$. The first term is locally in the range of $\Box_b$ (near p) by Cauchy-Kowaleski. Let $\varphi \in C_c^\infty$, $\varphi = 1$ near p. Then $(I-S)[(1-\varphi)g]$ is locally in the range of $\Box_b$ (near p) by (a) and Cauchy-Kowaleski, while $(I-S)(\varphi g)$ is locally in the range of $\Box_b$ (near p) by (e) and Cauchy-Kowalewski, for supp φ sufficiently small. This proves (iii) => (i).

Thus we need only prove (10.9), (a), (d), (e) and (f) of Theorem 10.2.

We shall examine several cases:

(1) $n > 1$, M strictly pseudoconvex;

(ii) $n = 1$; $M = \partial D$, D a smooth strictly convex domain in $\mathbb{C}^2$ which is analytic in a neighborhood of $\mathcal{U}$;

(iii) the general case.

In our discussions of case (i) and (ii), we shall seek to establish only some of the desired conclusions. All the conclusions will be established in our discussion of Case (iii).

Case (i). $n > 1$, M strictly pseudoconvex.

In this case, we establish (10.9), (a), (d) and most of (e).

All of these conclusions follow easily from Kohn's formula

$$I - S = \vartheta_b N_1 \bar{\partial}_b \tag{10.12}$$

where N_1 is as in #10 of Chapter 9. This is immediately verified on $C^\infty(M)$. (Indeed, say $f \in C^\infty$, $h = N_1\bar{\partial}_b f$, and $g = \vartheta_b N_1 \bar{\partial}_b f$. Then $\square_b^1 h = \bar{\partial}_b f$. Applying $\vartheta_b\bar{\partial}_b$ to both sides, we see $\vartheta_b\bar{\partial}_b h = 0$, so $\bar{\partial}_b h = 0$, so $\bar{\partial}_b g = \square_b^1 h = \bar{\partial}_b f$. Since $f = (f-g) + g$, $f - g \in \ker \bar{\partial}_b$, $g \perp \ker \bar{\partial}_b$, we must have $g = (I-S)f$.) Thus $S : C^\infty \to C^\infty$, proving (10.9); we extend $S : \mathcal{E}' \to \mathcal{E}'$, and (10.12) still holds on $\mathcal{E}'$. (a) and (d) now follow at once. Further, it follows at once from (10.12) that

$$I - S = \square_b B = B\square_b \tag{10.13}$$

with

$$B = \vartheta_b N_1^2 \bar{\partial}_b \tag{10.14}$$

Since $\vartheta_b N_1^2 \bar{\partial}_b$ is locally in $\mathcal{O}^{-2n}$ near p, after ψ, up to an analytic regularizing error, we have (e)-except that we have not yet shown that K, in (e), may be chosen to be self-adjoint on $C_c^\infty(U_0)$.

<u>Case (ii)</u>. $n = 1$; $M = \partial D$, D a smooth strictly convex domain in $\mathbb{C}^2$ which is analytic in a neighborhood of $\mathcal{U}$.

In this case, we carry through the analysis just far enough to lay the groundwork for the general case.

In this case, there is no "Hodge theory" for $\square_b^1$, and the approach through (10.12) is not available. In fact, if we merely assumed n = 1, M strictly pseudoconvex, we could not expect an analogue of (10.13) with B (everywhere) locally in the C^∞ analogue of $\mathcal{O}^{-2n}$ after a contact transformation. For this would imply that the range of $\bar{\partial}_b$ on $L_2^{0,1}(M)$ is closed. As is known [57], [11], if this holds then M is embeddable in $\mathbb{C}^N$ for some N, and this is not true in general [69].

We base our analysis instead on the work of Henkin [43].

We may write $D = \{\rho < 0\}$, ρ smooth in a neighborhood of $\bar{D}$ and analytic in a neighborhood of p in $\mathbb{C}^2$. We may assume ρ is analytic on $\mathcal{U}$. Further, for some $C > 0$,

$$\sum \frac{\partial^2 \rho}{\partial w_i \partial \bar{w}_j}\Big|_\xi w_i \bar{w}_j \geq C|w|^2 \qquad \text{for all } \xi \in M,\ w \in T_\xi(M) \qquad (10.15)$$

and $(\operatorname{grad} \rho)(\xi) \neq 0$ for any $\xi \in M$.

We let $\sigma = i(\bar{\partial} - \partial)\rho$, the natural contact form on M. We may assume $\mathcal{U}$ is small enough that there is an analytic (0,1) form $\bar{\omega}$ on $\mathcal{U}$, with $\langle \bar{\omega}, \bar{\omega} \rangle = 1$ at each point, and that there is an analytic contact transformation $\psi: \mathcal{U}' \to \mathcal{U}$, $\mathcal{U}' \subset \mathbb{H}^1$. We use the basis $\{\bar{\omega}\}$ in forming pullbacks to $\mathbb{H}^1$ (as in #8 of Chapter 9). We shall need the following Proposition.

<u>Proposition 10.4</u>. There exist operators R,H so that

(i) $R:\Lambda^{0,1}(M) \to C^\infty(M)$, $H:C^\infty(M) \to C^\infty(M)$.

(ii) $R:L_2^{0,1}(M) \to L^2(M)$ continuously, $H:L^2(M) \to L^2(M)$ continuously.

(iii) $\bar\partial_b H = 0$ on $C^\infty(M)$.

(iv) $R\bar\partial_b = I-H$ on $C^\infty(M)$.

(v) $H:L^2(M) \to L^2(M)$ is a projection onto $\ker \bar\partial_b$.

(vi) R, acting on $\Lambda^{0,1}$, is locally in $\mathcal{O}^{-3}$ near p, after ψ. H, acting on C^∞, is locally in $\mathcal{O}^{-4}$ near p, after ψ.

Proof. The main point is the verification of (vi). (i)-(v) were either proved by Henkin or follow from the methods of [20]. At the end of the proof we shall sketch arguments for (i)-(iii). Note that (v) is an immediate consequence of (i)-(iv).

We need the explicit formula for R. ρ, again, is smooth in an open neighborhood of $\bar D$; for ζ, z in this neighborhood, Henkin sets

$$\Phi(\zeta,z) = \sum_{k=1}^{2} P_k(\zeta)(\zeta_k - z_k),\ P(\zeta) = (P_1(\zeta), P_2(\zeta)),$$

$$\text{where } P_k(\zeta) = \frac{\partial\rho}{\partial z_k}(\zeta).$$

Then if $f \in \Lambda^{0,1}(M)$, Rf is the following function on M:

$$(Rf)(z) = (2\pi i)^{-2} \int_{\zeta\in M} K(z,\zeta) f(\zeta) \wedge d\zeta_1 \wedge d\zeta_2 \tag{10.16}$$

where

$$K(z,\zeta) = [P_1(z)P_2(\zeta) - P_1(\zeta)P_2(z)]/[\Phi(\zeta,z)\Phi(z,\zeta)] \qquad (10.17)$$

(This is equivalent to the formula right after equation (3.3) in [43].)

We may assume $\psi(0) = p$. We first seek to show that there exists a neighborhood U_0' of $0 \in \mathbb{H}^1$ with $\overline{U}_0' \subset \mathcal{U}'$, and there exist $r > 0$, $K_0^m \in AK^{-3+m}(m \in \mathbb{Z}^+)$ as follows: If

$$K_0(w) = \chi_{\{0<|w|<r\}}(w) \sum_{m=0}^{\infty} K_0^m(w), \qquad (10.18)$$

then the sum converges absolutely for $0 < |w| < r$, and for $f \in \Lambda^{0,1}(M)$ with supp $f \subset \psi(U_0')$,

$$(Rf)(p) = \int_{\mathbb{H}^1} K_0(w^{-1}) f_\psi(w)\,dw \quad \text{on} \quad \psi(U_0'). \qquad (10.19)$$

For (10.19), we may assume $p = 0$. For $w \in \mathcal{U}'$, we examine $K_1(w) = K(0,\psi(w))$. We examine the denominator in the expression for $K_1(w)$ obtained from (10.18). It will be enough to show that, if $w = (z,t) \in \mathcal{U}'$, then

$$\Phi(\psi(w),0) = [Q(z,z) + i(at+Q_1(z,z))] + O^3 \qquad (10.20)$$

$$\Phi(0,\psi(w)) = [Q(z,z) - i(at+Q_1'(z,z))] + O^3. \qquad (10.21)$$

Here Q is a positive definite quadratic form, and Q, Q' are other quadratic forms. In general, O^m denotes a smooth function in a neighborhood of 0, all of whose Taylor coefficients of weighted degree less than m vanish; here, of course, O^3 will be analytic on $\mathcal{U}'$. Once we know (10.20), (10.21), then since the expressions in square brackets are comparable to $|w|^2$, the denominator for $K_1(w)$ will have an expansion as $\sum_{m=0}^{\infty} K^m$, $K^m \in AK^{-4+m}$, for w small. Since the numerator arising from (10.17) vanishes to first order as $w \to 0$, and since the Jacobian arising from using ψ to change variable in (10.16) is analytic near 0, we shall have (10.19).

To see (10.20) and (10.21), for $\zeta \in M$, identify $\zeta \in \mathbb{C}^2$ with $\xi \in \mathbb{R}^4$ by $\zeta_1 = \xi_1 + i\xi_2$, $\zeta_2 = \xi_3 + i\xi_4$. Then $\mathrm{Re}\,\Phi(0,\zeta) = -\frac{1}{2}\nabla\rho(0)\cdot\xi$, and $\mathrm{Re}\,\Phi(\zeta,0) = -\frac{1}{2}\nabla\rho(\xi)\cdot\xi$ (explaining why the denominator in (10.17) cannot vanish, D being strictly convex). By Taylor's Theorem at 0,

$$0 = \rho(\xi) = 0 + \nabla\rho(0)\cdot\xi + \frac{1}{2}\sum\frac{\partial^2\rho}{\partial\xi_j\partial\xi_k}(0)\xi_j\xi_k + O(|\xi|^3) \qquad (10.22)$$

$$0 = \rho(0) = 0 - \nabla\rho(\xi)\cdot\xi + \frac{1}{2}\sum\frac{\partial^2\rho}{\partial\xi_j\partial\xi_k}(\xi)\xi_j\xi_k + O(|\xi|^3) \qquad (10.23)$$

Now ξ agrees with an element of $T_0(M)$ to second order in ξ, for $|\xi|$ small. From (10.15), (10.22) we now see $c_1|\zeta|^2 \geq \mathrm{Re}\,\Phi(0,\zeta) \geq c_2|\zeta|^2$ for ζ small, for some $c_1, c_2 > 0$. Thus, $c_1'|w|^2 \geq \mathrm{Re}\,\Phi(0,\psi(w)) \geq c_2'|w|^2$ for w small. This at once gives us the real part of (10.21). Also, by (10.22) and (10.23), $\mathrm{Re}\,\Phi(0,\zeta) - \mathrm{Re}\,\Phi(\zeta,0)$ vanishes to third order in ζ as $\zeta \to 0$, so the claim for the real part in (10.20) follows from that of (10.21).

As for the imaginary parts, it will suffice to show that they are of the form $\pm at$ $(a\neq 0)$, plus terms vanishing to second order at $w = 0$. It suffices to consider $\Phi(0,\psi(w))$, since $\Phi(0,\zeta) + \Phi(\zeta,0)$ vanishes to second order as $\zeta \to 0$. Define $J : \mathbb{R}^4 \to \mathbb{R}^4$ by $J(\xi) = (-\xi_2,\xi_1,-\xi_4,\xi_3)$ (equivalently, $J(\zeta) = i\zeta$). Let $v = \nabla\rho(0)$. Then $\operatorname{Im}\Phi(0,\zeta) = -\frac{1}{2}Jv\cdot\xi$. Put $b = \|v\|/2$. By making a unitary change of coordinates in $\mathbb{C}^2$ if necessary, we may assume $v = 2b(1,0,0,0)$; then $Jv = 2b(0,1,0,0)$. Since $\nabla\rho(0) = 2b(1,0,0,0)$, by the implicit function theorem we may use ξ_2, ξ_3, ξ_4 as coordinates on M near 0. By using ψ we may also use $(t,x,y) \in \mathbb{H}^1$ as coordinates on M near 0. Now, writing $w = (t,x,y)$, we have by Taylor's Theorem

$$\operatorname{Im}\Phi(0,\psi(w)) = -b\psi_2(w)$$

$$= -b\left[\frac{\partial\psi_2}{\partial t}(0)t + \frac{\partial\psi_2}{\partial x}(0)x + \frac{\partial\psi_2}{\partial y}(0)y\right] \text{ plus terms vanishing to second order at 0.}$$

We need to show $\frac{\partial\psi_2}{\partial t}(0) \neq 0$, $\frac{\partial\psi_2}{\partial x}(0) = \frac{\partial\psi_2}{\partial y}(0) = 0$. But - under the change of coordinates ψ- we have at 0 that

$$\frac{\partial}{\partial t} = \sum_{j=2}^{4} \frac{\partial\psi_j}{\partial t}\frac{\partial}{\partial\xi_j}, \quad \frac{\partial}{\partial x} = \sum_{j=2}^{4} \frac{\partial\psi_j}{\partial x}\frac{\partial}{\partial\xi_j}, \quad \frac{\partial}{\partial y} = \sum_{j=2}^{4} \frac{\partial\psi_j}{\partial y}\frac{\partial}{\partial\xi_j}. \quad (10.24)$$

At 0, $\tau(\frac{\partial}{\partial t}) = 1$, $\tau(\frac{\partial}{\partial x}) = \tau(\frac{\partial}{\partial y}) = 0$. Also $d\rho = 2bd\xi_1$ at 0, so that the contact form $\sigma = 2bd\xi_2$ there. Of course $d\xi_2(\frac{\partial}{\partial\xi_j}) = \delta_{2j}$. Applying these relations to (10.24), and using the fact that ψ is a contact transformation, we find the relations we need for the derivatives of ψ at 0. (10.20) and (10.21) follow, hence (10.19) follows. r in (10.18) clearly depends only on the radii of convergence of ψ near $0 \in \mathbb{H}^1$ and of ρ near $0 \in M$, and U_0' need only be small enough that $|w| < r$ for all $w \in U_0'$.

We can now prove (vi) for R. If $\eta \in \mathcal{U}$, note that the map ψ_η given by $\psi_\eta(w) = \psi(\psi^{-1}(\eta)w)$, for $\psi^{-1}(\eta)w \in \mathcal{U}'$, is a contact transformation taking 0 to η. By the same reasoning as that leading to (10.19), we have a K_η, r_η and a neighborhood U_η' of $0 \in \mathbb{H}^n$ so that for $f \in \Lambda^{0,1}(M)$, supp $f \subset \psi_\eta(U_\eta') = \psi(\psi^{-1}(\eta)U_\eta')$, we have

$$Rf(\eta) = \int f_{\psi_\eta}(w)K_\eta(w^{-1})\,dw = \int f_\psi(\psi^{-1}(\eta)w)K_\eta(w^{-1})\,dw =$$
$$= (f_\psi * K_\eta)(\psi^{-1}(\eta)),$$

which virtually gives (vi) for R. One has only to observe that, for η close enough to p, we may choose the number r_η, and hence the open set U_η', independently of η. Thus, for η close enough to p, there will be an open neighborhood U" of $0 \in \mathbb{H}^1$ contained in all the sets $\psi^{-1}(\eta)U_\eta'$, and (vi) follows for R. The required estimates (7.72)-(7.78) for a core in C^{-3} follow at once from (10.17), together with (10.20), (10.21) and the analogous equations appropriate for η, suitable complexifications of all these equations, and the method of majorants, comparing to the power series for $[1-s/(1-s)]^{-1}$ for small s. In effect, one verifies the condition (*) which follows (7.78). The "Q" in (7.73) may be taken to be zero. Note that no log terms can arise.

As for H, for $f \in C^\infty(M)$, we have that Hf is the boundary value of the following function, defined and holomorphic for $z \in D$:

$$Hf(z) = (2\pi i)^{-2} \int_{\zeta \in M} f(\zeta) K_H(z,\zeta) \wedge d\zeta_1 \wedge d\zeta_2 \qquad (10.25)$$

where

$$K_H(z,\zeta) = [P_1(\zeta)(\bar{\partial}P_2)(\zeta) - P_2(\zeta)(\bar{\partial}P_1)(\zeta)]/[\Phi(\zeta,z)]^2. \qquad (10.26)$$

Henkin [43] shows that the boundary values exist, and also shows (iv) on $C^\infty(M)$. (Actually, Henkin assumes, but does not use, that (10.15) holds for all $\xi \in \bar{D}$, $w \in \mathbb{R}^4$, not just for $\xi \in M$, $w \in T_\xi(M)$. One can, by the way, start with a ρ satisfying (10.15) for $\xi \in M$, $w \in T_\xi(M)$, and replace ρ by $\tilde{\rho} = (e^{\lambda\rho}-1)$ for λ large. $\tilde{\rho}$ satisfies (10.15) for all $\xi \in M$, $w \in \mathbb{R}^4$.) (vi) for H follows at once from (iv) and the product rule Theorem 7.11(c).

(i)-(iii) follow from the methods of [20] and [52], and would also follow if we had stopped to develop a C^∞ analogue of our calculus. To keep this book to a reasonable length, and since Taylor's C^∞ calculus [79] would be equivalent, we have not done so here. For completeness, we sketch those arguments needed to complete the proof.

We may as well work in a neighborhood of a point, say $0 \in M$, since the kernels of the operators are smooth off the diagonal. Let $\Psi : (U_0', T) \to (U_0, \sigma)$ be a smooth contact transformation, $0 \in U_0' \subset \mathbb{H}^1$ open, $\Psi(0) = 0$. We select a smooth (0,1) form ω_0 on U_0 with $\langle\omega_0,\omega_0\rangle = 1$ at each point, and use it in forming pullbacks by Ψ to $\mathbb{H}^1$. In a neighborhood of $0 \in M$, we may use coordinates $\zeta \in \mathbb{C}^2$, or alternatively $(w,h) \in \mathbb{H}^1 \times \mathbb{R}$, where $\zeta = \Psi(w) - h\nu(\Psi(w))$, ν being the outward unit normal to ∂D at ζ.

To analyze H, first note $\Phi(\Psi(w),-h\nu(0)) = \Phi(\Psi(w),0) - (h/2)(V+iJV)\cdot\nu(0) = \Phi(\Psi(w),0)+(b+O(w))h$; here $V = \nabla\rho(\Psi(w))$, $b = \|\nabla\rho(0)\|/2$. Thus, if $w = (t,z)$,

$$\Phi(\Psi(w),-h\nu(0)) = P_0(h,t,z) + O^3 + O^3h \tag{10.27}$$

where

$$P_0(h,t,z) = Q(z,z) + q(z,\bar{z})h + i((a+ch)t + Q_1(z,z)) \tag{10.28}$$

where $q(z,\bar{z})$ is a polynomial of degree 2 in z, $\bar{z}$ with positive constant term b, and $c \in \mathbb{C}$. Now $P_0 \neq 0$ for $(0,0) \neq (z,h)$ small, since $|zh| <$ (small const)$|z|^2$ + (large const)h^2. Now if $f \in C_c^\infty(U_0)$, we find

$$Hf(-h\nu(0)) = \int_{\mathbb{H}^1} K_h(w^{-1}) f_\Psi(w)\,dw, \text{ where}$$

$$K_h(w^{-1}) = \sum_{m=0}^{N-1} F_{m0}(w,h)P_0^{-2-m} + G_{N0}(w,h)(P_0+O^3+O^3h)^{-2-N} \tag{10.29}$$

for any N, if w is small. The F_{m0}, G_{N0} are polynomials in h whose coefficients are smooth in w. For F_{m0} the coefficients are O^{3m}; for G_{N0} they are O^{3N}. By reasoning we just saw, we find that for v near 0

$$Hf(\Psi(v)-h\nu(\Psi(v))) = f_\Psi * K_{v,h}(v) \tag{10.30}$$

where $K_{v,h}$ has an expansion like (10.29), but where all functions and constants in (10.29), (10.28) depend smoothly on v. We call the corresponding functions P_v, F_{mv}, G_{Nv}. Now, convolution with the Nth error term of the expansion clearly gives a C^M function of (h,v) as $(h,v) \to (0,0)$, where $M \to \infty$ as $N \to \infty$. Convolution with any of the terms in the sum appearing in the expansion give C^∞ functions of (h,v) as $(h,v) \to (0,0)$. To see this, one uses the identity $[A+Bt]^{-M} = (-1)^{M-1}[(M-1)!B^{M-1}]^{-1}T^{M-1}[A+Bt]^{-1}$ to integrate by parts in t M -1 times before taking the limit as $(h,v) \to (0,0)$.

We have therefore shown (i) for H, as well as (iii), since Hf is holomorphic in D. We prove (ii) for H soon.

As for R, write $P_v^0(t,z) = P_v(0,t,z)$, P_v as just above. Then for some open set V containing 0, for $v \in V$ and $f_\Psi \in C_c^\infty(V)$, $Rf(\Psi(v)) = (f_\Psi * K_v)(v)$ where $K_v(w) = A_v(w)K_v'(w)K_v''(w)$. Here $K_v' = \sum_{m=0}^{N-1} F_{mv}'(P_v^0)^{-m-1} + G_{Nv}'(P_v^0+O^3)^{-1-N}$; F_{mv}' is O^{3m}, G_{Nv}' is O^{3N}. K_v'' has a similar form, the only differences arising from the differences between (10.20) and (10.21). Also $A_v(0) = 0$, and we can assume, by shrinking V if necessary, that $K_v(w)$ is smooth in (v,w) for $w \neq 0$ and vanishes unless both v and w are sufficiently small. One now easily shows that the function $g(v,v') = (f_\Psi * K_v)(v')$ is smooth in (v,v') for supp $f_\Psi \subset V$; this then shows (i) for R.

Define $\mathcal{R} : C_c^\infty(\Psi^{-1}(V)) \to C^\infty(\Psi^{-1}(V))$ by $\mathcal{R}(f_\Psi) = (Rf)_\Psi$. Write $K(v,w) = K_v(w)$. The kernel of $\mathcal{R}$ is then $K(v,w^{-1}v)$. Since $|K(v,w)| < C/|w|^{-3}$, C independent of v, we have $\int|K(w,v^{-1}w)|dw < C'$, so the kernel of $\mathcal{R}^* : C_c^\infty(V) \to L^\infty(V)$ is $\overline{K(w,v^{-1}w)}$. Thus the kernel L of $\mathcal{R}$ satisfies, for some $C > 0$,

$$\int|L(v,w)|dw < C, \int|L(v,w)|dv < C. \tag{10.31}$$

This implies $R : L^2(V) \to L^2(V)$, which implies (ii) for R.

We return to H. Using (10.30), and (10.29) for N = 1, and an analysis similar to that for R, we see that modulo operators with kernels satisfying conditions like (10.31), we have $Hf(\Psi(v)) = q(v)f_\Psi * TP_v$ for v near 0 and f supported near 0. It is now easy to use the Cotlar-Stein lemma, and to make simple changes in the proof of L^2 boundedness of convolution with principal value distributions in [66], pages 18-20, to show (iii) for H. This completes the proof of Proposition 10.4.

On a strictly convex domain, one could now get directly at S, in order to prove theorem 10.2, by using this observation of Kerzman and Stein [52]: since S and H project onto the same subspace, we have SH = H, HS = S ⇒ SH* = S, so S(I+H-H*) = H, so

$$S = H(I+H-H^*)^{-1}. \tag{10.32}$$

I + H - H* is invertible on L^2 because H - H* is skew-adjoint. Further, one may use Theorem 8.1 to show $(I+H-H^*)^{-1}$ is locally in $\mathcal{O}^{-4}$ near p, after ψ, up to an analytic regularizing error. One may then show Theorem 10.2(a) and (d), by using (10.32). Further, $[(I-S)R]\bar\partial_b = I - S$, since $(I-S)(I-H) = I - S$; this is partway toward Theorem 10.2(e).

Rather than carry out this plan at this point, we will find it a little more convenient to work on the formal level, and to wait to carry out the construction until we deal with the general case of Theorem 10.2.

We will need the analogue of Proposition 10.4 for ϑ_b. We need another formula of Henkin [43]. We can find an ambient metric $\langle,\rangle_A$ on $\mathbb{C}T^*\mathbb{C}^2$ in a neighborhood of M in $\mathbb{C}^2$, compatible with the complex structure, so that whenever η_1, η_2 are cotangent vectors at a point $q \in M$, $\langle\eta_1,\eta_2\rangle_A = \langle\eta_1|_{TM}, \eta_2|_{TM}\rangle$. We may assume $\langle,\rangle_A$ is analytic in a neighborhood of $\mathcal{U}$ in $\mathbb{C}^2$. We identify $\mathbb{C}T^*M$ with the orthocomplement of dr with respect to $\langle,\rangle_A$, at each point.

By (10.15), for some $\varepsilon_0 > 0$, the domain $D_\varepsilon = \{\rho<\varepsilon\}$ is also strictly convex for all $0 < \varepsilon \leqq \varepsilon_0$. Then [43]

$$\bar\partial_b R = I - G_1 \quad \text{on} \quad \Lambda^{0,1}(M) \tag{10.33}$$

where if $f \in \Lambda^{0,1}(M)$, $z \in D_{\varepsilon_0} \setminus \bar{D}$, we define

$$G_{ext} f(z) = \int_{\zeta\in\partial D} K_G(\zeta,z) f(\zeta) \wedge d\zeta_1 \wedge d\zeta_2 \tag{10.34}$$

where

$$K_G(\zeta,z) = [P_1(z)\bar\partial_z P_2(z) - P_2(z)\bar\partial_z P_1(z)]/[\Phi(z,\zeta)]^2 \tag{10.35}$$

We shall verify momentarily that $G_{ext}f$ extends smoothly up to M. If g denotes the boundary value, we let $G_1 f = \mathcal{P}g$, where $\mathcal{P}$ is the orthogonal projection perpendicular to $\bar{\partial}r$ at each point, in the ambient metric in $\mathbb{C}T^*\mathbb{C}^2$.

The smoothness of $G_{ext}f$ up to M is verified in much the same manner as for H. Let $V_h = \nabla\rho(h\nu(0))$, $2b = \|\nabla\rho(0)\|$. Identify $\zeta \in \mathbb{C}^2$ with $\xi \in \mathbb{R}^4$. Observe $\Phi(h\nu(0),\zeta) = -\frac{1}{2}(V_h+iJV_h)\cdot(\xi-h\nu(0))$ $= -\frac{1}{2}(V_0+iJV_0)\cdot(\xi-h\nu(0)) + O(h)O(\zeta) + O(h^2) = \Phi(0,\zeta)+bh+O(h)O(\zeta)$ $+O(h^2)$. Using this with (10.21) we find analogues of (10.27), (10.28) for $\Phi(h\nu(0),\Psi(w))$. The error estimates are to be slightly modified; replace each occurence of $O^{3\ell}$ $(\ell \geq 1)$ in (10.29), (10.30) by $(O^3+O^3O(h)+O(h^2))^{\ell}$. One still sees $G_{ext}f$ extends smoothly up to M, and that $G_1 : L_2^{0,1} \to L_2^{0,1}$.

If η is the restriction to M of a form which is in $\Lambda^{2,0}$ on a neighborhood of M in $\mathbb{C}^2$, then $\int_M \bar{\partial}_b \eta = 0$. Indeed, if η is extended smoothly to $\bar{D}$, $\int_M \bar{\partial}_b \eta = \int_D d\bar{\partial}\eta = \int_D d^2\eta = 0$. Thus if $F(\zeta)$ is holomorphic on a neighborhood of $\bar{D}$, and $g \in C^{\infty}(M)$, then $\int_M F(\zeta)\bar{\partial}_b g \wedge d\zeta_1 \wedge d\zeta_2 = \int_M \bar{\partial}_b(Fg d\zeta_1 \wedge d\zeta_2) = 0$. From (10.34), (10.35) we now see $(G_{ext}\bar{\partial}_b g)(z) = 0$ for $z \in D_{\varepsilon_0} \setminus \bar{D}$, so by continuity,

$$G_1\bar{\partial}_b = 0 \quad \text{on } C^{\infty}(M). \tag{10.36}$$

As in Proposition 10.4, we see G_1 preserves $\Lambda^{0,1}(M)$, $(\mathcal{E}')^{0,1}(M)$ and $L_2^{0,1}(M)$; that (10.33) and (10.36) hold on $(\mathcal{E}')^{0,1}(M)$; that $I - G_1$ is a projection onto ran $\bar{\partial}_b \subset L_2^{0,1}(M)$, and that G_1

is locally in $\mathcal{O}^{-4}$ near p, after ψ.

Using $\vartheta_b R^* = I - G_1^*$, we could now show an analogue of Theorem 10.2(a) and (d) for ϑ_b, and the projection onto ker ϑ_b, in place of $\bar{\partial}_b$ and S, by a plan analogous to that we have sketched for S.

Indeed, however, we immediately turn to the general situation.

Case (iii). The general case.

We now prove (10.9), (a), (d), (e) and (f) of Theorem 10.2, in the general case.

We need the following well-known proposition.

Proposition 10.5. Let $U_0 \subset \mathbb{C}^{n+1}$ be open, and $r : U_0 \to \mathbb{R}$ be real analytic. Suppose $M_0 = \{z \in U_0 : r(z) = 0\}$, that $dr \neq 0$ on M_0, that $\Omega_0 = U_0 \cap \{z : r(z) < 0\}$, and that M_0, viewed locally as the boundary of Ω_0, is strictly pseudoconvex at each point. Say $p \in M_0$. Then there exist a biholomorphic map $\Phi : U \to U'$ for some open neighborhood U of p in $\mathbb{C}^n$, $U \subset U_0$, and a smooth, bounded, strictly convex domain D_0' so that $\Phi(M_0 \cap U) = \partial D_0' \cap U'$, $\Phi(\Omega_0 \cap U) = D_0' \cap U'$.

Proof. This is a simple version of Narasimhan's result [67]. To see it, we may assume p = 0. Note that by the usual E. E. Levi construction (see e.g. [20] pages 500-502), after a holomorphic change of coordinates, Ω_0 takes the form

$2\,\mathrm{Re}\, z_1 + \sum_{j,k=1}^{n+1} a_{j\bar{k}} z_j \bar{z}_k + f(z) < 0$ near $p = 0$, where f vanishes to at least third order at $p = 0$, and $a_{j\bar{k}} = \frac{1}{2}\frac{\partial^2 r}{\partial z_j \partial \bar{z}_k}(0)$. We may assume the summation here defines a strictly positive definite form on $\mathbb{C}^{n+1}$ (otherwise we may use the usual device of replacing r by $(e^{\lambda r}-1)/\lambda$, λ large). After the Cayley transform $z_1 = -(1+w_1)/(1-w_1)$, $z_k = iw_k/(1-w_1)\,(k>1)$, Ω_0 takes the form $2(|w_1|^2-1) + \sum_{j,k=1}^{n+1} a_{j\bar{k}} w_j' \bar{w}_k' + g(w) < 0$ near $p = (-1,0)$, where g vanishes to at least third order at p, and $w_j' = iw_j$ if $j > 1$, $w_1' = -(1+w_1)$. Choose $\varphi \in C_c^\infty(\mathbb{C}^{n+1})$, φ real-valued, with $\varphi = 1$ near 0. Then for sufficiently small $\varepsilon > 0$, we may choose D_0' to be $\{w : 2|w_1|^2 + \sum a_{j\bar{k}} w_j' \bar{w}_k' + \varphi((w+e_1)/\varepsilon)g(w) < 2\}$. (Here $e_1 = (1,0,\ldots,0)$.) ∎

It is now easy to see that (f) will follow once we have shown (10.9), (a), (d) and (e). Indeed, as we have seen, these statements imply the equivalence of (b)(ii) and (iii). Further, for (f), we may assume that $M = \partial D_0'$, D_0' as in Proposition 10.5. Indeed, since M is real analytic near p, we may find [1] a real analytic CR embedding Φ of a neighborhood U_1 of p into $\mathbb{C}^{n+1}$, and in fact, by Proposition 10.5 we may assume $\Phi : U_1 \to \partial D_0' = M'$. By choosing a suitable metric on M', we may clearly then assume $M = \partial D_0'$. Using the methods of [19], construct, for any $k \geq 1$, $F \in C^k(\bar{D}_0')$, $F \notin C^\infty$ near p, F holomorphic in D_0'. If $g = F|_M$, then $Sg = g$ and we cannot solve $\bar{\partial}_b \omega = g$ locally near p.

To get at S in the general case, we need to use an operator S' which is a "local analogue" of it, in the sense of the following lemma. Let "$\mathcal{Q}$" be as after (iii) of #8 of Chapter 9.

Lemma 10.6. In the situation of Theorem 10.2, say $p \in \mathcal{U}$, and $\mathcal{U}$ is sufficiently small that there is an analytic contact transformation $\psi : \mathcal{U}' \to \mathcal{U}$, $\mathcal{U}' \subset \mathbb{H}^n$.

On some open neighborhood $U_0 \subset \mathcal{U}$ of p, there exist operators R', S', Q_1, Q_2 with $R' : (\mathcal{E}')^{0,1}(U_0) \to \mathcal{D}'(U_0)$, S', $Q_1 : \mathcal{E}'(U_0) \to \mathcal{D}'(U_0)$, $Q_2 : \mathcal{E}'(U_0) \to (\mathcal{D}')^{0,1}(U_0)$, so that:

(i) R', S', Q_2 are in $\mathcal{O}^{-2n-1}(\psi^{-1}(U_0))$, $\mathcal{O}^{-2n-2}(\psi^{-1}(U_0))$, $\mathcal{Q}(\psi^{-1}(U_0))$ respectively, after ψ, and $Q_1 f$ is analytic on U_0 for all $f \in \mathcal{E}'(U_0)$;

(ii) $(R'\bar{\partial}_b)f = (I-S'+Q_1)f$, $(\bar{\partial}_b S')f = Q_2 f$ on U_0 for $f \in \mathcal{E}'(U_0)$;

(iii) $(S'f_1,f_2) = (f_1,S'f_2)$ for $f_1 \in \mathcal{E}'(U_0)$, $f_2 \in C_c^\infty(U_0)$;

(iv) $S'\vartheta_b\omega = Q_2^*\omega$ is analytic on U_0 for any $\omega \in (\mathcal{E}')^{0,1}(U_0)$.

Proof. Say first $n > 1$. Again we find a real analytic CR embedding $\Phi : U_1 \to \partial D_0' = M'$ (U_1 a neighborhood of p, D_0' as in Proposition 10.5.) We arrange that the metric on $\mathbb{C}TM'$ agrees with that pulled over by Φ on a neighborhood $\mathcal{U}_1'$ of $\Phi(p)$ ($\mathcal{U}_1' \subset \Phi(U_1)$). Using M', $\mathcal{U}_1'$ and our discussion in Case (i), we produce $\mathcal{K}$ as in (e) (though it might not be self-adjoint on C_c^∞

of an open set). We restrict $K\mathcal{N}_b$ and the Szegö projection on M' to a suitably small neighborhood of $\Phi(p)$, and use Φ to pull them back to M, to produce the needed R', S'. In this case, $Q_2 = 0$.

When n = 1 we use the same method of embedding and invoking Proposition 10.5, to locally embed into a suitable $\partial D_0'$. It follows then that p has an open neighborhood $U \subset\subset \mathcal{U}$, and there exist (new) operators $R : (\mathcal{E}')^{0,1}(U) \to (\mathcal{D}')^{0,1}(U)$ and $H : \mathcal{E}'(U) \to \mathcal{D}'(U)$ so that

$$\text{(I)} \qquad R\bar{\partial}_b = I - H \quad \text{and} \quad \text{(II)} \quad \bar{\partial}_b H = 0 \quad \text{on} \quad \mathcal{E}'(U),$$

and so that R, H are in $\mathcal{O}^{-3}(\psi^{-1}(U))$, $\mathcal{O}^{-4}(\psi^{-1}(U))$ (respectively) after ψ. The problem is that H does not satisfy the self-adjointness condition (iii). To produce S' we will use the Kerzman-Stein method, but at the <u>germ</u> level. It would be equally effective to pull over the Szegö projection from $\partial D_0'$; however we shall find it somewhat more convenient to work entirely with formal series.

We naturally denote the pullbacks to $\psi^{-1}(U) = U'$ of $\bar{\partial}_b$, R, H by $(\bar{\partial}_b)_\psi$, R_ψ, H_ψ respectively. These are in $\mathcal{O}^{-5}(U')$, $\mathcal{O}^{-3}(U')$, $\mathcal{O}^{-4}(U')$ respectively. Say that $\varkappa((\bar{\partial}_b)_\psi) \sim \sum D^m$, $\varkappa(R_\psi) \sim \sum K^m$, $\varkappa(H_\psi) \sim \sum H^m$. We abbreviate the formal sums $\sum D^m = d$, $\sum R^m = r$, $\sum H^m = h$, so that d, r, h are in $F\mathcal{C}^{-5}(U')$, $F\mathcal{C}^{-3}(U')$, $F\mathcal{C}^{-4}(U')$ respectively. Then

$$r\#d = \delta - h \quad \text{and } d\#h = 0 \quad \text{mod } FP\text{, on } U' \tag{10.37}$$

since these relations need only be proved in an open neighborhood of each point of U', and we can apply Theorem 7.11(c) and (i) and (iv) of #1 of Chapter 9. Also, by associativity, and (ii) and (v) of #1 of Chapter 9,

$$h\#h = h \qquad \text{mod } FP\text{, on } U'. \tag{10.38}$$

(By (8.4), we actually have $d\#h \equiv 0$ on U'; such "improvements" are irrelevant for our arguments.)

We wish to use * to refer to adjoints taken with respect to the inner product on $L^2(M)$. We must therefore change our previous notation. We use "adj" to denote that mapping on operators and formal sums on $\mathbb{H}^1$ which we previously denoted "*", for example in Theorem 7.11(a) and (vi) of #1 of Chapter 9.

Say, then, $K : \mathcal{D}'(V) \to \mathcal{D}'(V)$, where $V \subset\subset \mathcal{U}$ is an open neighborhood of p, and say K is in $\mathcal{O}^k(\psi^{-1}(V))$ after ψ, with pullback K_ψ. Then it is easy to see that there is an operator $K^* : C_c^\infty(V) \to C^\infty(V)$ so that $(Kf_1, f_2) = (f_1, K^*f_2)$ for $f_1, f_2 \in C_c^\infty(V)$, and where we use the $L^2(M)$ inner product. In fact, it is easy to see that for $f \in C_c^\infty(V)$, $(K^*f)_\psi = K_\psi^* f_\psi$, where $K_\psi^* = M_{\Phi^{-1}} \circ K_\psi^{\text{adj}} \circ M_\Phi$, where M_Φ denotes multiplication by Φ, a certain positive-valued analytic function on $\mathcal{U}'$. Here $\int_{\mathcal{U}} f = \int_{\mathcal{U}'} f_\psi \Phi$ for $f \in C_c^\infty(\mathcal{U})$. Put $V' = \psi^{-1}(V)$. Note $M_\Phi \in \mathcal{O}^{-4}(V')$, with core $\Phi(u)\delta(w)$. If now $K_1 \in C^{k_1}(V')$, define $(K_1^*)_u^m$ by

$\sum(K_1^*)_u^m = (\Phi^{-1}(u)\delta(w)) \# \sum(K_1^{adj})_u^m \# (\Phi(u)\delta(w))$. If also $K_2 \in C^{k_2}(V')$, and $\sum K_2^m \# \sum K_1^m = \sum K^m$, note $\sum(K^*)^m = \sum(K_1^*)^m \# \sum(K_2^*)^m$, by associativity and (vi) of #1 of Chapter 9.

By the product formula Theorem 7.11(c), the principal core of h^* is the same as the principal core of h^{adj}. By the adjoint formula Theorem 7.11(a), this is $\tilde{H}_u^o$. (Recall H_u^o is the principal core of h.) If $a = h - h^*$, its principal core A_u^o satisfies $\tilde{A}_u^o = -A_u^o$. Thus $\hat{A}_u^o(\lambda)$ is skew-adjoint for all $\lambda \neq 0$. We may therefore apply Theorem 8.1 and (viii) of #1 of Chapter 9 to $\delta + a$ and $\delta + a^*$, since the representation-theoretic criteria are met. Thus we find $k_1, k_2 \in FC^{-4}(U_1')$ so that

$$k_1\#(\delta+a) = \delta, \quad k_1\#(\delta+a^*) = \delta \quad \text{mod } F\mathcal{P} \text{ on } U_1'.$$

Thus $(\delta+a^*)\#k_1^* = \delta, \quad (\delta+a)\#k_2^* = \delta \quad \text{mod } F\mathcal{P} \text{ on } U_1'.$

By the associative law applied to $k_1\#(\delta+a)\#k_2^*$, and (vii) of #1 of Chapter 9,

$$k_1 = k_2^*, \; k_1^* = k_2 \qquad \text{mod } F\mathcal{P} \text{ on } U_1'. \tag{10.39}$$

Thus in fact $k_1\#(\delta+a) = (\delta+a)\#k_1 = \delta$, $k_2\#(\delta+a^*) = (\delta+a^*)\#k_2 = \delta$ mod $F\mathcal{P}$ on U_1'. We use the symbol $(\delta+a)^{-1}$, or $(\delta+h-h^*)^{-1}$, for k_1, and $(\delta+a^*)^{-1}$ or $(\delta+h^*-h)^{-1}$ for k_2.

Now put

$$s' = h\#(\delta+a)^{-1} \tag{10.40}$$

which is in $FC^{-4}(U_1')$. Note that, on U_1',

$$(\delta+h^*-h)\#h = h^*\#(\delta+h-h^*) = h^*\#h \quad \text{mod } F\mathcal{P} \qquad (10.41)$$

so $\quad h\#(\delta+a)^{-1} = (\delta+a^*)^{-1}\#h^* = (\delta+a^*)^{-1}\#h^*\#h\#(\delta+a)^{-1}$ mod $F\mathcal{P}$,

so

$$s' = (s')^* = s'\#s' \qquad (10.42)$$

by (10.39). Thus s' is a "formal orthogonal projection". Further, by (10.41),

$s'\#h = (\delta+a^*)^{-1}\#h^*\#h = (\delta+a^*)^{-1}\#(\delta+a^*)\#h$ mod $F\mathcal{P}$, so on U_1',

$$s'\#h = h, \qquad h\#\, s' = s' \qquad \text{mod } F\mathcal{P}. \qquad (10.43)$$

Thus also $(\delta-s')\#(\delta-h) = (\delta-s')$ mod $F\mathcal{P}$, on U_1'. Put

$$r' = (\delta-s')\#r \qquad \text{on } U_1'.$$

Then by (10.37) and (10.40),

$$r'\#d = \delta - s', \quad d\#s' = 0 \qquad \text{mod } F\mathcal{P}, \quad \text{on } U_1'. \qquad (10.44)$$

By (x) of #1 of Chapter 9, we may select $\mathcal{R}' \in \mathcal{O}^{-3}(U')$, $\mathcal{S}' \in \mathcal{O}^{-4}(U')$, with support radii as small as we please, so that $\varkappa(\mathcal{R}') \sim r'$, $\varkappa(\mathcal{S}') = s'$, and so that $(\mathcal{S}'f_1,f_2) = (f_1,\mathcal{S}'f_2)$ if one of f_1,f_2 is in $\mathcal{E}'(U_1')$ and the other is in $C_c^\infty(U_1')$, if the pairing (,) is the one induced from the inner product on $L^2(M)$.

We may clearly assume that the support radii of the $\mathcal{R}',\mathcal{S}'$ obtained through this construction are as small as we please, and different. Thus we can pull R', S' over to M, and Lemma 10.5(i), (ii), (iii) follow at once from Theorem 7.11(c) and (iii) of #1 of Chapter 9. Lemma 10.6(iv) follows at once from (iii) and (ii), since if U_0 is sufficiently small, the kernel of Q_2 is analytic on $U_0 \times U_0$. This proves the Lemma.

With this preparation, we can now begin the proof in the general case. We retain, from the proof of the lemma, the notation R', S' for the pullbacks of R',S' to $\psi^{-1}(U_0) = U_0'$, say. In the notation of #6 of Chapter 9, we may clearly find $0 < a < b$, φ_N as in (9.2), and open neighborhoods of $p' = \psi^{-1}(p)$, say $V' \subset\subset W' \subset\subset U_0'$, so that if $R_N' = R'_{N,a,b,\varphi_N}$, $S_N' = S'_{N,a,b,\varphi_N}$, then

(I) If $f \in E'(V')$, $R_N' f = R'f$ on V' and $S_N' f = S'f$ on V'.

(II) If $f \in E'(U_0')$, f = 0 on W', then $R_N' f = S_N' f = 0$ on V'.

(III) If $f \in E'(V')$, and f is analytic on an open set $V_0' \subset\subset V'$, then on V_0', $S_N' f = S'f$ is analytic.

(Choose a,b,W' first so that $R_N' f = S_N' f = 0$ near p' if f = 0 on W'; then choose V' small enough. Use Theorem 7.11(b).)

Pull R_N', S_N' over to U_0; call the pulled over operators $R_N', S_N' : E'(U_0) \to \mathcal{D}'(U_0)$, and let $V = \psi(V')$, $W = \psi(W')$.

We will always be restricting $R_N' f, S_N' f$ to V. Then we can actually define natural extensions of these operators, also denoted $R_N', S_N' : E'(M) \to \mathcal{D}'(V)$, simply by setting

$R_N'f = R_N'(\varphi f)$, $S_N'f = S_N'(\varphi f)$ where $\varphi \in C_c^\infty(U_0)$, $\varphi = 1$ on W. By property (II), this condition is independent of the choice of cutoff φ.

Finally, we may turn to S. For $f \in C^\infty(M)$, we may examine $S_N'Sf$ (on V, of course). Let $\zeta \in C_c^\infty(V)$, $\zeta = 1$ on V_1, an open neighborhood of p, and otherwise arbitrary. Using the partial global assumption (10.8), write $Sf = f - \vartheta_b u$, $u \in \mathcal{E}'(M)$; then, on V,

$$S_N'Sf = S_N'f - S_N'\vartheta_b u = S'(\zeta f) - S'\vartheta_b(\zeta u) + S_N'[(1-\zeta)f - \vartheta_b(1-\zeta)u]$$

$$= S'(\zeta f) + g_1 + S_N'[(1-\zeta)f - \vartheta_b(1-\zeta)u]$$

where g_1 is analytic on V, by Lemma 10.6(iv). At the same time, on V_1,

$$S_N'Sf = Sf + [(S_N'-I) + R_N'\bar{\partial}_b]Sf$$

$$= Sf + [(S'-I) + R'\bar{\partial}_b](\zeta Sf) + [S_N'+R_N'\bar{\partial}_b][(1-\zeta)Sf]$$

$$= Sf + g_2 + [S_N' + R_N'\bar{\partial}_b][(1-\zeta)Sf]$$

where g_2 is analytic on V_1, by Lemma 10.6(ii). Equating these, we see that on V_1,

$$Sf - S'(\zeta f) - g = S_N'h + R_N'\omega \qquad (10.45)$$

where g is analytic on V_1, $h \in \mathcal{E}'(M)$, $h = 0$ on V_1, and $\omega \in \mathcal{E}'(M)$, $\omega = 0$ on V_1. Let u be the left side of (10.45). Pulling back to $\mathbb{H}^n$, we see that on $V_1' = \psi^{-1}(V_1)$,

$$u_\psi = S'_N(\varphi h)_\psi + R'_N(\varphi\omega)_\psi$$

where $\varphi \in C_c^\infty(U_0)$, $\varphi = 1$ on W. Since $(\varphi h)_\psi$, $(\varphi\omega)_\psi = 0$ near p', p' has a neighborhood V_2' so that for some $L \geq 0$, we have <u>for all</u> <u>N</u>: $|D_1 \ldots D_M u_\psi(v)| < CR^M N^{M+L}$ for $M \leq N - L$, $v \in V_2'$, $D_1, \ldots, D_M \in \{T, X_1, \ldots, X_n, Y_1, \ldots, Y_n\}$. But then, by Proposition 9.3(b), u_ψ is analytic on V_2'. Thus Sf - S'(ζf) is analytic on $V_2 = \psi(V_2')$ for $f \in C^\infty(M)$.

In particular, since p is arbitrary in $\mathcal{U}$, $S : C^\infty(M) \to C^\infty(\mathcal{U})$. This map is continuous by the closed graph theorem for Frechet spaces, since $S : C^\infty(M) \to L^2(\mathcal{U})$ is continuous and the $L^2(\mathcal{U})$ topology on $C^\infty(\mathcal{U})$ is weaker than the usual topology. This proves (10.9).

If we now repeat the above arguments for $f \in \mathcal{D}(S)$, using the full global assumption (10.10), we see that Sf - S'(ζf) is analytic on V_2 for any $\zeta \in C_c^\infty(V)$, if $\zeta = 1$ on V_1. In particular, if f is analytic near p, so is ζf, and hence (by (III) above) so is Sf. This proves Theorem 10.2(a). Theorem 10.2(d) also follows; in fact:

for all $f \in \mathcal{E}'(V_1 \cap V_2)$, Sf - S'f is analytic on $V_1 \cap V_2$. (10.46)

That leaves only (e). If n > 1, we may proceed as at the beginning of the proof of Lemma 10.6, and pull back the "K" for M' to M, via φ. In this way we produce a new K with $\Box_b^0 Kf = (I-S')f + g_1$, $K\Box_b^0 f = (I-S')f + g_2$ on an open set

V_3, if $f \in \mathcal{E}'(V_3)$, and where g_1, g_2 are analytic on V_3. We then obtain (10.11) on U_0, for $f \in \mathcal{E}'(U_0)$, if $U_0 = V_1 \cap V_2 \cap V_3$.

We still must show that, when $n > 1$, U_0 can be shrunk and K modified so that $K = K^*$ on $C_c^\infty(U_0)$. We may assume that there is an orthonormal basis $\{\overline{\eta}_1, \ldots, \overline{\eta}_n\}$ for (0,1) forms on $\mathcal{U}$, such that all $\overline{\eta}_j$ are analytic, and such that the pullback of $f = \sum_{j=1}^{n} f_j \overline{\eta}_j$ $(f \in \Lambda^{0,1}(\mathcal{U}))$ has been defined to be the vector $(f_j \circ \psi)_{1 \leq j \leq n}$. As in the case $n = 1$, we select Φ so that $\int_{\mathcal{U}} f = \int_{\mathcal{U}'} f_\psi \Phi$ for $f \in C_c^\infty(\mathcal{U})$. Then the pullbacks $(\vartheta_b)_\psi$ and $(\overline{\partial}_b)_\psi$ are related by $(\vartheta_b)_\psi = M_{\Phi^{-1}} \circ (\overline{\partial}_b)_\psi^{\mathrm{adj}} \circ M_\Phi$, where adj is the adjoint operator on $\mathbb{H}^n$. With K, U_0 as just constructed, say $\kappa(K_\psi) \sim k$. Let $k^* = \Phi^{-1}(u)\delta(w) \# k \# \Phi(u)\delta(w)$. Then it is easy to see from (10.14), (9.21), and the analysis at the end of #8 of Chapter 9, that $k = k^*$ mod FP on $\psi^{-1}(U_0)$. But then, if K has sufficiently small support radius, $K - K^*$ is analytic regularizing in a neighborhood of p, by (iii) of #1 of Chapter 9. Therefore, $K - (K+K^*)/2$ is also analytic regularizing in a neighborhood of p. Thus we may replace K by $(K+K^*)/2$ as in (x) of #1 of Chapter 9. This completes the proof of the theorem if $n > 1$.

If $n = 1$, with U' as in (10.37), we invoke (10.33) and (10.36) to see that

$$d \# r = \delta - g \text{ and } g \# d = 0 \quad \text{mod } FP, \text{ on } U' \tag{10.47}$$

for some $g \in FC^{-4}(U')$. The proof of (10.47) is entirely analogous to that of (10.37).

We may assume that there is an analytic (0,1)-form $\overline{\eta}$ on $U = \psi(U')$, which is a unit vector at each point of U, and we may assume that the pullback of $f = f_0\overline{\eta}$ $(f_0 \varepsilon C^{\infty})$ has been defined to be $f_0 \circ \psi$. Then, again, the pullback of ϑ_b, namely $(\vartheta_b)_\psi$, is simply $[(\overline{\partial}_b)_\psi]^* = M_{\Phi^{-1}}(\overline{\partial}_b)_\psi M_\Phi$. Then

$$\varkappa((\vartheta_b)_\psi) \sim d^* \text{ and } \varkappa((\square_b^0)_\psi) \sim d^*\#d. \tag{10.48}$$

From (10.47) we see

$$r^*\#d^* = \delta - g^*, \; d^*\#g^* = 0 \mod F\mathcal{P}, \text{ on } U_1'. \tag{10.49}$$

These relations are entirely analogous to (10.37), with d^*, r^*, g^* in place of d, r, h. The same analysis that led to (10.42), (10.44) then shows we may find $r'' \varepsilon FC^{-3}(U')$, $t' \varepsilon FC^{-4}(U')$, so that, mod $F\mathcal{P}$, on U_1', we have

$$t' = (t')^* = t'\#t' \tag{10.50}$$

and

$$r''\#d^* = \delta - t', \; d^*\#t' = 0 \tag{10.51}$$

so also

$$t'\#d = 0. \tag{10.52}$$

We finally see that, mod $F\mathcal{P}$, on U',

$$\delta - s' = r'\#d = r'\#(\delta - t')\#d = r'\#r''\#d^*\#d$$

$$= r'\#r''\#(\delta - s')\#d^*\#d = k\#\ell \tag{10.53}$$

where $k = r'\#r''\#(\delta - s')$, $\ell = d^*\#d$. We used, in order, (10.44), (10.52), (10.51), (10.44). Note

$$k\#(\delta - s') = k. \tag{10.54}$$

In (10.53), we take * to see $\ell\#k^* = \delta - s'$ also; thus

$$k = k\#(\delta - s') = k\#\ell\#k^* = (\delta - s')\#k^* = k^* \tag{10.55}$$

by (10.54); thus

$$\delta - s' = k\#\ell = \ell\#k. \tag{10.56}$$

If we use (10.55), (10.56) and (x) of #1 of Chapter 9, we obtain (e) for $n = 1$. This completes the proof of Theorem 10.2.

References

[1] A. Andreotti and C. D. Hill, Complex characteristic coordinates and tangential Cauchy-Riemann equations, Ann. Scuola Norm. Sup. Pisa (3) 26 (1972), 299-324.

[2] A. Andreotti and C. D. Hill, E. E. Levi convexity and the Hans Lewy problem I, II, Ann. Scuola Norm. Super. Pisa 26 (1972), 325-363, ibid. 26 (1972), 747-806.

[3] M. S. Baouendi, C. H. Chang and F. Trèves, Microlocal hypo-analyticity and extension of CR functions, J. Diff. Geom. 18 (1983), 331-391.

[4] M. S. Baouendi, H. Jacobowitz and F. Trèves, On the analyticity of CR mappings, Ann. Math. 122 (1985), 365-400.

[5] M. S. Baouendi and L. P. Rothschild, Normal forms for generic manifolds and holomorphic extension of CR functions, J. Diff. Geom. 25 (1987), No. 3, 431-467.

[6] M. S. Baouendi, S. R. Bell and L. P. Rothschild, Mappings of three-dimensional CR manifolds and their holomorphic extension, Duke Math. J. 56 (1988), 503-530.

[7] R. Beals, Spatially inhomogeneous pseudodifferential operators II, Comm. Pure Appl. Math 27 (1974), 161-205.

[8] R. Beals and P. Greiner, Calculus on Heisenberg manifolds, Ann. Math Studies Number 119, Princeton University Press, Princeton (1988).

[9] L. Boutet de Monvel and P. Krée, Pseudo-differential operators and Gevrey classes, Ann. Inst. Fourier Grenoble 27 (1967), 295-323.

[10] L. Boutet de Monvel and J. Sjöstrand, Sur la singularité des noyaux de Bergman et de Szegö, Soc. Math. de France, Astérisque 34-35 (1976), 123-164.

[11] D. Burns, Global behavior of some tangential Cauchy-Riemann equations, Partial Differential Equations and Geometry (Proc. Conf., Park City, Utah, 1977), Dekker, New York (1979), 51-56.

[12] M. Christ and D. Geller, Singular integral characterizations of Hardy spaces on homogeneous groups, Duke Math. J. 51 (1984), 547-598.

[13] L. Corwin and L. P. Rothschild, Necessary conditions for local solvability of homogeneous left invariant differential operators on nilpotent Lie groups, Acta Math. 147 (1981), 265-288.

[14] T. Cummins, A. pseudodifferential calculus associated to 3-step nilpotent groups, Comm. PDE 14 (1989), 129-171.

[15] M. Derridj and D. S. Tartakoff, Local analyticity for $\Box_b$ and the $\bar{\partial}$-Neumann problem at certain weakly pseudoconvex points, Comm. PDE 13 (1988), 1521-1600.

[16] R. Douglas, Banach Algebra Techniques in Operator Theory, Academic Press, 1972.

[17] A. S. Dynin, Pseudodifferential operators on the Heisenberg group, Soviet Math. Dokl., Vol. 16 (1975), 1608-1612.

[18] G. B. Folland, Subelliptic estimates and function spaces on nilpotent groups, Arkiv für Math. 13 (1975), 161-207.

[19] G. B. Folland and J. J. Kohn, The Neumann problem for the Cauchy-Riemann complex, Annals of Math Studies, No. 75, Princeton Univ. Press, Princeton, NJ, 1972.

[20] G. B. Folland and E. M. Stein, Estimates for the $\bar{\partial}_b$ complex and analysis on the Heisenberg group, Comm. Pure Appl. Math. 27 (1974), 429-522.

[21] I. M. Gelfand and G. E. Šilov, Fourier transforms of rapidly increasing functions and questions of the uniqueness of the solution of Cauchy's problem, Uspehi. Math. Nauk (N.S.) 8, No. 6 (58), 3-54 (1953); American Mathematical Society Translations, Series 2, Volume 5 (1957), 221-274.

[22] I. M. Gelfand and G. E. Šilov, Generalized Functions, Vol. 2, Academic Press, 1968.

[23] I. M. Gelfand and G. E. Šilov, Generalized Functions, Vol. 3, Academic Press, 1967.

[24] D. Geller, Fourier Analysis on the Heisenberg Group, Princeton University thesis, 1977.

[25] D. Geller, Fourier analysis on the Heisenberg group I: Schwartz space, J. Func. Anal., Vol. 36, No. 2 (1980), 205-254.

[26] D. Geller, Local solvability and homogeneous distributions on the Heisenberg group, Comm. in P.D.E., 5 (5), (1980), 475-560.

[27] D. Geller, The Laplacian and the Kohn Laplacian for the sphere, J. Diff. Geom. 15 (1980), 417-435.

[28] D. Geller, Spherical harmonics, the Weyl transform and the Fourier transform on the Heisenberg group, Can. J. Math. 36, (1984), 615-684.

[29] D. Geller, Liouville's theorem for homogeneous groups, Comm. in P.D.E. 8 (1983), 1665-1677.

[30] D. Geller, Toward analytic pseudodifferential operators for the Heisenberg group, Contemp. Math.27 (1984), 205-229.

[31] D. Geller and P. Heller, Local solvability of left invariant differential operators on the Heisenberg group, to appear.

[32] P. Glowacki, The Rockland condition for non-differential convolution operators II, to appear.

[33] R. Goodman, Differential operators of infinite order on a Lie group, J. Math. Mech. 19 (1970), 879-894.

[34] R. Goodman, Holomorphic representations of nilpotent Lie groups, J. Func. Anal. 31 (1979), 115-137.

[35] P. C. Greiner, J. J. Kohn and E. M. Stein, Necessary and sufficient conditions for solvability of the Lewy equation, Proc. Natl. Acad. Sci. USA, 72 (1975), 3287-3289.

[36] A. Grigis and L. P. Rothschild, A criterion for analytic hypoellipticity of a class of differential operators with polynomial coefficients, Ann. of Math. 118 (1983), 443-460.

[37] A. Grossman, G. Loupias and E. M. Stein, An algebra of pseudodifferential operators and quantum mechanics in phase space, Ann. Inst. Fourier Grenoble 18 (1968), 343-368.

[38] V. V. Grušin, On a class of hypoelliptic operators, Mat. Sbornik, Tom 83 (125) (1970), 456-473; Math. USSR Sbornik, Vol. 12 (1970), No. 3, 458-476.

[39] V. V. Grušin, On a class of elliptic pseudodifferential operators degenerate on a submanifold, Mat. Sbornik, Tom 84 (126) (1971) No. 2, 163-195; Math. USSR Sbornik, Vol. 13 (1971), No. 2, 155-185.

[40] B. Helffer, Conditions nécessaries d'hypoanalyticité pour des operateurs invariants a gauche homogenes sur un groupe nilpotent gradué, J. Diff. Eq. 44 (1982), 460-481.

[41] B. Helffer and J. Nourrigat, Caracterization des operateurs hypoelliptiques homogenes invariant a gauche sur un groupe de Lie nilpotent gradué, Comm. P.D.E. 4 (1979), 899-958.

[42] P. N. Heller, Analyticity and regularity for nonhomogeneous operators on the Heisenberg group, Princeton University thesis, 1986.

[43] G. M. Henkin, The Lewy equation and analysis on pseudoconvex manifolds, Uspekhi Mat. Nauk. 32:3 (1977), 57-118=Russian Math. Surveys 32:3 (1977), 59-130.

[44] E. Hille, Contributions to the theory of Hermitian series, II. The representation problem, Trans. A.M.S. 47 (1940), 80-94.

[45] L. Hörmander, The Analysis of Linear Partial Differential Operators, Springer-Verlag, 1983.

[46] L. Hörmander, Pseudo-differential operators and non-elliptic boundary problems, Ann. of Math 83 (1966), 129-209.

[47] L. Hörmander, Uniqueness theorems and wave front sets for solutions of linear differential equations with analytic coefficients, Comm. Pure Appl. Math. 17 (1971), 671-704.

[48] L. Hörmander, The Weyl calculus of pseudo-differential operators, Comm. Pure Appl. Math. 32 (1979), 359-443.

[49] R. Howe, Quantum mechanics and partial differential equations, J. Funct. Anal. 38 (1980), 188-254.

[50] D. Jerison, The Dirichlet problem for the Kohn Laplacian on the Heisenberg group, I, J. Func. Anal. (1981), 97-142.

[51] M. Kashiwara and T. Oshima, Systems of differential equations with regular singularities and their boundary value problems, Ann. Math. 106 (1977), 145-200.

[52] N. Kerzman and E. M. Stein, The Szegö kernel in terms of Cauchy-Fantappie kernels, Duke Math. J. 45 (1978), 197-224.

[53] A. W. Knapp and E. M. Stein, Intertwining operators for semisimple groups, Ann. of Math. 93 (1971), 489-578.

[54] J. J. Kohn, Harmonic integrals on strongly pseudoconvex manifolds, I and II, Ann. Math. 78 (1963), 112-147; 79 (1964), 450-472.

[55] J. J. Kohn, Global regularity for $\bar{\partial}$ on weakly pseudoconvex manifolds, Trans. Amer. Math. Soc. 181 (1973), 273-292.

[56] J. J. Kohn, Methods of P.D.E. in complex analysis, Proc. in Pure Math 30 (1977), 215-237.

[57] J. J. Kohn, Estimates for $\bar{\partial}_b$ on pseudo-convex CR manifolds, Proc. Symp. Pure Math. 43 (1985), 207-217.

[58] J. J. Kohn, The range of the tangential Cauchy-Riemann operator, Duke Math. J. 53 (1986), 525-545.

[59] H. Lewy, An example of a smooth linear partial differential equation without solution, Ann. Math. 66 (1957), 155-158.

[60] A. Melin, Parametrix constructions for some classes of right invariant differential operators on the Heisenberg group, Comm. P.D.E. 6 (1981), 1363-1405.

[61] A. Melin, Parametrix constructions for right-invariant differential operators on nilpotent groups, Annals of Global Analysis and Geometry 1 (1983), 79-130.

[62] G. Métivier, Hypoellipticité analytique sur des groupes nilpotents de rang 2, Duke Math. J. 47 (1980), 195-221.

[63] G. Métivier, Analytic hypoellipticity for operators with multiple characteristics, Comm. P.D.E. 6 (1981), 1-90.

[64] K. Miller, Hypoellipticity on the Heisenberg group, J. Func. Anal. 31 (1979), 306-320.

[65] N. Moukaddem, Inversibilité d'opérateurs intégraux singuliers sur des groupes nilpotents de rang 3, Université de Rennes I, thèse, 1986.

[66] A. Nagel and E. M. Stein, Lectures on Pseudo-Differential Operators: Regularity Theorems and Applications to Non-Elliptic Problems, Princeton University Press, 1979.

[67] R. Narasimhan, Compact analytical varieties, Enseignement Math. (2) 14 (1968), 75-98.

[68] C. Rockland, Hypoellipticity on the Heisenberg group: representation theoretic criteria, Trans. A.M.S. 240 (1978), 1-52.

[69] H. Rossi, Attaching analytic spaces to a space along a pseudo-convex boundary, Proc. Conf. Complex Analysis (Minneapolis, 1965), Springer-Verlag (1965), 242-256.

[70] L. P. Rothschild and E. M. Stein, Hypoelliptic differential operators and nilpotent groups, Acta Math. 137 (1976), 247-320.

[71] L. Rothschild and D. Tartakoff, Analyticity of relative fundamental solutions and projections for left invariant operators on the Heisenberg group, Ann. Sci. Ec. Norm. Sup. 15 (1982), 419-440.

[72] J. Sjöstrand, Singularités Analytiques Microlocales, Asterisque 95, Societé Mathématique de France, 1982.

[73] E. M. Stein, Singular Integrals and Differentiability Properties of Functions, Princeton University Press, 1970.

[74] E. M. Stein, Boundary Behavior of Holomorphic Functions of Several Complex Variables, Mathematical Notes, Princeton University Press, Princeton, NJ, 1972.

[75] E. M. Stein, An example on the Heisenberg group related to the Lewy operator, Invent. Math. 69 (1982), 209-216.

[76] E. M. Stein and G. Weiss, Introduction to Fourier Analysis on Euclidean Spaces, Princeton University Press, 1971.

[77] D. Tartakoff, Local analytic hypoellipticity for $\Box_b$ on nondegenerate Cauchy-Riemann manifolds, Proc. Nat. Acad. Sci. 75 (1978), 3027-3028.

[78] D. Tartakoff, The local real analyticity of solutions to $\Box_b$ and the $\bar{\partial}$- Neumann problem, Acta. Math. 145 (1980), 177-204.

[79] M. Taylor, Noncommutative Microlocal Analysis, Part I, Memoirs of the A.M.S., Vol. 52, No. 313, 1984.

[80] F. Trèves, Topological Vector Spaces, Distributions and Kernels, Academic Press, 1967.

[81] F. Trèves, Analytic hypoellipticity of a class of pseudo-differential operators, Comm. in P.D.E., 3 (1978), 475-642.

[82] G. Tomassini, Trace delle functional olomorfe sulle sotto varietà analitiche reali d'una varietà complessa, Ann. Scuola Norm. Sup. Pisa Cl. Sci. 20 (1966), 31-43.

[illegible]